# Electromagnetic Compatibility Engineering

# Electromagnetic Compatibility Engineering

**Henry W. Ott**
Henry Ott Consultants

A JOHN WILEY & SONS, INC., PUBLICATION

For general information on our other products and services or for technical support, please contact our Customer Care Department within the United States at (800) 762-2974, outside the United States at (317) 572-3993 or fax (317) 572-4002.

Wiley also publishes its books in a variety of electronic formats. Some content that appears in print may not be available in electronic formats. For more information about Wiley products, visit our web site at www.wiley.com.

*Library of Congress Cataloging-in-Publication Data:*

Ott, Henry W., 1936-
  Electromagnetic compatibility engineering / Henry W. Ott. – Rev. ed.
       p. cm.
  Earlier ed. published under title: Noise reduction techniques in electronic systems, 1988.
  Includes bibliographical references and index.
  ISBN 978-0-470-18930-6
1.  Electronic circuits–Noise. 2.  Electromagnetic compatibility.
I.  Ott, Henry W., 1936- Noise reduction techniques in electronic systems. II.  Title.
  TK7867.5.O867 2009
  621.382′24—dc22                                                              2009006814

Printed in the United States of America

10  9  8  7  6  5

*To my parents, the late Henry and Virginia Ott.*
*The values they instilled in me as a child have served me well throughout my life.*

Everything should be made as simple as possible, but no simpler.
Albert Einstein, 1879–1955

# CONTENTS

# PREFACE

*Electromagnetic Compatibility Engineering* started out being a third edition to my previous book *Noise Reduction Techniques in Electronic Systems*, but it turned out to be much more than that, hence, the title change. Nine of the original twelve chapters were completely rewritten. In addition, there are six new chapters, plus two new appendices, with over 600 pages of new and revised material (including 342 new figures). Most of the new material relates to the practical application of the theory of electromagnetic compatibility (EMC) engineering, and it is based on experience gained from my EMC consulting work, and teaching of EMC training seminars over the last 20 plus years.

Some of the more difficult and frustrating problems faced by design engineers concerns electromagnetic compatibility and regulatory compliance issues. Most engineers are not well equipped to handle these problems because the subject is not normally taught in engineering schools. Solutions to EMC problems are often found by trial and error with little or no understanding of the theory involved. Such efforts are very time consuming, and the solutions are often unsatisfactory. This situation is unfortunate, because most of the principles involved are simple and can be explained by elementary physics. This book is intended to remedy that situation.

This book is intended primarily for the practicing engineer who is involved in the design of electronic equipment or systems and is faced with EMC and regulatory compliance issues. It addresses the practical aspects of electromagnetic compatibility engineering, covering both emission and immunity. The concepts presented in this book are applicable to both analog and digital circuits operating from below audio frequencies up to the GHz range. Emphasis is on cost-effective EMC designs, with the amount and complexity of the mathematics kept to a minimum. The reader should obtain the knowledge necessary to design electronic equipment that is compatible with the electromagnetic environment and compliant with national and international EMC regulations.

The book is written in such a way that it can easily be used as a textbook for teaching a senior level or continuing education course in electromagnetic compatibility. To this end, the book contains 251 problems for the student to work out, the answers to which are included in Appendix F.

The book is divided into two parts: Part 1, EMC Theory, includes Chapters 1 to 10. Part 2, EMC Applications, includes Chapters 11 to 18. In addition, the book contains six appendices with supplemental information.

The organization of the material is as follows. Chapter 1 is an introduction to electromagnetic compatibility and covers national and international EMC regulations, including the European Union, FCC, and U.S. Military. Chapter 2 covers both electric and magnetic field cable coupling and crosstalk, as well as cable shielding and grounding. Chapter 3 covers safety, power, signal, and hardware/systems grounding.

Chapter 4 discusses balancing and filtering as well as differential amplifiers, and low-frequency analog circuit decoupling. Chapter 5 is on passive components and covers the nonideal characteristics of components that affect their performance. In addition to resistors, capacitors, and inductors—ferrite beads, conductors and transmission lines are also included. Chapter 6 is a detailed analysis of the shielding effectiveness of metallic sheets as well as conductive coatings on plastic, and the effect of apertures on the shielding effectiveness.

Chapter 7 covers contact protection for relays and switches. Chapters 8 and 9 discuss internal noise sources in components and active devices. Chapter 8 covers intrinsic noise sources, such as thermal and shot noise. Chapter 9 covers noise sources in active devices.

Chapters 10, 11, and 12 cover electromagnetic compatibility issues associated with digital circuits. Chapter 10 examines digital circuit grounding, including ground plane impedance and a discussion on how digital logic currents flow. Chapter 11 is on digital circuit power distribution and decoupling, and Chapter 12 covers digital circuit radiation mechanisms, both common mode and differential mode.

Chapter 13 covers conducted emissions on alternating current (ac) and direct current (dc) power lines, as well as EMC issues associated with switching power supplies and variable-speed motor drives. Chapter 14 covers radio frequency(rf) and transient immunity, as well as a discussion of the electromagnetic environment. Chapter 15 covers electrostatic discharge protection in the design of electronic products. It focuses on the importance of a three-prong approach, which includes mechanical, electrical, and software design.

Chapter 16 covers printed circuit board layout and stackup, a subject not often discussed. Chapter 17 addresses the difficult problem of partitioning, grounding, and layout of mixed-signal printed circuit boards.

The final chapter (Chapter 18) is on precompliance EMC measurements, that is, measurements that can be performed in the product development laboratory, using simple and inexpensive test equipment, which relate to the EMC performance of the product.

At the end of each chapter, there is a summary of the most important points discussed as well as many problems for the reader to work out. For those desiring additional information on the subjects covered, each chapter has an extensive reference, and further reading section.

Supplemental information is provided in six appendices. Appendix A is on the decibel. Appendix B covers the 10 best ways to *maximize* the emission from your product. Appendix C derives the equations for multiple reflections of magnetic fields in thin shields.

Appendix D, "Dipoles for Dummies," is a simple, insightful, and intuitive discussion of how a dipole antenna works. If a product picks up or radiates electromagnetic energy, then it is an antenna, therefore, an understanding of some basic antenna theory would be helpful for all engineers, especially EMC engineers.

Appendix E explains the important, and not well understood, theory of partial inductance, and Appendix F provides answers to the problems contained at the end of each chapter.

I would like to express my gratitude and appreciation to all those who took the time to comment on *Noise Reduction Techniques in Electronic Systems* and to all those who encouraged me to write *Electromagnetic Compatibility Engineering*. In particular, I would especially like to thank John Celli, Bob German, Dr. Clayton Paul, Mark Steffka, and Jim Brown for their insightful review of major portions of the manuscript, as well as for their encouragement and the many fruitful discussions we had on the subject of EMC. *Electromagnetic Compatibility Engineering* is a better book because of them.

Portions of the manuscript were also used for an electromagnetic compatibility class taught by Mark Steffka at the University of Michigan–Dearborn, during the 2007 and 2008 semesters. My heartfelt thanks go out to the students in those two classes for the large number of comments and suggestions that I received (many of which have been incorporated into this book), in particular their suggestions for additional problems to be included in the book. I would also like to express my appreciation to James Styles who, Mark Steffka and I both agreed, submitted the most useful comments.

Finally, I would like to thank all my colleagues who took the time to review various portions of this manuscript and make useful comments and suggestions.

Additional technical information, updated information on EMC regulations, as well as an errata sheet for this book are on the Henry Ott Consultants website at www.hottconsultants.com.

*Livingston, New Jersey*                                                    HENRY W. OTT
*January 2009*

# PART I
# EMC Theory

# 1 Electromagnetic Compatibility

## 1.1 INTRODUCTION

The widespread use of electronic circuits for communication, computation, automation, and other purposes makes it necessary for diverse circuits to operate in close proximity to each other. All too often, these circuits affect each other adversely. Electromagnetic interference (EMI) has become a major problem for circuit designers, and it is likely to become even more severe in the future. The large number of electronic devices in common use is partly responsible for this trend. In addition, the use of integrated circuits and large-scale integration has reduced the size of electronic equipment. As circuitry has become smaller and more sophisticated, more circuits are being crowded into less space, which increases the probability of interference. In addition, clock frequencies have increased dramatically over the years—in many cases to over a gigahertz. It is not uncommon today for personal computers used in the home to have clock speeds in excess of 1 GHz.

Today's equipment designers need to do more than just make their systems operate under ideal conditions in the laboratory. Besides that obvious task, products must be designed to work in the "real world," with other equipment nearby, and to comply with government electromagnetic compatibility (EMC) regulations. This means that the equipment should not be affected by external electromagnetic sources and should not itself be a source of electromagnetic noise that can pollute the environment. Electromagnetic compatibility should be a major design objective.

## 1.2 NOISE AND INTERFERENCE

*Noise is any electrical signal present in a circuit other than the desired signal.* This definition excludes the distortion products produced in a circuit due to nonlinearities. Although these distortion products may be undesirable, they are not considered noise unless they are coupled into another part of the circuit. It follows that a desired signal in one part of a circuit can be considered to be noise when coupled to some other part of the circuit.

*Electromagnetic Compatibility Engineering,* by Henry W. Ott
Copyright © 2009 John Wiley & Sons, Inc.

Noise sources can be grouped into the following three categories: (1) intrinsic noise sources that arise from random fluctuations within physical systems, such as thermal and shot noise; (2) man-made noise sources, such as motors, switches, computers, digital electronics, and radio transmitters; and (3) noise caused by natural disturbances, such as lightning and sunspots.

*Interference is the undesirable effect of noise.* If a noise voltage causes improper operation of a circuit, it is interference. Noise cannot be eliminated, but interference can. Noise can only be reduced in magnitude, until it no longer causes interference.

## 1.3  DESIGNING FOR ELECTROMAGNETIC COMPATIBILITY

*Electromagnetic compatibility (EMC) is the ability of an electronic system to (1) function properly in its intended electromagnetic environment and (2) not be a source of pollution to that electromagnetic environment.* The electromagnetic environment is composed of both radiated and conducted energy. EMC therefore has two aspects, emission and susceptibility.

*Susceptibility* is the capability of a device or circuit to respond to unwanted electromagnetic energy (i.e., noise). The opposite of susceptibility is *immunity*. The immunity level of a circuit or device is the electromagnetic environment in which the equipment can operate satisfactorily, without degradation, and with a defined margin of safety. One difficulty in determining immunity (or susceptibility) levels is defining what constitutes performance degradation.

*Emission* pertains to the interference-causing potential of a product. The purpose of controlling emissions is to limit the electromagnetic energy emitted and thereby to control the electromagnetic environment in which other products must operate. Controlling the emission from one product may eliminate an interference problem for many other products. Therefore, it is desirable to control emission in an attempt to produce an electromagnetically compatible environment.

To some extent, susceptibility is self-regulating. If a product is susceptible to the electromagnetic environment, the user will become aware of it and may not continue to purchase that product. Emission, however, tends not to be self-regulating. A product that is the source of emission may not itself be affected by that emission. To guarantee that EMC is a consideration in the design of all electronic products, various government agencies and regulatory bodies have imposed EMC regulations that a product must meet before it can be marketed. These regulations control allowable emissions and in some cases define the degree of immunity required.

EMC engineering can be approached in either of two ways: one is the *crisis approach*, and the other is the *systems approach*. In the crisis approach, the designer proceeds with a total disregard of EMC until the functional design is finished, and testing—or worse yet—field experience suggests that a problem exists. Solutions implemented at this late stage are usually expensive and consist of undesirable "add ons." This is often referred to as the "Band Aid" approach.

As equipment development progresses from design to testing to production, the variety of noise mitigation techniques available to the designer decreases steadily. Concurrently, cost goes up. These trends are shown in Fig. 1-1. Early solutions to interference problems, therefore, are usually the best and least expensive.

The systems approach considers EMC throughout the design; the designer anticipates EMC problems at the beginning of the design process, finds the remaining problems in the breadboard and early prototype stages, and tests the final prototypes for EMC as thoroughly as possible. This way, EMC becomes an integral part of the electrical, mechanical, and in some cases, software/firmware design of the product. As a result, EMC is designed into— and not added onto—the product. This approach is the most desirable and cost effective.

If EMC and noise suppression are considered for one stage or subsystem at a time, when the equipment is initially being designed, the required mitigation techniques are usually simple and straightforward. Experience has shown that when EMC is handled this way, the designer should be able to produce equipment with 90% or more of the potential problems eliminated prior to initial testing.

A system designed with complete disregard for EMC will almost always have problems when testing begins. Analysis at that time, to find which of the many possible noise path combinations are contributing to the problem, may not be simple or obvious. Solutions at this late stage usually involve the addition of extra components that are not integral parts of the circuit. Penalties paid include the added engineering and testing costs, as well as the cost of the

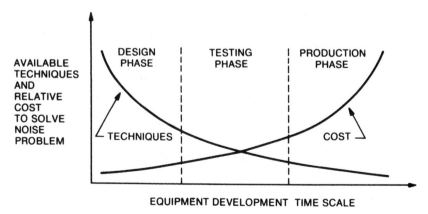

**FIGURE 1-1.** As equipment development proceeds, the number of available noise-reduction techniques goes down. At the same time, the cost of noise reduction goes up.

mitigation components and their installation. There also may be size, weight, and power dissipation penalties.

## 1.4 ENGINEERING DOCUMENTATION AND EMC

As the reader will discover, much of the information that is important for electromagnetic compatibility is not conveyed conveniently by the standard methods of engineering documentation, such as schematics, and so on. For example, a ground symbol on a schematic is far from adequate to describe where and how that point should be connected. Many EMC problems involve parasitics, which are not shown on our drawings. Also, the components shown on our engineering drawings have remarkably ideal characteristics.

The transmission of the standard engineering documentation alone is therefore insufficient. Good EMC design requires cooperation and discussion among the complete design team, the systems engineer, the electrical engineer, the mechanical engineer, the EMC engineer, the software/firmware designer, and the printed circuit board designer.

In addition, many computer-assisted design (CAD) tools do not include sufficient, if any, EMC considerations. EMC considerations therefore must often be applied manually by overriding the CAD system. Also, you and your printed circuit designer often have different objectives. Your objective is, or should be, to design a system that works properly and meets EMC requirements. Your printed circuit board (PCB) designer has the objective of doing what ever has to be done to fit all the components and traces on the board regardless of the EMC implications.

## 1.5 UNITED STATES' EMC REGULATIONS

Added insight into the problem of interference, as well as the obligations of equipment designers, manufacturers, and users of electronic products, can be gained from a review of some of the more important commercial and military EMC regulations and specifications.

The most important fact to remember about EMC regulations is that they are "living documents" and are constantly being changed. Therefore, a 1-year-old version of a standard or regulation may no longer be applicable. When working on a new design project, always be sure to have copies of the most recent versions of the applicable regulations. These standards may actually even change during the time it takes to design the product.

### 1.5.1 FCC Regulations

In the United States, the Federal Communications Commission (FCC) regulates the use of radio and wire communications. Part of its responsibility concerns the control of interference. Three sections of the FCC Rules and Regulations* have requirements that are applicable to nonlicensed electronic equipment. These requirements are contained in Part 15 for radio frequency devices; Part 18 for industrial, scientific, and medical (ISM) equipment; and Part 68 for terminal equipment connected to the telephone network.

Part 15 of the FCC Rules and Regulations sets forth technical standards and operational requirements for radio frequency devices. *A radio-frequency device is any device that in its operation is capable of emitting radio-frequency energy by radiation, conduction, or other means (§ 2.801).* The radio-frequency energy may be emitted intentionally or unintentionally. Radio-frequency (rf) energy is defined by the FCC as any electromagnetic energy in the frequency range of 9 kHz to 3000 GHz (§15.3(u)). The Part 15 regulations have a twofold purpose: (1) to provide for the operation of low-power transmitters without a radio station license and (2) to control interference to authorized radio communications services that may be caused by equipment that emits radio-frequency energy or noise as a by-product to its operation. Digital electronics fall into the latter category.

Part 15 is organized into six parts. Subpart A—General, Subpart B—Unintentional Radiators, Subpart C—Intentional Radiators, Subpart D—Unlicensed Personal Communications Devices, Subpart E—Unlicensed National Information Infrastructure Devices, and Subpart F—Ultra-Wide-band Operation. Subpart B contains the EMC Regulations for electronic devices that are not intentional radiators.

Part 18 of the FCC Rules and Regulations sets forth technical standards and operational conditions for ISM equipment. ISM equipment is defined as any device that uses radio waves for industrial, scientific, medical, or other purposes (including the transfer of energy by radio) and that is neither used nor intended to be used for radio communications. Included are medical diathermy equipment, industrial heating equipment, rf welders, rf lighting devices, devices that use radio waves to produce physical changes in matter, and other similar non-communications devices.

Part 68 of the FCC Rules and Regulations provides uniform standards for the protection of the telephone network from harm caused by connection of terminal equipment [including private branch exchange (PBX) systems] and its wiring, and for the compatibility of hearing aids and telephones to ensure that persons with hearing aids have reasonable access to the telephone network. Harm to the telephone network includes electrical hazards to telephone company workers, damage to telephone company equipment, malfunction of telephone company billing equipment, and degradation of service to persons other than the user of the terminal equipment, his calling or called party.

In December 2002, the FCC released a Report and Order (Docket 99-216) privatizing most of Part 68, with the exception of the requirements on hearing

---

*Code of Federal Regulations, Title 47, Telecommunications.

aid compatibility. Section 68.602 of the FCC rules authorized the Telecommunications Industry Association (TIA) to establish the Administrative Council for Terminal Attachments (ACTA) with the responsibility of defining and publishing technical criteria for terminal equipment connected to the U.S. public telephone network. These requirements are now defined in TIA-968. The legal requirement for all terminal equipment to comply with the technical standards, however, remains within Part 68 of the FCC rules. Part 68 requires that terminal equipment connected directly to the public switched telephone network meet both the criteria of Part 68 and the technical criteria published by ACTA.

Two approval processes are available to the manufacturer of telecommunications terminal equipment, as follows: (1) The manufacturer can provide a Declaration of Conformity (§68.320) and submit it to ACTA, or (2) the manufacturer can have the equipment certified by a Telecommunications Certifying Body (TCB) designated by the Commission (§68.160). The TCB must be accredited by the National Institute of Standards and Technology (NIST).

### 1.5.2  FCC Part 15, Subpart B

The FCC rule with the most general applicability is Part 15, Subpart B because it applies to virtually all digital electronics. In September 1979, the FCC adopted regulations to control the interference potential of digital electronics (at that time called "computing devices"). These regulations, "Technical Standards for Computing Equipment" (Docket 20780); amended Part 15 of the FCC rules relating to restricted radiation devices. The regulations are now contained in Part 15, Subpart B of Title 47 of the Code of Federal Regulations. Under these rules, limits were placed on the maximum allowable radiated emission and on the maximum allowable conducted emission on the alternating current (ac) power line. These regulations were the result of increasing complaints to the FCC about interference to radio and television reception where digital electronics were identified as the source of the interference. In this ruling the FCC stated the following:

> Computers have been reported to cause interference to almost all radio services, particularly those services below 200 MHz,* including police, aeronautical, and broadcast services. Several factors contributing to this include: (1) digital equipment has become more prolific throughout our society and are now being sold for use in the home; (2) technology has increased the speed of computers to the point where the computer designer is now working with radio frequency and electromagnetic interference (EMI) problems—something he didn't have to contend with 15 years ago; (3) modern production economics has replaced the steel cabinets which shield or reduce radiated emanations with plastic cabinets which provide little or no shielding.

---

* Remember this was 1979.

In the ruling, the FCC defined a digital device (previously called a computing device) as follows:

> An unintentional radiator (device or system) that generates and uses timing signals or pulses at a rate in excess of 9000 pulses (cycles) per second and uses digital techniques; inclusive of telephone equipment that uses digital techniques or any device or system that generates and uses radio frequency energy for the purpose of performing data processing functions, such as electronic computations, operations, transformations, recording, filing, sorting, storage, retrieval or transfer (§ 15.3(k)).

Computer terminals and peripherals, which are intended to be connected to a computer, are also considered to be digital devices.

This definition was intentionally broad to include as many products as possible. Thus, if a product uses digital circuitry and has a clock greater than 9 kHz, then it is a digital device under the FCC definition. This definition covers most digital electronics in existence today.

Digital devices covered by this definition are divided into the following two classes:

Class A: A digital device that is marketed for use in a commercial, industrial, or business environment (§ 15.3(h)).

Class B: A digital device that is marketed for use in a residential environment, notwithstanding use in commercial, business, and industrial environments (§ 15.3(i)).

Because Class B digital devices are more likely to be located in closer proximity to radio and television receivers, the emission limits for these devices are about 10 dB more restrictive than those for Class A devices.

Meeting the technical standards contained in the regulations is the obligation of the manufacturer or importer of a product. To guarantee compliance, the FCC requires the manufacturer to test the product for compliance before the product can be *marketed* in the United States. The FCC defines marketing as shipping, selling, leasing, offering for sale, importing, and so on (§ 15.803(a)). Until a product complies with the rules, it cannot legally be advertised or displayed at a trade show, because this would be considered an offer for sale. To advertise or display a product legally prior to compliance, the advertisement or display must contain a statement worded as follows:

> This device has not been authorized as required by the rules of the Federal Communications Commission. This device is not, and may not be, offered for sale or lease, or sold or leased, until authorization is obtained (§ 2.803(c)).

For personal computers and their peripherals (a subcategory of Class B), the manufacturer can demonstrate compliance with the rules by a Declaration of Conformity. A Declaration of Conformity is a procedure where the manufacturer makes measurements or takes other steps to ensure that the equipment complies with the applicable technical standards (§ 2.1071 to 2.1077). Submission of a sample unit or representative test data to the FCC is not required unless specifically requested.

For all other products (Class A and Class B—other than personal computers and their peripherals), the manufacturer must verify compliance by testing the product before marketing. *Verification* is a self-certification procedure where nothing is submitted to the FCC unless specifically requested by the Commission, which is similar to a declaration of conformity (§ 2.951 to 2.956). Compliance is by random sampling of products by the FCC. The time required to do the compliance tests (and to fix the product, and redo the test if the product fails) should be scheduled into the product's development timetable. Precompliance EMC measurements (see Chapter 18) can help shorten this time considerably.

Testing must be performed on a sample that is representative of production units. This usually means an early production or preproduction model. Final compliance testing must therefore be one of the last items in the product development timetable. This is no time for unexpected surprises! If a product fails the compliance test, then changes at this point are difficult, time consuming, and expensive. Therefore, it is desirable to approach the final compliance test with a high degree of confidence that the product will pass. This can be done if (1) proper EMC design principles (as described in this book) have been used throughout the design and (2) preliminary pre-compliance EMC testing as described in Chapter 18 was performed on early models and subassemblies.

It should be noted that the limits and the measurement procedures are interrelated. The derived limits were based on specified test procedures. Therefore, compliance measurements must be made following the procedure outlined by the regulations (§ 15.31). The FCC specifies that for digital devices, measurements to show compliance with Part 15, must be performed following the procedures described in measurement standard ANSI C63.4–1992 titled "Methods of Measurement of Radio-Noise Emissions from Low-Voltage Electrical and Electronic Equipment in the Range of 9 kHz to 40 GHz," excluding Section 5.7, Section 9, and Section 14 (§ 15.31(a)(6)).*

The test must be made on a complete system, with all cables connected and configured in a reasonable way that tends to maximize the emission (§ 15.31(i)). Special authorization procedures are provided in the case of central processor unit (CPU) boards and power supplies that are used in personal computers and sold separately (§ 15.32).

---

*Section 5.7 pertains to the use of an artificial hand to support handheld devices during testing. Section 9 pertains to measuring radio-noise power using an absorbing clamp in lieu of radiated emission measurements for certain restricted frequency ranges and certain types of equipment. Section 14 pertains to relaxing the radiated and/or conducted emission limits for short duration (≤200 ms) transients.

### 1.5.3  Emissions

The FCC Part 15 EMC Regulations limit the maximum allowable conducted emission, on the ac power line in the range of 0.150 to 30 MHz, and the maximum radiated emission in the frequency range of 30 MHz to 40 GHz.

*1.5.3.1  Radiated Emissions.*  For radiated emissions, the measurement procedure specifies an open area test site (OATS) or equivalent measurement made over a ground plane with a tuned dipole or other correlatable, linearly polarized antenna. This setup is shown in Fig. 1-2. ANSI C63.4 allows for the use of an alternative test site, such as an absorber-lined room, provided it meets specified site attenuation requirements. However, a shielded enclosure without absorber lining may not be used for radiated emission measurements.

The specified receive antenna in the 30- to- 1000-MHz range is a tuned dipole, although other linearly polarized broadband antennas may also be used. However, in case of a dispute, data taken with the tuned dipole will take precedence. Above 1000 MHz, a linearly polarized horn antenna shall be used.

Table 1-1 lists the FCC radiated emission limits (§ 15.109) for a Class A product when measured at a distance of 10 m. Table 1-2 lists the limits for a Class B product when measured at a distance of 3 m.

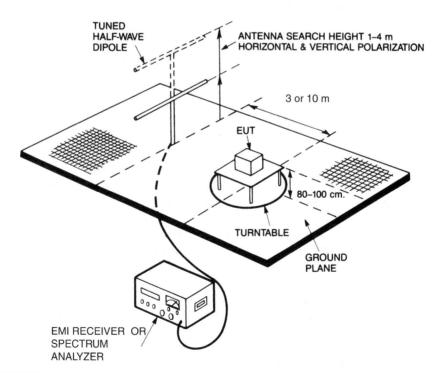

**FIGURE 1-2.**  Open area test site (OATS) for FCC radiated emission test. The equipment under test (EUT) is on the turntable.

**TABLE 1-1. FCC Class A Radiated Emission Limits Measured at 10 m.**

| Frequency (MHz) | Field Strength ($\mu$V/m) | Field Strength (dB $\mu$V/m) |
|---|---|---|
| 30–88 | 90 | 39.0 |
| 88–216 | 150 | 43.5 |
| 216–960 | 210 | 46.5 |
| >960 | 300 | 49.5 |

**TABLE 1-2. FCC Class B Radiated Emission Limits Measured at 3 m.**

| Frequency (MHz) | Field Strength ($\mu$V/m) | Field Strength (dB $\mu$V/m) |
|---|---|---|
| 30–88 | 100 | 40.0 |
| 88–216 | 150 | 43.5 |
| 216–960 | 200 | 46.0 |
| >960 | 500 | 54.0 |

**TABLE 1-3. FCC Class A and Class B Radiated Emission Limits Measured at 10 m.**

| Frequency (MHz) | Class A Limits (dB $\mu$V/m) | Class B Limit (dB $\mu$V/m) |
|---|---|---|
| 30–88 | 39.0 | 29.5 |
| 88–216 | 43.5 | 33.0 |
| 216–960 | 46.5 | 35.5 |
| >960 | 49.5 | 43.5 |

A comparison between the Class A and Class B limits must be done at the same measuring distance. Therefore, if the Class B limits are extrapolated to a 10-m measuring distance (using a $1/d$ extrapolation), the two sets of limits can be compared as shown in Table 1-3. As can be observed, the Class B limits are more restrictive by about 10 dB below 960 MHz and 5 dB above 960 MHz. A plot of both FCC Class A and Class B radiated emission limits over the frequency range of 30 MHz to 1000 MHz (at a measuring distance of 10 m) is shown in Fig. 1-5.

The frequency range over which radiated emission tests must be performed is from 30 MHz up to the frequency listed in Table 1-4, which is based on the highest frequency that the equipment under test (EUT) generates or uses.

*1.5.3.2 Conducted Emissions.* Conducted emission regulations limit the voltage that is conducted back onto the ac power line in the frequency range of 150 kHz to 30 MHz. Conducted emission limits exist because regulators believes

**TABLE 1-4.  Upper Frequency Limit for Radiated Emission Testing.**

| Maximum Frequency Generated or Used in the EUT (MHz) | Maximum Measurement Frequency (GHz) |
|---|---|
| <108 | 1 |
| 108–500 | 2 |
| 500–1000 | 5 |
| >1000 | 5th Harmonic or 40 GHz, whichever is less |

**TABLE 1-5.  FCC/CISPR Class A Conducted Emission Limits.**

| Frequency (MHz) | Quasi-peak (dB μV) | Average (dB μV) |
|---|---|---|
| 0.15–0.5 | 79 | 66 |
| 0.5–30 | 73 | 60 |

**TABLE 1-6.  FCC/CISPR Class B Conducted Emission Limits.**

| Frequency (MHz) | Quasi-peak (dB μV) | Average (dB μV) |
|---|---|---|
| 0.15–0.5 | 66–56[a] | 56–46[a] |
| 0.5–5 | 56 | 46 |
| 5–30 | 60 | 50 |

[a]Limit decreases linearly with log of frequency.

that at frequencies below 30 MHz, the primary cause of interference with radio communications occurs by conducting radio-frequency energy onto the ac power line and subsequently radiating it from the power line. Therefore, conducted emission limits are really radiated emission limits in disguise.

The FCC conducted emission limits (§ 15.107) are now the same as the International Special Committee on Radio Interference (CISPR, from its title in French) limits, used by the European Union. This is the result of the Commission amending its conducted emission rules in July 2002 to make them consistent with the international CISPR requirements.

Tables 1-5 and 1-6 show the Class A and Class B conducted emission limits, respectively. These voltages are measured common-mode (hot to ground and neutral to ground) on the ac power line using a 50-Ω/50-μH line impedance stabilization network (LISN) as specified in the measurement procedures.* Figure 1-3 shows a typical FCC conducted emission test setup.

* The circuit of an LISN is shown in Fig. 13-2.

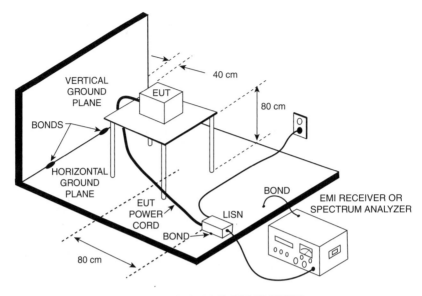

BOND METER, LISN AND GROUND PLANES TOGETHER

**FIGURE 1-3.**   Test setup for FCC conducted emission measurements.

A comparison between Tables 1-5 and 1-6 shows that the Class B quasi-peak conducted emission limits are from 13 dB to 23 dB more stringent than the Class A limits. Note also that both peak and average measurements are required. The peak measurements are representative of noise from narrowband sources such as clocks, whereas the average measurements are representative of broadband noise sources. The Class B average conducted emission limits are from 10 to 20 dB more restrictive than the Class A average limits.

Figure 1-4 shows a plot of both the average and the quasi-peak FCC/CISPR conducted emission limits.

### 1.5.4   Administrative Procedures

The FCC rules not only specify the technical standards (limits) that a product must satisfy but also the administrative procedures that must be followed and the measuring methods that must be used to determine compliance. Most administrative procedures are contained in Part 2, Subpart I (Marketing of Radio Frequency Devices), Subpart J (Equipment Authorization Procedures), and Subpart K (Importation of Devices Capable of Causing Harmful Interference) of the FCC Rules and Regulations.

Not only must a product be tested for compliance with the technical standards contained in the regulations, but also it must be labeled as compliant (§ 15.19), and information must be provided to the user (§ 15.105) on its interference potential.

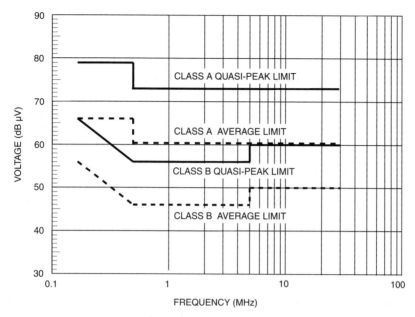

**FIGURE 1-4.**  FCC/CISPR conducted emission limits.

In addition to the technical standards mentioned above, the rules also contain a noninterference requirement, which states that if use of the product causes harmful interference, the *user* may be required to cease operation of the device (§ 15.5). Note the difference in responsibility between the technical standards and the noninterference requirement. Although meeting the technical standards (limits) is the responsibility of the manufacturer or importer of the product, satisfying the noninterference requirement is the responsibility of the user of the product.

In addition to the initial testing to determine compliance of a product, the rules also specify that the manufacturer or importer is responsible for the continued, or ongoing, compliance of subsequently manufactured units (§ 2.953, 2.955, 2.1073, 2.1075).

If a change is made to a compliant product, the manufacturer has the responsibility to determine whether that change has an effect on the compliance of the product. The FCC has cautioned manufacturers (Public Notice 3281, April 7, 1982) to note that:

Many changes, which on their face seem insignificant, are in fact very significant. Thus a change in the layout of a circuit board, or the addition or removal or even rerouting of a wire, or even a change in the logic will almost surely change the emission characteristics of the device. Whether this change in characteristics is enough to throw the product out of compliance can best be determined by retesting.

As of this writing (September 2008), the FCC has exempted eight subclasses of digital devices (§ 15.103) from meeting the technical standards of the rules. These are as follows:

1. Digital devices used exclusively in a transportation vehicle such as a car, plane, or boat.
2. Industrial control systems used in an industrial plant, factory, or public utility.
3. Industrial, commercial, or medical test equipment.
4. Digital devices exclusively used in an appliance such as a microwave oven, dishwasher, clothes dryer, air conditioner, and so on.
5. Specialized medical devices generally used at the direction or under the supervision of a licensed health care practitioner, whether used in a patient's home or a health care facility. Note, medical devices marketed through retail channels for use by the general public, are not exempted.
6. Devices with power consumption not exceeding 6 nW, for example, a digital watch.
7. Joystick controllers or similar devices (such as a mouse) that contain no digital circuitry. Note, a simple analog to digital converter integrated circuit (IC) is allowed in the device.
8. Devices in which the highest frequency is below 1.705 MHz and that does not operate from the ac power line, or contain provisions for operation while connected to the ac power line.

Each of the above exempted devices is, however, still subject to the noninterference requirement of the rules. If any of these devices actually cause harmful interference in use, the user must stop operating the device or in some way remedy the interference problem. The FCC also states, although not mandatory, it is strongly recommended that the manufacturer of an exempted device endeavor to have that device meet the applicable technical standards of Part 15 of the rules.

Because the FCC has purview over many types of electronic products, including digital electronics, design and development organizations should have a complete and current set of the FCC rules applicable to the types of products they produce. These rules should be referenced during the design to avoid subsequent embarrassment when compliance demonstration is required.

The complete set of the FCC rules is contained in the Code of Federal Regulations, Title 47 (Telecommunications)—Parts 0 to 300. They consist of five volumes and are available from the Superintendent of Documents, U.S. Government Printing Office. The FCC rules are in the first volume that contains Parts 0 to 19 of the Code of Federal Regulations. A new edition is published in the spring of each year and contains all current regulations codified as of October 1 of the previous year. The Regulations are also available online at the FCC's website, www.fcc.gov.

When changes are made to the FCC regulations, there is a transition period before they become official. This transition period is usually stated as x-number of days after the regulation is published in the Federal Register.

### 1.5.5    Susceptibility

In August 1982, the U.S. Congress amended the Communications Act of 1934 (House Bill #3239) to give the FCC authority to regulate the susceptibility of home electronics equipment and systems. Examples of home electronics equipment are radio and television sets, home burglar alarm and security systems, automatic garage door openers, electronic organs, and stereo/high-fidelity systems. Although this legislation is aimed primarily at home entertainment equipment and systems, it is not intended to prevent the FCC from adopting susceptibility standards for devices that are also used outside the home. To date, however, the FCC has not acted on this authority. Although it published an inquiry into the problem of Radio Frequency Interference to Electronic Equipment in 1978 (General Docket No. 78-369), the FCC relies on self-regulation by industry. Should industry become lax in this respect, the FCC may move to exercise its jurisdiction.

Surveys of the electromagnetic environment (Heirman 1976, Janes 1977) have shown that a field strength greater than 2 V/m occurs about 1% of the time. Because no legal susceptibility requirements exist for commercial equipment in the United States, a reasonable minimum immunity level objective might be 2 to 3 V/m. Clearly products with susceptibility levels of less than 1 V/m are not well designed and are very likely to experience interference from rf fields during their life span.

In 1982, the government of Canada released an Electromagnetic Compatibility Advisory Bulletin (EMCAB-1) that defined three levels, or grades, of immunity for electronic equipment, and stated the following:

1. Products that meet GRADE 1 (1 V/m) are likely to experience performance degradation.
2. Products that meet GRADE 2 (3 V/m) are unlikely to experience degradation.
3. Products that meet GRADE 3 (10 V/m) should experience performance degradation only under very arduous circumstances.

In June 1990, an updated version of EMCAB-1 was issued by Industry Canada. This updated version concludes that products located in populated areas can be exposed to field strengths that range from 1 V/m to 20 V/m over most of the frequency band.

### 1.5.6    Medical Equipment

Most medical equipment (other than what comes under the Part 18 Rules) is exempt from the FCC Rules. The Food and Drug Administration (FDA), not the FCC, regulates medical equipment. Although the FDA developed EMC standards, as early as 1979 (MDS-201-0004, 1979), they have never officially adopted them as mandatory. Rather, they depend on their inspectors' guideline document to assure that medical devices are properly designed to be immune to electromagnetic interference (EMI). This document, *Guide to Inspections of Electromagnetic Compatibility Aspects of Medical Devices Quality Systems*, states the following:

> At this time the FDA does not require conformance to any EMC standards. However, EMC should be addressed during the design of new devices, or redesign of existing devices.

However, the FDA is becoming increasingly concerned about the EMC aspects of medical devices. Inspectors are now requiring assurance from manufacturers that they have addressed EMC concerns during the design process, and that the device will operate properly in its intended electromagnetic environment. The above-mentioned Guide encourages manufacturers to use IEC 60601-1-2 Medical Equipment, Electromagnetic Compatibility Requirements and Tests as their EMC standard. IEC 60601-1-2 provides limits for both emission and immunity, including transient immunity such as electrostatic discharge (ESD).

As a result, in most cases, IEC 60601-1-2 has effectively become the unofficial, de facto, EMC standard that has to be met for medical equipment in the United States.

### 1.5.7   Telecom

In the United States, telecommunications central office (network) equipment is exempt from the FCC Part 15 Rules and Regulations as long as it is installed in a dedicated building or large room owned or leased by the telephone company. If it is installed in a subscriber's facility, such as an office or commercial building, the exemption does not apply and the FCC Part 15 Rules are applicable.

Telecordia's (previously Bellcore's) GR-1089 is the standard that usually applies to telecommunications network equipment in the United States. GR-1089 covers both emission and susceptibility, and it is somewhat similar to the European Union's EMC requirements. The standard is often referred to as the NEBS requirements. NEBS stands for New Equipment Building Standard. The standard is derived from the original AT&T Bell System internal NEBS standard.

These standards are not mandatory legal requirements but are contractual between the buyer and the seller. As such, the requirements can be waived or not applied in some cases.

**1.5.8   Automotive**

As stated, much (although not all) of the electronics built into transportation vehicles are exempt from EMC regulation, such as the FCC Part 15 Rules, in the United States (§ 15.103). This does not mean that vehicle systems do not have legal EMC requirements. In many regions of the world, there are legislated requirements for vehicle electromagnetic emissions and immunity. The legislated requirements are typically based on many internationally recognized standards, including CISPR, International Organization for Standardization (ISO), and the Society of Automotive Engineers (SAE). Each of these organizations has published several EMC standards applicable to the automotive industry. Although these standards are voluntary, the automotive manufacturers either rigorously apply them or use these standards as a reference in the development of their own corporate requirements. These developed corporate requirements may include both component and vehicle level items and are often based upon the customer satisfaction goals of the manufacturer—therefore, they almost have the effect of mandatory standards.

For example, SAE J551 is a vehicle-level EMC standard, and SAE J1113 is a component-level EMC standard applicable to individual electronic modules. Both standards cover emissions and immunity and are somewhat similar to the military EMC standards.

The resulting vehicle EMC standards cover both emissions and immunity and are some of the toughest EMC standards in the world, partly because of the combination of types of systems on vehicles and their proximity to each other. These systems include high-voltage discharges (such as spark ignition systems) located near sensitive entertainment radio receiver systems, wiring for inductive devices such as motors and solenoids in the same wiring harness as data communication lines, and with the newer "hybrid vehicles" high-current motor drive systems that operate at fast switching speeds. The radiated emission standards are typically 40 dB more stringent than the FCC Class B limits. Radiated immunity tests are specified up to an electric field strength of 200 V/m (or in some cases higher) as compared with 3 or 10 V/m for most non-automotive commercial immunity standards.

In the European Union, vehicles and electronic equipment intended for use in these vehicles are exempt from the EMC Directive (204/108/EC), but they do fall within the scope of the automotive directive (95/54/EC) that contains EMC requirements.

**1.6   CANADIAN EMC REQUIREMENTS**

The Canadian EMC regulations are similar to those of the United States. The Canadian regulations are controlled by Industry Canada. Table 1-7 lists the Canadian EMC standards applicable to various types of products. These standards can be accessed from the Industry Canada web page (www.ic.gc.ca).

**TABLE 1-7.  Canadian EMC Test Standards.**

| Equipment Type | Standard |
|---|---|
| Information technology equipment (ITE)[a] | ICES–003 |
| Industrial, Scientific & Medical Equipment (ISM) | ICES–001 |
| Terminal Equipment Connected to the Telephone Network | CS–03 |

[a]Digital Equipment.

The ITE and ISM standards can be accessed from the Industry Canada home page by following the following links: A-Z Index/Spectrum Management and Telecommunications/Official Publications/Standards/Interference-Causing Equipment Standards (ICES). The telecom standard can be accessed from the Industry Canada home page by following the following links: A-Z Index/ Spectrum Management and Telecommunications/Official Publications/Standards/Terminal Equipment-Technical Specifications List.

The methods of measurement and actual limits for ITE are contained in CAN/CSA-CEI/IEC CISPR 22:02, Limits and Methods of Measurement of Radio Disturbance Characteristics of Information Technology Equipment.

To reduce the burden on U.S. and Canadian manufacturers, the United States and Canada have a mutual recognition agreement whereby each country agrees to accept test reports from the other country for equipment authorization purposes (FCC Public Notice 54795, July 12, 1995).

## 1.7   EUROPEAN UNION'S EMC REQUIREMENTS

In May 1989, the European Union (EU) published a directive (89/336/EEC) relating to electromagnetic compatibility, which was to be effective January 1, 1992. However, the European Commission underestimated the task of implementing the directive. As a result, the European Commission amended the directive in 1992 allowing for a 4-year transition period and requiring full implementation of the EMC directive by January 1, 1996.

The European EMC directive differs from the FCC regulations by including immunity requirements in addition to emission requirements. Another difference is that the directive, without exception, covers all electrical/electronic equipment. There are no exemptions—the EMC directive even covers a light bulb. The directive does, however, exclude equipment that is covered by another directive with EMC provisions, such as the automotive directive. Another example would be medical equipment, which comes under the medical directive (93/42/EEC) not the EMC directive.

### 1.7.1   Emission Requirements

As stated, the EU's conducted emission requirements are now the same as the FCC's (see Tables 1-5 and 1-6 as well as Fig. 1-4). The radiated emission

**TABLE 1-8.   CISPR Radiated Emission Limits at 10 m.**

| Frequency (MHz) | Class A Limit (dB μV/m) | Class B Limit (dB μV/m) |
|---|---|---|
| 30–230 | 40 | 30 |
| 230–1000 | 47 | 37 |

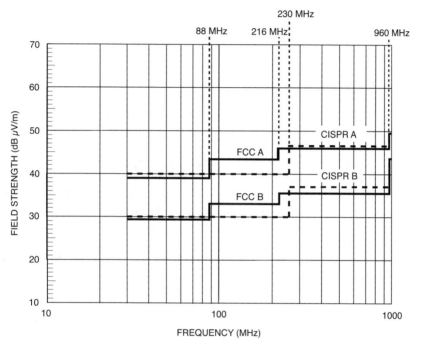

**FIGURE 1-5.** Comparison of FCC and CISPR radiated emission limits, measured at a distance of 10 m.

standards are similar but not exactly the same. Table 1-8 shows the European Union's Class A and Class B radiated emission limits when measured at 10 m.

Figure 1-5 compares the EU's radiated emission standard with the current FCC standard over the frequency range of 30 MHz to 1000 MHz. The FCC Class B limits have been extrapolated to a 10-m measuring distance for this comparison. As can be observed the European (CISPR) limits are more restrictive in the frequency range from 88 to 230 MHz. Below 88 MHz and above 230 MHz the CISPR and FCC limits are virtually the same (within 0.5 dB of each other). However, the EU has no radiated emission limit above 1 GHz, whereas the FCC limits, under some circumstances (see Table 1-4), go up to 40 GHz.

Table 1-9 is a simplified, composite worst-case combination of the FCC and CISPR radiated emission limits when measured at 10 m.

**TABLE 1-9.   Simplified, Composite Worst-Case Radiated Emission Limits for Commercial Products, Measured at a Distance of 10 m.**

| Frequency (MHz) | Class A Limit (dB μV/m) | Class B Limit (dB μV/m) |
|---|---|---|
| 30–230 | 39 | 29.5 |
| 230–1000 | 46.5 | 35.5 |
| >1000 | 49.5 | 43.5 |

### 1.7.2   Harmonics and Flicker

The EU has two additional emission requirements that relate to power quality issues—harmonics and flicker. These regulations apply to products that draw an input current of 16 A per phase or less and are intended to be connected to the public ac power distribution system. The FCC has no similar requirement.

The harmonic requirement (EN 61000-3-2) limits the harmonic content of the current drawn by the product from the ac power line, (see Table 18-3). The generation of harmonics is the result of the nonlinear behavior of the loads connected to the ac power line. Common nonlinear loads include switched-mode power supplies, variable-speed motor drives, and electronic ballasts for fluorescent lamps.

A major source of harmonics is a full-wave rectifier connected directly to the ac power line and followed by a large-value capacitor input filter. Under these circumstances, current is only drawn from the power line when the input voltage exceeds that on the filter capacitor. As a result, current is drawn from the power line only on the peaks of the ac voltage waveform (see Fig. 13-4). The resultant current waveshape is rich in odd harmonics (third, fifth, seventh, etc.). Total harmonic distortion (THD) values of 70% to 150% are not uncommon under these circumstances.

The number of harmonics present is determined by the rise and fall time of the current pulse, and their magnitude by the current wave shape. Most switching power supplies (the exception is very low-power supplies) and variable-speed motor drives cannot meet this requirement without some kind of passive or active power factor correction circuitry.

To alleviate this problem, the ac input current pulse must be spread out over a larger portion of a cycle to reduce the harmonic content. Normally the THD of the current pulse must be reduced to 25% or less to be compliant with the EU regulations.

The flicker requirements (EN 61000-3-3) limit the transient ac power line current drawn by the product; see Table 18-4. The purpose of this requirement is to prevent lights from flickering, because it is perceived as being disturbing to people. The regulations are based on not providing a noticeable change in the illumination of a 60-W incandescent lamp powered off the same ac power supply as the equipment under test.

Because of the finite source impedance of the power line, the changing current requirements of equipment connected to the line produces corresponding voltage fluctuations on the ac power line. If the voltage variation is large enough, it will produce a perceptible change in lighting illumination. If the load changes are of sufficient magnitude and repetition rate, the resulting flickering of lights can be irritating and disturbing.

To determine an applicable limit, many people were subjected to light flicker to determine the irritability threshold. When the flicker rate is low (<1 per minute), the threshold of irritability is when the ac line voltage changes by 3%. People are most sensitive to light flicker when the rate is around 1000 times per minute. At a rate of 1000 times per minute, a 0.3% voltage change is just as irritating as a 3% change at less than one change per minute. Above about 1800 changes per minute, light flicker is no longer perceived.

Most EMC emission requirements are based on the magnitude of a measured parameter not exceeding a specified amount (the limit). However, flicker tests are different in that they require many measurements to be made and then a statistical analysis to be performed on the measured data to determine whether the limit is exceeded.

For most equipment, this requirement is not a problem because they naturally do not draw large transient currents off the ac power line. However, the requirement can be a problem for products that suddenly switch on heaters that draw large currents, or motors under a heavy load. An example would be when an air conditioner compressor or a large heater in a copy machine is suddenly switched on.

### 1.7.3  Immunity Requirements

The EU's immunity requirements cover radiated and conducted immunity, as well as transient immunity that include ESD, electrical fast transient (EFT), and surge.

The EFT requirement simulates noise generated by inductively switched loads on the ac power line. As a contactor is opened to an inductive load, an arc is formed that extinguishes and restarts many times. The surge requirement is intended to simulate the effect of a nearby lightning pulse.

In addition, the EU has susceptibility requirements that cover ac voltage dips, sags, and interruptions.

For additional information on these transient immunity and power line disturbance requirements, see Sections 14.3 and 14.4.

### 1.7.4  Directives and Standards

The European regulations consist of directives and standards. The directives are very general and are the legal requirements. The standards provide one way, but not the only way, to comply with the directive.

The EMC Directive 2004/108/EC (which superceded the original EMC Directive 89/336/EEC) defines the *essential requirements* for a product to be marketed in the EU. They are as follows:

1. The equipment must be constructed to ensure that any electromagnetic disturbance it generates allows radio and telecommunication equipment and other apparatus to function as intended.
2. The equipment must be constructed with an inherent level of immunity to externally generated electromagnetic disturbances.

These are the *only legal requirements* with respect to EMC and the requirements are vague. The directive provides for two methods of demonstrating compliance with its requirements. The most commonly used is by a declaration of conformity; the other option is the use of a technical construction file.

If a product is tested to and complies with the applicable EMC standards it is presumed to meet the requirements of the directive, and the manufacturer can produce a declaration of conformity attesting to that fact.

A declaration of conformity is a self-certification process in which the responsible party, manufacturer or importer, must first determine the applicable standards for the product, test the product to the standards, and issue a declaration declaring compliance with those standards and the EMC directive. The declaration of conformity can be a single-page document but must contain the following:

- Application of which council directives (all applicable directives)
- Standards used (including date of standard) to determine conformity
- Product name and model number, also serial numbers if applicable
- Manufacturer's name and address
- A dated declaration that the product conforms to the directives
- A signature by a person empowered to legally bind the manufacturer

The technical construction file approach to demonstrating conformity is unique to the European Union. The technical construction file is often used where no harmonized standards exist for the product and the manufacturer does not think that the generic standards are appropriate. In this case, the manufacturer produces a technical file to describe the procedures and tests used to ensure compliance with the EMC directive. The manufacturer can develop its own EMC specifications and test procedures. The manufacturer can decide how, where, when, or if, the product is tested for EMC. An independent *competent body*, however, must approve the technical construction file. The competent bodies are appointed by the individual states of the European Union, and the European Commission publishes a list of them in the Official Journal of the European Union. The competent body must agree that, using the manufacturer's procedures and tests, the product satisfies the essential requirements of the EMC directive. This approach is acceptable, because in the European Union, the EMC directive is the legal document that must be satisfied, not the standards. In most other jurisdictions, the standards are the legal documents that must be complied with.

Products whose compliance with the EMC directive has been demonstrated by one of the above procedures shall be labeled with the *CE mark*. The CE mark consists of the lower case letters "ce" in a specified, distinctive font. Affixing the CE mark to a product indicates conformity to *all* applicable directives, not just the EMC directive. Other applicable directives might be, the safety directive, the toy directive, the machinery directive, and so on.

Two types of standards exist in the European Union: product specific and generic.* Product-specific standards always take precedence over generic standards. However, if no applicable product-specific standard exists for a product, the generic standards are then applicable. Emission and immunity requirements for a product are usually covered by different standards. Currently, over 50 different standards are associated with the EMC directive. Table 1-10 lists some of the more commonly applicable product-specific standards, as well as the four generic EMC standards. If a product-specific standard does not exist in a category, then the requirement defaults to the appropriate generic standard.

The EU's standards writing organization CENELEC (the European Committee for Electro-Technical Standardization) has been given the task of drawing up the corresponding technical specifications meeting the essential requirements of the EMC directive, compliance with which will provide a presumption of conformity with the essential requirements of the EMC

**TABLE 1-10.  European Union's EMC Test Standards.**

| Equipment Type | Emission | Immunity |
|---|---|---|
| **Product Specific Standards** | | |
| Information Technology Equipment (ITE) | EN 55022 | EN 55024 |
| Industrial, Scientific & Medical Equipment (ISM) | EN 55011 | – |
| Radio & Television Receivers | EN 55013 | EN 55020 |
| Household Appliances/Electric Tools | EN 55014-1 | EN 55014-2 |
| Lamps & Luminaries | EN 55015 | EN 61547 |
| Adjustable Speed Motor Drives | EN 61800-3 | EN 61800-3 |
| Medical Equipment[a] | EN 60601-1-2 | EN 60601-1-2 |
| **Generic Standards** | | |
| Residential, Commercial, Light Industrial Environment | EN 61000-6-3 | EN 61000-6-1 |
| Heavy Industrial Environment | EN 61000-6-4 | EN 61000-6-2 |

[a]Covered by the Medical Directive (93/42/EEC), not the EMC Directive

* A third type of standard also exists, which is a basic standard. Basic standards are usually test or measurement procedures and are referenced by the product-specific or generic standards.

directive. Such specifications are referred to as harmonized standards. Most CENELEC standards are derived from International Electro-Technical Committee (ITC) or CISPR standards—ITC for immunity standards and CISPR for emission standards. The CENELEC standards, or European Norms (EN), are not official until a reference to them is published in the "Official Journal of the European Union."

As new standards come into existence and existing standards are modified, as regularly happens, a transition period, usually of 2 years is specified in the standard. During the transition period, either the old standard or the new standard can be used to demonstrate compliance with the EMC directive.

The latest information on the EMC Directive 2004/108/EC and the harmonized standards can be obtained on the following website: http://europa.eu.int/comm/enterprise/newapproach/standardization/harmstds/reflist/emc.html.

In light of the large breadth and scope of the EMC Directive and the variety of products covered, the European Commission in 1997 felt it necessary to publish a 124-page guideline to the interpretation of the EMC directive to be used by manufacturers, test laboratories, and other parties affected by the directive (European Commission, 1997). This guideline was intended to clarify matters and procedures relating to the interpretation of the EMC Directive. It also clarified the application of the Directive to components, subassemblies, apparatus, systems, and installations, as well as the application of the Directive to spare parts, used, and repaired apparatus.

## 1.8 INTERNATIONAL HARMONIZATION

It would be desirable to have one international EMC standard for allowable emission and immunity of electronic products, instead of many different national standards. This would allow a manufacturer to design and test a product to one standard that would be acceptable worldwide. Figure 1-6 depicts a typical commercial product and shows the different types of EMC requirements, both emission and immunity, that it might have to meet in a harmonized world market.

Even more important than a single uniform EMC standard is a single uniform EMC test procedure. If the test procedure is the same, then an EMC test could be performed once and the results compared against many different standards (limits) to determine compliance with each regulation. When the test procedures are different, however, the product must be retested for each standard, which is a costly and time-consuming task.

The most likely vehicle for accomplishing harmonization is the European Union's EMC standards, which are based on the CISPR standards. CISPR was formed in 1934 to determine measurement methods and limits for radio-frequency interference to facilitate international trade. CISPR has no regulatory authority, but its standards, when adopted by governments, become

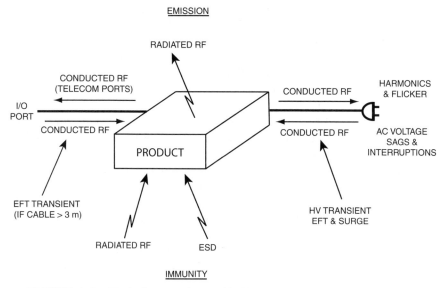

**FIGURE 1-6.** Typical composite worldwide commercial EMC requirements.

national standards. In 1985 CISPR adopted a new set of emission standards (Publication 22) for Information Technology Equipment (computer and digital electronics). The European Union has adopted the CISPR standard as the basis for their emission requirements. As a voting member of CISPR, the United States voted in favor of the new standard. This action puts considerable pressure on the FCC to adopt the same standards.

In 1996, the FCC modified its Part 15 Rules to allow manufacturers to use a Declaration of Conformity as a compliance procedure for personal computers and their peripherals, which is similar to that used by the EU's EMC regulations. As stated, the FCC also has adopted the CISPR limits for conducted emission.

## 1.9   MILITARY STANDARDS

Another important group of EMC standards are those issued by the U.S. Department of Defense and are applicable to military and aerospace equipment. In 1968, the Department of Defense consolidated the multitude of different EMC standards from the various branches of the service into two universally applicable standards. MIL-STD-461 specified the limits that had to be met, and MIL-STD-462 specified the test methods and procedures for making the tests contained in MIL-STD-461. These standards are more stringent than the FCC regulations, and they cover immunity as well as emissions in the frequency range of 30 Hz to 40 GHz.

Over the years, these standards have gone through revisions that ranged from MIL-STD-461A in 1968 to MIL-STD-461E in 1999. In 1999, MIL-STD-461D (Limits) and MIL-STD-462D (Test Procedures) were merged into one standard MIL-STD-461E that covered both limits and test procedures.*

Unlike commercial standards, MIL-STDs are not legal requirements; rather, they are contractual requirements. As such, test limits can be negotiated and waivers are possible. Earlier versions are still applicable to current products because the requirements are contractual, not legal. Normally whatever version the original procurement contract specified is still applicable.†

The test procedures specified in the military standards are often different than those specified by commercial EMC standards, which makes a direct comparison of the limits difficult. For radiated emissions the military standard specifies enclosed chamber (shielded room) testing, whereas the FCC and the EU rules require open-area testing. For conducted emission testing, the military standards originally measured current, whereas the commercial standards measure voltage.

As more was learned about EMC testing and its accuracy, the military has come under some criticism for some of its test procedures. As a result, the military has adopted some of the commercial test procedures. For example, MIL-STD-461E specifies the use of a LISN and the measurement of voltage rather than current for conducted emission testing. Also MIL-STD-461E requires that some absorber material must be used on the walls of chambers used for emission and immunity testing to make the chamber at least partially anechoic.

Table 1-11 is a list of the emission and immunity requirements established by MIL-STD-461E. Tests are required for both radiated and conducted emissions as well as for radiated, conducted, and high-voltage transient susceptibility.

The military standards are application specific, often with different limits for different environments (such as Army, Navy, aerospace, etc.). Some requirements listed in Table 1-11 are applicable to only certain environments and not to others. Table 1-12 lists the applicability of the requirements to the various environments.

## 1.10  AVIONICS

The commercial avionics industry has its own set of EMC standards, which are similar to those of the military. These standards apply to the entire spectrum of commercial aircraft, which includes light general aviation aircraft, helicopters,

---

* On December 10, 2007, MIL-STD 461F was released.
† By contrast, when a commercial standard is revised or modified, all newly manufactured products must comply with the new limits by the end of the specified transition period.

**TABLE 1-11.   Emission and Susceptibility Requirements of MIL-STD-461E.**

| Requirement | Description |
|---|---|
| CE101 | Conducted Emissions, Power Leads, 30 Hz to 10 kHz |
| CE102 | Conducted Emissions, Power Leads, 10 kHz to 10 MHz |
| CE106 | Conducted Emissions, Antenna Terminals, 10 kHz to 40 GHz |
| CS101 | Conducted Susceptibility, Power Leads, 30 Hz to 50 kHz |
| CS103 | Conducted Susceptibility, Antenna Port, Inter–modulation, 15 kHz to 10 GHz |
| CS104 | Conducted Susceptibility, Antenna Port, Rejection of Undesired Signals, 30 Hz to 20 GHz |
| CS105 | Conducted Susceptibility, Antenna Port, Cross-modulation, 30 Hz to 20 GHz |
| CS109 | Conducted Susceptibility, Structure Current, 60 Hz to 100 kHz |
| CS114 | Conducted Susceptibility, Bulk Current Injection, 10 kHz to 40 MHz |
| CS115 | Conducted Susceptibility, Bulk Current Injection, Impulse Excitation |
| CS116 | Conducted Susceptibility, Damped Sinusoidal Transients, Cables and Power Leads, 10 kHz to 100 MHz |
| RE101 | Radiated Emission, Magnetic Field, 30 Hz to 100 kHz |
| RE102 | Radiated Emission, Electric Field, 10 kHz to 18 GHz |
| RE103 | Radiated Emission, Antenna Spurious and Harmonic Outputs, 10 kHz to 40 GHz |
| RS101 | Radiated Susceptibility, Magnetic Field, 30 Hz to 100 kHz |
| RS103 | Radiated Susceptibility, Electric Field, 10 kHz to 40 GHz |
| RS105 | Radiated Susceptibility, Transient Electromagnetic Field |

**TABLE 1-12.   Requirement Applicability Matrix, MIL-STD-461E.**

| Equipment Installed In, On, or Launched From the Following Platforms or Installations | CE101 | CE102 | CE106 | CS101 | CS103 | CS104 | CS105 | CS109 | CS114 | CS115 | CS116 | RE101 | RE102 | RE103 | RS101 | RS103 | RS105 |
|---|---|---|---|---|---|---|---|---|---|---|---|---|---|---|---|---|---|
| Surface Ships | N | A | L | A | S | S | S | N | A | L | A | A | A | L | A | A | L |
| Submarines | A | A | L | A | S | S | S | L | A | L | A | A | A | L | A | A | L |
| Aircraft, Army, & Flight Line | A | A | L | A | S | S | S | N | A | A | A | A | A | L | A | A | L |
| Aircraft, Navy | L | A | L | A | S | S | S | N | A | A | A | L | A | L | L | A | L |
| Aircraft, Air Force | N | A | L | A | S | S | S | N | A | A | A | N | A | L | N | A | N |
| Space Systems & Launch Eq. | N | A | L | A | S | S | S | N | A | A | A | N | A | L | N | A | N |
| Ground, Army | N | A | L | A | S | S | S | N | A | A | A | N | A | L | L | A | N |
| Ground, Navy | N | A | L | A | S | S | S | N | A | A | A | N | A | L | A | A | L |
| Ground, Air Force | N | A | L | A | S | S | S | N | A | A | A | N | A | L | N | A | N |

A = applicable, L = limited applicability as specified in the standard, S = applicable only if specified in procurement document, N = not applicable.

and jumbo jets. The Radio Technical Commission for Aeronautics (RTCA) produces these standards for the avionics industry. The current version is RTCA/DO-160E *Environmental Conditions and Test Procedures For Airborne Equipment* and was issued in December 2004. Sections 15 through 23 and Section 25 cover EMC issues.

Like the military standard, DO-160E is a contractual, not legal, requirement, so its terms may be negotiable.

## 1.11   THE REGULATORY PROCESS

We are all probably familiar with the phrase *ignorance of the law is no defense.* How then do governments make their commercial EMC regulations public, so that we all presumably know of their existence? In most countries, regulations are made public by publication, or being referenced, in the "Official Journal" of that country. In the United States, the official journal is the *Federal Register*; in Canada, it is the *Canada Gazette*; and in the European Union, it is the *Official Journal of the European Union.*

Once a regulation is published, or referenced, in the official journal, *its official*, and everyone is presumed to know of its existence.

## 1.12   TYPICAL NOISE PATH

A block diagram of a typical noise path is shown in Fig. 1-7. As shown, three elements are necessary to produce an interference problem. First, there must be a *noise source.* Second, there must be a *receptor* circuit that is susceptible to the noise. Third, there must be a *coupling channel* to transmit the noise from the source to the receptor. In addition, the characteristics of the noise must be such that it is emitted at a *frequency* that the receptor is susceptible, an *amplitude* sufficient to affect the receptor, and a *time* the receptor is susceptible to the noise. A good way to remember the important noise characteristics is with the acronym FAT.

The first step in analyzing a noise problem is to define the problem. This is done by determining what is the noise source, what is the receptor, what is the coupling channel, and what are the FAT characteristics of the noise. It follows that there are three ways to break the noise path: (1) the characteristics of the noise can be changed at the source, (2) the receptor can be made insensitive to the noise, or (3) the transmission through the coupling channel

**FIGURE 1-7.**   Before noise can be a problem, there must be a noise source, a receptor, and a coupling channel.

can be eliminated or minimized. In some cases, the noise suppression techniques must be applied to two or to all three parts of the noise path.

In the case of an emission problem, we are most likely to attack the source of the emissions by changing its characteristics—its frequency, amplitude, or time. For a susceptibility problem, we are most likely to direct our attention to modifying the receptor to increase its immunity to the noise. In many cases, modifying the source or receptor is not practical, which then leaves us with only the option of controlling the coupling channel.

As an example, consider the circuit shown in Fig. 1-8. It shows a shielded direct current (dc) motor connected to its motor-drive circuit. Motor noise is interfering with a low-level circuit in the same equipment. Commutator noise from the motor is conducted out of the shield on the leads going to the drive circuit. From the leads, noise is radiated to the low-level circuitry.

In this example, the noise source consists of the arcs between the brushes and the commutator. The coupling channel has two parts: conduction on the motor leads and radiation from the leads. The receptor is the low-level circuit. In this case, not much can be done about the source or the receptor. Therefore, the interference must be eliminated by breaking the coupling channel. Noise conduction out of the shield or radiation from the leads must be stopped, or both steps may be necessary. This example is discussed more fully in Section 5.7.

## 1.13   METHODS OF NOISE COUPLING

### 1.13.1   Conductively Coupled Noise

One of the most obvious, but often overlooked, ways to couple noise into a circuit is on a conductor. A wire run through a noisy environment may pick up

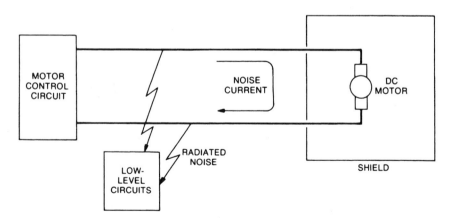

**FIGURE 1-8.**   In this example, the noise source is the motor, and the receptor is the low-level circuit. The coupling channel consists of conduction on the motor leads and radiation from the leads.

noise and then conduct it to another circuit. There it causes interference. The solution is to prevent the wire from picking up the noise or to remove the noise from it by filtering before it interferes with the susceptible circuit.

The major example in this category is noise conducted into a circuit on the power supply leads. If the designer of the circuit has no control over the power supply, or if other equipment is connected to the power supply, it becomes necessary to decouple or filter the noise from the wires before they enter the circuit. A second example is noise coupled into or out of a shielded enclosure by the wires that pass through the shield.

### 1.13.2   Common Impedance Coupling

Common impedance coupling occurs when currents from two different circuits flow through a common impedance. The voltage drop across the impedance observed by each circuit is influenced by the other circuit. This type of coupling usually occurs in the power and/or ground system. The classic example of this type of coupling is shown in Fig. 1-9. The ground currents 1 and 2 both flow through the common ground impedance. As far as circuit 1 is concerned, its ground potential is modulated by ground current 2 flowing in the common ground impedance. Some noise, therefore, is coupled from circuit 2 to circuit 1, and vice versa, through the common ground impedance.

Another example of this problem is illustrated in the power distribution circuit shown in Fig. 1-10. Any change in the supply current required by circuit 2 will affect the voltage at the terminals of circuit 1 because of the common impedances of the power supply lines and the internal source impedance of the power supply. A significant improvement can be obtained by connecting

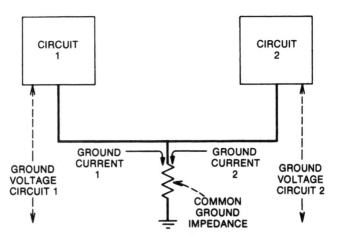

**FIGURE 1-9.**   When two circuits share a common ground, the ground voltage of each one is affected by the ground current of the other circuit.

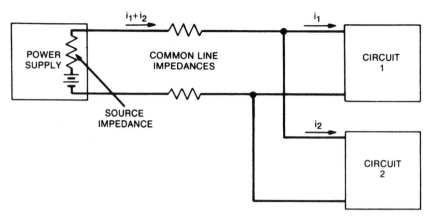

**FIGURE 1-10.** When two circuits share a common power supply, current drawn by one circuit affects the voltage at the other circuit.

the leads from circuit 2 directly to the power supply output terminals, thus bypassing the common line impedance. However, some noise coupling through the power supply's internal impedance will remain.

### 1.13.3   Electric and Magnetic Field Coupling

Radiated electric and magnetic fields provide another means of noise coupling. All circuit elements, including conductors, radiate electromagnetic fields whenever a charge is moved. In addition to this unintentional radiation, there is the problem of intentional radiation from sources such as broadcast stations and radar transmitters. When the receiver is close to the source (near field), electric and magnetic fields are considered separately. When the receiver is far from the source (far field), the radiation is considered as combined electric and magnetic or electromagnetic radiation.*

## 1.14   MISCELLANEOUS NOISE SOURCES

### 1.14.1   Galvanic Action

If dissimilar metals are used in the signal path in low-level circuitry, a noise voltage may appear from the galvanic action between the two metals. The presence of moisture or water vapor in conjunction with the two metals produces a chemical wet cell (galvanic couple). The voltage developed depends on the two metals used and is related to their positions in the galvanic series

---

* See Chapter 6 for an explanation of near field and far field.

**TABLE 1-13.   Galvanic Series.**

|  |  |  |  |
|---|---|---|---|
| **ANODIC END** |  |  |  |
| (Most susceptible to corrosion) |  |  |  |
| Group I | 1. Magnesium |  | 13. Nickel (active) |
|  |  |  | 14. Brass |
|  | 2. Zinc |  | 15. Copper |
|  | 3. Galvanized steel |  | 16. Bronze |
| Group II | 4. Aluminum 2S | Group IV | 17. Copper-nickel alloy |
|  | 5. Cadmium |  | 18. Monel |
|  | 6. Aluminum 17ST |  | 19. Silver solder |
|  |  |  | 20. Nickel (passive)$^a$ |
|  | 7. Steel |  | 21. Stainless steel |
|  | 8. Iron |  | (passive)$^a$ |
|  | 9. Stainless steel |  |  |
| Group III | (active) |  | 22. Silver |
|  | 10. Lead-tin solder | Group V | 23. Graphite |
|  | 11. Lead |  | 24. Gold |
|  | 12. Tin |  | 25 Platinum |
|  |  |  | **CATHODIC END** |
|  |  |  | (Least susceptibility to corrosion) |

$^a$Passivation by immersion in a strongly acidic solution.

shown in Table 1-13. The farther apart the metals are on this table, the larger the developed voltage. If the metals are the same, no potential difference can develop.

In addition to producing a noise voltage, the use of dissimilar metals can produce a corrosion problem. Galvanic corrosion causes positive ions from one metal to be transferred to the other one. This action gradually causes the anode material to be destroyed. The rate of corrosion depends on the moisture content of the environment and how far apart the metals are in the galvanic series. The farther apart the metals are in the galvanic series, the faster the ion transfer. An undesirable, but common, combination of metals is aluminum and copper. With this combination, the aluminum is eventually eaten away. The reaction slows down considerably, however, if the copper is coated with lead-tin solder because aluminum and lead-tin solder are closer in the galvanic series.

The following four elements are needed before galvanic action can occur:

1. Anode material (higher rank in Table 1-13)
2. Electrolyte (usually present as moisture)
3. Cathode material (lower rank in Table 1-13)
4. Conducting electrical connection between anode and cathode (usually present as a leakage path)

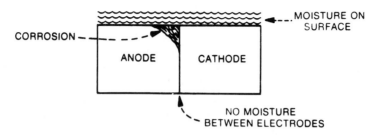

**FIGURE 1-11.** Galvanic action can occur if two dissimilar metals are joined and moisture is present on the surface.

Galvanic action can take place even if moisture does not get between the anode and the cathode. All that is needed is some moisture on the surface where the two metals come together, as shown in Fig. 1-11.

As observed in Table 1-13, the metals of the galvanic series are divided into five groups. When dissimilar metals must be combined, it is desirable to use metals from the same group. Usually metals from adjacent groups can be used together if the product is to be used in a fairly benign indoor environment.

Other methods of minimizing corrosion between two dissimilar metals are as follows:

- Keep the cathode material as small as possible.
- Plate one of the materials to change the group that the contact surface is in.
- Coat the surface, after joining to exclude surface moisture.

### 1.14.2 Electrolytic Action

A second type of corrosion is caused by electrolytic action. It is caused by a direct current flowing between two metals with an electrolyte (which could be slightly acidic ambient moisture) between them. This type of corrosion does not depend on the two metals used and will occur even if both are the same. The rate of corrosion depends on the magnitude of the current and on the conductivity of the electrolyte.

### 1.14.3 Triboelectric Effect

A charge can be produced on the dielectric material within a cable, if the dielectric does not maintain contact with the cable conductors. This is called the triboelectric effect. It is usually caused by mechanical bending of the cable. The charge acts as a noise voltage source within the cable. Eliminating sharp bends and cable motion minimizes this effect. A special "low-noise" cable is available in which the cable is chemically treated to minimize the possibility of charge buildup on the dielectric.

### 1.14.4    Conductor Motion

If a conductor is moved through a magnetic field, a voltage is induced between the ends of the wire. Because of power wiring and other circuits with high-current flow, stray magnetic fields exist in most environments. If a wire with a low-level signal is allowed to move through this field, then a noise voltage will be induced in the wire. This problem can be especially troublesome in a vibrational environment. The solution is simple: prevent wire motion with cable clamps and other tie-down devices.

## 1.15    USE OF NETWORK THEORY

For the exact answer to the question of how any electric circuit behaves, Maxwell's equations must be solved. These equations are functions of three space variables $(x, y, z)$ and of time $(t)$—a four-dimensional problem. Solutions for any but the simplest problems are usually complex. To avoid this complexity, an approximate analysis technique called "electric circuit analysis" is used during most design procedures.

Circuit analysis eliminates the spatial variables and provides approximate solutions as a function of time (or frequency) only. Circuit analysis assumes the following:

1. All electric fields are confined to the interiors of capacitors.
2. All magnetic fields are confined to the interiors of inductors.
3. Dimensions of the circuits are small compared with the wavelength(s) under consideration.

What is really implied is that external fields, even though actually present, can be neglected in the solution of the network. Yet these external fields may not necessarily be neglected where their effect on other circuits is concerned.

For example, a 100-W power amplifier may radiate 100 mW of power. These 100 mW are completely negligible as far as the analysis and operation of the power amplifier is concerned. However, if only a small percentage of this radiated power is picked up on the input of a sensitive circuit, it may cause interference.

Even though the 100 mW of radiated emission is completely negligible to the 100-W power amplifier, a sensitive radio receiver, under the right conditions, may be capable of picking up the signal thousands of miles away.

Whenever possible, noise-coupling channels are represented as equivalent lumped component networks. For instance, a time-varying electric field that exists between two conductors can be represented by a capacitor connecting the two conductors as shown in Fig. 1-12. A time-varying magnetic field that

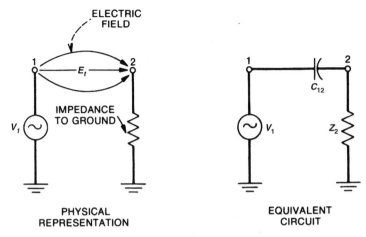

PHYSICAL
REPRESENTATION

EQUIVALENT
CIRCUIT

**FIGURE 1-12.**   When two circuits are coupled by an electric field, the coupling can be represented by a capacitor.

couples two conductors can be represented by a mutual inductance between the two circuits as shown in Fig. 1-13.

For this approach to be valid, the physical dimensions of the circuits must be small compared with the wavelengths of the signals involved. Wherever appropriate, this assumption is made throughout this book.

Even when this assumption is not truly valid, the lumped component representation is still useful for the following reasons:

1. The solution of Maxwell's equations is not practical for most "real-world" noise problems because of the complicated boundary conditions.
2. Although lumped component representation will not produce the most accurate numerical answer, it does clearly show how noise depends on the parameters of the system. On the other hand, the solution of Maxwell's equations, even if possible, does not clearly show such parameter dependence.
3. To solve a noise problem, a parameter of the system must be changed, and lumped circuit analysis clearly points out the parameter dependence.

In general, the numerical values of the lumped components are extremely difficult to calculate with any precision, except for certain special geometries. One can conclude, however, that these components exist, and as will be shown, the results can be very useful even when the components are only defined in a qualitative sense.

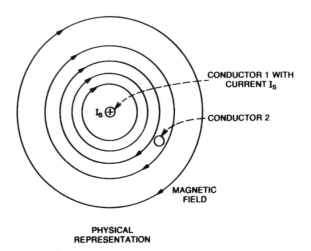

**PHYSICAL
REPRESENTATION**

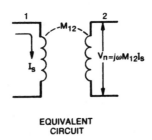

**EQUIVALENT
CIRCUIT**

**FIGURE 1-13.** When two circuits are coupled by a magnetic field, the coupling can be represented as a mutual inductance.

## SUMMARY

- Designing equipment that does not generate noise is as important as designing equipment that is not susceptible to noise.
- Noise sources can be grouped into the following three categories: (1) intrinsic noise sources, (2) man-made noise sources, and (3) noise caused by natural disturbances.
- To be cost effective, noise suppression must be considered early in the design.
- Electromagnetic compatibility is the ability of an electronic system to function properly in its intended electromagnetic environment.
- Electromagnetic compatibility has two aspects, emission and susceptibility.
- Electromagnetic compatibility should be designed into a product not added on at the end of the design.

- Most electronic equipment must comply with EMC regulations before being marketed.
- EMC regulations are not static but are continually changing.
- The three major EMC regulations are the FCC rules, the European Union's regulations, and the military standards.
- The following products are temporarily exempt from the FCC requirements:
  - Digital electronics in transportation vehicles
  - Industrial control systems
  - Test equipment
  - Home appliances
  - Specialized medical devices
  - Devices with power consumption not exceeding 6 nW
  - Joystick controllers or similar devices
  - Devices with clock frequencies less than 1.705 kHz, and which do not operate from the AC power line
- Virtually no products are exempt from the European Union's EMC requirements.
- Electromagnetic compatibility should be a major design objective.
- The following three items are necessary to produce an interference problem:
  - A noise source
  - A coupling channel
  - A susceptible receptor
- Three important characteristics of noise are as follows:
  - Frequency
  - Amplitude
  - Time (when does it occur)
- Metals in contact with each other must be galvanically compatible.
- Noise can be reduced in an electronic system using many techniques; a single unique solution to most noise reduction problems does not exist.

## PROBLEMS

1.1  What is the difference between noise and interference?

1.2  a. Does a digital watch satisfy the FCC's definition of a digital device?
     b. Does a digital watch have to meet the FCC's EMC requirements?

1.3  a. Does test equipment have to meet the technical standards of the FCC's Part 15 EMC regulations?
     b. Does test equipment have to meet the non-interference requirement of the FCC's Part 15 EMC regulations?

1.4   a. Who is responsible for meeting the technical standards of the FCC's EMC regulations?

     b. Who is responsible for meeting the non-interference requirement of the FCC's EMC regulations?

1.5   Are the FCC's or the European Union's Class B radiated emission limits more restrictive:

     a. In the frequency range of 30 to 88 MHz?

     b. In the frequency range of 88 to 230 MHz?

     c. In the frequency range of 230 to 960 MHz?

     d. In the frequency range of 960 to 1000 MHz?

1.6   a. Over what frequency range, below 500 MHz, does the maximum difference exist between the FCC's and the European Union's Class B radiated emission limits?

     b. What is the magnitude of the maximum difference over this frequency range?

1.7   a. Over what frequency range does the FCC specify conducted emission limits?

     b. Over what frequency range does the FCC specify radiated emission limits?

1.8   a. What are the essential requirements for a product to be marketed in the European Union?

     b. Where are the essential requirements defined?

1.9   By what process are commercial EMC regulations made public?

1.10  What is the *major* difference between the FCC's EMC requirement and the European Union's EMC requirements?

1.11  What additional emission requirements does the European Union have that the FCC does not?

1.12  Your company is in the process of designing a new electronic widget to be marketed in the European Union. The widget will be used in both residential and commercial environments. You review the most current list of harmonized product specific EMC standards, and none of them apply to widgets. What EMC standards (specifically) should you use to demonstrate EMC compliance?

1.13  To be legally marketed in the European Union, must an electronic product be compliant with the harmonized EMC standards?

1.14  In the European Union, what are the two methods of demonstrating compliance with the EMC directive?

1.15  Which of the following EMC standards are legal requirements and which are contractual?

- FCC Part 15 B
- MIL-STD-461E

- 2004/108/EC EMC Directive
- RTCA/DO-160E for avionics
- GR-1089 for telephone network equipment
- TIA-968 for telecom terminal equipment
- SAE J551 for automobiles

1.16 What are the official journals of the following countries: the United States, Canada, and the European Union?

1.17 In the United States, does medical equipment have to meet the FCC's EMC requirements?

1.18 What are the three necessary elements to produce an interference problem?

1.19 When analyzing the characteristics of a noise source, what does the acronym FAT stand for?

1.20 a. Which of the following metals is the most susceptible to corrosion: cadmium, nickel (passive), magnesium, copper, or steel?

   b. Which is the least susceptible to corrosion?

1.21 If a tin plate is bolted to a zinc casting, because of galvanic action, which metal will be corroded or eaten away?

## REFERENCES

2004/108/EEC. Council Directive 2004/108/EEC Relating to Electromagnetic Compatibility and Repealing Directive 89/336/EEC, *Official Journal of the European Union*, No. L 390 December 31, 2004, pp. 24–37.

ANSI C63.4-1992. *Methods of Measurement of Radio-Noise Emissions from Low-Voltage Electrical and Electronic Equipment in the Range of 9 kHz to 40 GHz*, IEEE, July 17, 1992.

CAN/CSA-CEI/IEC CISPR 22:02. *Limits an Methods of Measurement of Radio Disturbance Characteristics of Information Technology Equipment*, Canadian Standards Association, 2002.

*CISPR, Publication 22*. "Limits and Methods of Measurement of Radio Interference Characteristics of Information Technology Equipment," 1997.

Code of Federal Regulations, Title 47, Telecommunications (47CFR). Parts 1, 2, 15, 18, and 68, U.S. Government Printing Office, Washington, DC.

*EMCAB-1, Issue 2*. Electromagnetic Compatibility Bulletin, "Immunity of Electrical/ Electronic Equipment Intended to Operate in the Canadian Radio Environment (0.014–10,000 MHz)." Government of Canada, Department of Communications, August 1, 1982.

EN 61000-3-2. *Electromagnetic Compatibility (EMC)—Part 3-2: Limits—Limits for Harmonic Current Emissions (Equipment Input Current $\leq$ 16 A Per Phase)*, CENELEC, 2006.

EN 61000-3-3. *Electromagnetic Compatibility (EMC)—Part 3-3: Limits—Limitation of Voltage Changes, Voltage Fluctuations and Flicker in Public Low-Voltage Supply*

*Systems, for Equipment with Rated Current ≤ 16 A Per Phase and Not Subject to Conditional Connection*, CENELEC, 2006.

European Commission. *Guidelines on the Application of Council Directive 89/336/EEC on the Approximation of the Laws of the Member States Relating to Electromagnetic Compatibility*, European Commission, 1997.

FCC. "Commission Cautions Against Changes in Verified Computing Equipment." Public Notice No. 3281, April 7, 1982.

FCC. "United States and Canada Agree on Acceptance of Measurement Reports for Equipment Authorization," Public Notice No. 54795, July 12, 1995.

FDA. Guide to Inspections of Electromagnetic Compatibility Aspects of Medical Device Quality Systems, US Food and Drug Administration, Available at http://www.fda.gov/ora/inspect_ref/igs/elec_med_dev/emc1.html. Accessed September 2008.

GR-1089-CORE. *Electromagnetic Compatibility and Electrical Safety—Generic Criteria for Network Telecommunications Equipment*, Telcordia, November 2002.

Heirman, D. N. "Broadcast Electromagnetic Environment Near Telephone Equipment." IEEE National Telecommunications Conference, 1976.

Janes, D. E. et al. "Nonionizing Radiation Exposure in Urban Areas of the United States." *Proceedings of the 5th International Radiation Protection Association.* April 1977.

MDS-201-0004. *Electromagnetic Compatibility Standards for Medical Devices*, U.S. Department of Health Education and Welfare, Food and Drug Administration, October 1, 1979.

MIL-STD-461E. *Requirements For The Control of Electromagnetic Interference Characteristics of Subsystems and Equipment*, August 20, 1999.

MIL-STD-889B. *Dissimilar Metals*, Notice 3, May 1993.

RTCA/DO-160E. *Environmental Conditions and Test Procedures for Airborne Equipment*, Radio Technical Commission for Aeronautics (RTCA), December 7, 2004.

SAE J551. *Performance Levels and Methods of Measurement of Electromagnetic Compatibility of Vehicles and Devices (60 Hz to 18 GHz)*, Society of Automotive Engineers, June 1996.

SAE J1113. *Electromagnetic Compatibility Measurement Procedure for Vehicle Components (Except Aircraft) (60 Hz to 18 GHz)*, Society of Automotive Engineers, July 1995.

TIA-968-A. *Telecommunications Telephone Terminal Equipment Technical Requirements for Connection of Terminal Equipment to the Telephone Network*, Telecommunications Industry Association, October 1, 2002.

**FURTHER READING**

Cohen, T. J., and McCoy, L. G. "RFI–A New Look at an Old Problem." *QST*, March 1975.

Gruber, M. ed. *The ARRL RFI Book*, Newington, CT, American Radio Relay League, 2007.

Marshman, C. *The Guide to the EMC Directive 89/336/EEC*, IEEE Press, New York, 1992.

Wall, A. " Historical Perspective of the FCC Rules For Digital Devices and a Look to the Future," 2004 *IEEE International Symposium on Electromagnetic Compatibility*, August 9–13, 2004.

# 2 Cabling

Cables are important because they are usually the longest parts of a system and therefore act as efficient antennas that pick up and/or radiate noise. This chapter covers the coupling mechanisms that occur between fields and cables, and between cables (crosstalk), both unshielded and shielded cables are considered.

In this chapter, we assume the following:

1. Shields are made of nonmagnetic materials and have a thickness much less than a skin depth at the frequency of interest.*
2. The receptor is not coupled so tightly to the source that it loads down the source.
3. Induced currents in the receptor circuit are small enough not to distort the original field. (This does not apply to a shield around the receptor circuit.)
4. Cables are short compared with a wavelength.

Because cables are assumed short compared with a wavelength, the coupling between circuits can be represented by lumped capacitance and inductance between the conductors. The circuit can then be analyzed by normal network theory.

Three types of couplings are considered. The first is capacitive or electric coupling, which results from the interaction of electric fields between circuits. This type of coupling is commonly identified in the literature as electrostatic coupling, an obvious misnomer because the fields are not static.

The second is inductive, or magnetic, coupling, which results from the interaction between the magnetic fields of two circuits. This type of coupling is commonly described as electromagnetic, which again is misleading terminology because no electric fields are involved. The third is a combination of electric and magnetic fields and is appropriately called electromagnetic coupling or radiation. The techniques developed to cope with electric coupling are also

---

*If the shield is thicker than a skin depth, some additional shielding is present besides that calculated by methods in this chapter. The effect is discussed in more detail in Chapter 6.

---

appropriate for the electromagnetic case. For analysis in the near field, we normally consider the electric and magnetic fields separately, whereas the electromagnetic field case is considered when the problem is in the far field.* The circuit causing the interference is called the source, and the circuit being affected by the interference is called receptor.

## 2.1  CAPACITIVE COUPLING

A simple representation of capactive coupling between two conductors is shown in Fig. 2.1.† Capacitance $C_{12}$ is the stray capacitance between conductors 1 and 2. Capacitance $C_{1G}$ is the capacitance between conductor 1 and ground, $C_{2G}$ is the total capacitance between conductor 2 and ground, and $R$ is the resistance of circuit 2 to ground. The resistance $R$ results from the circuitry connected to conductor 2 and is not a stray component. Capacitance $C_{2G}$ consists of both the stray capacitance of conductor 2 to ground and the effect of any circuit connected to conductor 2.

The equivalent circuit of the coupling is also shown in Fig. 2-1. Consider the voltage $V_1$ on conductor 1 as the source of interference and conductor 2 as the affected circuit or receptor. Any capacitance connected directly across the source, such as $C_{1G}$ in Fig 2-1 can be neglected because it has no effect on the noise coupling. The noise voltage $V_N$ produced between conductor 2 and ground can be expressed as follows:

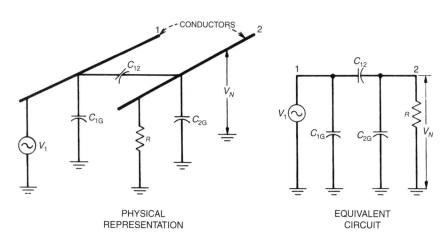

PHYSICAL
REPRESENTATION

EQUIVALENT
CIRCUIT

**FIGURE 2-1.** Capacitive coupling between two conductors.

---

* See Chapter 6 for definitions of near and far fields.
† The two conductors in Fig. 2-1 do not have to represent wires in a cable. They could be any two conductors in space. For example, they could just as well represent traces on a printed circuit board (PCB).

$$V_N = \frac{j\omega[C_{12}/(C_{12} + C_{2G})]}{j\omega + 1/R(C_{12} + C_{2G})} V_1. \tag{2-1}$$

Equation 2-1 does not show clearly how the pickup voltage depends on the various parameters. Equation 2-1 can be simplified for the case when $R$ is a lower impedance than the impedance of the stray capacitance $C_{12}$ plus $C_{2G}$. In most practical cases this will be true. Therefore, for

$$R \ll \frac{1}{j\omega(C_{12} + C_{2G})},$$

Eq. 2-1 can be reduced to the following:

$$V_N = j\omega R C_{12} V_1. \tag{2-2}$$

Electric field (capacitive) coupling can be modeled as a current generator, connected between the receptor circuit and ground, with a magnitude of $j\omega$ $C_{12}V_1$. This is shown in Fig. 2-9A.

Equation 2-2 is the most important equation to describe the capacitive coupling between two conductors, and it shows clearly how the pickup voltage depends on the parameters. Equation 2-2 shows that the noise voltage is directly proportional to the frequency ($\omega = 2\pi f$) of the noise source, the resistance $R$ of the affected circuit to ground, the mutual capacitance $C_{12}$ between conductors 1 and 2, and the magnitude of the voltage $V_1$.

Assuming that the voltage and frequency of the noise source cannot be changed, this leaves only two remaining parameters for reducing capacitive coupling. The receiver circuit can be operated at a lower resistance level, or the mutual capacitance $C_{12}$ can be decreased. Capacitance $C_{12}$ can be decreased by proper orientation of the conductors, by shielding (described in Section 2.2), or by physically separating the conductors. If the conductors are moved farther apart, $C_{12}$ decreases, thus decreasing the induced voltage on conductor 2.* The effect of conductor spacing on capacitive coupling is shown in Fig. 2-2. As a reference, 0 dB is the coupling when the conductors are separated by three times the conductor diameter. As can be observed in the figure, little additional attenuation is gained by spacing the conductors a distance greater than 40 times their diameter (1 in in the case of 22-gauge wire).

If the resistance from conductor 2 to ground is large, such that

$$R \gg \frac{1}{j\omega(C_{12} + C_{2G})},$$

---

* The capacitance between two parallel conductors of diameter $d$ and spaced $D$ apart is $C_{12} = \pi\varepsilon/ \cosh^{-1}(D/d)$, (F/m). For $D/d > 3$, this reduces to $C_{12} = \pi\varepsilon/\ln(2\ D/d)$, (F/m), where $\varepsilon = 8.5 \times 10^{-12}$ farads per meter (F/m) for free space.

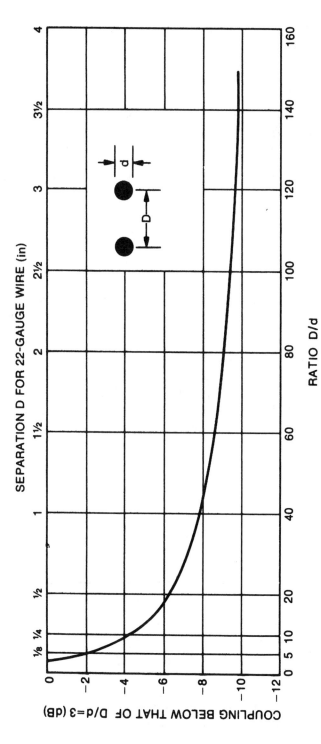

**FIGURE 2-2.** Effect of conductor spacing on capacitive coupling. In the case of 22-gauge wire, most attenuation occurs in the first inch of separation.

47

then Eq. 2-1 reduces to

$$V_N = \left(\frac{C_{12}}{C_{12} + C_{2G}}\right) V_1. \tag{2-3}$$

Under this condition, the noise voltage produced between conductor 2 and ground is the result of the capacitive voltage divider $C_{12}$ and $C_{2G}$. The noise voltage is independent of frequency and is of a larger magnitude than when $R$ is small.

A plot of Eq. 2-1 versus $\omega$ is shown in Fig. 2-3. As can be observed, the maximum noise coupling is given by Eq. 2-3. The figure also shows that the actual noise voltage is always less than or equal to the value given by Eq. 2-2. At a frequency of

$$\omega = \frac{1}{R(C_{12} + C_{2G})}, \tag{2-4}$$

Eq. 2-2 gives a value of noise that is 1.41 times the actual value. In almost all practical cases, the frequency is much less than this, and Eq. 2-2 applies.

## 2.2   EFFECT OF SHIELD ON CAPACITIVE COUPLING

First, consider the case of an ideal shielded conductor as shown in Fig. 2-4. An equivalent circuit of the capacitive coupling is also shown in the figure. This is an ideal case because of the following:

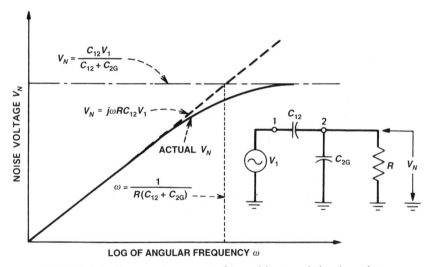

FIGURE 2-3. Frequency response of capacitive coupled noise voltage.

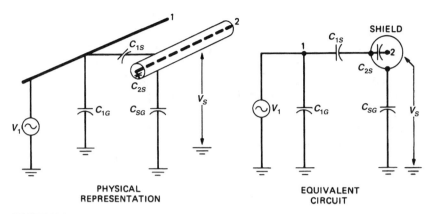

**FIGURE 2-4.** Capacitive coupling with shield placed around receptor conductor.

1. The shield completely encloses conductor 2—none of conductor 2 extends beyond the shield.
2. The shield is solid—there are no holes in the shield such as would be the case of a braided shield.
3. The shield is not terminated, and there is no terminating impedance on conductor 2.

The shield is an unshielded conductor exposed to conductor 1, and because there is no termination on the shield it has a high terminating impedance. Therefore Eq. 2-3 can be used to determine the voltage picked up by the shield. The noise voltage on the shield will be

$$V_S = \left( \frac{C_{1S}}{C_{1S} + C_{SG}} \right) V_1. \qquad (2\text{-}5)$$

From the equivalent circuit shown in Fig. 2-4, we recognize, that for this ideal case, the only impedance connected to conductor 2 is capacitance $C_{2S}$. Because no other impedances are connected to conductor 2, no current can flow through $C_{2S}$. As a result, there can be no voltage drop across $C_{2S}$, and voltage picked up by the conductor 2 will be

$$V_N = V_S. \qquad (2\text{-}6)$$

The shield therefore did not reduce the noise voltage picked up by conductor 2.

If, however, the shield is grounded, the voltage $V_S = 0$, and from Eq. 2-6, the noise voltage $V_N$ on conductor 2 is likewise reduced to zero. Therefore, we can conclude that the shield is not effective unless it is properly terminated (grounded). As we will observe, in many cases the shield termination is more important than the characteristics of the shield itself (see Section 2.15 on Shield Terminations).

In many practical cases, the center conductor does extend beyond the shield, and the situation becomes that of Fig. 2-5. There, $C_{12}$ is the capacitance between conductor 1 and the shielded conductor 2, and $C_{2G}$ is the capacitance between conductor 2 and ground. Both of these capacitances exist because the ends of conductor 2 extend beyond the shield and as the result of any holes in the shield. Even if the shield is grounded, there is now a noise voltage coupled to conductor 2. Its magnitude is expressed as follows:

$$V_N = \frac{C_{12}}{C_{12} + C_{2G} + C_{2S}} V_1. \tag{2-7}$$

The value of $C_{12}$, and hence $V_N$, in Eq. 2-7 depends primarily on the length of conductor 2 that extends beyond the shield and to a lesser extent on any holes present in the shield.

*For good electric field shielding, it is therefore necessary (1) to minimize the length of the center conductor that extends beyond the shield and (2) to provide a good ground on the shield.* A single ground connection makes a good shield ground, provided the cable is not longer than one twentieth of a wavelength. On longer cables, multiple grounds may be necessary.

If in addition the receiving conductor has finite resistance to ground, the arrangement is that shown in Fig. 2.6. If the shield is grounded, the equivalent circuit can be simplified as shown in the figure. Any capacitance directly across the source can be neglected because it has no effect on the noise coupling. The simplified equivalent circuit can be recognized as the same circuit analyzed in Fig. 2.1, provided $C_{2G}$ is replaced by the sum of $C_{2G}$ and $C_{2S}$. Therefore, if

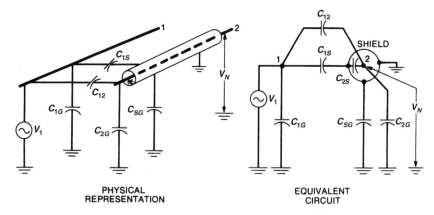

PHYSICAL
REPRESENTATION

EQUIVALENT
CIRCUIT

**FIGURE 2-5.** Capacitive coupling when center conductor extends beyond shield; shield grounded at one point.

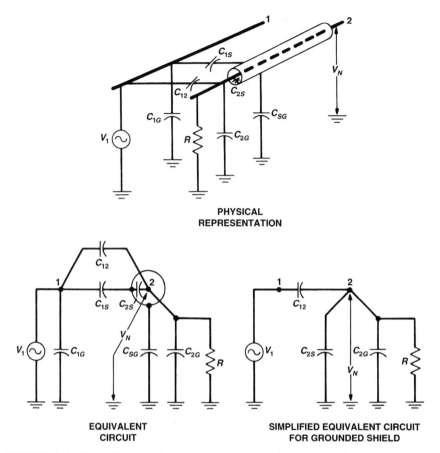

**FIGURE 2-6.** Capacitive coupling when receptor conductor has resistance to ground.

$$R \ll \frac{1}{j\omega(C_{12} + C_{2G} + C_{2S})},$$

which is normally true, then the noise voltage coupled to conductor 2 is

$$V_N = j\omega R C_{12} V_1. \tag{2-8}$$

This is the same as Eq. 2-2, which is for an unshielded cable, except that $C_{12}$ is greatly reduced by the presence of the shield. Capacitance $C_{12}$ now consists primarily of the capacitance between conductor 1 and the unshielded portions of conductor 2. If the shield is braided, any capacitance that exists from conductor 1 to 2 through the holes in the braid must also be included in $C_{12}$.

## 2.3   INDUCTIVE COUPLING*

When a current $I$ flows through a conductor, it produces a magnetic flux $\Phi$, which is proportional to the current. The constant of proportionality is the inductance $L$; hence, we can write

$$\phi_T = LI, \tag{2-9a}$$

where $\Phi_T$ is the total magnetic flux and $I$ is the current producing the flux. Rewriting Eq. 2-9a, we get for the *self-inductance* of a conductor

$$L = \frac{\phi_T}{I}. \tag{2-9b}$$

The inductance depends on the geometry of the circuit and the magnetic properties of the media containing the field.

When current flow in one circuit produces a flux in a second circuit, there is a *mutual inductance* $M_{12}$ between circuits 1 and 2 defined as

$$M_{12} = \frac{\phi_{12}}{I_1}. \tag{2-10}$$

The symbol $\phi_{12}$ represents the flux in circuit 2 because of the current $I_1$ in circuit 1.

The voltage $V_N$ induced in a closed loop of area $\bar{A}$ resulting from a magnetic field of flux density $\bar{B}$ can be derived from Faraday's law (Hayt, 1974, p. 331) and is

$$V_N = -\frac{d}{dt} \int_A \bar{B} \cdot d\bar{A}, \tag{2-11}$$

where $\bar{B}$ and $\bar{A}$ are vectors. If the closed loop is stationary and the flux density is sinusoidally varying with time but constant over the area of the loop, Eq. 2-11 reduces to[†]

$$V_N = j\omega BA \cos\theta. \tag{2-12}$$

As shown in Fig. 2-7, $A$ is the area of the closed loop, $B$ is the root mean square (rms) value of the sinusoidally varying flux density of frequency $\omega$ radians per second, and $V_N$ is the rms value of the induced voltage.

---

* A more detailed discussion of the concept of inductance is in Appendix E.
[†] Equation 2-12 is correct when the MKS system of units is being used, Flux density $B$ is in webers per square meter (or tesla), and area $A$ is in square meters. If $B$ is expressed in gauss and $A$ in square centimeters (the CGS system), then the right side of Eq. 2-12 must be multiplied by $10^{-8}$.

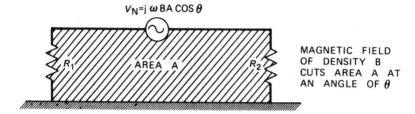

**FIGURE 2-7.** Induced noise depends on the area enclosed by the disturbed circuit.

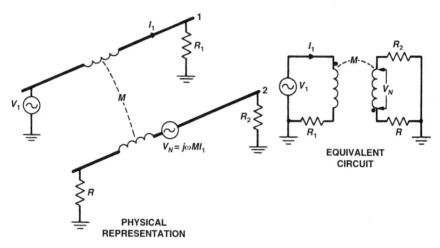

**FIGURE 2-8.** Magnetic coupling between two circuits.

Because $BA \cos \theta$ represents the total magnetic flux ($\phi_{12}$) coupled to the receptor circuit, Eqs. 2-10 and 2-12 can be combined to express the induced voltage in terms of the mutual inductance $M$ between two circuits, as follows:

$$V_N = j\omega M I_1 = M \frac{di_1}{dt}. \qquad (2\text{-}13)$$

Equations 2-12 and 2-13 are the basic equations describing inductive coupling between two circuits. Figure 2-8 shows the inductive (magnetic) coupling between two circuits as described by Eq. 2-13. $I_1$ is the current in the interfering circuit, and $M$ is the term that accounts for the geometry and the magnetic properties of the medium between the two circuits. The presence of $\omega$ in Eqs. 2-12 and 2-13 indicates that the coupling is directly proportional to frequency. To reduce the noise voltage, $B$, $A$, or $\cos \theta$ must be reduced. The $B$ term can be reduced by physical separation of the circuits or by twisting the source wires, provided the current flows in the twisted pair and not through the ground

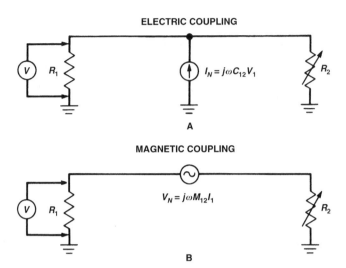

**FIGURE 2-9.** (A) Equivalent circuit for electric field coupling; (B) equivalent circuit for magnetic field coupling.

plane. Under these conditions, twisting causes the $B$ fields from each of the wires to cancel. The area of the receiver circuit can be reduced by placing the conductor closer to the ground plane (if the return current is through the ground plane) or by using two conductors twised together (if the return current is on one of the pair instead of the ground plane). The cos $\theta$ term can be reduced by proper orientation of the source and receiver circuits.

It may be helpful to note some differences between magnetic and electric field coupling. For magnetic field coupling, a noise voltage is produced in series with the receptor conductor (Fig. 2-9B), whereas for electric field coupling, a noise current is produced between the receptor conductor and ground (Fig. 2-9A). This difference can be used in the following test to distinguish between electric and magnetic coupling. Measure the noise voltage across the impedance at one end of the cable while decreasing the impedance at the opposite end of the cable (Fig. 2-9). If the measured noise voltage decreases, the pickup is electric, and if the measured noise voltage increases, the pickup is magnetic.

## 2.4   MUTUAL INDUCTANCE CALCULATIONS

To evaluate the expression in Eq. 2-13, the mutual inductance between the source and receptor circuit must be known. Most texts do not pay much attention to mutual inductance calculations for practical circuit geometries. Grover (1973), however, provides an extensive treatment of the subject, and Ruehli (1972) develops the useful concept of partial mutual inductance (also see

Appendix E). This concept of partial mutual inductance is further developed in Paul (1986).

Before the mutual inductance can be calculated, an expression must be determined for the magnitude of the magnetic flux density as a function of distance from a current-carrying conductor. Using the Biot-Savart law, one can write the magnetic flux density $B$ at a distance $r$ from a long current-carrying conductor as

$$B = \frac{\mu I}{2\pi r},$$

(2-14)

for $r$ greater than the radius of the conductor (Hayt, 1974, pp. 235–237). The flux density $B$ is equal to the flux $\phi$ per unit area. Therefore, the magnetic field is directly proportional to the current $I$ and inversely proportional to the distance $r$. Using Eqs. 2-14 and 2-10, one can determine the mutal inductance for any arbitrary configuration of conductors by calculating the magnetic flux coupled to the pickup loop from each current-carrying conductor individually, and then superimposing all the results to obtain the total flux coupling.

**Example 2.1**. Calculate the mutual inductance between the two nested coplaner loops shown in Fig. 2-10A, assuming that the sides of the loop are much longer than the ends (i.e., the coupling contributed by the end conductors can be neglected). Conductors 1 and 2 are carrying a current $I_1$ which induces a voltage $V_N$ into the loop formed by conductors 3 and 4. Figure 2-10B is a cross-sectional view showing the spacing between the conductors. The magnetic flux produced by the current in conductor 1 crossing the loop between conductors 3 and 4 is

$$\phi_{12} = \int_a^b \frac{\mu I_1}{2\pi r} dr = \frac{\mu I_1}{2\pi} \ln\left(\frac{b}{a}\right).$$

(2-15)

Conductor 2 also contributes an equal flux because of the symmetry of the conductors. This flux is in the same direction as the flux produced by the current in conductor 1. Therefore, the total flux coupled to the loop formed by conductors 3 and 4 is twice that given by Eq. 2-15, or

$$\phi_{12} = \left[\frac{\mu}{\pi} \ln\left(\frac{b}{a}\right)\right] I_1.$$

(2-16)

Dividing Eq. 2-16 by $I_1$ and substituting $4\pi \times 10^{-7}$ H/m for $\mu$, we obtain as the mutual inductance in H/m

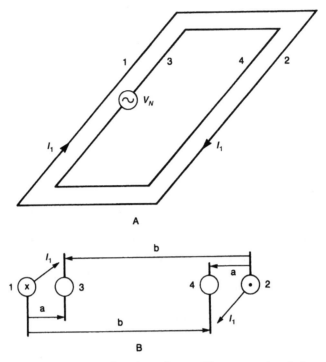

**FIGURE 2-10.** (A) Nested coplanar loops; (B) cross-sectional view of A.

$$M = 4 \times 10^{-7} \ln\left(\frac{b}{a}\right). \tag{2-17}$$

The voltage coupled between the two loops can be calculated by substituting the result from Eq. 2-17 into Eq. 2-13.

## 2.5   EFFECT OF SHIELD ON MAGNETIC COUPLING

If an ungrounded and nonmagnetic shield is now placed around conductor 2, the circuit becomes that of Fig. 2-11, where $M_{1s}$ is the mutual inductance between conductor 1 and the shield. Because the shield has no effect on the geometry or magnetic properties of the medium between circuits 1 and 2, it has no effect on the voltage induced into conductor 2. The shield does, however, pick up a voltage because of the current in conductor 1:

$$V_S = j\omega M_{1S} I_1, \tag{2-18}$$

A ground connection on one end of the shield does not change the situation. *It follows, therefore, that a nonmagnetic shield placed around a conductor and*

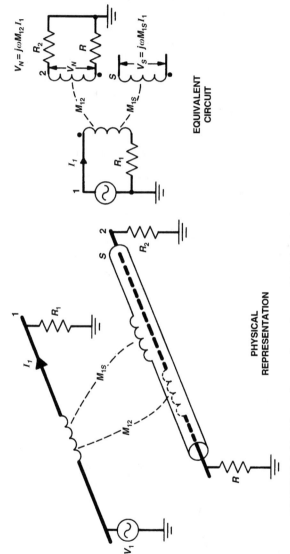

**FIGURE 2-11.** Magnetic coupling when a shield is placed around the receptor conductor.

*grounded at one end has no effect on the magnetically induced voltage in that
conductor.*

If, however, the shield is grounded at both ends, the voltage induced into the
shield from $M_{1S}$ in Fig. 2-11, will cause shield current to flow. The shield
current will induce a second noise voltage into conductor 2, and this must be
taken into account. Before this voltage can be calculated, the mutual inductance between a shield and its center conductor must be determined.

For this reason, it will be necessary to calculate the magnetic coupling
between a hollow conducting tube (the shield) and any conductor placed
inside the tube, before continuing the discussion of inductive coupling. This
concept is fundamental to a discussion of magnetic shielding and will be
needed later.

### 2.5.1   Magnetic Coupling Between Shield and Inner Conductor

First, consider the magnetic field produced by a tubular conductor carrying a
*uniform* axial current, as shown in Fig. 2-12. If the hole in the tube is concentric
with the outside of the tube, there is no magnetic field in the cavity, and the
total magnetic field is external to the tube (Smythe, 1924, p. 278).

Now, let a conductor be placed inside the tube to form a coaxial cable, as
shown in Fig. 2-13. All of the flux $\phi$ from the current $I_s$ in the shield tube
encircles the inner conductor. The inductance of the shield is equal to

$$L_S = \frac{\phi}{I_S}. \tag{2-19}$$

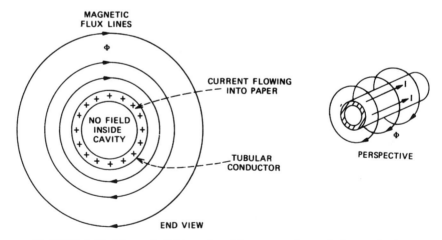

**FIGURE 2-12.** Magnetic field produced by current in a tubular conductor.

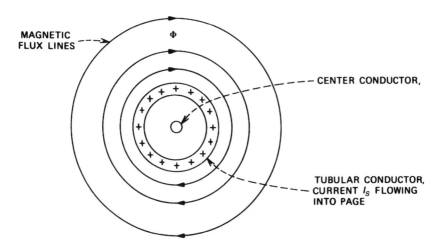

**FIGURE 2-13.** Coaxial cable with shield current flowing uniformly around the circumference of the shield.

The mutual inductance between the shield and the inner conductor is equal to

$$M = \frac{\phi}{I_S}. \qquad (2\text{-}20)$$

Because all the flux produced by the shield current encircles the center conductor, the flux $\phi$ in Eqs. 2-19 and 2-20 is the same. The mutual inductance between the shield and center conductor is therefore equal to the self-inductance of the shield

$$M = L_S. \qquad (2\text{-}21)$$

Equation 2-21 is a most important result and one that we will often have occasion to refer to. It was derived to show that the *mutual inductance between the shield and the center conductor is equal to the shield inductance.* Based on the reciprocity of mutual inductance (Hayt, 1974, p. 321), the inverse must also be true. That is, the mutual inductance between the center conductor and the shield is equal to the shield inductance.

The validity of Eq. 2-21 depends only on the fact that there is no magnetic field in the cavity of the tube because of shield current. The requirements for this to be true are that the tube be cylindrical and the current density be *uniform around the circumference of the tube* as shown in Fig. 2-12. Because there is no magnetic field inside the tube, Eq. 2-21 applies regardless of the position of the conductor within the tube. In other words, the two conductors do not have to be coaxial. Equation 2-21 is also applicable to the case of multiple conductors within a shield, in which case it will represent the mutual inductance between the shield and each conductor in the shield.

The voltage $V_N$ induced into the center conductor due to a current $I_S$ in the shield can now be calculated. Assume that the shield current is produced by a voltage $V_S$ induced into the shield from some other circuit. Figure 2-14 shows the circuit being considered; $L_S$ and $R_S$ are the inductance and resistance of the shield. The voltage $V_N$ is equal to

$$V_N = j\omega M I_S. \tag{2-22}$$

The current $I_S$ is equal to

$$I_S = \frac{V_S}{L_S}\left(\frac{1}{j\omega + R_S/L_S}\right). \tag{2-23}$$

Therefore

$$V_N = \left(\frac{j\omega M V_S}{L_S}\right)\left(\frac{1}{j\omega + R_S/L_S}\right). \tag{2-24}$$

Because $L_S = M$ (from Eq. 2-21),

$$V_N = \left(\frac{j\omega}{j\omega + R_S/L_S}\right)V_S, \tag{2-25}$$

A plot of Eq. 2-25 is shown in Fig. 2-15. The break frequency for this curve is defined as the shield cutoff frequency ($\omega_c$) and occurs at

$$\omega_c = \frac{R_S}{L_S}, \quad \text{or} \quad f_c = \frac{R_S}{2\pi L_S}. \tag{2-26}$$

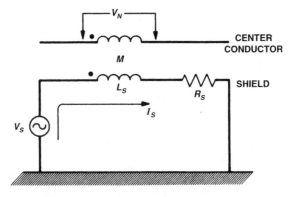

**FIGURE 2-14.** Equivalent circuis of shielded conductor.

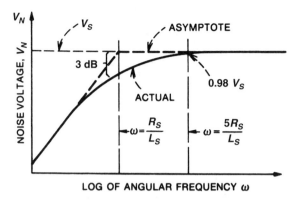

**FIGURE 2-15.** Noise voltage in center conductor of coaxial cable because of shield current.

The noise voltage induced into the center conductor is zero at dc and increases to almost $V_S$ at a frequency of $5R_S/L_S$ rad/s. Therefore, if shield current is allowed to flow, a voltage is induced into the center conductor that nearly equals the shield voltage at frequencies greater than five times the shield cutoff frequency.

This is a very important property of a conductor inside a shield. Measured values of the shield cutoff frequency and five times this frequency are tabulated in Table 2-1 for various cables. For most cables, five times the shield cutoff frequency is near the high end of the audio-frequency band. The aluminum-foil-shielded cable listed has a much higher shield cutoff frequency than any other, which is caused by the increased resistance of its thin aluminum-foil shield.

### 2.5.2   Magnetic Coupling—Open Wire to Shielded Conductor

Figure 2-16 shows the magnetic couplings that exist when a nonmagnetic shield is placed around conductor 2 and the shield is grounded at both ends. In this figure, the shield conductor is shown separated from conductor 2 to simplify the drawing. Because the shield is grounded at both ends, the shield current flows and induces a voltage into conductor 2. Therefore, here are two components to the voltage induced into conductor 2. The voltage $V_2$ from direct induction from conductor 1, and the voltage $V_c$ from the induced shield current. Note that these two voltages are of opposite polarity. The total noise voltage induced into conductor 2 is therefore

$$V_N = V_2 - V_c. \tag{2-27}$$

If we use the identity of Eq. 2-21 and note that mutual inductance $M_{1S}$ from conductor 1 to the shield is equal to the mutual inductance $M_{12}$ from conductor

**TABLE 2-1.**   Measured Values of Shield Cutoff Frequency $f_c$.

| Cable | Impedance ($\Omega$) | Cutoff Frequency (kHz) | Five Times Cutoff Frequency (kHz) | Remarks |
|---|---|---|---|---|
| Coaxial cable | | | | |
| RG-6A | 75 | 0.6 | 3.0 | Double shielded |
| RG-213 | 50 | 0.7 | 3.5 | |
| RG-214 | 50 | 0.7 | 3.5 | Double shielded |
| RG-62A | 93 | 1.5 | 7.5 | |
| RG-59C | 75 | 1.6 | 8.0 | |
| RG-58C | 50 | 2.0 | 10.0 | |
| Shielded twisted pair | | | | |
| 754E | 125 | 0.8 | 4.0 | Double shielded |
| 24 Ga. | – | 2.2 | 11.0 | |
| 22 Ga.[a] | – | 7.0 | 35.0 | Aluminum-foil shield |
| Shielded single | | | | |
| 24 Ga. | – | 4.0 | 20.0 | |

[a] One pair out of an 11-pair cable (Belden, 8775).

1 to conductor 2 (because the shield and conductor 2 are located in the same place in space with respect to conductor 1), then Eq. 2-27 becomes

$$V_N = j\omega M_{12}I_1\left[\frac{R_S/L_S}{j\omega + R_S/L_S}\right], \qquad (2\text{-}28)$$

If $\omega$ is small in Eq. 2-28, the term in brackets equals 1, and the noise voltage is the same as for the unshielded cable. Therefore, at low frequencies, a shield, even when grounded at both ends, provides no magnetic field shielding.

If $\omega$ is large, then Eq. 2-28 reduces to

$$V_N = M_{12}I_1\left(\frac{R_S}{L_S}\right). \qquad (2\text{-}29)$$

Equation 2-28 is plotted in Fig. 2-17. At low frequencies, the noise pickup in the shielded cable is the same as for an unshielded cable; however, at frequencies above the shield cutoff frequency, the pickup voltage stops increasing and remains constant. The shielding effectiveness (shown cross-hatched in Fig. 2-17) is therefore equal to the difference between the line for the unshielded cable and for the shielded cable.

From Eq. 2-29, we can conclude that to minimize the noise voltage coupled into conductor 2, the shield resistance $R_S$ should be minimized. This comes about because it is the induced shield current that produces the magnetic field that cancels a large percentage of the direct induction into conductor 2.

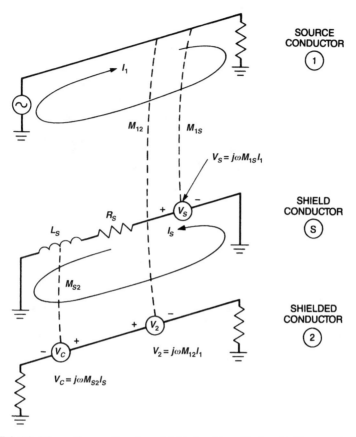

**FIGURE 2-16.** Magnetic coupling to a shielded cable with the shield grounded at both ends.

Because $R_S$ reduces the shield current, it decreases the magnetic shielding effectiveness.

From the middle diagram in Fig. 2-16, we can infer that $R_S$ represents not only the resistance of the shield but also all the resistance in the loop in which the shield current $I_S$ flows. Therefore, $R_S$ actually includes not only the shield resistance but also the termination resistance of the shield and any resistance in the ground. For maximum shielding effectiveness, all these resistances must be minimized. Therefore, the practice, which is sometimes recommended, of terminating a shield with a resistor instead of directly to ground will drastically reduce the magnetic field shielding effectiveness of the cable and should be avoided.

Figure 2-18 shows a transformer analogy equivalent circuit for the configuration of Fig. 2-16. As can be observed, the shield acts as a shorted turn in the transformer to short out the voltage in winding 2. Any resistance (such as shield resistance) in the shorted turn (shield) will decrease its effectiveness in shorting out the voltage in winding 2.

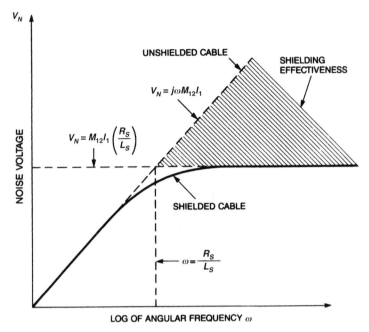

**FIGURE 2-17.** Magnetic field coupled noise voltage for an unshielded and shielded cable (shield grounded at both ends) versus frequency.

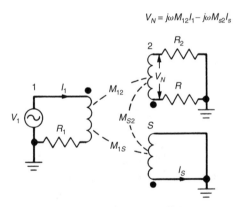

**FIGURE 2-18.** Transformer analogy of magnetic field coupling to a shielded cable when shield is grounded at both ends ($M_{S2}$ is much larger than $M_{12}$ or $M_{1S}$).

## 2.6   SHIELDING TO PREVENT MAGNETIC RADIATION

To prevent radiation, the source of the interference may be shielded. Figure 2-19 shows the electric and magnetic fields that surround a current-carrying conductor located in free space. If a non-magnetic shield is placed around the

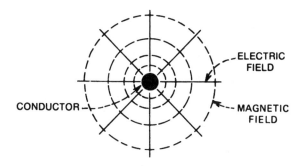

**FIGURE 2-19.** Fields surrounding a current-carrying conductor.

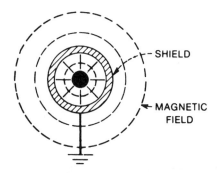

**FIGURE 2-20.** Fields around shielded conductor; shield grounded at one point.

**FIGURE 2-21.** Fields around shielded conductor; shield grounded and carrying a current equal to the conductor current but in the opposite direction.

conductor, then the electric field lines will terminate on the shield, but there will be very little effect on the magnetic field, as shown in Fig. 2-20. If a shield current equal and opposite to that in the center conductor is made to flow on the shield, it generates an equal but opposite external magnetic field. This field cancels the magnetic field caused by the current in the center conductor external to the shield, which results in the condition shown in Fig. 2-21, with no fields external to the shield.

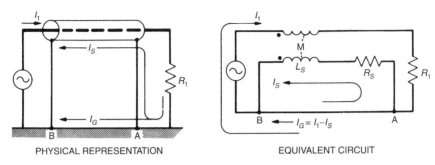

**FIGURE 2-22.** Division of current between shield and ground plane.

Figure 2-22 shows a circuit that is grounded at both ends and carries a current $I_1$. To prevent magnetic field radiation from this circuit, the shield must be grounded at both ends, and the return current must flow from $A$ to $B$ in the shield ($I_S$ in the figure) instead of in the ground plane ($I_G$ in the figure). But why should the current return from point $A$ to $B$ through the shield instead of through the zero-resistance ground plane? The equivalent circuit can be used to analyze this configuration. By writing a mesh equation around the ground loop $(A-R_S-L_S-B-A)$, the shield current $I_S$ can be determined:

$$0 = I_S(j\omega L_S + R_S) - I_1(j\omega M), \qquad (2\text{-}30)$$

where $M$ is the mutual inductance between the shield and center conductor and as previously shown (Eq. 2-21), $M = L_S$. Making this substitution and rearranging produces this expression for $I_S$:

$$I_S = I_1\left(\frac{j\omega}{j\omega + R_S/L_S}\right) = \left(\frac{j\omega}{j\omega + \omega_c}\right)I_1 \qquad (2\text{-}31)$$

As can be observed from the preceding equation, if the frequency is much above the shield cutoff frequency $\omega_c$, the shield current approaches the center conductor current. Because of the mutual inductance between the shield and center conductor, a coaxial cable acts as a common-mode choke (see Fig. 3-36), and the shield provides a return path with lower total circuit inductance than the ground plane at high frequency. As the frequency decreases below $5\omega_c$, the cable provides less and less magnetic shielding as more of the current returns via the ground plane.

*To prevent radiation of a magnetic field from a conductor grounded at both ends, the conductor should be shielded, and the shield should be grounded at both ends.* This approach provides good magnetic field shielding at frequencies considerably above the shield cutoff frequency. This reduction in the radiated magnetic field is not because of the magnetic shielding properties of the shield

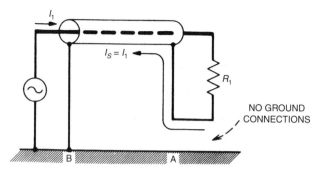

**FIGURE 2-23.** Without ground at far end, all return current flows through shield.

as such. Rather, the return current on the shield generates a field that cancels the center conductor's field.

If the ground is removed from one end of the *circuit*, as shown in Fig. 2-23, then the shield should not be grounded at that end because the return current must now all flow on the shield. This is true especially at frequencies less than the shield cutoff frequency. Grounding both ends of the shield, in this case, reduces the shielding because some current will return via the ground plane.

## 2.7   SHIELDING A RECEPTOR AGAINST MAGNETIC FIELDS

*The best way to protect against magnetic fields at the receptor is to decrease the area of the receptor loop.* The area of interest is the total area enclosed by current flow in the receptor circuit. An important consideration is the path taken by the current in returning to the source. Often, the current returns by a path other than the one intended by the designer, and therefore, the area of the loop changes. If a nonmagnetic shield placed around a conductor causes the current to return over a path that encloses a smaller area, then some protection against magnetic fields will have been provided by the shield. This protection, however, is caused by the reduced loop area and not by any magnetic shielding properties of the shield.

Figure 2-24 illustrates the effect of a shield on the loop area of a circuit. In Fig. 2-24A, the source $V_S$ is connected to the load $R_L$ by a single conductor, using a ground return path. The area enclosed by the current is the rectangle between the conductor and the ground plane. In Fig. 2-24B, a shield is placed around the conductor and grounded at both ends. If the current returns through the shield rather than the ground plane, then the area of the loop is decreased, and a degree of magnetic protection is provided. The current will return through the shield if the frequency is greater than five times the shield cutoff frequency as previously shown. A shield placed around the conductor and grounded at one end only, as shown in Fig. 2-24C, does not change the loop area and therefore provides no magnetic protection.

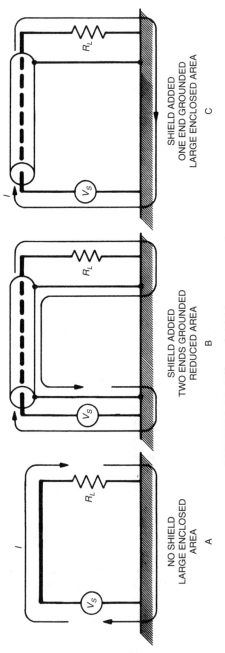

**FIGURE 2-24.** Effect of shield on receptor loop area.

The arrangement of Fig. 2-24B does not protect against magnetic fields at frequencies below the shield cutoff frequency because then most of the current returns through the ground plane and not through the shield. At low frequencies, this circuit also has two other problems, as follows: (1) Because the shield is one of the circuit conductors, any noise current in it will produce an *IR* drop in the shield and appear to the circuit as a noise voltage, and (2) if there is a difference in ground potential between the two ends of the shield then it will show up as a noise voltage in the circuit.

## 2.8   COMMON IMPEDANCE SHIELD COUPLING

When a coaxial cable is used at low frequencies and the shield is grounded at both ends, only a limited amount of magnetic field protection is possible because of the noise current induced into the shield. Because the induced current flows through the shield, which is also one of the signal conductors, a noise voltage is produced in the shield, that is equal to the shield current times the shield resistance. This is shown in Fig. 2-25. The current $I_S$ is the noise current caused by a ground differential or by external magnetic field coupling. If the voltages are summed around the input loop, then the following expression is obtained:

$$V_{IN} = -j\omega M I_S + j\omega L_S I_S + R_S I_S. \tag{2-32}$$

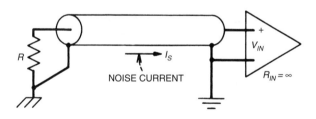

PHYSICAL REPRESENTATION

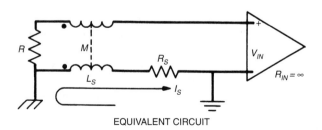

EQUIVALENT CIRCUIT

**FIGURE 2-25.** Effect of noise current flowing in the shield of a coaxial cable.

Because, as previously shown, $L_S = M$, Eq. 2-32 reduces to

$$V_{IN} = R_S I_S. \tag{2-33}$$

Notice that the two inductive noise voltages (the first and second terms in Eq. 2-32) cancel, which leaves only the resistive noise voltage term.

This example shows common impedance coupling and is the result of the shield having to serve two functions. First, it is the return conductor for the signal, and second it is a shield and carries the induced noise current. This problem can be eliminated, or at least minimized, by using a three-conductor cable (e.g., a shielded twisted pair). In this case, the two twisted pair conductors carry the signal and the shield only carries the noise current; therefore, the shield is not performing two functions.

Common impedance shield coupling is often a problem in consumer audio systems that use unbalanced interconnections, which usually consist of a cable with a center conductor and a shield terminated in a phono plug. The problem can be minimized by reducing the resistance of the cable shield or by using a balanced interconnection and a shielded twisted pair.

Even if the shield is grounded at only one end, noise currents may still flow in the shield because of electromagnetic field coupling (that is, the cable acts as an antenna and picks up radio frequency (rf) energy). This is often referred to as shield current induced noise (SCIN) (Brown and Whitlock, 2003).

This problem does not occur at high frequencies, because as the result of skin effect, a coaxial cable actually contains the following three isolated conductors: (1) the center conductor, (2) the inner surface of the shield conductor, and (3) the outer surface of the shield conductor. The signal return current flows only on the inside surface of the shield, and the noise current flows only on the outer surface of the shield. Therefore, the two currents do not flow through a common impedance, and the noise coupling discussed above does not occur.

## 2.9    EXPERIMENTAL DATA

The magnetic field shielding properties of various cable configurations were measured and compared. The test setup is shown in Fig. 2-26, and the test results are tabulated in Figs. 2-27 and 2-28. The frequency (50 kHz) is greater than five times the shield cutoff frequency for all the cables tested. The cables shown in Figs. 2-27 and 2-28 represent tests cables shown as $L_2$ in Fig. 2-26.

In circuits $A$ through $F$ (Fig. 2-27), both ends of the circuit are grounded. They provide much less magnetic field attenuation than do circuits $G$ through $K$ (Fig. 2-28), where only one end is grounded.

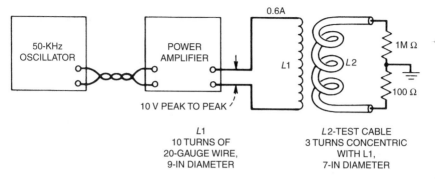

**FIGURE 2-26.** Test setup of inductive coupling experiment.

Circuit $A$ in Fig. 2-27 provides essentially no magnetic field shielding. The actual noise voltage measured across the 1 MΩ resistor in this case was 0.8 V. The pickup in configuration $A$ is used as a reference and is called 0 dB, to compare the performance of all the other ciucuits. In circuit $B$, the shield is grounded at one end; this has no effect on the magnetic shielding. Grounding the shield at both ends as in configuration $C$ provides some magnetic field protection because the frequency is above the shield cutoff frequency. The protection would be even greater if it were not for the ground loop formed by grounding both ends of the circuit. The magnetic field induces a large noise current into the low-impedance ground loop that consists of the cable shield and the two ground points. The shield noise current then produces a noise volatage in the shield, as was shown in the preceding section.

Use of a twisted pair as in circuit $D$ should provide much greater magnetic field noise reduction, but its effect is defeated by the ground loop formed by circuit grounds at both ends. This effect can clearly be observed by comparing the attenuation of circuit $H$ to that of circuit $D$. Adding a shield with one end grounded to the twisted pair as in $E$ has no effect. Grounding the shield at both ends as in $F$ provides additional protection, because the low-impedance shield shunts some of the magnetically induced ground-loop current away from the signal conductors. In general, however, none of the circuit configurations in Fig. 2-27 provide good magnetic field protection because of the ground loops. If the circuit must be grounded at both ends, then configurations $C$ or $F$ should be used.

Circuit $G$ shows a significant improvement in magnetic field shielding, which is caused by the small loop area formed by the coaxial cable and the fact that no ground loop is available to defeat the shielding. The coax provides a small loop area because the shield can be represented by an equivalent conductor located on its center axis. This effectively locates the shield at or near the axis of the center conductor.

It was expected that the twisted pair of circuit $H$ would provide considerably more shielding than the 55 dB shown. The reduced shielding is because some

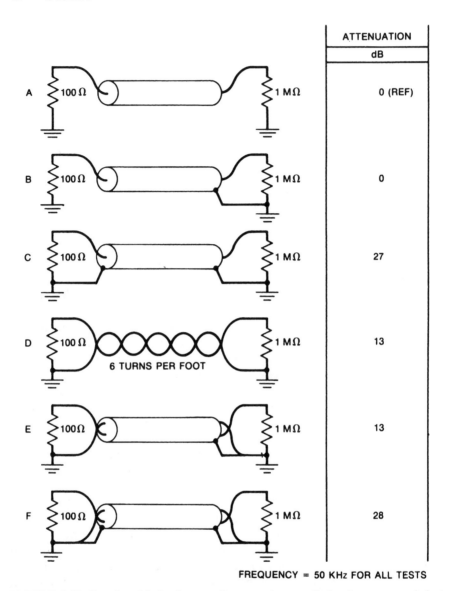

FIGURE 2-27. Results of inductive coupling experiment; all circuits are grounded at both ends.

electric field coupling is now beginning to show up because the twisted pair is unshielded and the termination is unbalanced (see Section 4.1). This can be seen in circuit *I*, where attenuation increases to 70 dB by placing a shield around the twisted pair. The fact that attenuation in circuit *G* is better than in *I* indicates that in this case the particular coaxial cable presents a smaller loop area to the magnetic field than does the twisted pair. This, however, is not necessarily true

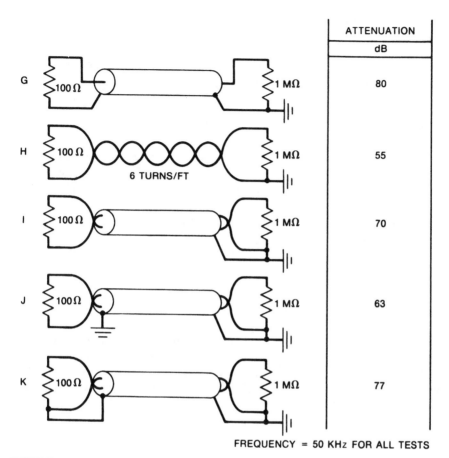

|  | ATTENUATION |
|---|---|
|  | dB |
| G | 80 |
| H | 55 |
| I | 70 |
| J | 63 |
| K | 77 |

FREQUENCY = 50 KHz FOR ALL TESTS

**FIGURE 2-28.** Results of inductive coupling experiment; all circuits are grounded at one end only.

in general. Increasing the number of turns per unit length for either of the twisted pairs (*H* or *I*) would reduce the pickup. In general, circuit *I* is preferred to circuit *G* for low-frequency magnetic shielding because in *I* the shield is not also one of the signal conductors.

Grounding both ends of the shield as in circuit *J* decreases the shielding slightly. This reduction is because of the high shield current in the ground loop formed by the shield inducing unequal voltages in the two center conductors. Circuit *K* provides more shielding than *I* because it combines the features of the coax *G* with those of the twisted pair *I*. Circuit *K* is not normally desirable because any noise voltages or currents that do get on the shield can flow down the signal conductor. It is almost always better to connect the shield and the signal conductors together at just one point. That point should be such that noise current from the shield does not have to flow down the signal conductor to get to ground.

Keep in mind that these experimental results are for relatively low-frequency (50 kHz) magnetic field shielding only, and no difference in ground potential existed between the two ends of the cable in the test setup.

## 2.10   EXAMPLE OF SELECTIVE SHIELDING

The shielded loop antenna is an example where the electric field is selectively shielded, whereas the magnetic field is unaffected. Such an antenna is useful in radio direction finders and as a magnetic field probe for precompliance *EMC* measurements (see Section 18.4). It can also decrease the antenna noise pickup in broadcast receivers. The latter effect is significant because most local noise sources generate a predominantly electric field. Figure 2-29A shows the basic loop antenna. From Eq. 2-12, the magnitude of the voltage produced in the loop by the magnetic field is

$$V_m = 2\pi f B A \cos \theta. \qquad (2\text{-}34)$$

The angle $\theta$ is measured between the magnetic field and a perpendicular to the plane of the loop. The loop, however, also act as a vertical antenna and will pick up a voltage from an incident electric field. This voltage is equal to the $E$ field times the effective height of the antenna. For a circular single-loop antenna, the effective height is $2\pi \ A/\lambda$ (ITT, 1968, p. 25-6). The induced voltage from the electric field becomes

$$V_c = \frac{2\pi A E}{\lambda} \cos \theta' \qquad (2\text{-}35)$$

The angle $\theta'$ is measured between the electric field and the plane of the loop.

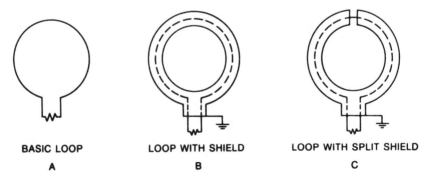

| BASIC LOOP | LOOP WITH SHIELD | LOOP WITH SPLIT SHIELD |
| A | B | C |

**FIGURE 2-29.** Split shield on loop antenna selectively reduces electric field while passing magnetic field.

To eliminate pickup from the electric field, the loop could be shielded as shown in Fig. 2-29B. However, this configuration allows shield current to flow, which will cancel the magnetic field as well as the electric field. To preserve the magnetic sensitivity of the loop, the shield must be broken to prevent the flow of shield current. This can be done as shown in Fig. 2-29C by breaking the shield at the top. The resulting antenna responds only to the magnetic filed component of an applied wave.

## 2.11   SHIELD TRANSFER IMPEDANCE

In 1934, Schelkunoff first proposed the concept of transfer impedance as a means of measuring the shielding effectiveness of cable shields. The shield transfer impedance is a property of the shield that relates the open circuit voltage (per unit length) developed between the center conductor and the shield to the shield current. The shield transfer impedance can be written as

$$Z_T = \frac{1}{I_S} \left( \frac{dV}{dl} \right), \tag{2-36}$$

where $Z_T$ is the transfer impedance in ohms per unit length, $I_S$ is the shield current, $V$ is the voltage induced between the internal conductors and the shield, and $l$ is the length of the cable. The smaller the transfer impedance, the more effective the shielding.

At low frequencies, the transfer impedance is equal to the direct current (dc) resistance of the shield. This result is equivalent to the results obtained in Eq. 2-33. At higher frequencies (above 1 MHz for typical cables), the transfer impedance of a solid tubular shield decreases because of skin effect, and the shielding effectiveness of the cable increases. Skin effect causes the noise current to remain on the outside surface of the shield, and the signal current on the inside; therefore, it eliminates the common impedance coupling between the two currents.

Figure 2-30 is a plot of the magnitude of the transfer impedance (normalized to the value of the dc resistance $R_{dc}$) for a *solid* tubular shield. If the shield is braided, then the transfer impedance will increase with frequency above about 1 MHz, as shown in Fig. 2-34.

## 2.12   COAXIAL CABLE VERSUS TWISTED PAIR

When comparing coaxial cable with twisted pair, it is important to first recognize the usefulness of both types of cables from a propagation point of view. This is shown in Fig. 2-31. Prior to the 1980s, the normal useful frequency for twisted pair cables was considered to be about 100 kHz, with special

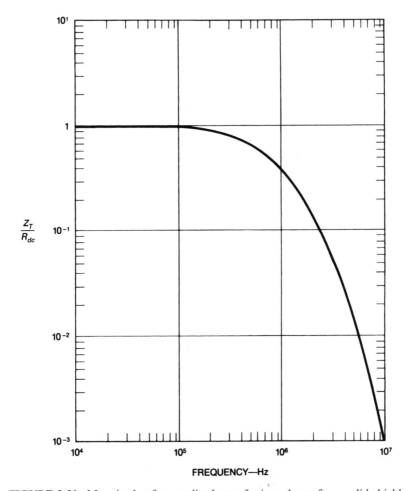

**FIGURE 2-30.** Magnitude of normalized transfer impedance for a solid shield.

applications going up to as high as 10 MHz. However, realizing the economic advantage of twisted pair over coaxial cable in many applications, today's cable designers and manufacturers have found ways to overcome this limitation.

Twisted pair cables do not have as uniform characteristic impedance as coaxial cables, which is the result of the two conductors not maintaining a constant position with respect to each other and is especially true when the cable is flexed or bent. Today's cable designers have been able to extend the normal useful frequency of twisted pairs up to 10 MHz with some applications [e.g., Ethernet and high-definition multimedia interface (HDMI)] extending up to hundreds of megahertz. These high-performance cables have less capacitance and are more tightly and uniformly twisted. In addition, in some cases, they have the two wires of the pair bonded together so they remain in the exact same relationship to each other over the length of the cable. Bonded twisted pair

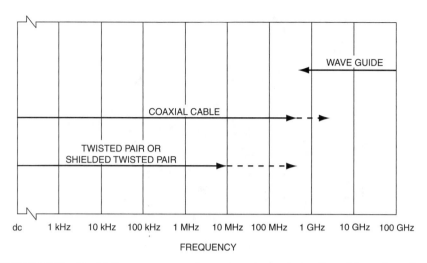

**FIGURE 2-31.** Useful frequency range for various transmission lines. Normal applications (solid line) and special applications (dashed line).

cables provide a more uniform characteristic impedance and are more immune to noise and produce much less radiation.

Many modern-day unshielded twisted pair (UTP) cables perform as well or better than older shielded twisted pair (STP) cables. A twisted pair cable is inherently a balanced structure and effectively rejects noise as discussed in Chapter 4—many of today's cables are extremely well balanced. Excellent examples of this are category (Cat) 5 and Cat 6 Ethernet cables. ANSI/TIA/ EIA 568B-2.1 defines performance specifications, such as impedance, cable loss, crosstalk, and radiation, for Cat 5 Ethernet cables. The Cat 5 cable, (see Fig. 2-32) consists of four UTPs made from 24-gauge solid wire. The nominal pitch of the twist is 1 per cm (2.5/in). Each pair in the cable, however, has a slightly different pitch to minimize the crosstalk between pairs. The Ethernet cable terminations are balanced. Cat 5e UTP cables are designed to perform well up to 125 MHz, Cat 6 cables up to 250 MHz, and in the future, Cat 7 cables should be useful up to about 600 MHz.

Coaxial cable has a more uniform characteristic impedance with lower losses. It is useful from dc up to very high frequency (VHF), with some applications extending up to ultra high frequency (UHF). Above 1 GHz, the losses in coaxial cable become large, and a waveguide often becomes a more practical transmission medium.

A coaxial cable grounded at one end provides a good degree of protection from capacitive (electric field) pickup. But if a noise current flows in the shield, then a noise voltage is produced as was discussed in Section 2.8. Its magnitude is equal to the shield current times the shield resistance. Because in a coax the shield is also part of the signal path, this voltage appears as noise in series

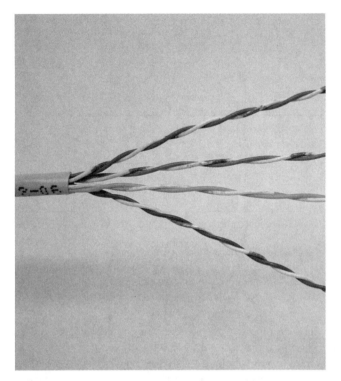

**FIGURE 2-32.** Cat 5 Ethernet cable.

with the input signal. A double-shielded, or triaxial, cable with insulation between the two shields can eliminate the noise voltage produced by the shield resistance. The noise current flows in the outer shield, and the signal return current flows in the inner shield. The two currents (signal and noise), therefore, do not flow through a common impedance.

Unfortunately, triaxial cables are expensive and awkward to use. A coaxial cable at high frequencies, however, acts as a triaxial cable because of skin effect. For a typical coaxial cable, the skin effect becomes important at about 1 MHz. The noise current flows on the outside surface of the shield, whereas the signal current flows on the inside surface. For this reason, a coaxial cable behaves better at high frequency.

A shielded twisted pair has characteristics similar to a triaxial cable and is not as expensive or awkward. The signal current flows in the two inner conductors, and any induced noise current flows in the shield. Common-impedance coupling is therefore eliminated. In addition, any shield current present is equally coupled (ideally), by mutual inductance, to both inner conductors and the two equal noise voltages cancel.

An unshielded twisted pair, unless its terminations are balanced (see Section 4.1), provides very little protection against capacitive (electric field) pickup, but it is very good for protection against magnetic field pickup. The effectiveness of twisting increases as the number of twists per unit length increases. When terminating a twisted pair, the more the two wires are separated, the less the noise suppression. Therefore, when terminating a twisted pair, shielded or unshielded, do not untwist the conductors any more than necessary to make the termination.

Twisted pair cables, even when unshielded, are very effective in reducing magnetic field coupling. Only two conditions are necessary for this to be true. First, the signal must flow equally and in opposite directions on the two conductors. Second, the pitch of the twist must be less than one twentieth of a wavelength at the frequencies of concern. (One twist per inch will be effective up to about 500 MHz.) The above is true whether the terminations are balanced or not. In addition, if the terminations are balanced, then twisted pair cables will also be effective in reducing electric field coupling (see Section 4.1). Do not confuse twisted pair wiring with balancing, because they are two completely different concepts, although often used together.

## 2.13  BRAIDED SHIELDS

Most cables are actually shielded with braid rather than with a solid conductor; see Fig. 2-33. The advantage of braid is flexibility, durability, strength, and long flex life. Braids typically provide only 60% to 98% coverage and are less effective as shields than solid conductors. Braided shields usually provide just slightly reduced electric field shielding (except at UHF frequencies) but greatly reduced magnetic field shielding. The reason is that braid distorts the uniformity of the longitudinal shield current. A braid is typically from 5 to 30 dB less effective than a solid shield for protecting against magnetic fields.

At higher frequencies, the effectiveness of the braid decreases even more as a result of the holes in the braid. Multiple shields offer more protection but at

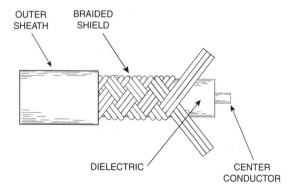

**FIGURE 2-33.** Cable with a braid shield.

higher costs and less flexibility. Premium cables with double and even triple shields, as well as silver-plated copper braid wires, are used in some critical military, aerospace, and instrumentation applications.

Figure 2-34 (Vance, 1978, Fig. 5-14) shows the transfer impedance for a typical braided-shielded cable normalized to the dc resistance of the shield. The decrease in transfer impedance around 1 MHz is because of the skin effect of the shield. The subsequent increase in transfer impedance above 1 MHz is caused by the holes in the braid. Curves are given for various percentages of coverage of the braid. Loose-weave braid (lower percentage shield coverage) provides more flexibility, whereas a tighter weave braid (higher percentage shield coverage) provides better shielding and less flexibility. As can be observed, for the best shielding, the braid should provide at least 95% coverage.

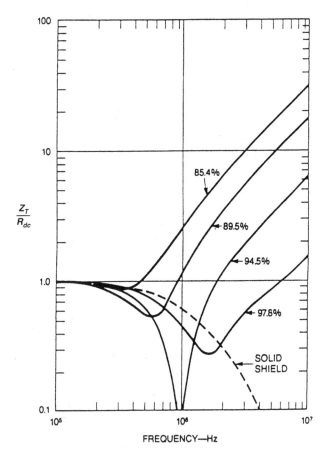

**FIGURE 2-34.** Normalized transfer impedance of a braided-wire shield, as a function of percent braid coverage (Vance, 1978, reprinted with permission of John Wiley & Sons, Inc.).

Cables with thin, solid aluminum-foil shields are available; these cables provide almost 100% coverage and more effective electric field shielding. They are not as strong as braid, have a higher shield cutoff frequency because of their higher resistance, and are difficult (if not impossible) to terminate properly. Shields are also available that combine a foil shield with a braid. These cables are intended to take advantage of the best properties of both foil and braid while minimizing the disadvantages of both. The braid allows proper 360° termination of the shield, and the foil covers the holes in the braid. The shielding effectiveness of braid over foil, or double-braid, cable does not start to degrade until about 100 MHz.

## 2.14  SPIRAL SHIELDS

A spiral shield (Fig. 2-35) is used on cables for one of three reasons, as follows: reduced manufacturing costs, ease of termination, or increased flexibility. It consists of a belt of conductors wrapped around the cable core (dielectric). The belt usually consists of from three to seven conductors.

Let us consider the differences between a spiral shield cable and an ideal, solid, homogeneous shield cable. In the solid homogeneous shield cable, the shield current is longitudinal along the axis of the cable, and the magnetic field produced by the shield current is circular and external to the shield as was shown in Fig. 2-12.

In the case of a spiral shield cable, the shield current follows the spiral and is at an angle $\phi$ with respect to the longitudinal axis of the cable, where $\phi$ is the pitch angle of the spiral as shown in Fig. 2-36.*

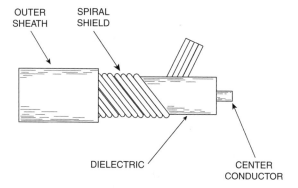

OUTER      SPIRAL
SHEATH     SHIELD

DIELECTRIC          CENTER
                    CONDUCTOR

**FIGURE 2-35.**  Cable with a spiral shield.

* This assumes that, as is usually the case in practice, poor electrical conductivity occurs between the individual conductors forming the spiral.

The total current $I$ in the shield can be decomposed into two components, one longitudinal along the axis of the cable and the other circular around the circumference of the cable as shown in Fig. 2-37. The longitudinal current $I_L$ along the cable axis is equal to

$$I_L = I \cos \phi, \qquad (2\text{-}37)$$

where $I$ is the total shield current and $\phi$ is the pitch angle. The circular current $I_c$ perpendicular to the cable axis and around the circumference of the cable is equal to

$$I_c = I \sin \phi. \qquad (2\text{-}38)$$

The longitudinal current $I_L$ behaves the same as the shield current does on a solid, homogenous, shielded cable and produces a circular magnetic field external to the cable. In the case of a long thin cable, the circular current behaves as a solenoid (coil of wire or inductor) along the axis of the cable and produces a longitudinal magnetic field $H$ inside the shield and no magnetic field outside the shield, as shown in Fig. 2-38. This is just the opposite of the magnetic field produced by the longitudinal shield current, which exists outside the shield with no magnetic field inside the shield. The longitudinal magnetic field produced by the circular component of the shield current has the detrimental effect of increasing the shield inductance. Therefore, the cable behaves as a normal coaxial cable as a result of the longitudinal current component, with additional inductance as a result of the circular current component.

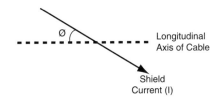

**FIGURE 2-36.** Direction of shield current flow on a spiral shield cable; $\phi$ is the pitch angle of the spiral.

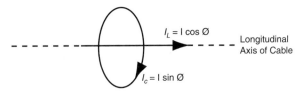

**FIGURE 2-37.** The shield current on a spiral shield cable can be decomposed into two components, one longitudinal ($I_L$) and one circular ($I_c$).

A braided shield cable can be thought of as having two, or more, interwoven spiral belts of conductor sets woven in opposite direction, such that each belt alternately passes over then under the other belt as shown in Fig. 2-33. One belt is applied in a clockwise direction, and the other belt is applied in a counter-clockwise direction. Because of the opposite direction of the lay of the two belts, the circular components of shield current tend to cancel each other, which leaves only the longitudinal component of the shield current—hence the much better high-frequency performance of braided shield cables compared with spiral shield cables.

As previously stated (Section 2.11), the shielding effectiveness of a shielded cable can be expressed in terms of the shield transfer impedance. For a spiral shield cable, the transfer impedance contains two terms, one from the long-itudinal component of the shield current and the other from the circular component of the shield current.

The transfer impedance that results from the longitudinal component of shield current decreases with frequency (which is good) as was shown in Fig. 2-30, whereas the component from the circular component of the shield current increases with frequency (which is bad). The net effect is that for spiral shield cables, the transfer impedance increases with frequency above about 100 kHz. The high-frequency transfer impedance is a strong function of the pitch angle $\phi$ of the spiral; the larger this angle is the larger will be the transfer impedance and the less shielding effectiveness the cable will have.

Normal braided shield cables also have shield transfer impedances that increase at high frequency, but the increase is much less than for the case of spiral shield cables. Figure 2-39 shows measured values of transfer impedance for cables that have various types of shields (Tsaliovich, 1995, Fig. 3-9).

Spiral shields are basically coils of wire that exhibit inductive effects with large high-frequency transfer impedances. Therefore, spiral shield cables should not be used in applications where signals are above about 100 kHz; basically, their use should be limited to audio frequencies and below. A more detailed discussion of spiral, as well as braided, shields is contained in *Cable Shielding For Electromagnetic Compatibility* (Tsaliovich, 1995).

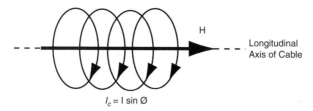

**FIGURE 2-38.** The circular component of shield current on a spiral shield cable produces a magnetic field along the longitudinal axis of the cable.

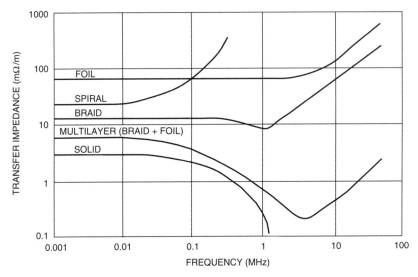

**FIGURE 2-39.** Measured values of shield transfer impedance, versus frequency, for various types of cable shields. The lower the transfer impedance, the better the shielding (from Tsaliovich, 1995 © AT&T).

## 2.15 SHIELD TERMINATIONS

Most shielded cable problems are the result of improper shield terminations. The maximum benefit of a well-shielded cable will only be realized if the shield is properly terminated. Requirements of a proper shield termination are as follows:

1. Termination at the proper end, or ends, and to the proper point or points
2. A very low impedance termination connection
3. A 360° contact with the shield

### 2.15.1 Pigtails

The magnetic shielding previously discussed depends on a uniform distribution of the longitudinal shield current around the circumference of the shield. Therefore, the magnetic shielding effectiveness near the ends of the cable depends strongly on the method by which the shield is terminated. A pigtail connection (Fig. 2-40) causes the shield current to be concentrated on one side of the shield. For maximum protection, the shield should be terminated uniformly around its circumference. This can be accomplished by using coaxial connectors such as BNC, UHF, or Type N. Such a connector, shown in Fig. 2-41, provides a 360° electrical contact to the shield. A coaxial termination also provides complete coverage of the inner conductor, which preserves the integrity of electric field shielding.

A 360° contact is important not only between the shield and the connector but also between the two mating halves of the connector. Screw type connectors, such as Type N and UHF, perform the best in this respect. Figure 2-42 shows an EMC version of a nonscrew-type XLR connector that contains spring fingers around its circumference to provide 360° shield contact between the mating halves. Figure 2-43 shows another method of providing a 360° shield termination, in this case without using a connector.

The use of a pigtail termination whose length is only a small fraction of the total cable length can have a significant effect on the noise coupling to the cable at frequencies above 100 kHz. For example, the coupling to a 3.66-m (12 ft) shielded cable with the shield grounded at both ends with 8-cm pigtail terminations is shown in Fig. 2-44 (Paul, 1980, Fig. 8a). The terminating

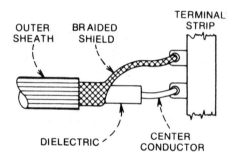

**FIGURE 2-40.** Pigtail shield connection concentrates current on one side of the shield.

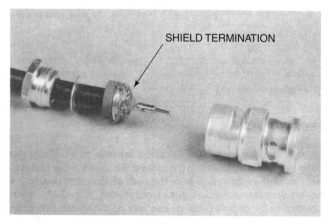

**FIGURE 2-41.** Disassembled BNC connector showing a 360° contact to the shield.

impedance of the shielded conductor was 50 Ω. This figure shows the individual contributions of the magnetic coupling to the shielded portion of the cable, the magnetic coupling to the unshielded (pigtail) portion of the cable, and the electric coupling to the unshielded portion of the cable. The capacitive (electric field) coupling to the shielded portion of the cable was negligible because the

**FIGURE 2-42.** A female XLR connector (used in pro-audio installations) with spring fingers around its circumference to provide a 360° contact between mating halves of the connector backshell.

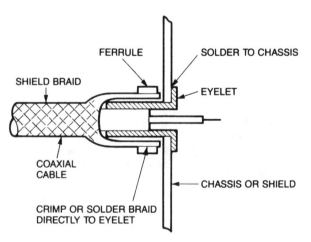

**FIGURE 2-43.** One method of terminating a cable, without a connector, and still providing a 360° contact to the shield.

shield was grounded and the shielded conductor's terminating impedance was low (50 $\Omega°$). As shown in Fig. 2-44, above 100 kHz, the primary coupling to the cable is from the inductive coupling to the pigtail.

If the terminating impedance of the shielded conductor is increased from 50 to 1000 $\Omega$, the result is as shown in Fig. 2-45 (Paul, 1980, Fig. 8b). In this case, the capacitive coupling to the pigtail is the predominant coupling mechanism above 10 kHz. Under these conditions, the coupling at 1 MHz is 40 dB greater than what it would have been if the cable had been completely shielded (no pigtail).

As can be observed from Figs. 2-44 and 2-45, although short pigtail terminations may be acceptable for low-frequency (<10 kHz) cable shield grounding, they are unacceptable for high-frequency shield grounding.

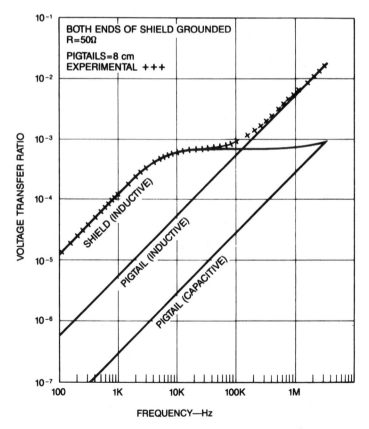

**FIGURE 2-44.** Coupling to a 3.7-m shielded cable with an 8-cm pigtail termination. Circuit termination equals 50 $\Omega$ (from Paul, 1980, © IEEE).

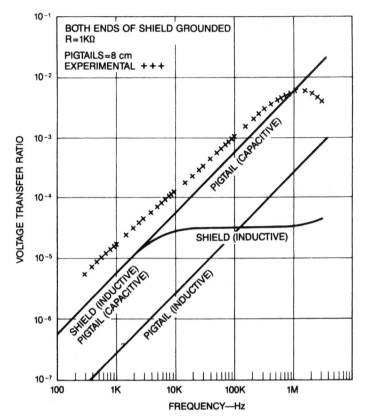

**FIGURE 2-45.** Coupling to a 3.7-m shielded cable with an 8-cm pigtail termination. Circuit termination equals 1000 Ω (from Paul, 1980, © IEEE).

### 2.15.2   Grounding of Cable Shields

The most common questions asked with respect to cable shield grounding are, where should the shield be terminated? At one end or at both ends and to what should it be connected? The simple answer is that it depends!

***2.15.2.1 Low-Frequency Cable Shield Grounding.***   The main reason to shield cables at low frequency is to protect them against electric field coupling primarily from 50/60-Hz power conductors. As was discussed in Section 2.5.2, a shield provides no magnetic field protection at low frequency. This points out the advantage of using shielded twisted pair cables at low frequency: The shield protects against the electric field coupling and the twisted pair protects against the magnetic field coupling. Many low-frequency circuits contain high-impedance devices that are susceptible to electric field coupling, hence, the importance of low-frequency cable shielding.

At low frequency, shields on multiconductor cables where the shield is not the signal return conductor are often grounded at only one end. If the shield is grounded at more than one end, then noise current may flow in the shield because of a difference in ground potential at the two ends of the cable. This potential difference, and therefore the shield current, is usually the result of 50/60-Hz currents in the ground. In the case of a coaxial cable, the shield current will produce a noise voltage whose magnitude is equal to the shield current times the shield resistance, as was shown in Eq. 2-33. In the case of a shielded twisted pair, the shield current may inductively couple unequal voltages into the twisted pair signal conductors and be a source of noise (see Section 4.1 on balancing). But if the shield is to be grounded at only one end, then which end should it be grounded, and to what ground?

Usually, it is better to ground the shield at the source end, because that is the reference for the signal voltage. If the signal source is floating (not grounded), however, then it is better to ground the cable shield at the load end.

Preferred low-frequency shielding schemes for both the shielded twisted pair and the coaxial cable are shown in Fig. 2-46. Circuits $A$ through $D$ are cases where either the source or the amplifier (load) circuit is grounded, but not both. In these four cases, the cable shield is also shown grounded at only one end, and it is the same end at which the circuit is grounded. When the signal circuit is grounded at both ends as shown in circuits $E$ and $F$, the amount of noise reduction possible is limited by the difference in ground potential and by the susceptibility of the ground loop to magnetic fields. In the case of circuit $E$, the shielded twisted pair cable is also grounded at both ends to force some ground-loop current to flow through the lower impedance shield, rather than on the signal return conductor. In circuit $F$, the shield of the coaxial cable must be grounded at both ends because it is also the return conductor for the signal, and the signal is grounded at both ends. In this case, the noise coupling can be decreased by decreasing the resistance of the cable shield, as this will reduce the common impedance coupling as was discussed in Section 2.8. If additional noise immunity is required, then the ground loop must be broken. This can be done by using transformers, optical couplers, or common-mode chokes as discussed in Section 3.4.

An indication of the type of performance to be expected from the configurations shown in Fig 2-46 can be obtained by referring to the results of the magnetic coupling experiment presented in Figs. 2-27 and 2-28.

Grounding the cable shield at only one end to eliminate power line frequency noise coupling, however, allows the cable to act as a high-frequency antenna and be vulnerable to rf pickup. AM and FM radio transmitters can induce high-frequency rf currents into the cable shield. If the cable shield is connected to the circuit ground, then these rf currents will enter the equipment and may cause interference. Therefore, the proper way to terminate the cable shield is to the equipment's shielded enclosure, not to the circuit ground. This connection should have the lowest impedance possible, and the connection should be made to the outside of the shielded enclosure. In this way, any rf noise current on the

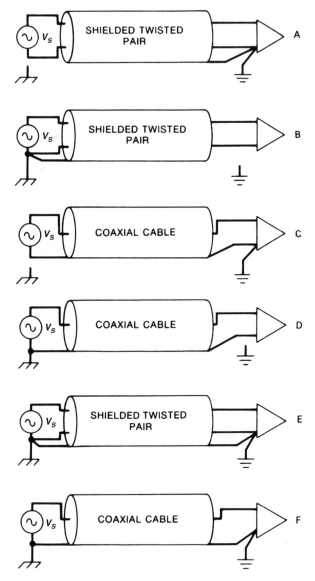

**FIGURE 2-46.** Preferred low-frequency shield grounding schemes for shielded twisted pair and coaxial cables.

shield will flow harmlessly on the outside surface of the enclosure and through the enclosure's parasitic capacitance to ground, bypassing the sensitive electronics inside the box.

   If you think of a cable shield as being an extension of the enclosure's shield, then it becomes clear that *the shield should be terminated to the enclosure not to the circuit ground.*

In the professional audio world, a noise problem that occurs from connecting a cable shield to circuit ground, instead of to chassis ground, is referred to as the "*Pin 1* Problem." Neil Muncy coined this term in 1995 in his classic paper on the subject, *Noise Susceptibility in Analog and Digital Signal Processing Systems*. The term *Pin 1* refers to the connector pin that is connected to the cable shield in XLR connectors, which are commonly used in professional audio systems. For a phone jack, *Pin 1* would refer to the sleeve; for a phono plug or BNC connector, it would refer to the outside shell of the connector.

Because of the many documented cases of interference caused by terminating cable shields to circuit ground, the Audio Engineering Society, in 2005, published a standard on cable shield grounding in audio equipment that says: *The cable shield and the shell of the equipment connector(s) shall have a direct connection to the shielding enclosure via the lowest impedance path possible (AES48, 2005)*.

Shields on coaxial cables, where the shield is the signal return conductor, must be grounded at both ends, and this ground, for functionality, must be to the circuit ground. However, for noise considerations, as discussed above, the shield should first be terminated to the enclosure. This can be accomplished easily by terminating the cable shield to the enclosure and then connecting the circuit ground to the enclosure at the same point.

The single-ended shield ground is effective at low frequencies (audio and below), because it prevents power-frequency currents from flowing on the shield and possibly introducing noise into the signal circuit. The single-point ground also eliminates the shield ground loop* and its associated possible magnetic field pickup. As frequency increases, however, the single-point ground becomes less and less effective. As the cable length approaches one-quarter wavelength, a cable shield grounded at one end becomes a very efficient antenna. Under these circumstances, grounding both ends of the shield is normally required.

### 2.15.2.2 High-Frequency Cable Shield Grounding.

At frequencies above about 100 kHz, or where cable length exceeds one twentieth of a wavelength, it becomes necessary to ground the shield at both ends. This is true for either multiconductor or coaxial cables. Another problem develops at high frequency; stray capacitance tends to complete the ground loop, as shown in Fig. 2-47, which makes it difficult or impossible to maintain ground isolation at the unterminated end of the shield.

It is therefore common practice at high frequency, and with digital circuits, to ground the cable shield at both ends. Any small noise voltage caused by a difference in ground potential that may couple into the circuit (primarily at power line frequencies and its harmonics) will not affect digital circuits and can usually be filtered out of rf circuits, because of the large frequency difference

---

* The loop formed between the cable shield and the external ground.

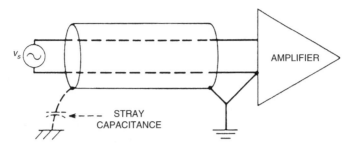

**FIGURE 2-47.** At high frequencies, stay capacitance completes the ground loop.

between the noise and the signal. At frequencies above 1 MHz, the skin effect reduces the common impedance coupling from the signal and noise currents that flow in the shield. The skin effect causes the noise current to flow on the outside surface of the shield and the signal current to flow on the inside surface of the shield. The multiple grounds also provide magnetic field shielding at frequencies above the shield cutoff frequency.

### 2.15.2.3 Hybrid Cable Shield Grounding.

Although single-point grounds are effective at audio frequencies and below, and multipoint grounds are effective at high frequency, what does one do when the signal contains both high- and low-frequency components, for example, a video signal? Most audio equipment today also contains digital circuitry for signal processing, so even in audio equipment, high-frequency signals are often present. Although, in this later case, the high-frequency signals may not be intentionally in the cable, they can unintentionally be coupled as a common-mode signal to the cable, and the cable shield may then be needed to prevent the radiation of these high-frequency signals.

In these situations, the circuit shown in Fig. 2-47 can be taken advantage of by replacing the stray capacitance with an actual capacitor (i.e., 47 nF), which forms a combination or hybrid ground. At low frequency, a single-point ground exists because the impedance of the capacitor is large. However, at high frequency, the capacitor becomes a low impedance, which converts the circuit to one that is grounded at both ends.

The actual implementation of an effective hybrid cable shield ground may, however, be difficult, because any inductance in series with the capacitor will decrease its effectiveness. Ideally, the capacitor should be built into the connector. Recently, some audio connector manufacturers have begun to understand the advantage of the hybrid shield ground approach and have designed connectors with an effective shield-terminating capacitor built in. Figure 2-48 shows an XLR connector with 10 radial surface-mount capacitors connected between the cable shield termination and the connector backshell, which effectively breaks the ground loop at low frequency while maintaining a

**FIGURE 2-48.** XLR connector with 10 SMT capacitors between the cable shield and the *connector backshell.*

low impedance cable-shield-to-connector-shell termination at high frequency. The paralleling of 10 capacitors decreases the inductance in series with any one capacitor by a factor of 10 and produces a hybrid shield termination that can be effective up to about 1 GHz.

*2.15.2.4 Double Shielded Cable Grounding.* Two reasons to use a double-shielded cable are as follows: One is to increase the high-frequency shielding effectiveness; the other is when you have both high-frequency and low-frequency signals in the same cable. In the first case, the two shields can be in contact with each other; in the second case, the two shields must be isolated from each other (often referred to as a triaxial cable).

   Having two shields that are isolated from each other allows the designer the option of terminating the two shields differently. The outer shield can be terminated at both ends to provide effective high-frequency as well as magnetic field shielding. The outer shield is often also used to prevent radiation from the cable, which results from high-frequency common-mode currents on the cable. The inner shield can then be terminated at only one end, thus avoiding the ground-loop coupling that would occur if grounded at both ends. Effectively, the inner shield is low-frequency terminated, whereas the outer shield is high-frequency terminated, which thereby solves the problem caused by a signal containing both high- and low-frequency components. The outer shield should be connected to the enclosure, and the inner shield should be connected to the enclosure or to the circuit ground, whichever performs best in that circumstance.

Another interesting possibility when using a triaxial cable is to terminate both shields at one end only—but at opposite ends. Therefore, no low-frequency ground loop is present, but the inter-shield capacitance closes the loop at high frequency. This can often be effective in cases that have very long cable runs where there might be a large ground voltage differential between the two ends, and where there will be a lot of inter-shield capacitance because of the long cable run.

## 2.16  RIBBON CABLES

A major cost associated with the use of cables is the expense related to the termination of the cable. The advantage of ribbon cables is that they allow low-cost multiple terminations, which is the primary reason for using them.

Ribbon cables have a second advantage. They are "controlled cables" because the position and orientation of the wires within the cable is fixed, like the conductors on a printed wiring board. However, a normal wiring harness is a "random cables" because the position and orientation of the wires within the cable is random and varies from one harness to the next. Therefore, the noise performance of a "random cable" can vary from one unit to the next.

The major problem associated with the use of ribbon cables relates to the way the individual conductors are assigned with respect to signal leads and grounds.

Figure 2-49A shows a ribbon cable where one conductor is a ground and all the remaining conductors are signal leads. This configuration is used because it minimizes the number of conductors required; however, it has three problems. First, it produces large loop areas between the signal conductors and their ground return, which results in radiation and susceptibility. The second problem is the common impedance coupling produced when all the signal conductors use the same ground return. The third problem is crosstalk between the individual conductors—both capacitive and inductive; therefore, this configuration should seldom be used. If it is used, the single ground should be assigned to one of the center conductors to minimize the loop areas.

Figure 2-49B shows a better configuration. In this arrangement, the loop areas are small because each conductor has a separate ground return next to it. Because each conductor has a separate ground return, common impedance coupling is eliminated, and the crosstalk between leads is minimized. This is the preferred configuration for a ribbon cable, even though it does require twice as many conductors as Fig 2-49A. In applications where crosstalk between cables is a problem, two grounds may be required between signal conductors.

A configuration that is only slightly inferior to Fig. 2-49B, and one that used 25% fewer conductors is shown in Fig. 2-49C. This configuration also has a ground conductor next to every signal conductor and therefore has small loop areas. Two signal conductors share one ground, so some common impedance

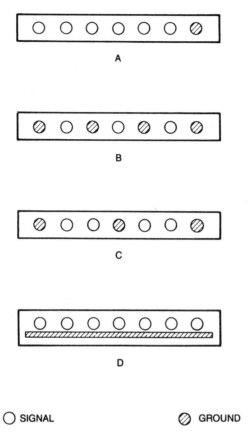

FIGURE 2-49. Ribbon cable configurations: (*A*) single ground; (*B*) alternate grounds; (*C*) ground/signal/signal/ground; (*D*) signal over ground plane.

coupling occurs, and the crosstalk is higher than in Fig. 2-49B because there is no ground between some of the adjacent signal conductors. This configuration may provide adequate performance in many applications and has the lowest cost-to-performance ratio.

Ribbon cables are also available with a ground plane across the full width of the cable as shown in Fig. 2-49D. In this case, the loop areas are determined by the spacing between the signal conductor and the ground plane under it. Because this dimension is usually less than the lead-to-lead spacing in the cable, the loop areas are smaller than in the alternate ground configuration of Fig. 2-49B. If allowed to do so, the ground current will flow under the signal conductor, for the same reason that the current returned on the shield in Fig. 2-22. However, unless the cable is terminated with a full-width electrical contact to the ground plane, the return currents will be forced out from under the signal leads, and the loop area will increase. Because it is difficult to terminate this kind of cable properly, it is not often used.

Shielded ribbon cables are also available; however, unless the shield is properly terminated with a 360° connection (a difficult thing to do), their effectiveness is considerably reduced. The effect of shield termination on the radiation from ribbon cables was discussed by Palmgren (1981). Palmgren points out that the outside conductors in a shielded ribbon cable are not as well shielded as the conductors located closer to the center of the cable (typically 7 dB less shielding). This effect is caused by the nonuniformity of the shield current at the outside edge of the shield. Therefore, critical signals should not be placed on the outside conductors of shielded ribbon cables.

## 2.17   ELECTRICALLY LONG CABLES

The analysis presented in this chapter has assumed that the cables were short compared with a wavelength. What this really means is that all the current on the cables is in phase. Under these circumstances, the theory predicts that both the electric and the magnetic field coupling increase with frequency indefinitely. In practice, however, the coupling levels off above some frequency.

As cables approach a quarter wavelength in length, some of the current in the cable is out of phase. When the cable is a half-wavelength long, the out-of-phase currents will cause the external coupling to be zero because of cancellation of effects. This does not alter the dependence of the coupling on the various other parameters of the problem; it only changes the numerical result. Therefore, the parameters that determine the coupling remain the same, regardless of the length of the cables.

Figure 2-50 shows the coupling between two cables with and without the assumption that the cables are short. The results are almost exact up to the point where the phasing effects start to occur, about one tenth of a wavelength. The short cable approximation results are, however, still useful up to about one quarter wavelength. Above this point, the actual coupling decreases because some of the current is out of phase, whereas the short cable approximation predicts an increase in the coupling. If the rise in coupling predicted by the short cable approximation is truncated at a quarter wavelength, it provides an approximation to the actual coupling. Note that the nulls and peaks produced by the phasing the currents are not taken into account under these circumstances. However, unless one is planning to take advantage of these nulls and peaks in the design of equipment—a dangerous thing to do—their location is not important.

For more information on analyzing long cables see Paul (1979) and Smith (1977).

## SUMMARY

- Electric field coupling can be modeled by inserting a noise current generator in *shunt* with the receptor circuit.

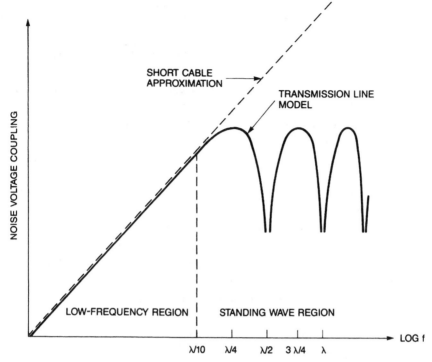

**FIGURE 2-50.** Electric field coupling between cables using the short cable approximation (dashed line) and the transmission line model (solid line).

- Magnetic field coupling can be modeled by inserting a noise voltage generator in *series* with the receptor
- Electric fields are much easier to guard against than magnetic fields.
- A shield grounded at one or more points shields against electric fields.
- The key to reducing magnetic coupling is to decrease the area of the pickup loop.
- For a coaxial cable grounded at both ends, virtually all the return current flows in the shied at frequencies above audio.
- To prevent magnetic field radiation or pickup, a shield grounded at both ends is useful above audio frequencies.
- Any shield in which noise currents flow should not be part of the signal path.
- Because of skin effect, at high frequencies, a coaxial cable behaves as a triaxial cable.
- The shielding effectiveness of a twisted pair increases as the number of twists per unit length increases.

- The magnetic shielding effects listed in this chapter require a cylindrical shield with uniform distribution of shield current over the circumference of the shield.
- For a solid-shield cable, the shielding effectiveness increases with frequency.
- For a braid-over-foil or double-braid cable, the shielding effectiveness begins to decrease above about 100 MHz.
- For a braided-shield cable, the shielding effectiveness begins to decrease above about 10 MHz.
- For a spiral shield cable, the shielding effectiveness begins to decrease above about 100 kHz.
- Most cable shielding problems are caused by improper shield terminations.
- At low frequency, cable shields may be grounded at one end only.
- At high frequency, cable shields should be grounded at both ends.
- Hybrid shield terminations can be used effectively when both low- and high-frequency signals are involved.
- Cable shields should be terminated to the equipment enclosure not to the circuit ground.
- The major problem with ribbon cables relates to how individual conductors are assigned between signals and grounds.

## PROBLEMS

2.1 In Fig. P2-1 the stray capacitance between conductors 1 and 2 is 50 pF. Each conductor has a capacitance to ground of 150 pF. Conductor 1 has a 10-V alternating current (ac) signal at a frequency of 100 kHz on it. What is the noise voltage picked up by conductor 2 if its termination $R_T$ is:

   a. An infinite resistance?
   b. A 1000-$\Omega$ resistance?
   c. A 50-$\Omega$ resistance?

2.2 In Fig. P2-2, a grounded shield is placed around conductor 2. The capacitance from conductor 2 to the shield is 100 pF, The capacitance between conductors 2 and 1 is 2 pF, and the capacitance between conductor 2 and ground is 5 pF. Conductor 1 has a 10-V ac signal at a frequency of 100 kHz on it. For this configuration, what is the noise voltage picked up by conductor 2 if its termination $R_T$ is:

   a. An infinite resistance?
   b. A 1000-$\Omega$ resistance?
   c. A 50-$\Omega$ resistance?

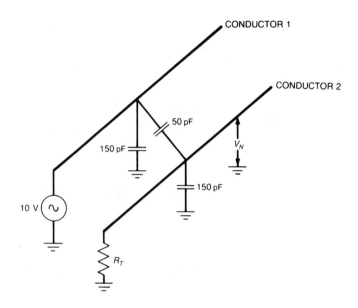

**FIGURE P2-1.**

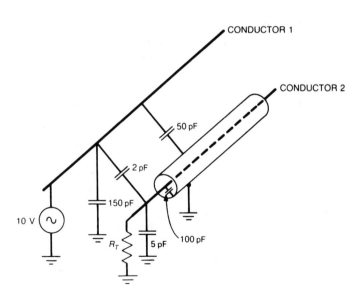

**FIGURE P2-2.**

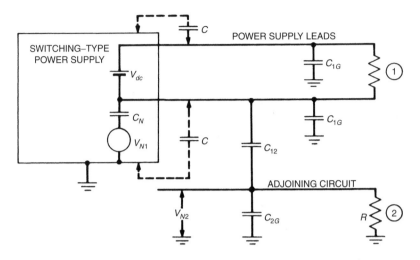

**FIGURE P2-3.**

2.3 Because of the switching action of power transistors, a noise voltage is usually introduced in switching-type power supplies between the power-supply output leads and the case. This is represented by $V_{N1}$ in Fig. P2-3. This noise voltage can capacitively couple into adjoining Circuit 2 as illustrated. $C_N$ is the equivalent coupling capacitance between the case and the output power leads. Assume $C_{12} \ll C_{10}$.

a. For this circuit configuration, determine and sketch the ratio $V_{N2}/V_{N1}$ as a function of frequency. (Neglect the capacitors $C$, shown dotted.)

Next, capacitors $C$ are added between the output power leads and the case, as indicated.

b. How does this affect the noise coupling?

c. How would shielding of the power-supply leads improve the noise performance?

2.4 Two conductors, each 10-cm long and space 1-cm apart, form a circuit. This circuit is located where there is a 10-gauss magnetic field at 60 Hz.

What is the maximum noise voltage coupled into the circuit from the magnetic field?

2.5 Figure P2-5A is a partial schematic for a low-level transistor amplifier. A printed circuit layout for the circuit is shown in Fig. P2-5B. The circuit is located within a strong magnetic field.

What is the advantage of the alternate layout shown in Fig. P2-5C over that of Fig. P2-5B?

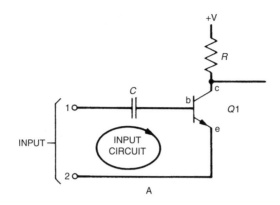

A

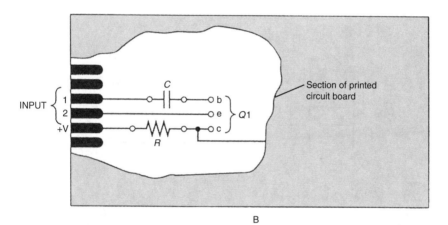

B

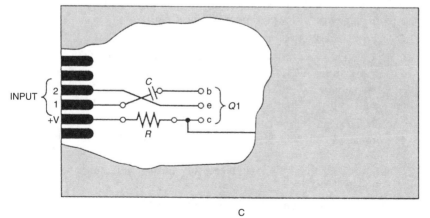

C

**FIGURE P2-5.**

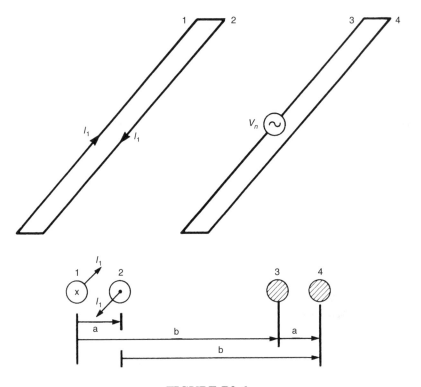

**FIGURE P2-6.**

2.6 Calculate the mutual inductance per unit length between the two coplaner parallel loops shown in Fig. P2-6.

2.7 Using the results of Problem 2.6:

   a. Calculate the mutual inductance per unit length between adjacent pairs (first and second pair) of a ribbon cable having a conductor spacing of 0.05 in. Also calculate the mutual inductance between the first and third pair, and between the first and fourth pair.

   b. If the signal in one pair is a 10-MHz, 5-V sine wave and the cable is terminated with 500 $\Omega$, what is the voltage induced into the adjacent pair?

2.8 What is the maximum value of the mutual inductance between two circuits?

2.9 How does the magnitude of the magnetic field vary versus distance from:

   a. A single isolated conductor?

   b. Closely spaced parallel conductors carrying the signal and return current?

2.10 A receptor circuit consists of a 1-m long wire, located 5 cm above a ground plane. Each end of the circuit is terminated with a 50-Ω resistor. An electric field induces a noise current of 0.5 mA into the circuit. The magnetic field from the same noise source induces a noise voltage of 25 mV into the circuit.

    a. If the noise voltage is measured across each of the terminating resistors, what will the two readings be?

    b. What general conclusion can you draw from the above results?

    c. What will happen if the polarity of the magnetic-field induced voltage is reversed?

2.11 Explain why an unshielded twisted pair will only provide protection against capacitive pickup when its terminations are balanced (i.e., both have the same impedance to ground)?

2.12 In a spiral shield cable, what percentage of the magnetic field $H$ produced by

    a. The longitudinal component of the shield current is inside the shield. What percentage is outside the shield?

    b. The circular component of the shield current is inside the shield. What percentage is outside the shield?

# REFERENCES

AES48-2005, *AES Standard on Interconnections—Grounding and EMC Practices—Shields of Connectors in Audio Equipment Containing Active Circuitry*, Audio Engineering Society, 2005.

ANSI/TIA/EIA-568-B.2.1, *Commercial Building Telecommunications Cabling Standard—Part 2: Balanced Twisted Pair Components—Addendum I—Transmission Performance Specifications for 4-Pair 100 Ohm Category 6 Cabling*, 2002.

Brown, J. and Whitlock, B. "Common-Mode to Differential-Mode Conversion in Shielded Twisted-Pair Cables (Shield Current Induced Noise)." *Audio Engineering Society 114th Convention*, Amsterdam, The Netherlands, 2003.

Grover, F. W. "Inductance Calculations—Working Formulas and Tables." *Instrument Society of America*, 1973.

Hayt, W. H., Jr. *Engineering Electromagnetics*. 3rd ed. McGraw-Hill, New York, 1974.

ITT. *Reference Data for Radio Engineers*. 5th ed. Howard W. Sams & Co., New York, 1968.

Muncy, N. "Noise Susceptibility in Analog and Digital Signal Processing Systems." *Journal of the Audio Engineering Society*, June 1995.

Palmgren, C. "Shielded Flat Cables for EMI & ESD Reduction," IEEE Symposium on EMC, Boulder, Co, 1981.

Paul, C. R. "Prediction of Crosstalk Involving Twisted Pairs of Wires—Part I: A Transmission-Line Model for Twisted-Wire Pairs," *IEEE Transactions on EMC*, May 1979.

Paul, C. R. "Prediction of Crosstalk Involving Twisted Pairs of Wires—Part II: A Simplified Low-Frequency Prediction Model." *IEEE Transactions on EMC*, May 1979.

Paul, C. R. "Effect of Pigtails on Crosstalk to Braided-Shield Cables." *IEEE Transactions on EMC*, August 1980.

Paul, C. R. "Modelling and Prediction of Ground Shift on Printed Circuit Boards." *1986 IERE Symposium on EMC*. York, England, October 1986.

Ruehli, A. E. "Inductance Calculations in a Complex Integrated Circuit Environment." *IBM Journal of Research and Development*, September 1972.

Schelkunoff, S. A. "The Electromagnetic Theory of Coaxial Transmission Lines and Cylindrical Shields." *Bell System Technical Journal*, Vol. 13, October 1934, pp. 532–579.

Smith, A. A *Coupling of External Electromagnetic Fields to Transmission Lines*. Wiley, New York, 1977.

Smythe, W. R. *Static and Dynamic Electricity*. McGraw-Hill, New York, 1924.

Tsaliovich, A. *Cable Shielding for Electromagnetic Compatibility*. New York, Van Nostrand Reinhold, 1995.

Vance, E. F. *Coupling to Shielded Cables*. Wiley, New York, 1978.

## FURTHER READING

Buchman, A. S. "Noise Control in Low Level Data Systems." *Electromechanical Design*, September 1962.

Cathy, W., and Keith, R. "Coupling Reduction in Twisted Wires." *IEEE International Symposium on EMC*, Boulder, August 1981.

Ficchi, R. O. *Electrical Interference*. Hayden Book Co., New York, 1964.

Ficchi, R. O. *Practical Design For Electromagnetic Compatibility*. Hayden Book Co., New York, 1971.

Frederick Research Corp, *Handbook on Radio Frequency Interference*. Vol. 3 (Methods of Electromagnetic Interference Suppression). Frederick Research Corp., Wheaton, MD, 1962.

Hilberg. W. *Electrical Characteristics of Transmission Lines*. Artech House, 1979.

Lacoste, R., "Cable Shielding Experiments," *Circuit Cellar* October, 2008.

Mohr, R. J. "Coupling between Open and Shielded Wire Lines over a Ground Plane," *IEEE Transactions on EMC*, September 1976.

Morrison, R. *Grounding and Shielding*. Wiley, 2007.

Nalle, D. "Elimination of Noise in Low Level Circuits." *ISA Journal*, vol. 12, August 1965.

Ott, H. W. *Balanced vs. Unbalanced Audio System Interconnections*, Available at www.hottconsultants.com/tips.html. Accessed December 2008.

Paul, C. R. "Solution of the Transmission-Line Equations for Three-Conductor Lines in Homogeneous Media." *IEEE Transactions on EMC,* February 1978.

Paul, C. R. "Prediction of Crosstalk in Ribbon Cables: Comparison of Model Predictions and Experimental Results." *IEEE Transactions on EMC,* August 1978.

Paul, C. R. *Introduction to Electromagnetic Compatibility*, 2nd ed. Chapter 9 (Crosstalk) New York Wiley, 2006.

Rane Note 110. *Sound System Interconnection.* Rane Corporation, 1995.

Timmons, F. *"Wire or Cable Has Many Faces*, Part 2." *EDN,* March 1970.

Trompeter, E. "Cleaning Up Signals with Coax." *Electronic Products Magazine,* July 16, 1973.

# 3 Grounding

*The search for a "good ground" is very similar to the search for the Holy Grail in many respects—tales abound about its existence and we all say we want and need it, but we cannot seem to find it.*

—*Warren H. Lewis**

Of all the conductors used in interconnecting electronics, the most complex is ironically the one that generally gets the least attention—the ground. Grounding is one of the primary ways of minimizing unwanted noise and of producing a safe system. That said, a noise-free system is not necessarily a safe system, and conversely, a safe system is not necessarily a noise-free system. It is the responsibility of the designer to provide both a safe and a noise-free system. *A good ground system must be designed;* it is wishful thinking to expect a ground system to perform well if little or no thought has been given to its design. It is sometimes difficult to believe that expensive engineering time should be devoted to sorting out the minute details of circuit grounding, but in the end, not having to solve mysterious noise problems once the equipment is built and tested saves both time and money.

One advantage of a well-designed ground system is that it can often provide protection against unwanted interference and emission, without any additional per-unit cost to the product. The only cost is the engineering time required to design the system. In comparison, an improperly designed ground system may be a primary source of interference and emission and therefore require considerable engineering time to eliminate the problem. Hence, a properly designed ground system can be truly cost effective.

Grounding is an important and often misunderstood aspect of noise and interference control. One problem with grounding is the word itself. The word "ground" can mean many different things to many different people. It could mean an 8-ft long rod driven into the earth for lightning protection, it could mean the green safety wire used in an alternating current (ac) power distribution system, it could mean a ground plane on a digital logic printed

* Lewis, 1995, p.301.

---

*Electromagnetic Compatibility Engineering,* by Henry W. Ott
Copyright © 2009 John Wiley & Sons, Inc.

circuit board (PCB), or it could refer to a narrow trace on a PCB that provides the return path for a low-frequency analog signal on a satellite orbiting the earth. In all the above cases, the requirements for the ground are different.

Grounds fall into two categories as follows: (1) safety grounds and (2) signal grounds. The second category probably should not be called grounds at all, but rather *returns*, and it could be further subdivided into either signal or power returns. If they are called grounds, they should be referred to as "signal grounds" or "power grounds" to define the type of current they carry and to distinguish them from "safety grounds." Common usage, however, often just refers to them all as grounds.

In most cases, safety grounds do not carry current, except during a fault, as discussed later in this chapter. This distinction is important, because signal grounds do carry current during normal operation. Therefore, another way to classify grounds would be (1) those that carry current during normal operation (such as signal or power returns) and (2) those that do not carry current during normal operation (such as safety grounds).

In addition, if a ground is connected to the enclosure or chassis of the equipment, then it is often called a *chassis ground*. If a ground is connected to the earth through a low-impedance path, then it may be called an *earth ground*. Safety grounds are usually connected to the earth or some conducting body that serves in place of the earth such as the airframe of an aircraft or the hull of a ship. Signal grounds may or may not be connected to the earth. In many cases, a safety ground is required at a point that is unsuitable for a signal ground, which can complicate the design problem. However, the basic objective of grounding must always be to *first make it safe and then make it work properly without compromising the safety*. In all cases, grounding techniques are available to do this.

## 3.1 AC POWER DISTRIBUTION AND SAFETY GROUNDS

In the power industry, grounding usually means a connection to the earth. In the United States, ac power distribution, grounding, and wiring standards for facilities are contained in the National Electric Code (NEC), which is often just referred to as "the code." Changes to the code are made every 3 years. The basic purpose of power system grounding is to protect personnel, animals, structures, and buildings from harm because of electrical shock or fire. In facility wiring, this is usually accomplished by the following:

1. Insuring the operation of a protective device (fuse or circuit breaker) in the event of a fault (i.e., contact between the hot conductor and the equipment enclosure)

2. Minimizing the potential difference between conductive enclosures and other metal objects

3. Providing lightning protection

### 3.1.1 Service Entrance

Electrical power enters a facility (residential or commercial/industrial) at the service entrance. The utility is responsible for the wiring and grounding on the supply side of the service entrance, and the user is responsible for the wiring and grounding on the load side of the service entrance. The service entrance is the interface between the utility and the user; it is where metering takes place, where power can be disconnected from the facility, and where lightning is likely to enter the facility. With few exceptions, the NEC allows only one service entrance per building (Article 230.2); the few exceptions to this are listed in Articles 230.2(A) through 230.2(D).

A three-phase, high-voltage (typically 4160 or 13,800 V) system is usually used to feed power to a neighborhood. The service drop to the facility can either be supplied aerially or underground. In industrial areas, all three phases are available for the service drop. In residential areas, however, only one phase is available for the service drop—the high-voltage being stepped down using a single-phase distribution transformer. A very common configuration is a center-tapped secondary power distribution transformer as shown in Fig. 3-1. This arrangement provides the facility with single-phase power at 120 and 240 V. The center tap of the transformer secondary (neutral conductor) is *solidly grounded* to earth both at the distribution transformer and at the service entrance panel.

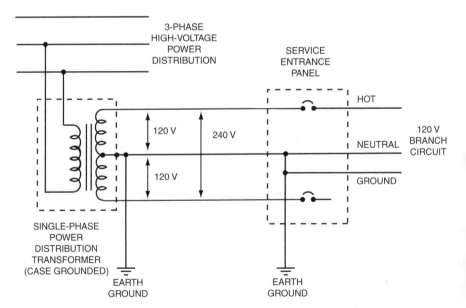

**FIGURE 3-1.** A single-phase residential service providing 120/240 V single-phase power.

The voltage between either of the outer terminals of the transformer secondary and the neutral is 120 V. The voltage between the two outer terminals of the transformer is 240 V. Two-hundred-forty volts is often used to power large appliances such as electric ovens, large air conditioners, and electric clothes dryers. The two outer conductors are "hot" (at a voltage with respect to the grounded neutral) and must contain overcurrent protection devices such as fuses or circuit breakers. Fuses or circuit breakers are not allowed in the neutral conductor (230.90(B)). The code also specifies that not more than six manually operated breakers or disconnects be required to completely disconnect power from a facility (230.71), and these disconnects must all be at the same location, preferably in the same panel box.

The NEC does not imply that grounding is the only way to make electrical installations safe; isolation, insulation, and guarding are also viable alternatives in some situations. For example, power distribution systems above 1000 V are normally not grounded. The NEC, however, provides one method of providing a safe ac power distribution system, and it is often codified into law by local and state governmental authorities.

In the early days of ac electrical power distribution, there was much discussion amongst "experts" as to whether the power distribution system in buildings should be grounded or not (International Association of Electrical Inspectors, Appendix A, 1999). The term "grounded," in this case, refers to connecting one of the current carrying conductors, usually called the neutral, to the earth. In 1892, the New York Board of Fire Underwriters issued a report on *Grounding of Electrical Wires*, which stated: "The New York Board of Fire Underwriters has condemned the practice of grounding the neutral as dangerous and orders it to be stopped. There can be no doubt that the practice of grounding the neutral is not as safe as a completely insulated system."

The only mandatory grounding requirement in the first (1897) edition of the National Electrical Code was that for lightning arrestors. It was not until 16 years later in the 1913 code that the neutral conductor of the ac power distribution transformer secondary was required to be grounded.

To this day, in many parts of the world, ungrounded ac power distribution systems are in common use. Over 100 years of experience with these ungrounded systems have shown that such systems can be safe. On naval ships where reliability is critical and seawater an ever-present problem, power distribution systems are usually not grounded. The United States, as well as many other nations, however, have adopted a grounded system—an approach many feel is safer. The above discussion, however, clearly demonstrates that there is more than one acceptable way to achieve electrical safety.

### 3.1.2  Branch Circuits

The grounding requirements for ac power distribution systems are contained in Article 250 of the code. In addition to the grounded neutral, an additional safety ground is required to protect against shock hazards. A 120-V branch

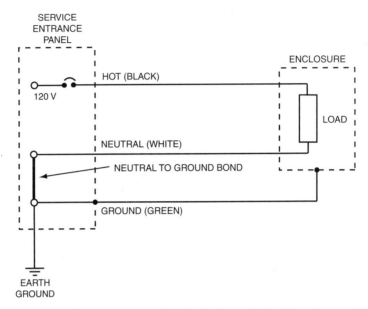

**FIGURE 3-2.** Standard 120-V ac power distribution branch circuit has three conductors.

circuit, therefore, must be a three-wire system as shown in Fig. 3-2. Load current flows through the hot (black) conductor, which contains overcurrent protection and returns through the neutral (white) conductor. The NEC refers to the neutral as the "grounded conductor." The safety ground (green, green with yellow stripe, or bare) conductor must be connected to all non-current-carrying metallic equipment enclosures and hardware, and it must be contained in the same cable or conduit as the black and white current-carrying conductors. The NEC refers to the safety ground as the "grounding conductor." In this book, however, we will just call the three conductors hot, neutral, and ground (or safety ground).

The only time the ground conductor carries current is during a fault, and then only momentarily, until the overcurrent protection device (fuse or circuit breaker) opens the circuit removing voltage and rendering the equipment safe. Because no current normally flows in the ground, it has no voltage drop, and the hardware and enclosures connected to it are all at the same potential. The NEC specifies that the neutral and ground wires shall be connected together at *one and only one point*, and that point shall be at the main service entrance panel (250.24(A)(5)). To do otherwise would allow some of the load current to return on the ground conductor, thereby producing a voltage drop in the conductor. In addition, the NEC requires that the ground conductor also be connected to the earth through a ground rod or some other means at the service entrance panel. The NEC refers to the ground rod as a "grounding electrode." We will just refer to it as a ground rod. Metal water pipes and building steel must also be bonded* to the ground rod to form a single "grounding electrode system" for the building. A properly wired branch circuit is shown in Fig. 3-3.

A combination 120/240-V branch circuit is similar to the 120-V circuit, except an additional hot (often red) conductor is added, as shown in Fig. 3-4. If the load requires only 240 V, the neutral (white) wire shown in Fig 3-4 is not required.

Figure 3-5 shows what happens when a fault occurs in a piece of electrical equipment connected to a properly installed ac power line. A low-impedance path that consists of the hot conductor, the ground conductor, and the neutral-to-ground bond exists, which draws a large fault current and quickly blows the overcurrent protection device, thereby removing power from the load and rendering the equipment safe.

Note that the ground-to-neutral bond is required for the overcurrent protection device to operate; however, the ground-to-earth bond is not required. Neither is the earth connection required for all the metal connected to the ground conductor to be at the same potential. Why then does the NEC require the earth connection? Overhead power lines are very susceptible to lightning strikes. The earth connection is used to divert the lightning current to ground and to limit the voltage imposed on the power system by lightning, as well as any voltage induced from power line surges or unintentional contact with higher voltage power lines.

### 3.1.3  Noise Control

It should be kept in mind that the NEC says very little about the issue of noise and interference control, it is mostly concerned with electrical safety and fire protection. The system designer must find a way to satisfy the code and still produce a low-noise system. In addition, the NEC is only concerned with a frequency of 50/60 Hz and its harmonics. An acceptable 60-Hz ground will not be an acceptable 1-MHz ground. A good resource on satisfying the NEC requirements while producing a low-noise system is IEEE Std. 1100-2005, *IEEE Recommended Practice for Powering and Grounding Electronic Equipment*, which is referred to as the "Emerald Book."

Noise can be differential mode (hot to neutral) or common mode (neutral to ground) as shown in Fig. 3-6. *Grounding, however, will only have an effect on the common-mode noise.*

To control noise and interference, one needs to create a low-impedance ground system that is effective not only at 50/60 Hz but also at much higher frequencies (tens if not hundreds of megahertz). To achieve this objective,

---

*The NEC defines bonding as follows: The permanent joining of metallic parts to form an electrically conductive path..."

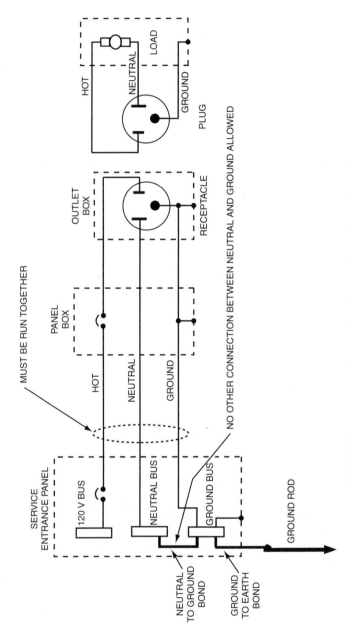

**FIGURE 3-3.** A properly wired and grounded ac branch circuit.

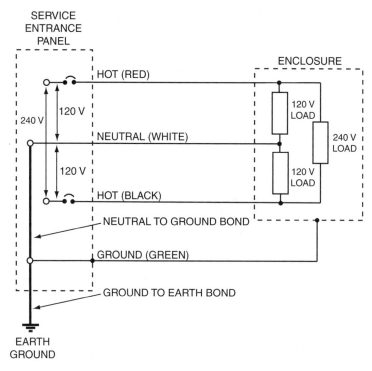

**FIGURE 3-4.** Combination 120/240-V ac power distribution circuit has four conductors.

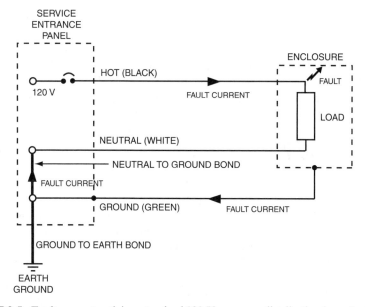

**FIGURE 3-5.** Fault current path in a standard 120-V ac power distribution branch circuit.

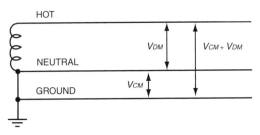

**FIGURE 3-6.** Differential-mode and common-mode noise.

supplemental ground conductors such as ground straps, ground planes, ground grids, and so on may be required. The NEC allows such supplemental ground conductors provided they meet the following two conditions:

1. They must be *in addition to* the ground conductors required by NEC, not in place of them.
2. They must be bonded to the NEC required ground conductors.

The most effective method of obtaining a low-impedance ground over the widest range of frequencies for physically separated units of equipment is to connect them to a solid ground plane as shown in Fig. 3-26. This plane is often referred to as a zero signal reference plane (ZSRP). A ZSRP has several orders of magnitude less impedance than a single ground wire of any practical dimensions, and this low impedance exists over a frequency range that spans many orders of magnitude (i.e., dc to hundreds of megahertz or more). *A ZSRP is without a doubt the optimum low-impedance, wide-bandwidth ground structure.*

In many cases, a solid ZSRP is not practical; in which case, a grid can be used to simulate a solid plane with good results. A grid can be thought of as a plane with holes in it. As long as the maximum dimension of the holes is kept to less than one twentieth of a wavelength, a grid will be almost as effective as a plane. This approach, which is a gridded ZSRP, is typically used in large computer room in the form of a cellular raised floor usually with a grid conductor spacing of 60 cm (approximately 2 ft).

### 3.1.4   Earth Grounds

It is a myth that connecting equipment to an earth ground reduces noise and interference. Some equipment designers, in an attempt to control noise and interference, have proposed connecting their equipment to a "quiet ground" that consists of an isolated separate ground rod. Such a system is shown in Fig. 3-7. If, however, a fault should occur between the hot conductor and the enclosure, the fault-current path, as indicated by the arrows in the figure, now includes the earth. The problem is that the earth is not a good conductor. Seldom is the earth resistance less than a few ohms, and often it is in the 10-to 15-$\Omega$ range. The NEC allows the resistance between the ground rod and the

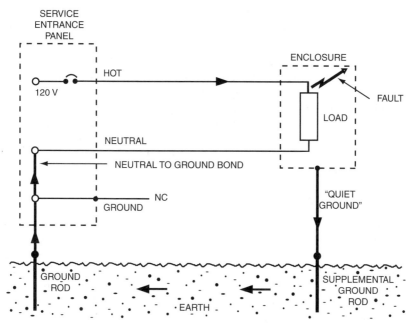

**FIGURE 3-7.** Fault current path when the load is connected to an isolated or separate "quiet" ground. This configuration is dangerous and violates the NEC.

earth to be as high as 25 Ω (250.56). If the resistance to earth is greater than 25 Ω, then a second ground rod must be driven at least 6 feet away from the first rod and electrically bonded to the first ground rod. The fault current, therefore, may be less than 5A, which is far too small to cause the circuit breaker to trip. This configuration is very dangerous, and it should never be used. It also violates the requirements contained in the NEC. The NEC states that: *The earth shall not be considered as an effective fault-current path* (250.4(B)(4)).

If multiple ground rods are used, the NEC requires that they all be bonded together to produce a single low-impedance fault-current path (250.50). Communications systems [cable television (CATV), telephone, etc.] ground rods are also required to be bonded to the building grounding electrode system (810.20(J)). This is necessary to reduce differences in potential between the ground rods during a lightning strike.

In addition to the fact that the configuration of Fig. 3-7 is unsafe, it seldom reduces noise or interference. The ac power earth ground system is an array of long wires, which act as antennas, to pick up all kinds of noise and interference. It is also heavily polluted with noisy power currents from other equipment and the power utility. The earth is not a low impedance, and it is far from being an equipotential. It is more likely the cause of noise and interference than the solution to these problems.

In practice, a properly installed ac power distribution system, in a building or home, will develop small, perfectly safe, voltage difference between the grounds of the various outlets. Leakage currents, magnetic field induction, and currents that flow through electromagnetic interference (EMI) filter capacitors connected to the equipment ground cause these voltages. The voltage measured between two ground points is typically less than 100 mV, but in some cases, it is as high as a few volts. These noise voltages, although safe, if coupled into most low-level signal circuits are clearly excessive. Therefore, the ac power ground is of little practical value as a signal reference. Connection should be made to the ac power, or earth, ground only when required for safety.

### 3.1.5 Isolated Grounds

When required for the reduction of noise (electromagnetic interference) on the ground of sensitive electronic equipment, the NEC allows the use of a receptacle in which the ground terminal is insulated from the receptacle mounting structure (250.146(D)).* A similar section (250.96(B)) allows for the use of an isolated ground with directly wired equipment enclosures. These examples are exceptions to the normal NEC requirement for a solidly grounded circuit. It is important to understand that, in this case, the term isolated refers to the method by which the receptacle is grounded, not if it is grounded.† An isolated ground (IG) receptacle is one in which there is no direct electrical connection between the ground terminal on the receptacle and any other metal part of the assembly. Isolated ground receptacles are usually colored orange; however, the only NEC requirement is that they be identified by an orange Δ on the face of the receptacle (406.2(D)). The use of an IG receptacle does not relieve the requirement that the metal outlet box and all other metallic structures be grounded.

The wiring of an isolated ground receptacle is shown in Fig. 3-8. The receptacle's ground pin is connected to the safety ground by running a separate insulated conductor to the service entrance panel or source of a separately derived system (see Section. 3.1.6). The insulated ground conductor must pass through any intermediate panel boxes without making an electrical connection to the boxes. The normal safety-ground conductor is still required and is connected to all the outlet and panel boxes. The system now has two grounding conductors. The isolated grounding conductor connects to the receptacle's ground pin and only grounds the equipment plugged into the receptacle. The normal safety ground conductor grounds all other hardware as well as the outlet box and any intermediate panel boxes. The ac power wiring is now a four-conductor system, which includes hot, neutral, isolated ground, normal

---

* The NEC does not address the issue of exactly what circumstances would justify the use of an isolated ground.

† Some prefer to use the term *insulated ground* instead of *isolated ground*, because it more accurately describes how the system is wired and does not imply that the circuit is isolated from ground.

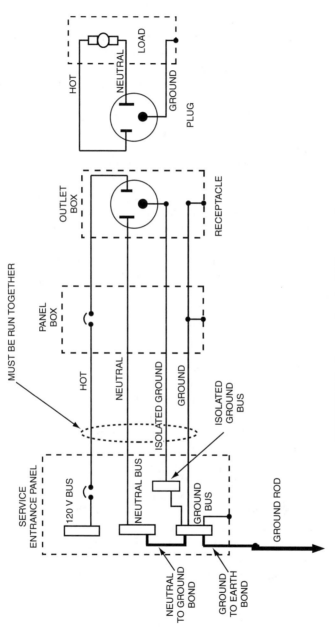

**FIGURE 3-8.** A properly wired IG receptacle.

hardware (safety) ground.* The isolated ground conductor and hardware ground conductor must all be run in the same cable or conduit as the associated current-carrying conductors (hot and neutral).

Comparing Figs. 3-3 and 3-8 shows that the only differences are (1) the removal of the bonding jumper between the receptacle's ground pin and the outlet box and (2) the addition of an insulated ground conductor run back to the service entrance panel.

Any intentional or accidental connection, other than the one at the service entrance panel, between the isolated ground conductor and the safety ground conductor will compromise the whole purpose behind the isolated ground configuration. Not only will it compromise the circuit where the extraneous connection occurs, but also it will compromise all isolated ground circuits supported by the isolated ground bus in the service entrance panel.

The benefits, if any, of using an IG receptacle is not universally agreed on and is not discussed in the code. Results range from no improvement, to some improvement, to increased noise. Any noise improvement that does occur, however, will only be with respect to common-mode noise, not to differential-mode noise.

In many cases, a separate branch circuit, a ZSRP, or the use of an isolation transformer to produce a separately derived system will give better noise performance. These approaches can also be used in combination with each other as well as with an isolated ground to reduce noise even more.

### 3.1.6   Separately Derived Systems

A separately derived system is a wiring system in which the hot and neutral conductors have no *direct* electrical connection to the main electrical service. Examples of a separately derived system are those provided by a generator, a battery or, a transformer—provided there is no direct electrical connection to another power source. Basically, in the case of a separately derived system, we start all over again, as if it was the main service entrance panel, and we create a new single-point neutral-to-ground bond.

An isolation transformer can often be used to reduce common-mode noise. Because the isolation transformer creates a separately derived system, a new neutral-to-ground bond point can be established. At this point, there will be no common-mode noise voltage between the ground and the neutral conductor. If the voltage derived from the transformer is used solely to feed sensitive electronic loads, then noise will be greatly reduced. Figure 3-9 shows the wiring between an isolation transformer and a load. Considering the transformer secondary as a separately derived (new) source of power, we observe that the right-hand portion of Fig. 3-9 (panel box to load) and Fig. 3-2 are wired the same.

An isolation transformer will not reduce any differential-mode (hot-to-neutral) noise present on the main power source, as this will directly couple through the

---

* A metallic conduit may also be used as the safety ground in place of a separate ground conductor.

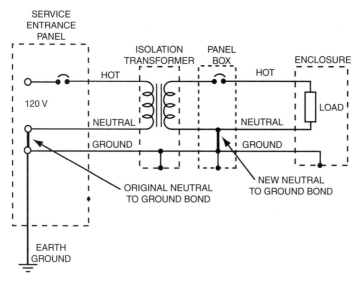

**FIGURE 3-9.** An isolation transformer used to create a separately derived system. A new neutral-to-ground bond point is established at the transformer or first panel box after the transformer.

transformer. An isolation transformer can be used with a solidly grounded receptacle or with an IG receptacle. This technique is probably the best use of an IG receptacle, when combined with an isolation transformer.

### 3.1.7 Grounding Myths

More myths exist relating to the field of grounding than in any other area of electrical engineering. The more common of these are as follows:

1. The earth is a low-impedance path for ground current. False, the impedance of the earth is orders of magnitude greater than the impedance of a copper conductor.
2. The earth is an equipotential. False, this is clearly not true as a result of (1).
3. The impedance of a conductor is determined by its resistance. False, what happened to the concept of inductive reactance?
4. To operate with low noise, a circuit or system must be connected to an earth ground. False, because airplanes, satellites, cars, and battery-powered laptop computers all operate fine without a ground connection. As a matter of fact, an earth ground is more likely to be the cause of a noise problem. More electronic system noise problems are resolved by removing (or isolating) a circuit from earth ground than by connecting it to earth ground.

5. To reduce noise, an electronic system should be connected to a separate "quiet ground" by use of a separate, isolated ground rod. False, in addition to being untrue, this approach is dangerous and violates the requirements of the NEC.

6. An earth ground is unidirectional, with current only flowing into the ground. False, because current must flow in loops, any current that flows into the ground must also flow out of the ground somewhere else.

7. An isolated receptacle is not grounded. False, the term "isolated" refers only to the method by which a receptacle is grounded, not if it is grounded.

8. A system designer can name ground conductors by the type of the current that they should carry (i.e., signal, power, lightning, digital, analog, quiet, noisy, etc.), and the electrons will comply and only flow in the appropriately designated conductors. Obviously false.

## 3.2  SIGNAL GROUNDS

A ground is often defined as an equipotential* point or plane that serves as a reference potential for a circuit or system. I like to refer to this as the *voltage definition* of ground. This definition, however, is not representative of practical ground systems, because they, in reality, are not equipotentials; also, this definition does not emphasize the importance of the actual path taken by the ground current. It is important for the designer to know the actual return current path to evaluate the radiated emission or susceptibility of a circuit. To understand the limitations and problems of "real-world" ground systems, it is better to use a definition more representative of the actual situation. A better definition for a signal ground is a low-impedance path for current to return to the source (Ott, 1979). This *current definition* of a ground emphasizes the importance of current flow. It implies that because current is flowing through some finite impedance, a difference in potential will exist between any two physically separated ground points. The voltage definition defines what a ground ideally should be, whereas the current definition more closely defines what a ground actually is. Keep in mind another important difference between the voltage and current concepts of ground. Voltage is always relative, and we must ask, "With respect to what?" What potential should our ground be at, and respect to what should it be measured? Current, on the other hand, is definitive—the current always wants to return to the source.

Three basic objectives of signal grounding are as follows:

1. Not to interrupt the ground return path
2. Return the current through the smallest loop possible[†]
3. Be aware of possible common impedance coupling in the ground

---

* A point where the voltage does not change, regardless of the current applied to it or drawn from it.
[†] This will be the lowest inductance path.

The most important characteristic of a ground conductor is its impedance. The impedance of any conductor can be written as

$$Z_g = R_g + j\omega L_g \tag{3-1}$$

Equation 3-1 clearly shows the effect that frequency has on ground impedance. At low frequency, the resistance $R_g$ will be dominant. At high frequency, the inductance $L_g$ will be the dominant impedance. At frequencies greater than 13 kHz, a straight length of 24-gauge wire, 1 in above a ground plane, has more inductive reactance than resistance; see Fig. 3-10.

In designing a ground, it is important to ask: How does the ground current flow? The path taken by ground current must be determined. Then, because any conductor carrying current will have a voltage drop, the effect of this voltage drop on the performance of all circuits connected to the ground must be evaluated. Ground voltage, just like all other voltage, obeys Ohms law; therefore,

$$V_g = I_g Z_g \tag{3-2}$$

Equation 3-2 points out two ways to minimize the ground noise voltage $V_g$.

1. Minimize the ground impedance $Z_g$
2. Decrease $I_g$ by forcing the ground current to flow through a different path

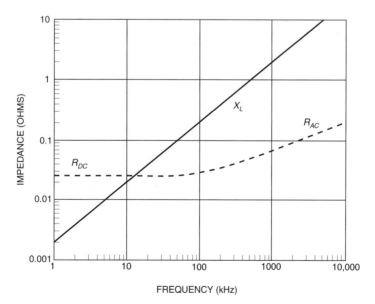

**FIGURE 3-10.** Resistance and inductive reactance versus frequency for a straight, 1-ft length of 24-gauge wire, located 1-in. above a ground plane.

The first approach is commonly used at high frequency and with digital circuits, by using a ground plane or grid. The latter approach is commonly used with low-frequency analog circuits by using single-point grounding. With single-point grounding, we can direct the ground current to flow where we want it to flow. Equation 3-2 also clearly demonstrates the all-important point that, assuming current is in the ground, two physically separated points will never be at the same potential.

Consider the case of the double-sided PCB as shown in Fig. 3-11. It consists of a trace routed as shown on the topside of the board and a solid ground plane on the bottom of the board. At points A and B, vias pass through the board connecting the topside trace to the ground plane to complete the current loop. The question is, exactly how does the current flow in the ground plane between points A and B?

At low frequencies, the ground current will take the path of least resistance, which is directly between points A and B as shown in Fig. 3-12A. However, at high frequencies, the ground current takes the path of least inductance, which is directly under the trace as shown in Fig. 3-12B, because this represents the smallest loop area. Therefore, the current return paths are different at low frequency and at high frequency. In this case, the distinction between low frequency and high frequency is typically a few hundred kilohertz.

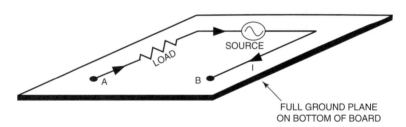

**FIGURE 3-11.** A double-sided PCB with a single trace on the topside and a full ground plane on the bottom side. How does the ground plane current flow between points A and B?

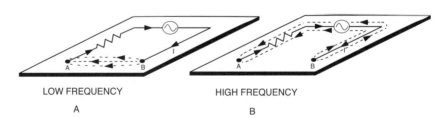

**FIGURE 3-12.** Ground plane current path, (A) at low frequency the return current takes the path of least resistance and (B), at high frequency the return current takes the path of least inductance.

Notice that for the low-frequency case (Fig. 3-12A), the current flows around a very large loop, which is undesirable. However in the high-frequency case (Fig. 3-12B), the current flows around a small loop (the length of the signal trace times the thickness of the board). One can therefore conclude that high-frequency ground currents do what we want them to (i.e., flow through a small loop), and as designers all we have to do is not to interrupt them or prevent them from flowing as they desire. Low-frequency ground currents, however, may or may not flow as we want them to (i.e., flow through a small loop), so we often must direct the current (or force the current) to flow where we want.

The proper signal ground system is determined by many things, such as the type of circuitry, the frequency of operation, the size of the system, whether it is self-contained or distributed, as well as other constraints such as safety and electrostatic discharge (ESD) protection. It is important to understand that no single ground system is proper for all applications.

Another factor to keep in mind is that grounding always involves compromise. All ground systems have advantages as well as disadvantages. The designer's job is to maximize the advantages and to minimize the disadvantages of the ground for the application at hand.

Also, grounding problems have more than one acceptable solution. Therefore, although two different engineers often will come up with two different solutions to the same grounding problem, both solutions may be acceptable.

Last, grounding is hierarchical. Grounding is done at the integrated circuit (IC) level, at the circuit board level, as well as at the system or equipment level. Each level is often performed by different people; each one does not usually know what will happen at the next level. For example, the IC designer does not know every use or application of the device, every piece of equipment that the device will go into, or what grounding strategy the final equipment designer will take.

Signal grounds can be divided into the following three categories:

1. Single-point grounds
2. Multipoint grounds
3. Hybrid grounds

Single-point and multipoint grounds are shown in Figs. 3-13 and 3-14, respectively.* Hybrid grounds are shown in Figs. 3-21 and 3-23. Two subclasses of single-point grounds are as follows: those with series connections and those with parallel connections, as shown in Fig. 3-13. The series connection is also called a common or daisy chain, and the parallel connection is often called a separate or star ground system.

---

* Because grounding is hierarchical, the boxes labeled circuit 1, 2, and so on in these figures can represent anything you desire—a large rack of electronic equipment, a small electronic module, a PCB, or even an individual component or IC.

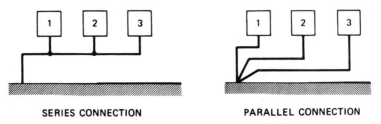

FIGURE 3-13. Two types of single-point grounding connections.

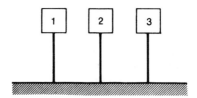

FIGURE 3-14. Multipoint grounding connections.

In general, it is desirable that the topology of the power distribution system follows that of the ground. Usually, the ground structure is designed first, and then the power is distributed in a similar manner.

### 3.2.1 Single-Point Ground Systems

Single-point grounds are most effectively used at low frequency, from dc up to about 20 kHz. They should usually not be used above 100 kHz, although sometimes this limit can be pushed as high as 1 MHz. With single-point grounding, we control the ground topology to direct the ground current to flow where we want it to flow, which decreases $I_g$ in the sensitive portions of the ground. From Eq. 3-2, we observe that decreasing $I_g$, decreases the voltage drop in that portion of the ground. In addition, single-point grounding can be used effectively to prevent ground loops.

The most undesirable single-point ground system is the common or daisy chain ground system shown in Fig. 3-15. This system is a series connection of all the individual circuit grounds. The impedances $Z$ shown* represent those of the ground conductors, and $I_1$, $I_2$, and $I_3$ are the ground currents of circuits 1, 2, and 3, respectively. Point A is not a zero potential but is at a potential of

$$V_A = (I_1 + I_2 + I_3)Z_1, \qquad (3\text{-}3)$$

---

* Although resistors are shown in the figure, they are intended to represent impedances in general and could just as well be inductors.

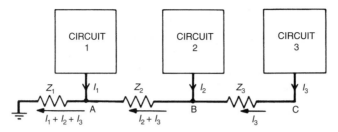

**FIGURE 3-15.** Common, or daisy chain, single-point ground system is a series ground connection and is undesirable from a noise standpoint, but it has the advantage of simple wiring.

and point C is at a potential of

$$V_C = (I_1 + I_2 + I_3)Z_1 + (I_2 + I_3)Z_2 + I_3Z_3. \tag{3-4}$$

Although this circuit is the least desirable single-point grounding system, it is commonly used because of its simplicity. In noncritical applications, it may be perfectly satisfactory. The configuration should not be used between circuits that operate at widely different current levels, because the high current stages will adversely affect the low-level circuits through the common ground impedance. When the system is used, the most critical circuit should be the one nearest the primary ground point. Note that point A in Fig. 3-15 is at a lower potential than points B or C.

The separate or parallel ground system shown in Fig. 3-16 is a more desirable single-point ground system. That is because no cross coupling occurs between ground currents from different circuits. The potentials at points A and C, for example, are as follows:

$$V_A = I_1Z_1, \tag{3-5}$$

$$V_C = I_3Z_3. \tag{3-6}$$

The ground potential of a circuit is now a function of the ground current and impedance of that circuit only. This system can be mechanically cumbersome, however, because in a large system an unreasonable number of ground conductors may be necessary.

Most practical single-point ground systems are actually a combination of the series and parallel connection. Such a combination is a compromise between the need to meet the electrical noise criteria and the goal of avoiding more wiring complexity than necessary. The key to balancing these factors success-fully is to group ground leads selectively, so that circuits of widely varying power and noise levels do not share the same ground return wire. Thus, several

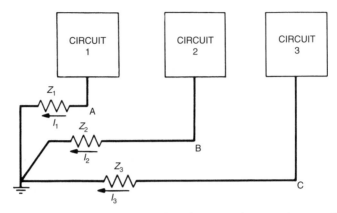

**FIGURE 3-16.** Separate, or parallel, single-point ground system is a parallel ground connection and provides good low-frequency grounding, but it may be mechanically cumbersome in large systems.

low-level circuits may share a common ground return, whereas other high-level circuits could share a different ground return conductor.

The NEC mandated ac power ground system is actually a combination of a series and parallel connected single-point ground. Within a branch circuit (connected to one circuit breaker), the grounds are series connected, and the various branch circuit grounds are parallel connected. The single, or star, point is at the service entrance panel, as shown in Fig. 3-17.

At high frequencies, the single-point ground system is undesirable because the inductance of the ground conductors increase the ground impedance. At still higher frequencies, the impedance of the ground conductors can be very high, if the length coincides with odd multiples of a quarter wavelength. Not only will these grounds have large impedance, but also they will act as antennas and radiate, as well as pick up energy effectively. To maintain a low impedance and to minimize radiation and pickup, ground leads should always be kept shorter than one twentieth of a wavelength.

At high frequency, there is no such thing as a single-point ground. Figure 3-18 shows what happens when a single-point ground configuration is attempted at high frequencies. Because of their inductance, the ground conductors represent high impedances. However at high frequency, the impedance of the stray capacitance between the circuits and ground is low. The ground current therefore flows through the low impedance of the stray capacitance and not the high impedance that results from the inductance of the long ground conductors. The result is a multipoint ground at high frequency.

### 3.2.2   Multipoint Ground Systems

Multipoint grounds are used at high frequency (above 100 kHz) and in digital circuitry. Multipoint ground systems minimize the ground noise voltage $V_g$ in

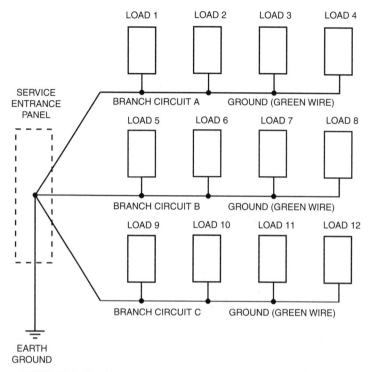

**FIGURE 3-17.** A single-point ac power ground, as per the NEC.

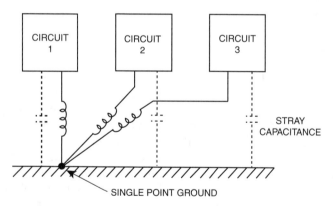

**FIGURE 3-18.** At high frequency, single-point grounds become multipoint grounds because of stray capacitance.

Eq. 3-2 by minimizing the ground impedance $Z_g$. From Eq. 3-1, we observe that at high frequency, this means minimizing the ground inductance, which can be done by the use of ground planes or grids. Where possible, use multiple connections between the circuits and the plane to reduce the inductance. In the

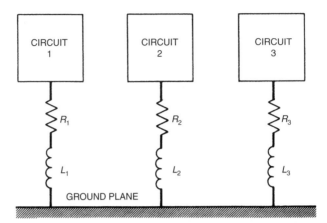

**FIGURE 3-19.** Multipoint ground system is a good choice at frequencies above about 100 kHz. Impedances $R_1 - R_3$ and $L_1 - L_3$ must be minimized at the frequency of interest.

multipoint system shown in Fig. 3-19, circuits are connected to the nearest available low-impedance ground plane. The low ground impedance is primarily the result of the low inductance of the ground plane. The connections between each circuit and the ground plane should be kept as short as possible to minimize their impedance. In many high-frequency circuits, the length of these ground leads may have to be kept to a small fraction of an inch. All single-point grounds become multipoint grounds at high frequency because of stray capacitance, as was shown in Fig. 3-18.

Increasing the thickness of the ground plane has no effect on its high-frequency impedance because (1) it is the inductance not the resistance of the ground that determines its impedance and (2) high-frequency currents only flow on the surface of the plane because of the skin effect (see Section 6.4).

A good low-inductance ground is necessary on any PCB that contains high-frequency or digital logic circuits. The ground can be either a ground plane; or on a double-sided board, a ground grid. The ground plane provides a low-inductance return for signal currents and allows for the possibility of using constant impedance transmission lines for signal interconnections.

Although the ground on a digital logic board should be multipoint, that does not mean that the power supplied to the board must also be multipoint grounded. Because the high-frequency digital logic currents should be confined to the board and not flow through the power supply conductors that feed the board, and because the power is dc, it can be wired as a single-point ground even though the logic board ground is multipoint.

### 3.2.3 Common Impedance Coupling

Many ground system problems occur as the result of common impedance coupling. An example of common impedance coupling is shown in Fig. 3-20,

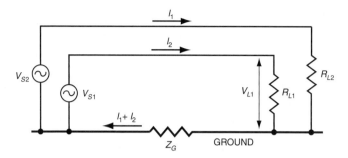

**FIGURE 3-20.** An example of common impedance coupling.

which shows two circuits that share the same ground return. The voltage $V_{L1}$ across the load impedance $R_{L1}$ of circuit 1 will be

$$V_{L1} = V_{S1} + Z_G(I_1 + I_2), \qquad (3\text{-}7)$$

where $Z_G$ is the common ground impedance and $I_1$ and $I_2$ are the signal currents in circuits 1 and 2, respectively. Notice that in this situation the signal voltage across the load $R_{L1}$ of circuit 1 is no longer a function of just the current in circuit 1, but also it is a function of the current in circuit 2. The term $I_1 Z_G$ in Eq. 3-7 represents an intracircuit noise voltage, and the term $I_2 Z_G$ represents an intercircuit noise voltage.

Common impedance coupling becomes a problem when two or more circuits share a common ground and one or more of the following conditions exist:

1. A high-impedance ground (at high frequency, this is caused by too much inductance; at low-frequency this is caused by too much resistance).
2. A large ground current.
3. A very sensitive, low-noise margin circuit, connected to the ground.

Single-point grounds overcome these problems by separating ground currents that are likely to interfere with each other and by forcing them to flow on different conductors, effectively controlling $I_g$ in Eq. 3-2. This approach is effective at low frequency. However, the signal current paths and long lead lengths associated with single-point grounds increase the inductance, which is detrimental at high frequencies. In addition, at high frequencies, single-point grounds are almost impossible to achieve because parasitic capacitance closes the ground loop as was shown in Fig. 3-18.

Multipoint grounds overcome these problems by producing a very low ground impedance, effectively controlling the $L_g$ term in Eq. 3-1.

Normally, at frequencies below 100 kHz, a single-point ground system may be preferable; above 100 kHz, a multipoint ground system is best.

### 3.2.4  Hybrid Grounds

When the signal frequency covers a wide range both above and below 100 kHz, a hybrid ground may be a solution. A video signal is a good example of this; the signal frequencies can range from 30 Hz to tens of megahertz. A hybrid ground is one in which the system-grounding configuration behaves differently at different frequencies. Figure 3-21 shows a common type of hybrid ground system that acts as a single-point ground at low frequency and as a multipoint ground at high frequency.

A practical application of this principle is the cable-shielding configuration shown in Fig. 3-22. At low frequency, the capacitor C is a high impedance and the cable shield is single-point grounded at the load end only. At high frequency, the capacitor C is a low impedance and the cable shield is effectively grounded at both ends. This type of hybrid shield ground was previously discussed in Section 2.15.2.3.

A different type of hybrid ground is shown in Fig. 3-23. This hybrid ground, although not very common, can be used when many equipment enclosures must be grounded to the power system ground, but it is desirable to have a single-point signal ground for the circuitry. The ground inductors provide a

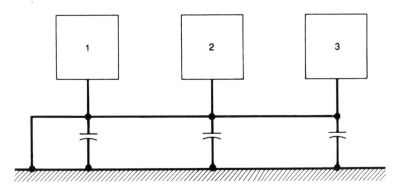

**FIGURE 3-21.** A hybrid ground connection that acts as a single-point ground at low frequencies and a multipoint ground at high frequencies.

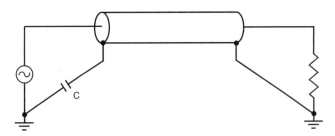

**FIGURE 3-22.** Example of a hybrid grounded cable shield.

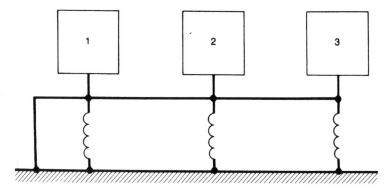

**FIGURE 3-23.** A hybrid ground connection that acts as a multipoint ground at low frequencies and a single-point ground at high frequencies.

low-impedance safety ground at 50/60 Hz, and ground isolation at higher frequencies. Another application might be if the equipment is conducting a noise current out on the ground conductor, which causes the power cable to radiate, thereby failing regulatory electromagnetic compatibility (EMC) requirements. If the ground conductor is removed, the product passes EMC, but that is a safety violation. An inductor or choke (e.g., 10 to 25 μH) added in series with the ground wire will provide a low impedance at 50/60 Hz, while providing a high impedance at the much higher noise frequencies.

### 3.2.5   Chassis Grounds

Chassis ground is any conductor that is connected to the equipment's metal enclosure. Chassis ground and signal ground are usually connected together at one or more points. The key to minimizing noise and interference is to determine *where* and *how* to connect the signal ground to the chassis. Proper circuit grounding will reduce the radiated emissions from the product as well as increase the product's immunity to external electromagnetic fields.

Consider the case of a PCB, with an input/output (I/O) cable, mounted inside a metallic enclosure as shown in Fig. 3-24. Because the circuit ground carries current and has a finite impedance, there will be a voltage drop $V_G$ across it. This voltage will drive a common-mode current out on the cable, and will cause the cable to radiate. If the circuit ground is connected to the chassis at the end of the PCB opposite the cable, then the full voltage $V_G$ will drive the current onto the cable. If, however, the circuit ground is connected to the enclosure at the I/O connector, the voltage driving common-mode current out onto the cable will ideally be zero. The full ground voltage will now appear at the end of the PCB without the cable connection. It is, therefore, important to establish a low-impedance connection between the chassis and the circuit ground in the I/O area of the board.

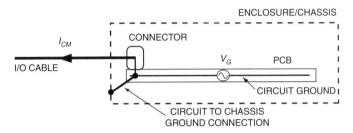

**FIGURE 3-24.** Circuit ground should be connected to the enclosure (chassis) in the I/O area of the PCB.

Another way to visualize this example is to assume that the ground voltage produces a common-mode noise current that flows toward the I/O connector. At the connector, there will be a current division between the cable and the PCB ground-to-chassis connection. The lower the value of the board ground to chassis impedance, the smaller the common-mode current on the cable will be. The key to the effectiveness of this approach is achieving a low impedance (at the frequencies of interest) in the PCB-to-chassis connection. This method is often easier said than done, especially when the frequencies involved can be in the range of hundreds of megahertz or more. At high frequency, this implies low inductance and usually requires multiple connections.

Establishing a low-impedance connection between the circuit ground and the chassis in the I/O area is also advantageous with respect to radio frequency (rf) immunity. Any high-frequency noise currents induced into the cable will be conducted to the enclosure, instead of flowing through the PCB ground.

## 3.3  EQUIPMENT/SYSTEM GROUNDING

Electronic circuits for many systems are mounted in large equipment racks or cabinets. A typical system will consist of one or more of these equipment enclosures. The equipment enclosures can be large or small; can be located adjacent to each other or dispersed; can be located in a building, ship, or aircraft; and can be powered by ac or direct current (dc) power systems. Equipment grounding objectives include electrical safety, lightning protection, EMC control, and signal integrity. The following discussion emphasizes ac-powered land-based systems; however, analogous grounding methods apply to systems in vehicles, and aircraft, as well as to dc-powered systems.

Although the examples in this section show the grounding of large equipment enclosures, the same principles are equally applicable to systems that contain many small circuit modules such as those that might be found in cars or aircraft. In fact, the principles are even applicable to a system that contains a collection of printed circuit boards. Remember, grounding is hierarchal and the same principles apply regardless of the scale.

The following three types of systems (originally categorized by Denny, 1983) will be considered: (1) isolated systems, (2) clustered systems, and (3) distributed systems.

### 3.3.1 Isolated Systems

An isolated system is one in which all functions are contained within a single enclosure with no external signal connections to other grounded systems. Examples of isolated systems are vending machines, television sets, component stereo system (with all the components mounted in a single rack),* desktop computer, and so on. This system is the simplest of all systems and the easiest to ground properly. The optimum way to minimize the potential difference between interconnected items of electronics is to have them all contained in a single six-sided metallic enclosure of the smallest size possible.

The NEC requires that all exposed metallic enclosures of electronic equipment be connected to the ac power ground conductor. Only one safety ground connection is required between the enclosure and the earth, structure, airframe, hull, and so on. This enclosure ground can be provided by (1) the ac power ground (green wire) when powered by single-phase ac (see Fig. 3-2) or (2) a separate ground conductor run with the power cable (or a metallic conduit) when powered by three-phase ac.

Internal signals should be grounded as appropriate for the type of circuitry and frequency of operation. Because an isolated system does not have any I/O to other grounded equipment, the grounding of I/O signals does not have to be considered.

### 3.3.2 Clustered Systems

A clustered system has multiple equipment enclosures (cabinets, racks equipment frames, etc.) located in a small area such as in an equipment closet or a single room as shown in Fig. 3-25. Multiple interconnecting I/O cables may exist, not shown in Fig. 3-25, between individual elements of the system, but not with any other grounded system. Examples of clustered systems are a small data-processing center, a mini computer with many large peripherals, or a component stereo system with the components scattered around the room.

*3.3.2.1 Safety Grounding of Clustered Systems.* For safety, the equipment enclosures must be connected to the ac power ground. This connection can be accomplished in many ways. The racks could be single-point grounded, or they could be multipoint grounded. If single-point grounded, they could be connected in a series (or daisy chain) fashion, or they could be connected in a parallel (or star) pattern. If a series connection, as was shown in Fig 3-15, is

---

* The fact that speakers may be mounted remote from the system does not change the fact that this is an isolated system, because the speakers are not grounded.

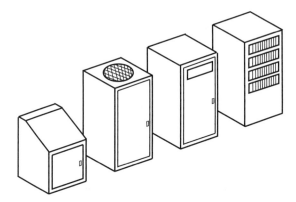

**FIGURE 3-25.** Four equipment enclosures forming a clustered system.

used, the potential of an enclosure will be a function of the ground current associated with another rack. As discussed in Section 3.2.1, the series connected single-point ground is the least desirable, and it is normally unsatisfactory for the interconnection of equipment racks that contain sensitive electronic equipment.

A better approach is to use a parallel or star ground as was shown in Fig. 3-16. In this case, each equipment rack is connected to the main grounding bus with a separate ground conductor. Because each equipment ground conductor carries less current, than in the case of the series connection, the voltage drops in the ground conductors are reduced and the ground potential of an enclosure is only a function of its own ground current. In many cases, this approach provides more than adequate noise performance.

***3.3.2.2 Signal Grounding of Clustered Systems.*** Internal signals should be grounded as appropriate for the type of circuitry and frequency of operation. Signal-ground referencing between elements may be single point, multipoint, or hybrid, whatever is appropriate for the characteristics of the signals involved. If single point, the signal ground reference is usually provided by the existing NEC-required equipment grounding conductors. If multipoint, the signal ground may be provided by cable shields (poorest), auxiliary ground conductors or wide metal straps (better), or with a wire grid or solid metal plane (best). For the case of nonsensitive electronic equipment in a benign environment, signal grounding with cable shields or auxiliary ground conductors is often acceptable. For sensitive electronic equipment in a harsh environment, wide metal straps, grids, or ground planes should be seriously considered.

By far the best way to obtain a low-impedance signal ground connection, over the broadest range of frequencies between separated units, is by interconnecting them with a solid metallic ground plane as shown in Fig. 3-26. The impedance of the plane will be three to four orders of magnitude less than that of any single wire. This produces a true multipoint grounding system and is

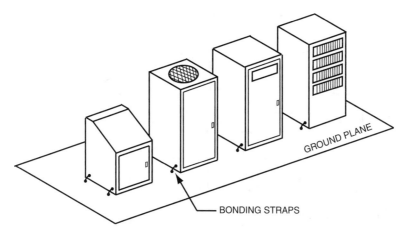

GROUND PLANE

BONDING STRAPS

**FIGURE 3-26.** A ZSRP is the optimum way to provide a low-impedance ground connection between individual equipment enclosures that is effective over the widest frequency range.

often referred to as a ZSRP. The second best approach is to use a ground grid. A grid can be considered to be a plane with holes in it. As long as the holes are small compared with a wavelength ($< \lambda/20$) at the highest frequency of interest, the grid will closely approximate the performance of a plane and is often easier to implement.

In a multipoint ground system, each electronic equipment enclosure or rack must first be grounded using the NEC-mandated equipment-grounding conductor. In addition, each rack or enclosure is connected to another ground, in this case, the ZSRP. Figure 3-27 shows this. Remember, a grid could be used in place of the plane. This approach is effective from dc to high frequency, and it is without a doubt the optimum configuration.

Although ZSRPs perform extremely well, they are not perfect. They, as well as all conductor configurations, have resonances. When a conductor or a current path becomes a quarter wavelength (or odd multiples thereof) long, it will present a high impedance. However, in the case of a ZSRP, or grid, when one current path becomes a quarter wavelength long, other parallel paths will exist that are not a quarter wavelength long. Therefore, the ground current will follow these lower impedance parallel paths, instead of the high-impedance quarter-wavelength path. Therefore, even when resonances are considered, the impedance of a ZSRP will be less than any that of any single-conductor alternative.

For lightning protection, the ZSRP should *not* be isolated from any objects or structures. It should be bonded to each item that penetrates it, such as pipes, metallic conduit, building steel, and so on. The ZSRP should also be bonded to any metal object located within 6 ft of it. This rule is important to prevent lightning flashover.

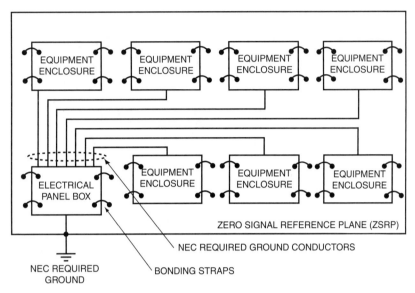

**FIGURE 3-27.** Implementation of a ZSRP ground structure while still retaining the NEC-mandated equipment-grounding conductors.

***3.3.2.3 Ground Straps.*** To minimize inductance, equipment enclosures should be bonded to the ZSRP at multiple points (a minimum of four is a good rule to follow) using short straps with a length-to-width ratio of 3:1 or less. Increasing the diameter of a round conductor does not reduce its inductance significantly because of the logarithmic relationship between the diameter and the inductance (see Section 10.5.1).

Instead of a round conductor, a short, flat rectangular strap should be used to bond the enclosures to the ZSRP. The inductance of a flat rectangular conductor is (Lewis, 1995, p. 316)

$$L = 0.002l \left[ 2.303 \log\left(\frac{2l}{w+t}\right) + 0.5 + 0.235\left(\frac{w+t}{l}\right) \right], \qquad (3\text{-}8)$$

where $L$ is the inductance in $\mu$H, $l$ is the length, $w$ is the width, and $t$ is the thickness of the flat-strap (all in centimeters). In the case of a rectangular conductor, the length-to-width ratio can have a significant effect on the inductance of the strap.* For a strap of a given length, the inductance decreases as the strap's width is increased. Figure 3-28 shows the inductance of a 10-cm-long, 0.1-cm-thick conductor as a function of the length-to-width ratio, as a percentage of the inductance at a length-to-width ratio of 100:1.

---

* Actually, this ratio is the length-to-width plus thickness ratio. However, because the thickness $t$ is usually much less than the width $w$, we usually just consider the length-to-width ratio.

Equation 3-8 also shows that as the strap length is increased, the inductance will increase regardless of the length-to-width ratio because of the $l$ term in front of the brackets. However, for a fixed-length strap, the inductance will decrease as a function of the length-to-width ratio as shown in Fig. 3-28 and as listed in Table 3-1. Therefore, for good high-frequency performance, a short-strap with the smallest length-to-width ratio possible should be used.

The ground strap can be constructed from solid metal or braid. Braid is often preferred over solid metal where flexibility is needed. Braid, however, is susceptible to corrosion between the individual strands that will increase its impedance. Copper braid, if used, should be tinned or silver plated to reduce this problem. A tinned or plated braid, in good condition, should perform as well as a solid strap in this application.

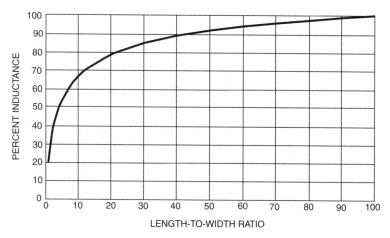

**FIGURE 3-28.** Inductance of a rectangular cross-section grounding strap as a function of its length-to-width ratio. Inductance is plotted as a percentage of the inductance of a 100:1 length-to-width ratio strap.

**TABLE 3-1.** **Percent decrease in inductance as a function of the length-to-width ratio of a fixed-length, rectangular cross-section conductor.**

| Length-to Width Ratio | Percent Decrease in Inductance[a] |
|---|---|
| 100:1 | 0 |
| 50:1 | 8 |
| 20:1 | 21 |
| 10:1 | 33 |
| 5:1 | 45 |
| 3:1 | 54 |
| 2:1 | 61 |
| 1:1 | 72 |

[a] The decrease is with respect to the inductance existing with a 100:1 length-to-width ratio.

Additional reduction in inductance can be obtained by using multiple grounding straps. The inductance of two straps in parallel will be half the inductance of one strap, provided they are separated so that the mutual inductance is negligible. This decrease in inductance is linear with the number of straps used. Four widely separated ground straps will have one quarter the inductance of one strap.

Therefore, for the lowest impedance, high-frequency equipment grounding, multiple ground straps, with the shortest length possible, and with the smallest length-to-width ratio possible should be used.

Equipment cabinets are often intentionally or unintentionally (e.g., as the result of the painted surface of the cabinet) insulated from the ground plane that they are sitting on. Therefore, the only electrical connection to the ground plane is through the bonding straps. This situation produces a resonance problem, as shown in Fig. 3-29, which is caused by the parasitic capacitance between the enclosure and the ground plane. This capacitance will resonate with the ground strap inductance at a frequency of

$$f_r = \frac{1}{2\pi\sqrt{LC}}, \tag{3-9}$$

where $L$ is the ground strap inductance and $C$ is the parasitic capacitance between the enclosure and the ground plane. Because this is a parallel resonance, the impedance at resonance will be very large, thus effectively disconnecting the enclosure from the plane. It is not unusual for this resonance to be in the 10- to 50-MHz frequency range. It is desirable to keep this ground strap resonance at a frequency above the operating frequencies of the system. To raise the resonant frequency, the capacitance and/or inductance must be decreased. The inductance can be decreased by using a multiplicity of short,

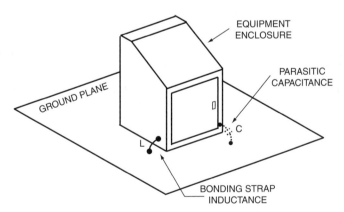

**FIGURE 3-29.** Bonding strap resonance can occur from the parallel combination of the bonding strap inductance and the parasitic capacitance of the enclosure.

wide grounding straps. Raising the cabinet, with an insulator, farther above the ground plane, will decrease the capacitance.

### 3.3.2.4 Inter-Unit Cabling.

*3.3.2.4 Inter-Unit Cabling.* Signal as well as power cables between the individual elements should be routed close to the ZSRP as shown in Fig. 3-30B, not in the air above the enclosures as shown in Fig. 3-30A. This approach minimizes the area between the cables and the reference plane, which will minimize the common-mode noise coupling into the cables. In some cases (e.g., computer rooms), the ZSRP or grid is part of a raised floor; in these cases, the cables can be run below the ZSRP, as shown as option 2 in Fig. 3-30B.

To understand how these equipment grounding concepts can be applied to things other than large racks of electronics, let us apply the concepts for clustered systems to a product in a small enclosure the size of a "bread box," which consists of several individual modules and printed circuit boards all interconnected by cables. There might be a power supply module, a disk drive module, a liquid crystal display (LCD) module, and many printed circuit boards. What would be the optimum way to mount, ground, and interconnect the modules and PCBs?

Remember, grounding is hierarchical, and although a single enclosure is an isolated system on a macroscale, if we only consider the elements inside the enclosure, they could be considered to be a clustered system on a microscale.

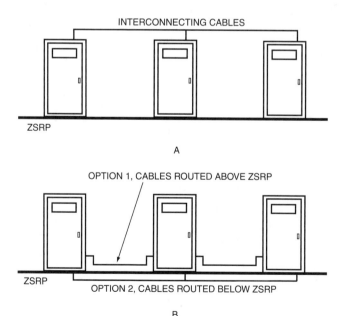

**FIGURE 3-30.** (A) Poor interunit cabling routing and (B) preferred interunit cabling routing.

The optimum configuration then would be to have all the modules and PCBs mounted and grounded to a single ground plane or metal chassis, acting as a ZSRP, and all the cables routed as close to the plane as possible. If each PCB were grounded to the plane with four metal $\frac{1}{4}$ in diameter stand offs, each stand off should have a length no greater than $\frac{3}{4}$ in to not exceed a length-to-width ratio of 3:1.

### 3.3.3   Distributed Systems

A distributed system has multiple equipment enclosures (cabinets, racks equipment frames, etc.) that are physically separated such as in different rooms, buildings, and so on as shown in Fig. 3-31. There will also be multiple interconnecting I/O cables between individual elements of the system, and these cables are often long, in excess of one twentieth of a wavelength at the

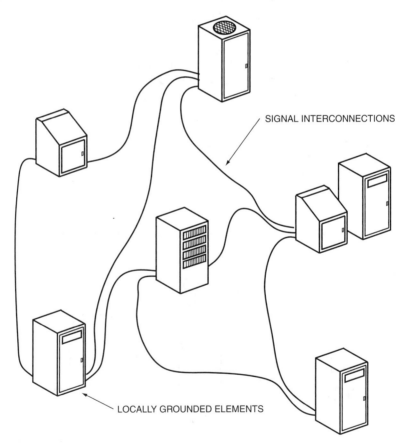

**FIGURE 3-31.** A distributed system that consists of a multiplicity of widely separated elements.

frequencies of interest. Examples of distributed systems are industrial process control equipment and large mainframe computer networks. The elements of the system are fed power from different sources, for example different branch circuits, if within a single building, or even different transformer banks if the elements are located in different buildings.

### 3.3.3.1 Grounding of Distributed Systems.

In a distributed system, the different elements usually have separate ac power, safety, and lightning protection grounds. The element, or elements, located at each location can, however, be considered to be either an isolated or clustered system, and they are safety grounded appropriately for that type of system.

Internal signals should be grounded as appropriate for the type of circuitry and frequency of operation. The primary problem associated with a distributed system is the treatment afforded to the signals that must interconnect between the individual elements of the system. All signal ports and interconnecting cables should be viewed as existing in a harsh (noisy) environment and treated appropriately using the principles covered in Chapter 2 on Cabling and Chapter 4 on Balancing and Filtering.

The main considerations involved in determining the applicable I/O treatment are as follows: What are the characteristics of the signal? What type of cabling and/or filtering will be used? Is the signal analog or digital? What is its frequency and amplitude? Is the signal balanced or unbalanced? (Balanced signals are more immune to noise than unbalanced signals.) Will the cabling be individual wires, twisted pairs (shielded or unshielded), ribbon cable, or coax? If shields are used, should they be grounded at one end, both ends, or hybrid grounded? Another important consideration is can or will some form of isolation or filtering be used? For example, can the signal be transformer or optically coupled to the cable? Filters and common-mode chokes can also be used to treat the I/O signals and to minimize the noise coupling.

In this situation, ground loops may be a problem. Section 3.4 covers this subject and its mitigation. All interconnecting signals should be analyzed to determine the degree of protection required. Some interconnects, depending on frequency, amplitude, signal characteristic, and so on, may not require any special protective measures.

### 3.3.3.2 Common Battery Systems.

A common battery distributed system is one in which the structure (chassis, hull, airframe, etc.) is used for the dc power return. Such a configuration can often be found in automobiles and aircraft. In this type of system, the structure often becomes the signal reference. Because all the ground currents (power and signal) flow through the structure, common impedance coupling can be a major problem. Any voltage differential that exists between various points of the structure, as a result of this common impedance coupling in the ground, will appear in series with all ground-referenced (single-ended) signal interconnections. From a noise and interference perspective, such a system is not very desirable, and it can at best be problematic.

The primary problem associated with a common battery distributed system is the treatment afforded to the signals that must interconnect between the individual elements of the system. All interconnecting signals should be analyzed to determine the degree of protection, if any, required. The magnitude of the noise voltage compared with the signal level in the circuit is important. If the signal-to-noise ratio is such that circuit operation is likely to be affected, then steps must be taken to provide adequate protection from the common-mode ground noise. Sensitive signals should be treated as discussed in Section 3.4 for ground loops. Because balanced interconnections are more immune to noise than single-ended interconnections, they should be seriously considered in this situation.

Providing power to equipment using a twisted pair is always better than using a chassis return. However, if a twisted pair is used in parallel with a chassis return, then it will not be as effective as without the chassis return. Using a twisted pair will still provide some benefits, however. It will reduce any magnetic field differential-mode pickup in the power leads. Although the low-frequency return currents will still return via the chassis (assuming that it is a lower resistance than the return conductor in the twisted pair), high-frequency noise currents will return via the twisted pair (because this provides a lower inductance path) and will not radiate as efficiently and possibly interfere with other equipment. In addition, if chassis ground is poor (high impedance) or degrades with time, the power current can return on the twisted pair, which provides increased reliability.

### 3.3.3.3 Central System With Extensions.

A special case of a distributed system is a central system with extensions. Such a system shown in Fig. 3-32 consists of a central element, or elements, connected, usually in a star arrangement, to remote elements that may extend out long distances from the central element. What distinguishes this from a distributed system is that the remote elements, which are usually small, are not powered or grounded locally; rather, they get their power from the central element. Usually, these are low-frequency systems. The best example of this type of system is the telephone network; another example would be a programmable logic controller (PLC) in a factory with remote ungrounded sensors and/or actuators.

The central element should be grounded as an isolated or clustered system, whichever is appropriate for the configuration. In addition, the cables to the remote elements should be treated to prevent noise pickup and/or radiation as was discussed in Chapter 2.

## 3.4  GROUND LOOPS

Ground loops, at times, can be a source of noise and interference. This is especially true when multiple ground points are separated by a large distance and are connected to the ac power ground, or when low-level analog circuits are

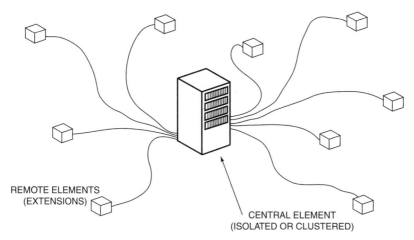

**FIGURE 3-32.** A central system with extensions. The extensions are not locally grounded and are fed power from the central system.

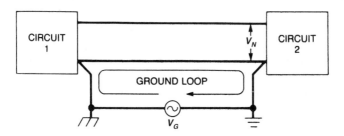

**FIGURE 3-33.** A ground loop between two circuits.

used. In these cases, it may be necessary to provide some form of discrimination or isolation against the ground path noise.

Figure 3-33 shows a system grounded at two different points. Two different ground symbols are shown in the figure to emphasize the fact that the two physically separated ground points are likely to be at different potentials. This configuration has three potential problems, as follows:

1. A difference in ground potential $V_G$ between the two grounds may couple a noise voltage $V_N$ into the circuit as shown in Fig. 3-33. The ground potential is usually the result of other currents flowing through the ground impedance.
2. Any strong magnetic fields can induce a noise voltage into the loop formed by the signal conductors and the ground, which is designated as "ground loop" in Fig. 3-33.

3. The signal current has multiple return paths and may, especially at low-frequency, flow through the ground connection and not return on the signal return conductor.

Item 3 is seldom a problem at high frequency because the larger loop associated with the ground return path will have much more inductance than the smaller loop if the current returns on the signal return conductor. Hence, the high-frequency signal current will return on the signal return conductor, not in the ground.

The magnitude of the noise voltage compared with the signal level in the circuit is important. If the signal-to-noise ratio is such that circuit operation is affected, then steps must be taken to remedy the situation. In many cases, however, nothing, special has to be done.

All ground loops are not bad, and a designer should not get paranoid about the presence of a ground loop. Most ground loops are benign. Most actual ground-loop problems occur at low frequency, under 100 kHz, and they are usually associated with sensitive analog circuits, such as audio or instrumentation systems. The classic example of this is 50/60-Hz hum coupling into an audio system. Ground loops are seldom a problem at high frequency, above 100 kHz, or in digital logic systems. Experience has shown that more problems are caused by trying to avoid ground loops than those caused by the ground loops themselves. Some ground loops are actually helpful, for example, in the case of a cable shield being grounded at both ends in order to provide magnetic field shielding, as was discussed in Section 2.5.

If ground loops are a problem, then they can be dealt with in one of three ways as follows:

1. Avoid them by using single-point or hybrid grounds. This technique is usually only effective at low frequencies, and often it makes the situation worse when attempted at high frequency.
2. Tolerate them by minimizing ground impedance (e.g., by using a ZSRP) and/or by increasing the circuit noise margin (e.g, by increasing the signal voltage level or by using a balanced circuit).
3. Break them by using one of the techniques discussed below.

The ground loop shown in Fig. 3-33 can be broken by one of the following:

1. Transformers
2. Common-mode chokes
3. Optical couplers

Figure 3-34 shows circuits 1 and 2 isolated with a transformer. The ground noise voltage now appears between the transformer's primary and secondary windings and not at the input to the circuit. The remaining noise coupling, if

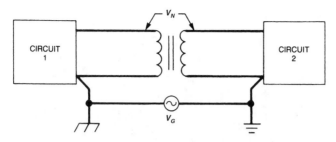

**FIGURE 3-34.** A transformer used to break a ground loop ground between two circuits.

any, is now primarily a function of the parasitic capacitance between the transformer windings, as discussed in Section 5.3 on transformers, and it can be reduced even more by placing a Faraday shield between the transformer windings. Although transformers provide excellent results, they do have some disadvantages. They are often large, have limited frequency response, provide no dc continuity, and are costly. In addition, if multiple signals are connected between the circuits, then multiple transformers are required.

Most professional audio equipment uses a balanced interface to minimize susceptibility to interference and ground loops. Most consumer audio equipment, however, uses a less expensive unbalanced interface. An unbalanced interface is much more susceptible to common-impedance coupling (see Fig. 2-25) resulting from any ground loops that may be formed by interconnecting various pieces of equipment. Isolation transformers are often used in the interconnecting signal leads to eliminate or break the ground loop. Figure 3-35 shows a dual, high-quality audio isolation transformer unit that can be used to eliminate hum and buzz caused by ground loops in such audio applications. The module shown is a dual unit intended for stereo interconnections that use RCA phono plugs. The transformers have mumetal external shields to reduce magnetic field pickup, as well as having Faraday shields between the primary and secondary windings to reduce the interwinding capacitance. The unit shown has a frequency response of 10 Hz to 10 MHz with an insertion loss of less than 0.5 dB and a common-mode (noise) rejection ratio (CMRR) of 120 dB at 60 Hz and 70 dB at 20 kHz.

Note that when isolation is used to break a ground loop, the isolation should be applied to the signal interconnections—not achieved by isolating (breaking) the ac power safety ground. The latter approach is a violation of the NEC and can be very dangerous.

In Fig. 3-36, the two circuits are isolated with a transformer connected as a common-mode choke that will transmit dc and differential-mode signals while rejecting common-mode ac signals. This is basically a transformer rotated 90° and connected in series with the signal conductors. The common-mode noise voltage now appears across the windings of the transformer (choke) and not at the input to the circuit. Because the common-mode choke has no effect on the

**FIGURE 3-35.** Dual channel (stereo) audio isolation transformer unit used to eliminate hum caused by a ground loop in an audio system.

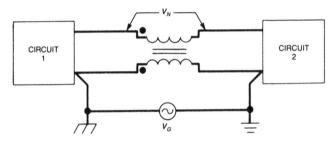

**FIGURE 3-36.** A common-mode choke used to break the ground loop between two circuits.

differential-mode signal being transmitted, multiple signal leads can be wound on the same core without crosstalk. The operation of the common-mode choke is analyzed in Sections 3.5 and 3.6.

Optical coupling (optical isolators or fiber optics) as shown in Fig. 3-37 is another very effective method of eliminating common-mode noise because it breaks the metallic path between the two grounds. It is most useful when a large difference in voltage exists between the two grounds, even in some cases hundreds of volts. The undesired common-mode noise voltage appears across the optical coupler and not across the input to the circuit.

Optical couplers, as shown in Fig. 3-37, are especially useful in digital circuits. They are not as useful in analog circuits because linearity through the coupler is not always satisfactory. Analog circuits have been designed, however, using optical feedback techniques to compensate for the inherent non-linearity of the coupler (Waaben, 1975). Isolation amplifiers (with internal

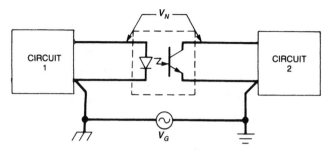

**FIGURE 3-37.** An optical coupler used to break the ground loop between two circuits.

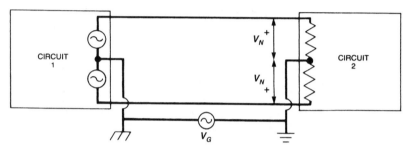

**FIGURE 3-38.** A balanced circuit can be used to cancel out the effect of a ground loop.

transformer or optical coupling), however, are also available for use in sensitive analog circuits.

Balanced circuits, as shown in Fig. 3-38, provide a way to increase the noise immunity of the circuit because they can discriminate between the common-mode noise voltage and the differential-mode signal voltage. In this case, the common-mode voltage induces equal currents in both halves of the balanced circuit, and the balanced receiver responds only to the difference between the two inputs. The better the balance, the larger the amount of common-mode rejection. As frequency increases, it becomes more and more difficult to achieve a high degree of balance. Balancing is discussed in Chapter 4.

When the common-mode noise voltages are at a frequency different from the desired signal, frequency-selective hybrid grounding can often be used to avoid the ground loop at the troublesome frequency.

## 3.5  LOW-FREQUENCY ANALYSIS OF COMMON-MODE CHOKE

A transformer can be used as a common-mode choke (also called a longitudinal choke, neutralizing transformer, or balun) when connected, as shown in Fig. 3-39. A transformer connected in this manner presents a low impedance to the signal

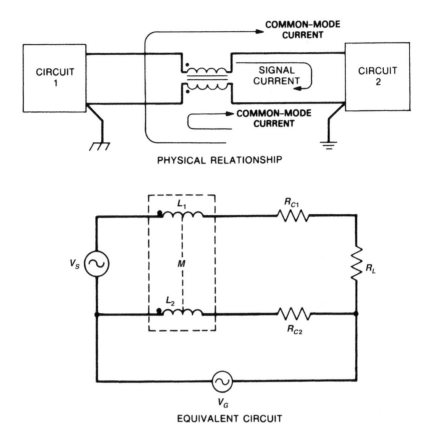

**FIGURE 3-39.** When dc or low-frequency continuity is required, a common-mode choke can be used to break a ground loop.

current and allows dc coupling. To any common-mode noise current, however, the transformer is a high impedance.

The signal current shown in Fig. 3-39 flows equally in the two conductors but in opposite directions. This is the desired current, and it is also known as the differential circuit current or metallic circuit current. The noise currents flow in the same direction along both conductors and are called common-mode currents.

Circuit performance for the common-mode choke of Fig. 3-39. may be analyzed by referring to the equivalent circuit in Fig. 3-39. Voltage generator $V_s$ represents a signal voltage that is connected to the load $R_L$ by conductors with resistance $R_{C1}$ and $R_{C2}$. The common-mode choke is represented by the two inductors $L_1$ and $L_2$ and the mutual inductance $M$. If both windings are identical and closely coupled on the same core, then $L_1$, $L_2$, and $M$ are equal. Voltage generator $V_G$ represents a common-mode voltage either from magnetic

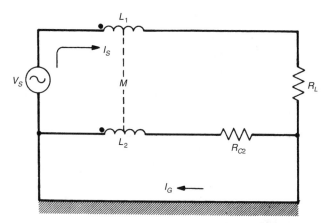

**FIGURE 3-40.** Equivalent circuit for Fig. 3-39 for analysis of response to the signal voltage $V_S$.

coupling in the ground loop or from a ground differential voltage. Because the conductor resistance $R_{C1}$ is in series with $R_L$ and of much smaller magnitude, it can be neglected.

The first step is to determine the response of the circuit to the signal voltage $V_S$, neglecting the effect of $V_G$. The circuit of Fig. 3-39 can be redrawn, as shown in Fig. 3-40. This figure is similar to the circuit of Fig. 2-22. There, it was shown that at frequencies greater than $\omega = 5R_{C2}/L_2$, virtually all the current $I_S$ returned to the source through the second conductor and not through the ground plane. If $L_2$ is chosen such that the lowest signal frequency is greater than $5R_{C2}/L_2$ rad/s, then $I_G = 0$. Under these conditions the voltages around the top loop of Fig. 3-40 can be summed as follows:

$$V_s = j\omega(L_1 + L_2)I_s - 2j\omega M I_s + (R_L + R_{C2})I_s. \qquad (3\text{-}10)$$

Remembering that $L_1 = L_2 = M$ and solving for $I_S$ gives

$$I_s = \frac{V_s}{R_L + R_{C2}} = \frac{V_s}{R_L}, \qquad (3\text{-}11)$$

provided $R_L$ is much greater than $R_{C2}$. Equation 3-11 is the same that would have been obtained if the choke had not been present. It, therefore, has no effect on the signal transmission so long as the choke inductance is large enough that the signal frequency $\omega$ is greater than $5R_{C2}/L_2$.

The response of the circuit of Fig. 3-39 to the common-mode voltage $V_G$ can be determined by considering the equivalent circuit shown in Fig. 3-41. If the choke were not present, the complete noise voltage $V_G$ would appear across $R_L$.

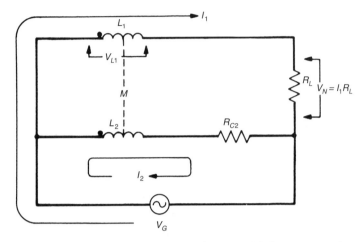

**FIGURE 3-41.** Equivalent circuit for Fig. 3-39 for analysis of response to the common-mode voltage $V_G$.

When the choke is present, the noise voltage developed across $R_L$ can be determined by writing equations around the two loops shown in the illustration. Summing voltages around the outside loop gives

$$V_G = j\omega L_1 I_1 + j\omega M I_2 + I_1 R_L \tag{3-12}$$

The sum of the voltages around the lower loop is

$$V_G = j\omega L_2 I_2 + j\omega M I_1 + R_{c2} I_2 \tag{3-13}$$

Equation 3-13 can be solved for $I_2$, giving the following result:

$$I_2 = \frac{V_G - j\omega M I_1}{j\omega L_2 + R_{C2}} \tag{3-14}$$

Remembering that $L_1 = L_2 = M = L$, and substituting Eq. 3-14 into Eq. 3-12 and solving for $I_1$, gives

$$I_1 = \frac{V_G R_{C2}}{j\omega L (R_{C2} + R_L) + R_{C2} R_L} \tag{3-15}$$

The noise voltage $V_N$ is equal to $I_1 R_L$, and because $R_{C2}$ is normally much less than $R_L$, we can write

$$V_N = \frac{V_G R_{C2}/L}{j\omega + R_{C2}/L} \tag{3-16}$$

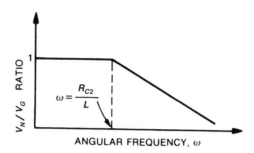

**FIGURE 3-42.** Noise voltage may be significant if $R_{C2}$ is large.

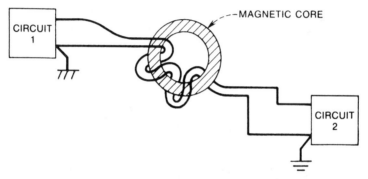

**FIGURE 3-43.** An easy way to place a common-mode choke in the circuit is to wind both conductors around a torodial magnetic core. A coaxial cable may also be used in place of the conductors shown.

An asymptotic plot of $V_N/V_G$ is shown in Fig. 3-42. To minimize this noise voltage, $R_{C2}$ should be kept as small as possible, and the choke inductance $L$ should be such that

$$L \gg \frac{R_{C2}}{\omega} \tag{3-17}$$

where $\omega$ is the frequency of the noise. The choke also must be large enough that any unbalanced dc current flowing in the circuit does not cause saturation.

The common-mode choke shown in Fig. 3-39 can be easily made; simply wind the conductors connecting the two circuits around a magnetic core, as shown in Fig. 3-43. At frequencies above 30MHz, a single turn choke is often effective. The signal conductors from more than one circuit may be wound around the same core without the signal circuits interfering (crosstalking). In this way, one core can be used to provide a common-mode choke for many circuits.

## 3.6   HIGH-FREQUENCY ANALYSIS OF COMMON-MODE CHOKE

The preceding analysis of the common-mode choke was a low-frequency analysis and neglected the effect of parasitic capacitance. If the choke is to be used at high frequencies ($>$ 10MHz), then the stray capacitance across the windings must be considered. Figure 3-44 shows the equivalent circuit of a two-conductor transmission line that contains a common-mode choke ($L_1$ and $L_2$). $R_{C1}$ and $R_{C2}$ represent the resistance of the windings of the choke plus the cable conductors, and $C_S$ is the stray capacitance across the windings of the choke. $Z_L$ is the common-mode impedance of the cable and $V_{CM}$ is the common-mode voltage driving the cable. In this analysis, $Z_L$ is not the differential-mode impedance, but the impedance of the cable acting as an antenna and may vary from about 35 to 350 $\Omega$.

The insertion loss (IL) of the choke can be defined as the ratio of the common-mode current without the choke to the common-mode current with the choke. For $R_{C1} = R_{C2} = R$ and $L_1 = L_2 = L$, the IL of the choke can be written as

$$IL = Z_L \sqrt{\frac{[2R(1 - \omega^2 LC_s)]^2 + R^4(\omega C_s)^2}{[R^2 + 2R(Z_L - \omega^2 LC_s Z_L)]^2 + [2R\omega L + \omega C_s R^2 Z_L]^2}} \qquad (3\text{-}18)$$

Figures 3-45 and 3-46 are plots of Eq. 3-18 for the case where $R_{C1} = R_{C2} = 5$ $\Omega$, and $Z_L = 200\ \Omega$. Figure 3-45 shows the insertion loss for a 10-$\mu$H choke for various values of shunt capacitance, and Fig. 3-46 shows the insertion loss for a choke with 5 pF of shunt capacitance and various values of inductance. As can be observed from these two figures, the insertion loss above 70 MHz does not

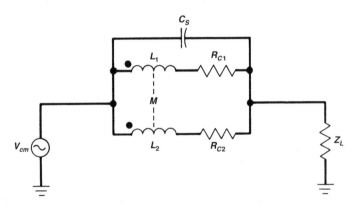

**FIGURE 3-44.** Equivalent circuit of a common-mode choke with parasitic shunt capacitance $C_S$.

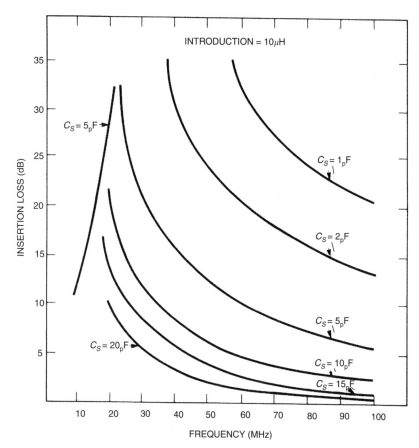

**FIGURE 3-45.** Insertion loss of a 10-μH common-mode choke, with various values of shunt capacitance.

vary much with the inductance of the choke; however, it varies considerably as a function of the shunt capacitance. *Therefore, the most important parameter in determining the performance of the choke is the shunt capacitance and not the value of inductance.* Actually, most chokes used in these applications are beyond self-resonance. The presence of the parasitic capacitance severely limits the maximum insertion loss possible at high frequencies. It is difficult to obtain more than 6 to 12 dB insertion loss at frequencies above 30 MHz by this technique.

At these frequencies, the choke can be thought of as an open circuit to the common-mode noise currents. The total common-mode noise current on the cable is therefore determined by the parasitic capacitance, not the inductance of the choke.

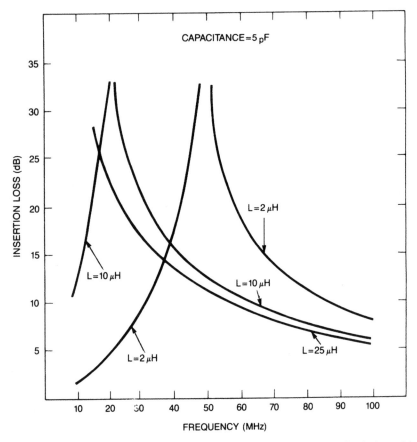

**FIGURE 3-46.** Insertion loss of various value inductance common-mode chokes with 5 pF of shunt capacitance.

## 3.7  SINGLE GROUND REFERENCE FOR A CIRCUIT

The way to make an antenna, dipole or monopole, is to have an rf potential between two pieces of metal; see Appendix D. The capacitance between the two pieces of metal will then provide the path for the rf current. Antennas radiate energy as well as pick up energy very efficiently. As a matter of fact, at some frequencies, the efficiency of a dipole antenna can be in excess of 98%. By the way, it does not matter what potential the halves of the antenna are at; all that matters is that a potential difference exists between them. The way to prevent the radiation is to connect the two pieces of metal together so that they are at the same potential and, therefore, cannot radiate—because without a voltage, there will be no current flow.

Many systems contain multiple ground planes, for example, separate analog and digital ground planes, which are only connected together at one point, possibly only at the power supply, as shown in Fig. 17-2. Having separate ground or reference planes for a system is a good way to make sure that efficient antennas are designed into the system. *In almost all cases, a system will perform better, both functionally and EMC wise, with a single reference plane.*

Terrell and Keenan best stated this in their book, *Digital Design for Interference Specifications*, where they said (Terrell and Keenan, 1983, p. 3-18): *"Thou shalt have but one ground before thee."*

## SUMMARY

- All conductors, including ground conductors, have finite impedance, which consists of both resistance and inductance.
- A ground conductor longer than 1/20 wavelength is not a low impedance.
- Grounds fall into two categories, safety grounds and signal grounds.
- The ac power ground is of little practical value as a signal ground.
- The earth is not a very low impedance and is polluted with noisy power currents; it is far from an equipotential.
- Make connections to earth ground only when required for safety.
- Do not look to an earth ground as a solution to your EMC problems.
- Single-point grounds should only be used at low frequencies, typically at frequencies below 100 kHz.
- Multipoint grounds should be used at high frequencies, typically above 100 kHz, and with digital circuits.
- One purpose of a good ground system is to minimize the noise voltage produced when two or more ground currents flow through a common ground impedance.
- The best way to make a low-impedance ground connection over the widest range of frequencies, between separate pieces of equipment, is by connecting them with a plane or grid.
- To minimize ground noise voltage,
    - At low-frequency, control the ground topology (direct the current).
    - At high-frequency, control the ground impedance.
- Ground loops can be controlled by:
    - Avoiding them
    - Tolerating them
    - Breaking them
- Three common ways to break a ground loop are:
    - Isolation transformers
    - Common-mode chokes
    - Optical couplers

## PROBLEMS

3.1 What type of ground does not carry current during normal operation?

3.2 Proper ac power grounding can be very effective in controlling differential- and common-mode noise on the power distribution system. True or false?

3.3 What is the optimum way to obtain a low-impedance ground that is effective over the widest frequency range?

3.4 To minimize ground noise voltage, what term of Eq.3-2 do we usually control:
   a. In the case of a low-frequency circuit?
   b. In the case of a high-frequency circuit?

3.5 a. What is the typical impedance of an earth ground?
   b. What is the maximum earth ground impedance allowed by the NEC, before a second ground electrode must be used?

3.6 What is the potential of point B in Fig. 3-15?

3.7 At high frequency, why should multiple ground-bonding straps be used on an electronic equipment enclosure?

3.8 What is the inductance of an equipment grounding strap that is 20 cm long, 5 cm wide, and 0.1 cm thick?

3.9 A large equipment rack is standing on a ground plane, but it is insulated from it because of the paint on the rack. The rack has 1000 pF of capacitance to the plane. It is grounded to the plane with four grounding straps, one located in each corner of the rack. Each strap has an inductance of 120 nH. At what frequency will the rack-to-ground-plane impedance be the largest?

3.10 In a distributed system, what is the primary problem that must be dealt with?

3.11 What is the major difference between a distributed system and a central system with extensions?

3.12 What are the three basic ways of dealing with a problematic ground loop?

3.13 Name three different components that can be used to break a ground loop.

3.14 A person has a problem with hum in a home stereo system and discovers that by disconnecting the ac power, green wire, ground to the amplifier the hum goes away. Why is this not an acceptable solution to the problem?

3.15 A common-mode choke is placed in series with a transmission line connecting a low-level source to a 900-$\Omega$ load. The transmission line conductors each have a resistance of 1 $\Omega$. Each winding of the common-mode choke has an inductance of 0.044 H and a resistance of 4 $\Omega$.
   a. Above what frequency will the choke have a negligible effect on the signal transmission?

b. How much attenuation (in decibels) does the choke provide to a ground differential noise voltage at 60, 180, and 300 Hz?

## REFERENCES

Denny, H. W. *Grounding For the Control of EMI*. Gainsville, VA, Don White Consultants, 1983.

IEEE Std. 1100–2005, *IEEE Recommended Practices for Powering and Grounding Electronic Equipment* (*The Emerald Book*).

Lewis, W. H. *Handbook of Electromagnetic Compatibility*, (Chapter 8, Grounding and Bonding), Perez, R. ed., New York, Academic Press, 1995.

*NEC 2008, NFPA 70: National Electric Code*. National Fire Protection Association (NFPA), Quincy, MA, 2008. (This code is revised every three years.)

Ott, H. W. "Ground—A Path for Current Flow." *1997 IEEE International Symposium on EMC*. San Diego, CA, October 9-11, 1979.

International Association of Electrical Inspectors. *SOARES Book on Grounding and Bonding*, 10th ed., Appendix A (The History and Mystery of Grounding, 2008). Available at http://www.iaei.org/products/pdfs/historyground.pdf. Accessed September 22, 2008.

Terrell, D. L. and Keenan, R. K. *Digital Design for Interference Specifications*, 2nd ed., Pinellas Park, FL, The Keenan Corporation, 1983.

Waaben, S. "High-Performance Optocoupler Circuits," *International Solid State Circuits Conference*, Philadelphia, PA, February 1975.

## FURTHER READING

Brown, H."Don't Leave System Grounding to Chance." *EDN/EEE*, January 15, 1972.

Cushman, R. H. "Designers Guide to Optical Couplers." *EDN*, July 20, 1973.

DeDad, J. "The Pros and Cons of IG Wiring." *EC&M*, November 2007.

*Integrating Electronic Equipment and Power Into Rack Enclosures*, Middle Atlantic Products, 2008.

Morrison, R. *Grounding and Shielding*, 5th ed., New York, Wiley, 2007.

Morrison, R. and Lewis, W. H. *Grounding and Shielding in Facilities*, New York, Wiley, 1990.

*The Truth—Power Distribution & Grounding in Residential AV Installations*. Parts 1 to 3, Middle Atlantic Products, 2008. Available at http://exactpower.com. Accessed February 2009.

# 4 Balancing and Filtering

## 4.1 BALANCING

A balanced circuit is a two-conductor circuit in which both signal conductors, and all circuits connected to them, have the same nonzero impedance with respect to a reference (usually ground) and all other conductors; The purpose of balancing is to make the noise pickup equal in both conductors; in which case, it will be a common-mode signal, which can be made to cancel in the load. If the impedances of the two signal conductors to ground are unequal, then the system is unbalanced. A circuit with a grounded return conductor therefore is unbalanced, and sometimes it is referred to as a single-ended circuit.

Balancing is an often overlooked—although in many cases cost-effective—noise reduction technique, which may be used in conjunction with shielding, when noise must be reduced below the level obtainable with shielding alone. In addition, it can be used, in some applications, in place of shielding as the primary noise-reduction technique.

For a balanced circuit to be most effective in reducing common-mode noise, not only must the terminations be balanced, but also the interconnection (cable) must be balanced. Using transformers or differential amplifiers are two possible approaches to providing a balanced termination.

An excellent example of the effectiveness of a balanced system in reducing noise is the telephone system, where signal levels are typically a few hundred millivolts. Telephone cables, which consist of unshielded twisted pairs, often run parallel to high-voltage (4 to 14 kV) ac power lines for many miles, and it is seldom that any 50/60-Hz hum is heard in the telephone system. This is the result of the telephone system being a balanced system; both the source and the load are balanced. On the rare occasion that hum is heard, it is because something has caused an unbalance (e.g., water getting into the cable) to occur to the lines, and the problem will go away once the balance is restored.

Consider the circuit shown in Fig. 4-1. If $R_{s1}$ equals $R_{s2}$ then the source is balanced, and if $R_{L1}$ equals $R_{L2}$ then the load is balanced. Under these conditions, the circuit will be balanced because both signal conductors have the same impedance to ground. Notice, that it is not necessary for $V_{s1}$ to

*Electromagnetic Compatibility Engineering*, by Henry W. Ott
Copyright © 2009 John Wiley & Sons, Inc.

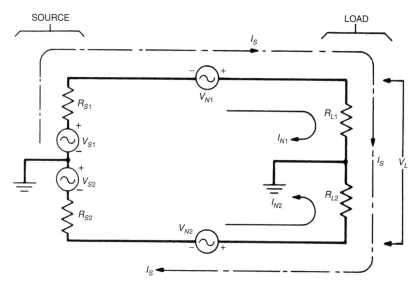

**FIGURE 4-1.** For balanced condition: $R_{s1} = R_{s2}$, $R_{L1} = R_{L2}$, $V_{N1} = V_{N2}$, and $I_{N1} = I_{N2}$.

be equal to $V_{s2}$ for the circuit to be balanced. One or both of these generators may even be equal to zero, and the circuit is still balanced.

In Fig. 4-1, the two common-mode noise voltages $V_{N1}$ and $V_{N2}$ are shown in series with the conductors. These noise voltages produce noise currents $I_{N1}$ and $I_{N2}$. The sources $V_{s1}$ and $V_{s2}$ together produce the signal current $I_s$. The total voltage $V_L$ produced across the load is then equal to the following:

$$V_L = I_{N1}R_{L1} - I_{N2}R_{L2} + I_s(R_{L1} + R_{L2}). \qquad (4\text{-}1)$$

The first two terms represent noise voltages, and the third term represents the desired signal voltage. If $I_{N1}$ is equal to $I_{N2}$ and $R_{L1}$ is equal to $R_{L2}$, then the noise voltage across the load is equal to zero. Equation 4-1 then reduces to

$$V_L = I_s(R_{L1} + R_{L2}), \qquad (4\text{-}2)$$

which represents a voltage resulting only from the signal current $I_s$.

Figure 4-1 just shows resistive terminations to simplify the discussion. In reality, both resistive and reactive balance are important. Figure 4-2 is more general in that it shows both resistive and capacitive terminations.

In the balanced circuit shown in Fig. 4-2, $V_1$ and $V_2$ represent inductive pickup voltages, and current generators $I_1$ and $I_2$ represent noise that is capacitively coupled into the circuit. The difference in ground potential between source and load is represented by $V_{cm}$. If the two signal conductors 1 and 2 are

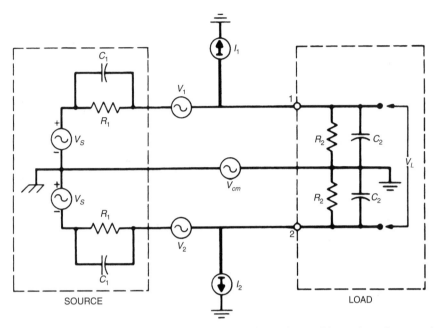

**FIGURE 4-2.** A balanced circuit that shows inductive and capacitive noise voltages and a difference in ground potential between source and load.

located adjacent to each other, or better yet twisted together, the two inductively coupled noise voltages $V_1$ and $V_2$ should be equal and cancel at the load.

The noise voltage produced between load terminals 1 and 2, resulting from to capacitive coupling, can be determined by referring to Fig. 4-3. Capacitors $C_{31}$ and $C_{32}$ represent capacitive coupling from the noise source, in this case conductor 3. Impedances $R_{c1}$ and $R_{c2}$ represent the total resistance to ground from conductors 1 and 2, respectively.*

The capacitive coupled noise voltage $V_{N1}$ induced into conductor 1 because of the voltage $V_3$ on conductor 3 is (Eq. 2-2)

$$V_{N1} = j\omega R_{c1} C_{31} V_3. \tag{4-3}$$

The noise voltage induced into conductor 2 because of $V_3$ is

$$V_{N1} = j\omega R_{c2} C_{32} V_3. \tag{4-4}$$

If the circuit is balanced, then resistances $R_{c1}$ and $R_{c2}$ are equal. If conductors 1 and 2 are located adjacent to each other, or better yet are twisted together, capacitance $C_{31}$ should be nearly equal $C_{32}$. Under these conditions,

---

* $R_{c1}$ and $R_{c2}$ are both equal to the parallel combination of $R_1$ and $R_2$ (see Fig 4-2)

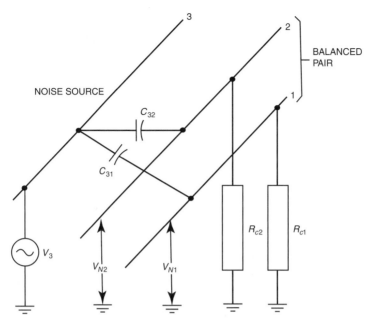

**FIGURE 4-3.** Capacitive pickup in balanced conductors.

$V_{N1}$ approximately equals $V_{N2}$, and the capacitively coupled noise voltages cancel in the load. If the terminations are balanced, then a twisted pair cable can provide protection against capacitive coupling. Because a twisted pair also protects against magnetic fields, whether the terminations are balanced or not (see Section 2.12), a balanced circuit using a twisted pair will protect against both magnetic and electric fields, even without a shield over the conductors. Shields may still be desirable, however, because it is difficult to obtain perfect balance, and additional protection may be required.

It should be noted in Fig. 4-2 that the difference in ground potential $V_{cm}$ between source and load produces equal voltages at terminals 1 and 2 at the load. These voltages cancel and produce no new noise voltage across the load.

### 4.1.1  Common-Mode Rejection Ratio

The common-mode rejection ratio (CMRR) is a metric that can be used to quantify the degree of balance or the effectiveness of a balanced circuit in rejecting common-mode noise voltages.

Figure 4-4 shows a balanced circuit with a common-mode voltage $V_{cm}$ applied to it. If the balance were perfect, then no differential-mode voltage $V_{dm}$ would appear across the input of the amplifier. Because of slight unbalances present in the system, however, a small differential-mode noise voltage $V_{dm}$ will

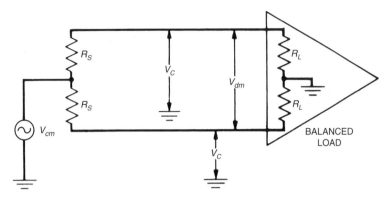

**FIGURE 4-4.** Circuit used to define CMRR.

appear across the input terminals of the amplifier as a result of the common-mode voltage $V_{cm}$. The CMRR, or balance, in dB, is defined as

$$\text{CMRR} = 20 \log \left( \frac{V_{cm}}{V_{dm}} \right) \text{dB.} \tag{4-5}$$

The better the balance, the higher the CMRR and the greater the common-mode noise reduction obtainable. Typically, 40 to 80 dB of CMRR is reasonable to expect from a well-designed circuit. CMRRs better than this range are possible, but individual circuit trimming and special cables may be necessary.

If the source resistance $R_s$ is small compared with the load resistance $R_L$, which is usually the case, then the voltage $V_c$ from either conductor to ground at the input to the amplifier will be almost equal to $V_{cm}$, and it can be used in place of $V_{cm}$ in Eq. 4-5, which gives the following alternative definition of CMRR:

$$\text{CMRR} = 20 \log \left( \frac{V_c}{V_{dm}} \right). \tag{4-6}$$

If the source and load are physically separated by an appreciable distance (e.g., a telephone system), then the definition in Eq. 4-6 is normally preferred because the measurement of both $V_c$ and $V_{dm}$ can be made at the same end of the circuit.

In an ideal balanced system, no common-mode noise will couple into the circuit. In the real world, however, small unbalances limit the noise suppression possible; these include source unbalance, load unbalance, and cable unbalance, as well as the balance of any stray or parasitic impedance present. Both resistive and reactive balances must be considered. Reactive balance becomes more important as the frequency increases.

In many practical applications, the load is balanced but the source is not. The CMRR caused by an *unbalanced source resistance* $\Delta R_s$ can be determined by referring to Fig. 4-5. In this case, the CMRR will be equal to

$$\text{CMRR} = 20\log\left[\frac{(R_L + R_s + \Delta R_s)(R_L + R_s)}{R_L \Delta R_s}\right]. \qquad (4\text{-}7)$$

If $R_L$ is much greater than $R_s + \Delta R_s$, which is usually the case, then Eq. 4-7 can be rewritten as

$$\text{CMRR} = 20\log\left[\frac{R_L}{\Delta R_s}\right]. \qquad (4\text{-}8)$$

If an unbalanced source (one end of the source grounded) is used with a balanced load, then $\Delta R_s$ will be equal to the total source resistance $R_s$.

For example, if $R_L$ equals 10 kΩ and $\Delta R_s$ equals 10 Ω, then the CMRR will be 60 dB.

The detrimental effect of source unbalance on the noise performance of the circuit shown in Fig. 4-5 can be reduced by the following:

- Reducing the common-mode voltage
- Reducing the source unbalance $\Delta R_s$
- Increasing the common-mode load impedance $R_L$

The CMRR caused by an *unbalanced load resistance* can be determined by referring to Fig. 4-6, where $\Delta R_L$ represents the unbalance in the load resistors.

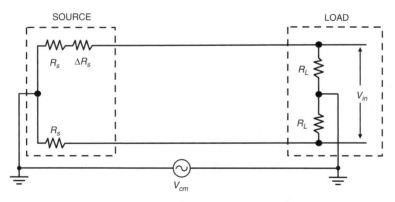

**FIGURE 4-5.** Circuit used to demonstrate the effect of an unbalanced source resistance on the CMRR of a balanced circuit.

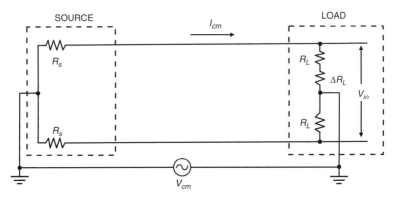

**FIGURE 4-6.** Circuit used to demonstrate the effect of an unbalanced load resistance on the CMRR on a balanced circuit.

The CMRR will be equal to

$$\text{CMRR} = 20\log\left[\frac{(R_L + R_s + \Delta R_L)(R_L + R_s)}{R_s \Delta R_L}\right]. \qquad (4\text{-}9)$$

For $R_L$ much greater than $R_s$, Eq. 4-9 reduces to

$$\text{CMRR} = 20\log\left[\left(\frac{R_L}{R_s}\right)\left(\frac{R_L + \Delta R_L}{\Delta R_L}\right)\right]. \qquad (4\text{-}10)$$

For example, if $R_s$ equals 100 $\Omega$, $R_L$ equals 10 k$\Omega$, and $\Delta R_L$ equals 100 $\Omega$, then the CMRR will be 80 dB. From Eq. 4-10, we observe that the CMRR is a function of the $R_L/R_s$ ratio; the larger this is the greater the noise rejection, regardless of the value of $\Delta R_L$. Therefore, a low source impedance with a high load impedance will provide the largest CMRR. Ideally, we would like a zero source impedance and an infinite load impedance.

From Eqs. 4-8 and 4-10, we can conclude that a large load resistance will maximize the CMRR for the case of both source unbalance and load unbalance. This is true because if the load resistance was infinite then no current would flow and no noise voltage drop would occur across the unbalanced source or load resistance to couple into the circuit.

If the load resistors in Fig. 4-6 are actually discrete resistors with a tolerance of $x\%$, then in the worst case, one load resistor could be $x\%$ high and the other load resistor could be $x\%$ low. If $p$ is the resistor tolerance (expressed as a numeric, not a percent), $\Delta R_L = 2\,p\,R_L$ or $\Delta R_L/R_L = 2\,p$. Therefore, for the case of $R_L \gg \Delta R_L$, the $\Delta R_L/(R_L + \Delta R_L)$ term in Eq. 4-10 represents twice the resistor tolerance. Equation 4-10 can then be rewritten as

$$\text{CMRR} = 20 \log \left( \frac{1}{2p} \right) \left( \frac{R_L}{R_s} \right), \tag{4-11}$$

where $p$ is the tolerance of the load resistors expressed as a numeric. For example, if $R_s$ equals 500 $\Omega$ and the load resistors $R_L$ are each 10 k$\Omega$, 1% resistors, then the CMRR could be as low as 60 dB, in the worst case.

### 4.1.2  Cable Balance

With respect to the interconnecting cable, both resistive and reactive balances must be maintained between the two conductors. Therefore, the resistance and reactance of each conductor must be equal. In many cases, the circuit unbalances are greater than the cable unbalances. However, when large amounts of common-mode rejection are required, greater than 100 dB, or very long cables are used, the cable imperfections must be considered.

The resistive unbalance of most cables is negligible and can usually be ignored. Capacitive unbalance is typically in the 3% to 5% range. At low frequency, this unbalance can usually be ignored because the capacitive reactances will be so much greater than the other impedances in the circuit. At high frequency, however, the capacitive unbalance may have to be considered.

Inductive unbalances are virtually nonexistent for braid shield cables if properly terminated. Improper termination of cable shields, that is non-360° contact to the shield, can be a problem, however.

Foil shield cables that contain a drain wire have significantly more inductive unbalance than braid shield cables because of the presence of current in the drain wire. The drain wire is usually physically closer; to and hence more tightly coupled; to, one of the signal conductors than to the other. Because the resistance of the drain wire is more than an order of magnitude less than that of the aluminum foil shield, at low frequency, nearly all the shield current flows on the drain wire instead of on the shield. This can cause a significant unbalance in the inductive coupling to the signal conductors.

Above about 10 MHz, the skin effect causes the shield current to flow on the foil, and the inductive unbalance is significantly reduced. However, if a foil-shielded cable is terminated via the drain wire, which is usually the case, the inductive unbalance will reappear in the vicinity of the ends of the cable—because the shield current will no longer be uniformly distributed around the cross section of the foil shield. Because of the two above-mentioned problems, inductive unbalance that results from current in the drain wire and/or improper shield termination, foil shielded cables with drain wires should not be used in sensitive circuits that require large amounts of common-mode noise suppression.

The effects of balancing and shielding are additive. The shielding can be used to reduce the amount of common-mode pickup in the signal conductors, and the balancing reduces the portion of the common-mode voltage that is converted to differential-mode voltage and coupled into the load.

Let us assume that a circuit is built with 60 dB of balance and that the cable is not shielded. Let us also assume that each conductor picks up a common-mode noise voltage of 300 mV from electric field coupling. Because of the balancing, the noise coupled into the load will be 60 dB below this, or 300 μV. If a grounded shield with 40 dB of shielding effectiveness is now placed around the conductors, then the common-mode pickup voltage in each conductor will be reduced to 3 mV. The noise coupled into the load will be 60 dB below that because of the balance, or 3 μV. This represents a total noise reduction of 100 dB—40 dB from the shielding, and 60 dB from the balancing.

Circuit balance depends on frequency. Normally, the higher the frequency, the harder it is to maintain good balance, because stray capacitance has more effect on circuit balance at high frequency.

### 4.1.3   System Balance

Knowing the CMRR provided by the individual components that make up a system does not necessarily allow predictions of the overall system CMRR when the components are combined. For example, unbalance in two of the components may complement each other such that the combined CMRR is greater than the CMRR of either of the individual components. However, the component balances may be such that the combined CMRR is less than that of either of the individual components.

One way to guarantee good system balance is to specify the CMRR for each component higher than the desired system CMRR. This method, however, may not produce the most economical system. When multiple individual components are used to construct a system, one way to estimate the overall CMRR of the system is to assume it is equal to the CMRR of the worst component. This method is especially useful if the CMRR of the worst component is 6 dB or more lower than that of the other components.

Because a twisted pair cable is inherently a balanced configuration, twisted pair or shielded twisted pair cables are often used as the interconnecting cables in a balanced system. A coaxial cable (coax); however, is inherently an unbalanced configuration. If a coaxial cable is to be used in a balanced system, then two cables can be used, as shown in Fig. 4-7. An example of this method is the balanced differential voltage probe described in Chapter 18, and shown in Fig. 18-8.

### 4.1.4   Balanced Loads

*4.1.4.1 Differential Amplifiers.*   Differential amplifiers are often used as the loads in balanced systems. Figure 4-8 shows the circuit of a basic differential amplifier. It consists of an operational amplifier (or op-amp), shown in the triangular block, surrounded by feedback and several resistors that determine

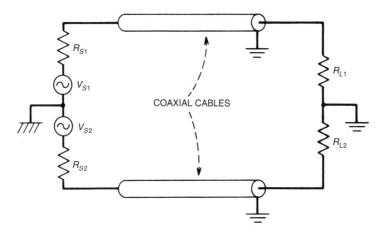

**FIGURE 4-7.** Use of coaxial cable in a balanced circuit.

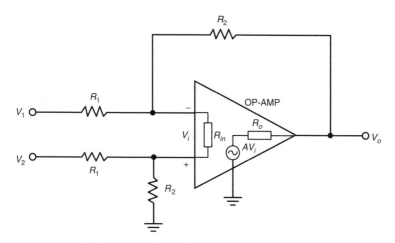

**FIGURE 4-8.** Basic differential amplifier circuit.

the performance of the circuit. The op-amp has a differential input with a single-ended output. The characteristics of an *ideal* op-amp are as follows:

- A very large voltage gain $A$ that ideally approaches infinity
- An infinite input resistance $R_{in}$ between the + and − input terminals
- A zero output resistance $R_o$
- An infinite CMRR

These ideal op-amp characteristics are never realized in practice, but they are approached. A typical op-amp may have a direct current (dc) gain of 100,000, an input resistance of a few megohms, and an output resistance of a few ohms. The CMRR of the op-amp alone, without any additional feedback resistors, will typically be in the 70-to-80-dB range. When resistors are added to form a practical amplifier, the assumption of an ideal gain unit (op-amp) greatly simplifies the analysis. By adding feedback and additional resistors, the op-amp can be configured as a single-ended amplifier, a differential amplifier, an inverting amplifier, or a noninverting amplifier.

The amplifier shown in Fig. 4-8 is a differential input, inverting amplifier. The differential-mode voltage gain of the amplifier is equal to

$$A_{dm} = -\frac{R_2}{R_1}. \tag{4-12}$$

The feedback connected to the negative input terminal drives the voltage between the $+$ and $-$ input terminals to a low value, theoretically zero. Because the voltage between the positive and negative terminals of the amplifier is small (ideally zero), the differential-mode input impedance, between the $V_1$ and $V_2$ input terminals, is just equal to

$$R_{in(dm)} = 2R_1. \tag{4-13}$$

Note that no current actually flows through the input of the op-amp, because ideally its impedance is infinite; the input current actually flows through the feedback resistor to the output, and back through the resistors $R_2$ and $R_1$ connected to the positive terminal of the amplifier.

The common-mode input impedance, between the two input terminals ($V_1$ and $V_2$ tied together) and ground, is equal to

$$R_{in(em)} = \frac{(R_1 + R_2)}{2}. \tag{4-14}$$

Figure 4-6 can be made to represent a differential amplifier driven from a balanced source if we let $R_L$ in Fig. 4-6 be equal to $R_1 + R_2$ in Fig. 4-8. The equivalent differential input voltage $V_{in}$ (in Fig. 4-6) is then equivalent to the voltage between $V_1$ and $V_2$ (in Fig. 4-8). The input voltage $V_{in}$, which is produced by the common-mode voltage $V_{cm}$ (as shown in Fig. 4-6), will be equal to the common-mode input current $I_{cm}$ times the unbalance in the load resistance $\Delta R_L$, or $V_{in} = I_{cm} \Delta (R_1 + R_2)$. The current $I_{cm}$ is equal to $V_{cm} / [R_s + \Delta(R_1 + R_2) + (R_1 + R_2)]$. For the case of $R_2 \gg R_1$, $R_2 \gg \Delta (R_1 + R_2)$, and $R_2 \gg R_s$, the equivalent differential-mode input voltage to the amplifier is equal to

$$V_{in} = \frac{\Delta R_2}{R_2}(V_{cm}). \tag{4-15}$$

The amplifier output voltage $V_{out}$ produced by the common-mode voltage $V_{cm}$ will be $V_{in}$ from Eq. 4-15 times the differential gain $A_{dm}$ of the amplifier (Eq. 4-12) or

$$V_{out} = \frac{\Delta R_2}{R_2}(A_{dm}V_{cm}).\qquad(4\text{-}16)$$

Therefore, the common-mode voltage gain will be

$$A_{cm} = \frac{V_{out}}{V_{cm}} = \frac{\Delta R_2}{R_2}(A_{dm}) = \frac{\Delta R_2}{R_1}.\qquad(4\text{-}17\text{a})$$

The term $\Delta R_2/R_2$ in Eq. 4-17a represents, in the worst case, twice the tolerance of resistor $R_2$. If the resistor tolerance is $p$ (expressed as a numeric, not a percentage), then Eq. 4-17a can be rewritten as

$$A_{cm} = 2pA_{dm} = \frac{2pR_2}{R_1}.\qquad(4\text{-}17\text{b})$$

The CMRR of the differential amplifier can be determined by substituting $V_{in}$ (Eq. 4-15) for $V_{dm}$ in Eq. 4-5 (reference Fig. 4-4), which gives

$$\mathrm{CMRR} = 20\log\left(\frac{V_{cm}}{V_{in}}\right) = 20\log\left(\frac{R_2}{\Delta R_2}\right) = 20\log\left(\frac{1}{2p}\right).\qquad(4\text{-}18)$$

If the differential amplifier shown in Fig. 4-8 is built using 0.1% resistors for $R_1$ and $R_2$ the CMRR, from Eq. 4-18, will be equal to 54 dB

Many books (e.g., Frederiksen, 1988; and Graeme et. al., 1971) define the CMRR of a differential amplifier as

$$\mathrm{CMRR} = 20\log\left(\frac{A_{dm}}{A_{cm}}\right).\qquad(4\text{-}19)$$

Substituting Eq. 4-17b into Eq. 4-19 for $A_{cm}$ gives CMRR $= 20\log(1/2p)$, which agrees with Eq. 4-18.

When using matched resistors, and driven from a balanced source, the differential amplifier can provide high CMRR. However, when driven from an unbalanced source as shown in Fig. 4-9, which is often the case, the CMRR is degraded considerably as the result of the source unbalance.

The CMRR of a balanced circuit driven from an unbalanced source was given in Eq. 4-8. This equation can be rewritten for the differential amplifier circuit shown in Fig. 4-9 as

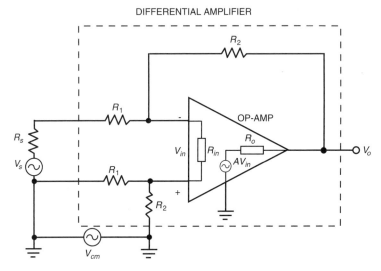

**FIGURE 4-9.** Differential amplifier driven from an unbalanced source.

$$\text{CMRR} = 20 \log \frac{R_2}{R_s}, \qquad (4\text{-}20)$$

for the case of $R_2 \gg R_1$.

In Fig 4-9, if $R_s = 500\ \Omega$, and $R_2 = 100\ \text{k}\Omega$, then from Eq 4-20, the CMMR = 46 dB. From Eq. 4-20, it is obvious that increasing the input impedance of the differential amplifier increases the CMRR proportionately. Decreasing the source resistance $R_s$ also will increase the CMRR of the amplifier.

Therefore, the best way to improve the CMRR of a differential amplifier driven from an unbalanced source is to increase the amplifier's common-mode input impedance to a value of several megohms or more. If in the previous example, $R_2$ were equal to 2 M$\Omega$ instead of 100 k$\Omega$, then the CMRR would have been 72 dB, which is a 26-dB improvement. Such large value resistors, however, are usually not practical, and will be a source of thermal noise on the input. The magnitude of the thermal noise in a resistor is a function of the square root of the magnitude of the resistance (see Section 8.1).

**4.1.4.2. Instrumentation Amplifiers.**   An alternative approach is to add two high-impedance buffer amplifiers to the inputs of a standard differential amplifier. A high-impedance, noninverting buffer can be made from a standard op-amp by feeding the input to the positive terminal and the feedback to the negative terminal. This method produces the classic instrumentation amplifier as shown in Fig. 4-10.

The instrumentation amplifier configuration also has the advantage of a single-resistor ($kR_f$) gain control, instead of having to change the ratio of two

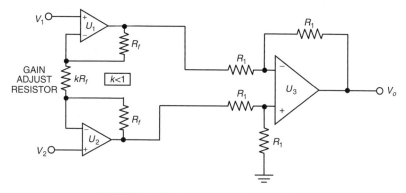

**FIGURE 4-10.** Instrumentation amplifier.

pairs of resistors. In this case, the input impedance of the instrumentation amplifier is equal to the input impedance of the op-amp because no feedback is applied to the positive input terminal of the op-amps. Input impedances in excess of a megohm are possible with this configuration.

If the inputs to the instrumentation amplifier are altenating current (ac) coupled, additional shunt resistors must be added to the circuit of Fig. 4-10, between the buffer inputs and ground, to provide a path for the input transistor bias currents to flow. Therefore, extremely low bias current transistors or FET op-amps should be used. The common-mode input impedance in this case will be the parallel combination of these input resistors and the op-amp input impedance.

In an instrumentation amplifier, the gain is in the buffer amplifiers ($U_1$ and $U_2$), and the differential amplifier ($U_3$) is set to have a gain of unity. The buffers have a common-mode gain of 1 and a differential-mode gain of $A_{dm} = 1 + (2/k)$, where $k$ is a constant less than one.

Because the buffers have a common-mode gain of unity, all the common-mode rejection occurs in the differential amplifier $U_3$. Table 4-1 summarizes the gains of the two stages of the instrumentation amplifier.

From Eq. 4-19, the CMRR of the instrumentation amplifier is

$$\text{CMRR} = 20 \log \left( \frac{A_{dm}}{A_{cm}} \right) = 20 \log \left( \frac{A_{dm}}{2p} \right) = 20 \log \left( \frac{1 + \frac{2}{k}}{2p} \right). \qquad (4\text{-}21)$$

Comparing Eq. 4-21 with Eq. 4-18 we observe that for the same tolerance resistors, an instrumentation amplifier has a CMRR, 20 log ($A_{dm}$) greater than an equivalent differential amplifier. For an instrumentation amplifier with a gain of 100, using 0.1% resistors, the CMRR will be 94 dB. A differential amplifier with a gain of 100, using 0.1% resistors, will have (from Eq. 4-18) a CMRR of only 54 dB.

**TABLE 4-1. Differential-Mode and Common-Mode Gains of the Instrumentation Amplifier Stages Shown in Fig. 4-10.**

| Gain | Buffer Amplifiers ($U_1$ and $U_2$) | Differential Amplifier ($U_3$) | Total for Instrumentation Amplifier |
|---|---|---|---|
| $A_{dm}$ | $1 + (2/k)$ | 1 | $1 + (2/k)$ |
| $A_{cm}$ | 1 | $2p$ | $2p$ |

Note: $p$ is the tolerance of resistors $R_1$, expressed as a numeric.

**TABLE 4-2. CMRR, in dB, for an Instrumentation Amplifier.**

| Resistor Tolerance | $A_{dm} = 1$ | $A_{dm} = 10$ | $A_{dm} = 100$ | $A_{dm} = 1000$ |
|---|---|---|---|---|
| 1% | 34 | 54 | 74 | 94 |
| 0.1% | 54 | 74 | 94 | 114 |
| 0.01% | 74 | 94 | 114 | 134 |

In a differential amplifier, a common-mode to differential-mode conversion occurs on the input of the amplifier as the result of the resistor tolerances. This converted common-mode signal (now a differential-mode signal) is amplified by the differential-mode gain of the amplifier. In an instrumentation amplifier, a similar conversion occurs at the input of the differential amplifier ($U_3$ in Fig 4-10) but that amplifier has a differential-mode gain of unity, so no amplification of the converted common-mode signal occurs.

Therefore the improved CMRR of an instrumentation amplifier results from the fact that the gain is in the buffers, $U_1$ and $U_2$ prior to the common-mode to differential-mode conversion that occurs at the input of the unity gain differential amplifier $U_3$.

Table 4-2 lists the CMRR for an instrumentation amplifier as a function of resistor tolerance and differential-mode gain.

***4.1.4.3 Transformer Coupled Inputs.*** Another approach to obtaining high common-mode input impedance is to use a transformer. The transformer can be used with a differential amplifier or even with a single-ended amplifier as shown in Fig. 4-11. With a transformer, the low-frequency common-mode input impedance will be determined by the insulation resistance (which is extremely large) between primary and secondary of the transformer. At high frequency, the transformer interwinding capacitance may also affect the common-mode input impedance. Transformers also provide galvanic isolation between the source and the load. Transformers, however, tend to be large and costly but perform well.

***4.1.4.4 Input Cable Shield Termination.*** As discussed, cable shields should normally be grounded at both ends. However, when using high common-mode input impedance amplifier circuits, such as instrumentation amplifiers, the input

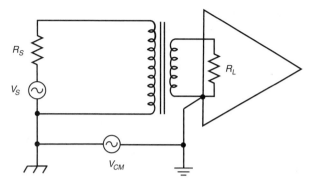

**FIGURE 4-11.** A transformer can be used to increase the common-mode load impedance and to provide galvanic isolation.

cable shield is often connected only to the source ground, not to the load ground. If the shield is connected to the load, or amplifier ground, then the high input impedance of the amplifier will be shunted by the cable capacitance, which will lower the input impedance of the amplifier and decrease the CMRR of the system. If however, the cable shield is grounded at the source, the cable capacitance shunts the source resistance, which is already low and does not reduce the CMRR. This approach however, will often increase the emissions from the product if high-frequency or digital circuits are also present. Therefore, a trade-off must be made between maximum CMRR and radiated emissions. To minimize the emissions, the shield should be grounded at both ends.

**Example 4-1.** The input to a high-impedance differential amplifier is fed from a 600-$\Omega$ unbalanced source through a shielded cable that has a capacitance of 30 pF/ft. If the cable is 100 ft. long, the total cable capacitance will be 3 nF. If the cable's shield is grounded at the amplifier end, then the cable capacitance will shunt the amplifier's input impedance, and the input impedance cannot exceed the capacitive reactance of the cable. At a frequency of 1000 Hz, the CMRR will be equal to or less than (from Eq. 4-8) the following:

$$\text{CMRR} \leq 20 \log \left[ \frac{1}{2\pi f C \Delta R_s} \right].$$

Therefore,

$$\text{CMRR} \leq 20 \log \left[ \frac{1}{2\pi (1000)(3 \times 10^{-9})(600)} \right] = 40 \, \text{dB}.$$

The CMRR of the system will be limited, by the cable capacitance, to no more than 40 dB, regardless of the actual input impedance of the amplifier. If the actual amplifier input impedance was 2 M$\Omega$, and the cable shield had been grounded at the source end, then the cable capacitance would shunt the source impedance, and the CMRR would have been 70 dB.

## 4.2    FILTERING

Filters are used to change the characteristics of, or in some cases eliminate, signals. Filters can be differential mode or common mode. Signal line, or differential-mode, filters are well understood. A multitude of books and articles exist on their design. common-mode filters, however, are often thought of as mysterious and are not well understood.

### 4.2.1    Common-Mode Filters

Common-mode filters are usually used to suppress noise on cables while allowing the intended differential-mode signal to pass undisturbed. Why are common-mode filters more difficult to design than differential-mode filters? Basically there are three reasons:

- We usually do not know the source impedance.
- We usually do not know the load impedance.
- The filter must not distort the intentional signal (the differential-mode signal) on the cable.

The effectiveness of a filter depends on the source and load impedances between which the filter is working. For a differential-mode filter, it is usually easy to find information on the output impedance of the driver and on the input impedance of the load. However, for common-mode signals, the source is the noise generated by the circuit (not documented anywhere), and the load is usually some cable that acts as an antenna, the impedance of which is not generally known and varies with frequency, cable length, conductor diameter, and cable routing.

In the case of a common-mode filter the source impedance is usually the printed circuit board (PCB) ground impedance (which is small and increases with frequency, because it is inductive), and the load is the impedance of a cable acting as an antenna (which except in the vicinity of cable resonance is large). So, although we may not know the source and load impedances exactly, we do have a handle on their magnitude and frequency characteristics.

For a common-mode filter not to distort the intentional differential-mode signal, the differential-mode pass band of the filter must be such that it satisfies the following:

- For narrow-band signals, the highest frequency present
- In the case of wide band digital signals, the $1/\pi\ tr$ frequency of the signal (where $tr$ is the rise time).

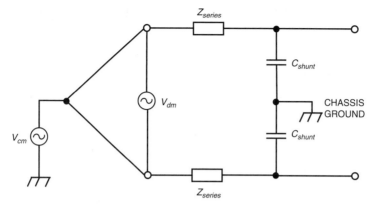

**FIGURE 4-12.** A two-element common-mode filter in both the signal and return conductors.

Differential-mode or signal line filters (e.g., clock line filters, etc.) should be placed as close to the source or driver as possible. Common-mode filters, however, should be located as close to where the cable enters or leaves the enclosure as possible.

Figure 4-12 shows a simple two-element, low-pass, common-mode filter that consists of a series element and a shunt element. The filter is inserted both in the signal conductor and in its return conductor. The figure also shows a common-mode (noise) and a differential-mode (signal) voltage source connected to the filter.

To the common-mode voltage source the two shunt capacitors are in parallel for a total capacitance of $2\,C_{shunt}$. To the differential-mode voltage source, the two capacitors are in series for a total capacitance of $C_{shunt}/2$. Therefore, the common-mode source observes four times the capacitance that the differential-mode source observes. This result is good, because we want the shunt capacitance of the filter to have more effect on the common-mode signal than on the differential-mode signal.

However, to the common-mode source the two series impedances are in parallel for a total impedance of $Z_{series}/2$. To the differential-mode source, the two series impedances are in series for a total capacitance of $2\,Z_{series}$. Therefore, the differential-mode source observes four times the series impedance that the common-mode source observes. This result is bad, because we want the series impedance of the filter to have more effect on the common-mode voltage than on the differential mode voltage. As a result, the series element in a common-mode filter is usually configured as a common-mode choke (see Section 3.5), in which case the differential-mode impedance is zero, and the series impedance only affects the common-mode signal and not the differential-mode signal.

Common-mode filters are usually low-pass filters that consist of from one to three elements arranged in one of the following topologies:

- Single-element filters
  - A single series element
  - A single shunt element
- Multielement filters
  - An L-filter (one series and one shunt element)
  - A T-filter (two series elements and one shunt element)
  - A π-filter (two shunt elements and one series element)

The advantage of single-element filters is that they only require one component. The advantage of multielement filters is that they will often be effective where single element filters are not and that they often can provide more attenuation than a single-element filter.

The shunt element in the filter is almost always a capacitor, the value of which is determined by the frequency range over which the filter is to be effective. The series element can be a resistor, an inductor, or a ferrite. If the dc voltage drop can be tolerated, a resistor can be used. If the dc voltage drop cannot be tolerated, then an inductor or ferrite should be used, both of which will have zero or very small dc voltage drop. At low frequency ($< 10$ to $30$ MHz), an inductor should be used; at high frequency a ferrite should be used. Inductors, being high-Q components, may have resonance problems as discussed in Section 4.3.1. Sometimes, a small value resistor can be added in series with the inductor to lower its Q. A ferrite configured as a common-mode choke has the added advantage of not affecting the differential-mode signal.

Filter attenuation occurs as a result of impedance mismatches. From the above discussion we know that the source impedance is usually low, and except at resonance the load impedance is usually high. Therefore, an L-filter with its high-impedance element (series element) facing the low source impedance, and its low-impedance element (a shunt capacitor) facing the high load impedance should be the most effective.

For a series impedance filter element to be effective, it must have an impedance larger than the sum of the source and load impedances. For a shunt filter element to be effective, it must have an impedance less than the parallel combination of the source and load impedances.

Therefore, three cases are possible, as follows (1) Both source and load impedance are low, in which case a series element will be effective; (2) both source and load impedance are high, in which case a shunt capacitor will be effective; and (3) one of the impedances, source or load, is low and the other is high (and it does not matter which is which), in which case no single element filter will be effective, and a multielement filter must be used.

Series elements will be most effective in the vicinity of cable resonance (where both source and load impedances are low), whereas shunt capacitors will be most effective above the resonant frequency of the cable (where the source impedance is moderate to high and the load impedance is high). A cable will have multiple resonance points where the impedance dips low and a series element will be effective. For the case of a cable acting as a dipole, at the first resonance the cable will have an impedance around 70 Ω, and for a monopole, this will be around 35 Ω.

With respect to multistage filters, the more stages a filter has, the less its attenuation is dependent on the terminating impedances. Most of the mismatch can be made to occur between the elements of the filter itself, independent of the actual source and load impedances.

Shunt capacitors need a low-impedance connection to ground, but what ground? When controlling common-mode noise produced in the PCB ground, the filter capacitors need to be connected to the enclosure or chassis ground. If the circuit ground and the chassis ground are tied together in the input/output (I/O) area of the PCB, which is recommended, then the chassis ground and the circuit ground will be the same at this point.

Note that common-mode filters need to be applied to all conductors leaving or entering the equipment enclosure, which include the circuit ground conductors. When shunt capacitors are used, one capacitor must be connected between each conductor, which include the ground conductor. In the case of a series resistor or inductor, one component must be placed in series with each of the conductors, which includes the ground conductor. In the case of a ferrite core, however, one component can be used to treat all the conductors in a cable, by running all the conductors through the single ferrite core. This is a major advantage of ferrites; one component can treat many conductors.

When a series element and a shunt capacitor are used together in an L-filter, the series element should be placed on the circuit side of the filter, (because this is the low-impedance side), and the capacitor on the cable side (because this is the high-impedance side). This configuration is necessary because the capacitor cannot work against the low-source impedance of the circuit ground. If the high-impedance series element is placed on the circuit side of the shunt capacitor, the series element effectively raises the source impedance to a point where the capacitor can be effective. If the ferrite is placed on the cable side of the capacitor, then it increases the already large cable impedance and has very little effect on the filter.

### 4.2.2 Parasitic Effects in Filters

Consider the low-pass π-filter shown in Fig. 4-13. The filter consists of a series impedance and two shunt capacitors. Parasitic capacitance $C_p$ is shown across the series impedance $Z_1$, and parasitic inductances $L_{p1}$ and $L_{p2}$ are shown in series with the two capacitors $C_1$ and $C_2$ respectively. As frequency increases, a point is reached where the series element becomes capacitive and the two shunt elements become inductive. At this point, the low-pass filter becomes a high-pass filter. The designer must make sure that this transformation from a low-pass to a high-pass filter does not occur in the frequency range of interest.

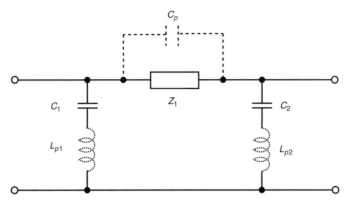

**FIGURE 4-13.** A low-pass π-filter with parasitics.

The frequency at which this transformation occurs is a function of the layout of the filter. In the case of a poor filter layout, this frequency can occur in the tens of megahertz range or less. For a good filter layout, this frequency can be hundreds of megahertz or more. The layout makes all the difference! Above some frequency, all low-pass filters will become high-pass filters. Conversely, above some frequency, all high-pass filters will become low-pass filters because of parasitics.

In many cases, at high frequency, the control of the parasitics is more important to the filter's performance than the value of the intended elements.

## 4.3  POWER SUPPLY DECOUPLING

In most electronic systems, the dc power-supply distribution system is common to many circuits. It is important, therefore, to design the dc power distribution system so that it is not a channel for noise coupling between the circuits connected to the system. The object of a power distribution system is to supply a nearly constant dc voltage to all loads under conditions of varying load currents. In addition, any ac noise signals generated by the load should not generate an ac voltage across the dc power bus.

Ideally, a power supply is a zero-impedance source of voltage. Unfortunately, practical supplies do not have zero impedance, so they represent a source of noise coupling between the circuits connected to them. Not only do the supplies have finite impedance, but also the conductors used to connect them to the circuits add to this impedance. Figure 4-14 shows a typical power distribution system as it might appear on a schematic. The dc source—a battery, power supply, or converter—is fused and connected to a variable load $R_L$ by a pair of conductors. A local decoupling capacitor $C$ may also be connected across the load.

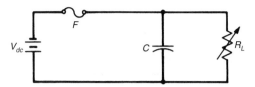

**FIGURE 4-14.** A dc power distribution system as it might appear on schematic.

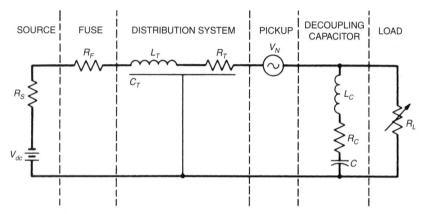

**FIGURE 4-15.** The actual circuit for a dc power distribution system, including parasitics.

For a detailed analysis, the simplified circuit of Fig 4-14 can be expanded into the circuit of Fig. 4-15. Here, $R_s$ represents the source resistance of the power supply and is a function of the power-supply regulation. Resistor $R_F$ represents the resistance of the fuse. Components $R_T$, $L_T$, and $C_T$ represent the distributed resistance, inductance, and capacitance, respectively, of the transmission line used to connect the power source to the load. Voltage $V_N$ represents noise coupled into the wiring from other circuits. The decoupling capacitor $C$ has resistance $R_c$ and inductance $L_c$ associated with it. Resistor $R_L$ represents the load.

The noise pickup $V_N$ can be minimized by the techniques previously covered in Chapters 2 and 3. The effect of the decoupling capacitor is discussed in Section 4.3.1. When the filter capacitor and $V_N$ are eliminated from Fig. 4-15, the circuit of Fig. 4-16 remains. This circuit can be used to determine the performance of the power distribution system. The problem can be simplified by dividing the analysis of Fig. 4-16 into two parts. First, determine the static or dc performance of the system, and second, determine the transient or noise performance of the system.

The static voltage drop is determined by the maximum load current and the resistances, $R_s$, $R_F$, and $R_T$. The source resistance $R_s$ can be decreased by improving the regulation of the power supply. The resistance $R_T$ of the power

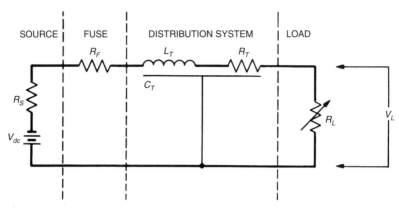

**FIGURE 4-16.** Circuit of Fig. 4-15, less the decoupling capacitor and noise pickup voltage.

distribution line is a function of the cross sectional area $A$ and length $l$ of the conductors and the resistivity ($\rho$) of the conductor material,

$$R_T = \rho \frac{l}{A}. \tag{4-22}$$

The resistivity $\rho$ equals 1.724 x $10^{-8}$ $\Omega$-meters for copper. The minimum dc load voltage is

$$V_{L(\min)} = V_{dc(\min)} - I_{L(\max)}(R_s + R_F + R_T)_{\max}. \tag{4-23}$$

Transient noise voltages on the power distribution circuit are produced by sudden changes in the current demand of the load. If the current change is assumed to be instantaneous, then the magnitude of the resulting voltage change is a function of the characteristic impedance ($Z_0$) of the transmission line:

$$Z_0 = \sqrt{\frac{L_T}{C_T}}. \tag{4-24}$$

The instantaneous voltage change $\Delta V_L$ across the load will then be

$$\Delta V_L = \Delta I_L Z_0. \tag{4-25}$$

The assumption of an instantaneous change in current is realistic for digital circuits, but not necessarily so for analog circuits. Even in the case of analog circuits, however, the characteristic impedance of the dc power distribution transmission line can be used as a figure of merit for comparing the noise performance of various power distribution systems. For best noise performance, a power distribution system transmission line with as low a characteristic impedance as possible is desired—typically 1 $\Omega$ or less. Equation 4-24 shows that the line should therefore have high capacitance and low inductance.

The inductance can be reduced by using a rectangular cross-sectional conductor instead of a round conductor and by having the supply and return conductors as close together as possible. Both of these efforts also increase the capacitance of the line, as does insulating the conductors with a material that has a high dielectric constant. Figure 4-17 gives the characteristic impedance for various conductor configurations. These equations can be used even if the inequalities listed in the figure are not satisfied. Under these conditions, however, these equations give higher values of $Z_0$ than the actual value because they neglect fringing. The typical values for the relative dielectric constant ($\varepsilon_r$) for various materials are listed in Table 4-3. The optimum power distribution line would be one with parallel flat conductors, as wide as possible, placed one on top of the other, and as close together as possible.

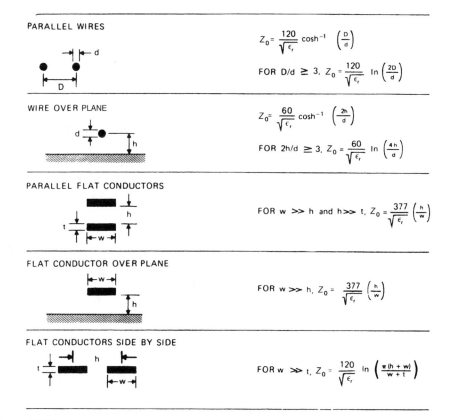

**PARALLEL WIRES**

$$Z_0 = \frac{120}{\sqrt{\varepsilon_r}} \cosh^{-1}\left(\frac{D}{d}\right)$$

FOR $D/d \geq 3$, $Z_0 = \frac{120}{\sqrt{\varepsilon_r}} \ln\left(\frac{2D}{d}\right)$

**WIRE OVER PLANE**

$$Z_0 = \frac{60}{\sqrt{\varepsilon_r}} \cosh^{-1}\left(\frac{2h}{d}\right)$$

FOR $2h/d \geq 3$, $Z_0 = \frac{60}{\sqrt{\varepsilon_r}} \ln\left(\frac{4h}{d}\right)$

**PARALLEL FLAT CONDUCTORS**

FOR $w \gg h$ and $h \gg t$, $Z_0 = \frac{377}{\sqrt{\varepsilon_r}}\left(\frac{h}{w}\right)$

**FLAT CONDUCTOR OVER PLANE**

FOR $w \gg h$, $Z_0 = \frac{377}{\sqrt{\varepsilon_r}}\left(\frac{h}{w}\right)$

**FLAT CONDUCTORS SIDE BY SIDE**

FOR $w \gg t$, $Z_0 = \frac{120}{\sqrt{\varepsilon_r}} \ln\left(\frac{\pi(h+w)}{w+t}\right)$

**FIGURE 4-17.** Characteristic impedance for various conductor configurations.

TABLE 4-3.   Relative Dielectric Constants of Various Materials.

| Material | $\varepsilon_r$ |
|---|---|
| Air | 1.0 |
| Styrofoam | 1.03 |
| Polyethylene foam | 1.6 |
| Cellular polyethylene | 1.8 |
| Teflon[a] | 2.1 |
| Polyethylene | 2.3 |
| Polystyrene | 2.5 |
| Nylon | 3.0 |
| Silicon rubber | 3.1 |
| Polyester | 3.2 |
| Polyvinylchloride | 3.5 |
| Epoxy resin | 3.6 |
| Delrin[TM] | 3.7 |
| Getek[b] | 3.9 |
| Epoxy glass | 4.5 |
| Mylar[a] | 5.0 |
| Polyurethane | 7.0 |
| Glass | 7.5 |
| Ceramic | 9.0 |

[a] Registered trademark of DuPont, Wilmington, DE.
[b] Registered trademark of General Electric, Fairfield, CT.

To demonstrate the difficulty involved in providing power distribution systems with very low characteristic impedance, it is helpful to work some numerical examples. First consider two round parallel wires spaced 1.5 times their diameter apart with Teflon® dielectric. The characteristic impedance is as follows:

$$Z_0 = \frac{120}{\sqrt{2.1}} \cosh^{-1}(1.5) = 80 \ \Omega.$$

If the dielectric had been air, the impedance would be 115 Ω. The actual impedance is between the two values because part of the field is in Teflon® and part in air. A value of 100 Ω is reasonable in this case.

As a second example, take two flat conductors 0.0027-in. thick by 0.02-in. wide placed side by side on the surface of an epoxy glass printed circuit board. If they are spaced 0.04-in apart, the characteristic impedance is

$$Z_0 = \frac{120}{\sqrt{4.5}} \ln\left(\frac{\pi(0.06)}{0.0227}\right) = 120 \ \Omega.$$

For an air dielectric, the impedance would be 254 Ω. The actual impedance is somewhere between these two values, because for a surface trace on a printed circuit board, part of the field is in air and part is in the epoxy glass. A value of 187 Ω is reasonable in this case.

Both of the preceding examples are common configurations, and neither one produced a low characteristic impedance transmission line. If, however, two flat conductors 0.25-in. wide are placed on top of one another and separated by a thin (0.005-in.) sheet of Mylar[®], the characteristic impedance becomes

$$Z_0 = \frac{377}{\sqrt{5}} \left( \frac{0.005}{0.25} \right) = 3.4 \ \Omega.$$

For an air dielectric, the impedance would be 7.6 Ω. The actual impedance is between the two values because part of the field is in Mylar[®] and part in air. A value of 5.5 Ω is reasonable in this case. Such a configuration makes a much lower impedance transmission line than the previous examples, but still it is not a very low impedance.

The above examples point out the difficulty of obtaining a power distribution system with a characteristic impedance of 1 Ω or less. This result makes it necessary to place decoupling capacitors across the power bus at the load end to achieve the desired low impedance. Although this is a good approach, a discrete capacitor will not maintain its low impedance at high frequencies, because of series inductance. A properly designed transmission line, however, maintains its low impedance even at high frequencies. Low-frequency analog decoupling is discussed in the next section. For information on high-frequency digital logic decoupling see Chapter 11.

### 4.3.1  Low-Frequency Analog Circuit Decoupling

Because the power supply and its distribution system are not an ideal voltage source, it is good practice to provide some decoupling at each circuit or group of circuits to minimize noise coupling through the supply system. This is especially important when the power supply and its distribution system are not under the control of the designer of the power-consuming circuit.

Resistor-capacitor and inductor-capacitor decoupling networks can be used to isolate circuits from the power supply, to eliminate coupling between circuits, and to keep power-supply noise from entering the circuit. Neglecting the dashed capacitor, Fig. 4-18 shows two such arrangements. When the R-C filter of Fig. 4-18A is used, the voltage drop in the resistor causes a decrease in power-supply voltage. This drop usually limits the amount of filtering possible with this configuration.

The L–C filter of Fig. 4-18B provides more filtering—especially at high frequencies—for the same loss in power-supply voltage. The L–C filter, however, has a resonant frequency,

$$f_r = \frac{1}{2\pi\sqrt{LC}}, \tag{4-26}$$

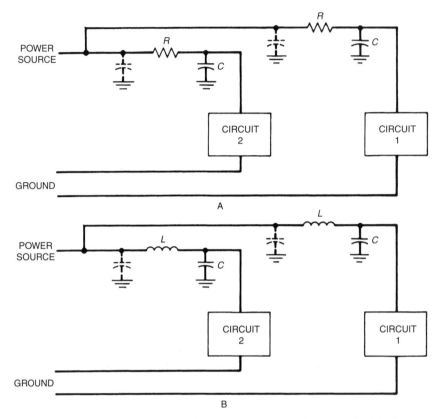

**FIGURE 4-18.** Circuit decoupling with (*A*) resistance-capacitance and (*B*) inductance-capacitance networks.

at which the signal transmitted through the filter may be greater than if no filter was used. Care must be exercised to ensure that this resonant frequency is well below the passband of the circuit connected to the filter. The amount of gain in an *L–C* filter at resonance is inversely proportional to the damping factor

$$\zeta = \frac{R}{2}\sqrt{\frac{C}{L}},$$  (4-27)

where *R* is the resistance of the inductor. The response of an *L–C* filter near resonance is shown in Fig. 4-19. To limit the gain at resonance to less than 2 dB, the damping factor must be greater than 0.5. Additional resistance can be added in series with the inductor, if required, to increase the damping. The inductor used must also be able to pass the direct current required by the circuit without saturating. A second capacitor, such as those shown dashed in

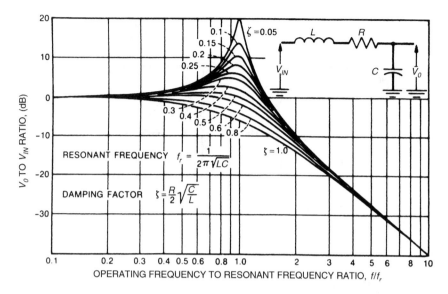

**FIGURE 4-19.** Effect of damping factor on filter response.

Fig. 4-18, can be added to each section to increase filtering to noise being fed back to the power supply from the circuit. This turns the filter into a pi-network.

When considering noise, a dissipative filter such as the $R$–$C$ circuit shown in Fig. 4-18A is preferred to a reactive filter, such as the $L$–$C$ circuit of Fig. 4-18B. In a dissipative filter, the undesirable noise voltage is converted to heat and eliminated as a noise source. In a reactive filter, however, the noise voltage is just moved around. Instead of appearing across the load, the noise voltage now appears across the inductor, where it may be radiated and become a problem in some other part of the circuit. It might then be necessary to shield the inductor to eliminate the radiation.

## 4.3.2  Amplifier Decoupling

Even if only a single amplifier is connected to a power supply, consideration of the impedance of the power supply is usually required. Figure 4.20 shows a schematic of a typical two-stage transistor amplifier. When this circuit is analyzed, it is assumed that the ac impedance between the power supply lead and ground is zero. This is hard to guarantee (because the power supply and its wiring has inductance and resistance) unless a decoupling capacitor is placed between the power supply and ground at the amplifier. This capacitor should serve as a short circuit across the frequency range over which the amplifier is capable of producing gain. This frequency range may be much wider than that of the signal being amplified. If this short circuit is not provided across the power-supply terminals of the amplifier, the circuit can produce an ac voltage gain to the

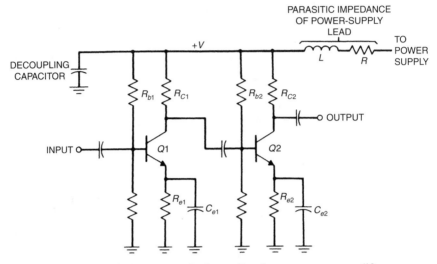

**FIGURE 4-20.** Power supply decoupling for a two-stage amplifier.

power-supply lead. This signal voltage on the power-supply lead can then be fed back to the amplifier input through resistor $R_{b1}$ and possibly cause oscillation.

## 4.4   DRIVING CAPACITIVE LOADS

An emitter follower, which feeds a capacitive load such as a transmission line, is especially susceptible to high-frequency oscillation caused by inadequate power-supply decoupling.* Figure 4-21 shows such a circuit. The collector impedance $Z_c$, consisting of the parasitic inductance of the power supply leads increases with frequency, and the emitter impedance $Z_e$ decreases with frequency because of the cable capacitance. At high frequency, the transistor therefore has a large voltage gain to its collector (point $A$ in Fig. 4-21),

$$\text{Voltage gain} \approx \frac{Z_c}{Z_e}. \tag{4-28}$$

This provides an ac feedback path around the transistor and through the bias resistor $R_b$, thus creating the possibility of oscillation. If previous stages of the same amplifier are connected to the same power supply, the feedback can propagate back through the preceding stages and the possibility of oscillation is greater. The oscillation is often a function of the presence or absence of the

---

* Even with zero-impedance power supply, an emitter follower with a capacitive load can oscillate if improperly designed. See Joyce and Clarke (1961, pp. 264-269) and article by Chessman and Sokol (1976).

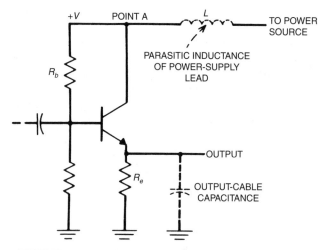

**FIGURE 4-21.** Emitter follower driving a capacitive load.

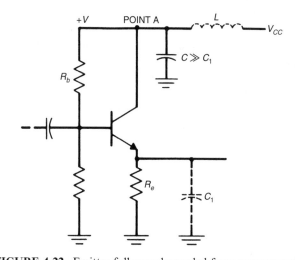

**FIGURE 4-22.** Emitter follower decoupled from power supply.

output cable, because the cable affects the emitter capacitance and hence the high-frequency gain and phase shift through the transistor.

To eliminate the effect of the parasitic lead inductance, a good high-frequency ground must be placed at the power terminal of the amplifier (point $A$).

This can be accomplished by connecting a capacitor between point $A$ and the amplifier's ground, as shown in Fig. 4-22. The value of this capacitor should be greater than the maximum value of the emitter capacitance $C_1$. This guarantees that the high-frequency gain to the collector of the transistor is always less than one.

Placing a capacitor across the amplifier's power-supply terminals will not guarantee zero ac impedance between power and ground. Therefore, some signal will still be fed back to the input circuit. In amplifiers with gains less than 60 dB, this feedback is usually not enough to cause oscillation. In higher gain amplifies, however, this feedback will often cause oscillation. The feedback can be decreased even more by adding an R–C filter in the power supply to the first stage, as shown in Fig. 4-23. The dc voltage drop across the filter resistor is usually not detrimental because the first stage operates at a low signal level and therefore does not require as much dc supply voltage.

A similar oscillation problem may occur when an operational amplifier (differential or single ended) is driving a heavy capacitive load as shown in Fig. 4-24A. This problem often occurs when the op-amp drives a long shielded cable, in which case the load capacitance is the capacitance of the shielded cable. This capacitance can range from a few nanofarads to a few microfarads. If the amplifier had zero output impedance, the problem would not exist. The amplifier output resistance $R_0$ and the load capacitance $C_L$ form a low-pass filter that adds phase shift to the output signal. As frequency increases, the phase shift of this filter increases. The pole, or break frequency, of this filter will be at a frequency of $f = 1/(2 \pi R_0 C_L)$. At this frequency, the phase shift will be equal to 45°. If the phase shift that results from the internal compensation capacitor plus the output filter reaches 180°, the negative feedback through resistor $R_2$ becomes positive feedback. If this occurs at a frequency where the amplifier still has a gain greater than unity, then the circuit will oscillate. The higher the break frequency of this output filter, the more stable the amplifier will be. For more discussion see Graeme (1971, pp. 2191–222).

This problem has many different solutions. One is to use an amplifier with a very low output impedance. Another possibility is to add an additional capacitor $C_2$ and resistor $R_3$ to the circuit as shown in Fig. 4-24B (Franco, 1989). The resistor $R_3$ (usually set equal to $R_0$) isolates the load capacitor from the amplifier, and the small feedback capacitor $C_2$ (10 to 100 pF) introduces a phase lead (a zero) to compensate for the phase lag (a pole) of capacitor $C_L$, which reduces the net phase shift and restores stability to the circuit.

## 4.5   SYSTEM BANDWIDTH

One simple but often overlooked method of minimizing noise in a system is to limit the system bandwidth to only that required by the intended signal. Use of a circuit bandwidth greater than that required by the signal just allows additional noise to enter the system. System bandwidth can be thought of as an open window, the wider the window is open, the more leaves and other debris (noise) can blow in.

The same principal also applies in the case of digital logic circuits. High-speed logic (fast rise time) is much more likely to generate and be susceptible to high-frequency noise than its lower speed counterpart (see Chapter 12).

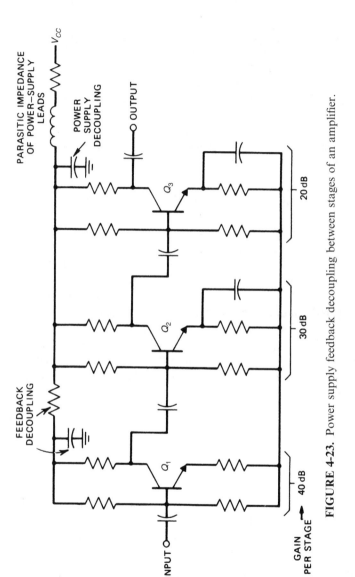

**FIGURE 4-23.** Power supply feedback decoupling between stages of an amplifier.

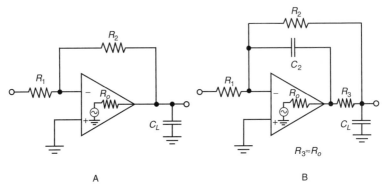

**FIGURE 4-24.** (*A*) Operational amplifier driving a capacitive load. (*B*). $C_2 R_3$ compensation network used to stabilize the amplifier of *A*.

## 4.6   MODULATION AND CODING

The susceptibility of a system to interference is a function not only of the shielding, grounding, cabling, and so on, but also of the coding or modulating scheme used for the signal. Modulation systems such as amplitude, frequency, and phase each have their own inherent immunity or lack thereof. For example, amplitude modulation is very sensitive to amplitude disturbances, whereas frequency modulation is very insensitive to amplitude disturbances. Digital modulation techniques such as pulse amplitude, pulse width, and pulse repetition rate coding may also be used to increase noise immunity. The noise advantages of various coding and modulation schemes are adequately covered in the literature (Panter, 1965; Schwartz, 1970; and Schwartz et al. 1966) and are not repeated here.

## SUMMARY

- In a balanced system, both resistive and reactive balance must be maintained.
- In a balanced system, the greater the degree of balance, or CMRR, the less noise that will couple into the system.
- Balancing can be used with shielding, to provide additional noise reduction.
- When the source impedance is low and the load impedance is high (or vice versa), no single element filter will be effective, and a multielement filter must be used.
- As the result of parasitics, all low pass filters become high pass filters above some frequency.

- The lower the characteristic impedance of a dc power distribution circuit, the less the noise coupling over it.
- Because most dc power distribution systems do not provide a low impedance, decoupling capacitors should be used at each load.
- From a noise point of view, a dissipative filter is preferred to a reactive filter.
- Some amplifier circuits will oscillate when driving a capacitive load, unless properly compensated and/or decoupled.
- To minimize noise, the bandwidth of a system should be no more than that necessary to transmit the desired signal.

## PROBLEMS

4.1   Derive Eq. 4-7.

4.2   If the balanced circuit of Fig. 4-4 has a CMRR of 60 dB, and a 300 mV common-mode ground voltage, what will be the noise voltage across the balanced load?

4.3   The circuit shown in Fig 4-5 has a source unbalance of 5 $\Omega$. Assume $R_L \gg R_s + \Delta R_s$.

    a.  What will be the CMRR if the load resistance is 5 k$\Omega$?
    b.  What will be the CMRR if the load resistance is 150 k$\Omega$?
    c.  What will be the CMRR if the load resistance is 1 M$\Omega$?

4.4   In a balanced circuit, if the tolerance of the load resistors is halved, by how much will the worst-case CMRR increase?

4.5   To maximize the CMRR of a balanced circuit, what should be the ratio of load resistance to the source resistance?

4.6   For the differential amplifier shown in Fig. 4-8, $R_1$ and $R_2$ are 1% resistors with values of 4.7 k$\Omega$ and 270 k$\Omega$, respectively.

    a.  What is the differential-mode input impedance?
    b.  What is the differential-mode gain?
    c.  What is the common-mode input impedance?
    d.  What is the common-mode gain?
    e.  What is the CMRR?

4.7   For the instrumentation amplifier shown in Fig. 4-10, $R_f = 1$ k$\Omega$, $kRf = 100$ $\Omega$, and $R_1 = 10$ k$\Omega$, all 1 % resistors.

a. What is the differential mode gain?
b. What is the CMRR?

4.8   A differential amplifier, similar to that shown in Fig. 4-8, and an instrumentation amplifier, similar to that shown in Fig. 4-10, are both designed to have a differential-mode gain of 50; both amplifiers using resistors having the same tolerance. Which amplifier will have the larger CMRR, and by how many dB?

4.9   Under what conditions will a single element filter not be effective?

4.10 a. To be effective, the impedance of a series filter element must be what?
   b. To be effective, the impedance of a shunt filter element must be what?

4.11 Name three reasons that common-mode filters are harder to design than differential-mode filters?

4.12 The shunt capacitor of a common-mode filter must be connected to where?

4.13 How can you minimize the parasitics in your common-mode filters?

4.14 What parameter can be used as a figure of merit for a dc power distribution system?

4.15 The power bus arrangement shown in Fig. P4.15 is used to transmit 5 V dc to a 10-A load. The bus bar is 5 m long.

a. What is the dc voltage drop in the distribution system?
b. What is the characteristic impedance of the power bus?
c. If the load current suddenly increases by 0.5-A, what is the magnitude of the transient voltage on the power bus?

## REFERENCES

Chessman, M. and Sokol, N. "Prevent Emitter Follower Oscillation." *Electronic Design*, June 21, 1976.

Franco, S. "Simple Techniques Provide Compensation for Capacitive Loads." *EDN*, June 8, 1989.

Frederiksen, T. M. *Intuitive Operational Amplifiers*. New York, McGraw Hill, 1988.

Graeme, J. G., Tobey, G. E., and Huelsman, L. P. *Operational Amplifiers*. New York, McGraw Hill, 1971.

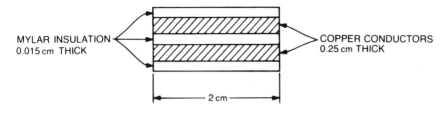

**FIGURE** P4-15.

Joyce M. V. and Clarke, K. K. *Transistor Circuit Analysis*, Reading, MA, Addison-Wesley, 1961.

Panter, P. F. *Modulation, Noise, and Spectral Analysis*, New York, McGraw Hill 1965.

Schwartz, M. *Information Transmission, Modulation and Noise*, 2nd ed. New York, McGraw Hill, 1970.

Schwartz, M., Bennett, W. R., and Stein, S. *Communications Systems and Techniques*, New York, McGraw Hill, 1966.

## FURTHER READING

Feucht, D. L. *Why Circuits Oscillate Spuriously*. Part 1: BJT Circuits, Available at http://www.analogzone.com/col_1017.pdf. Accessed April 2009.

Feucht, D. L. Why Circuits Oscillate Spuriously. Part 2: Amplifiers, Available at http://www.analogzone.com/col_1121.pdf. Accessed April 2009.

Nalle, D. "Elimination of Noise in Low Level Circuits." *ISA Journal*, Vol. 12, August 1965.

Siegel, B. L. "Simple Techniques Help You Conquer Op-Amp Instability." *EDN*, March 31, 1988.

# 5 Passive Components

Actual components are not "ideal"; their characteristics deviate from those of the theoretical components (Whalen and Paludi, 1977). Understanding these deviations is important in determining the proper application of these components. This chapter is devoted to those characteristics of passive electronic components that affect their performance, and/or their use in noise reduction circuitry.

## 5.1 CAPACITORS

Capacitors are most frequently categorized by the dielectric material from which they are made. Different types of capacitors have characteristics that makes them suitable for certain applications but not for others. An actual capacitor is not a pure capacitance; it also has both resistance and inductance, as shown in the equivalent circuit in Fig. 5-1. $L$ is the equivalent series inductance (ESL) and is from the leads as well as from the capacitor structure. Resistance $R_2$ is the parallel leakage and a function of the volume resistivity of the dielectric material. $R_1$ is the equivalent series resistance (ESR) of the capacitor and a function of the dissipation factor of the capacitor.

*Operating frequency is one of the most important considerations in choosing a capacitor.* The maximum useful frequency for a capacitor is usually limited by the inductance of the capacitor structure as well as by its leads. At some frequency, the capacitor becomes self-resonant with its own inductance. Below self-resonance, the capacitor looks capacitive and has an impedance that decreases with frequency. Above self-resonance, the capacitor looks inductive and has an impedance that increases with frequency. Figure 5-2 shows how the

**FIGURE 5-1.** Equivalent circuit for a capacitor.

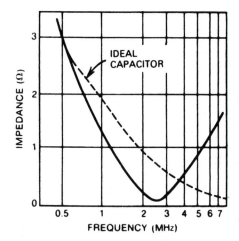

FIGURE 5-2. Effect of frequency on the impedance of a 0.1-μF paper capacitor.

impedance of a 0.1-μF paper capacitor varies with frequency. As can be observed, this capacitor is self-resonant at about 2.5 MHz. Any external leads or PCB traces will lower this resonant frequency.

Surface-mount capacitors, because of their small size and absence of leads, have significantly less inductance than leaded capacitors; therefore, they are more effective high-frequency capacitors. In general, the smaller the capacitor's package or case, the lower the inductance. Typical surface-mount, multilayer ceramic capacitors have inductances in the 1- to 2- nH range. A 0.01-μF surface mount capacitor with 1 nH of series inductance will have a self-resonant frequency of 50.3 MHz. Special package designs, which include multiple interdigitated leads, can decrease the capacitor's equivalent inductance to a few hundred pico-henries.

Figure 5-3 shows the approximate usable frequency range for various capacitor types. The high-frequency limit is caused by self-resonance or by an increase in the dielectric absorption. The low-frequency limit is determined by the largest practical capacitance value available for that type of capacitor.

### 5.1.1 Electrolytic Capacitors

The primary advantage of an electrolytic capacitor is the large capacitance value that can be put in a small package. The capacitance-to-volume ratio is larger for an electrolytic capacitor than for any other type.

An important consideration when using electrolytic capacitors is the fact that that they are polarized and that a direct current (dc) voltage of the proper polarity must be maintained across the capacitor. A nonpolarized capacitor can be made by connecting two equal value and equal voltage rated electrolytics in

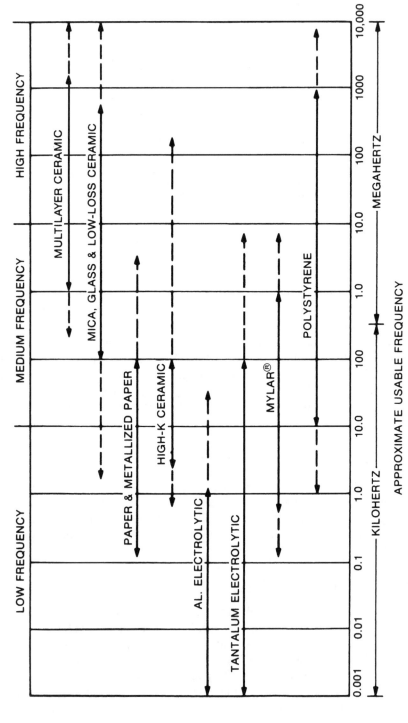

**FIGURE 5-3.** Approximate usable frequency ranges for various types of capacitors.

series, but poled in opposite directions. The resulting capacitance is one half that of each capacitor, and the voltage rating is equal to that of one of the individual capacitors. If unequal voltage-rated capacitors are connected in series, then the voltage rating of the combination will be that of the lowest rated capacitor.

Electrolytic capacitors can be divided into two categories, aluminum and tantalum.

An aluminum electrolytic capacitor may have 1 Ω or more of series resistance. Typical values are a few tenths of an ohm. The series resistance increases with frequency—because of dielectric losses—and with decreasing temperature. At −40°C, the series resistance may be 10 to 100 times the value at 25°C. Because of their large size, aluminum electrolytics also have large inductances. They are therefore low-frequency capacitors and should not normally be used at frequencies above 25 kHz. They are most often used for low-frequency filtering, bypassing, and coupling. For maximum life, aluminum electrolytic capacitors should be operated at between 80% and 90% of their rated voltage. Operating at less than 80% of their rated voltage does not provide any additional reliability.

When aluminum electrolytics are used in alternating current (ac) or pulsating (dc) circuits, the ripple voltage should not exceed the maximum-rated ripple voltage; otherwise, excessive internal heating may occur. Normally, the maximum ripple voltage is specified at 120 Hz, which is typical of operation as a filter capacitor in a full-wave bridge rectifier circuit. Temperature is the primary cause of aging, and electrolytic capacitors should never be operated outside their maximum temperature rating.

Solid tantalum electrolytic capacitors have less series resistance and a higher capacitance-to-volume ratio than aluminum electrolytics, but they are more expensive. Tantalum capacitors may have series resistance values that are an order of magnitude less that that of an equal value aluminum capacitor. Solid tantalum capacitors have lower inductance and can be used at higher frequencies than aluminum electrolytics. They often can be used up to a few megahertz. In general, they are more stable with respect to time, temperature, and shock than aluminum electrolytics. Unlike aluminum electrolytics, the reliability of solid tantalum capacitors is improved by voltage derating; typically they should be operated at 70% or less of their rated voltage. When used in ac or pulsating dc applications, the ripple voltage should not exceed the maximum rated ripple voltage; otherwise, the reliability of the capacitor may be affected as a result of internal heating. Tantalum capacitors are available in both leaded and surface-mount versions.

### 5.1.2 Film Capacitors

Film and paper capacitors have series resistances considerably less than electrolytics but still have moderately large inductances. Their capacitance-to-volume ratio is less than electrolytics, and they are usually available in values up to a few microfarads. They are medium-frequency capacitors useful up to a few megahertz. In most modern-day applications, film capacitors [Mylar* (polyester), polypropylene, polycarbonate, or polystyrene] are used in place of paper capacitors. These capacitors are typically used for filtering, bypassing, coupling, timing, and noise suppression in circuits operating under 1 MHz.

Polystyrene film capacitors have extremely low series resistance; very stable capacitance versus frequency characteristics, and excellent temperature stability. Although medium-frequency capacitors, they are in all other respects the closest to an ideal capacitor of all the types discussed. They are usually used in precision applications, such as filters, where stability with respect to time, and temperature, as well as a precise capacitance value, are required.

Paper and film capacitors are usually rolled into a tubular shape. These capacitors often have a band around one end, as shown in Fig. 5-4. Sometimes the band is replaced by just a dot. The lead connected to the banded or dotted end is connected to the outside foil of the capacitor. Even though the capacitors are not polarized, the banded end should be connected to ground, or to a common reference potential whenever possible. In this way, the outside foil of the capacitor can act as a shield to minimize electric field coupling to or from the capacitor.

### 5.1.3  Mica and Ceramic Capacitors

Mica and ceramic capacitors have low series resistance and inductance. They are therefore high-frequency capacitors and are useful up to about 500 MHz— provided the leads are kept short. Some surface-mount versions of these capacitors are useful up into the gigahertz range. These capacitors are normally used in radio frequency (rf) circuits for filtering, bypassing, coupling, timing and frequency discrimination, as well as decoupling in high-speed digital circuits. With the exception of high-K ceramic capacitors, they are normally very stable with respect to time, temperature, and voltage.

**FIGURE 5-4.** Band on tubular capacitor indicated the lead connected to the outside foil. This lead should be connected to ground.

---

* Mylar is a registered trademark of DuPont; Wilmungton, DE.

Ceramic capacitors have been used in high-frequency circuits for almost 100 years. The original ceramic capacitors were "disc capacitors." However, because of the large advancement in ceramic technology in the last few decades, ceramic capacitors now are available in many different styles, shapes, and formats. They are the "work horses" of high-frequency capacitors.

Mica has a low dielectric constant; therefore, mica capacitors tend to be large relative to their capacitance value. Combining the large advances in ceramic capacitor technology and the low capacitance-to-volume ratio of mica capacitors, ceramic has replaced mica in most low-voltage, high-frequency applications. Because of mica's high dielectric breakdown voltage, often in the kilovolt range, mica capacitors are still used in many high-voltage rf applications, such as radio transmitters.

Multilayer ceramic capacitors (MLCCs) are composed of multiple layers of ceramic material, often barium titanite, separated by interdigitated metal electrodes as shown in Fig. 5-5. Contact to the electrodes is made at the ends of the structure. This construction effectively places many capacitors in parallel. Some MLCCs contain hundreds of ceramic layers, each layer only a few micrometers thick.

This type of construction has the advantage of multiplying up the capacitance of each layer such that the total capacitance is equal to the capacitance of one layer multiplied by the number of layers, while at the same time dividing down the inductance of each layer such that the total inductance is equal to the inductance of one layer divided by the number of layers. Multilayer capacitor construction when combined with surface mount technology can produce almost ideal high-frequency capacitors. Some small-value (e.g., tens of picofarads) surface mount MLCCs can have self-resonant frequencies in the multiple gigahertz range.

Most MLCCs have capacitance values of 1 μF or less with voltage ratings of 50 V or less. The voltage rating is limited by the small spacing of the layers. However, the small spacing combined with the large number of layers has allowed manufacturers to produce larger value MLCC with capacitance values in the 10 to 100 μF range. MLCCs are excellent high-frequency capacitors and

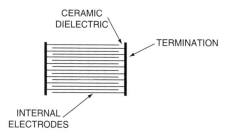

**FIGURE 5-5.** Multilayer ceramic capacitor construction.

are commonly used for high-frequency filtering as well as digital logic decoupling applications.

High-K ceramic capacitors are only medium-frequency capacitors. They are relatively unstable with respect to time, temperature, and frequency. Their primary advantage is a higher capacitance-to-volume ratio, compared with that of standard ceramic capacitors. They are usually used in noncritical applications for bypassing, coupling, and blocking. Another disadvantage is that they can be damaged by voltage transients. It is therefore not recommended that they be used as bypass capacitors directly across a low-impedance power supply.

Table 5-1 shows the typical failure modes for various capacitor types under normal use and when subjected to overvoltage.

### 5.1.4   Feed-Through Capacitors

Table 5-2 shows the effect of lead length on the resonant frequency of small ceramic capacitors. To keep the resonant frequency high, it is preferable to use the smallest value capacitor that will do the job.

If the resonant frequency cannot be kept above the frequency of interest, which is many times the case, then the impedance of the capacitor above

**TABLE 5-1.   Typical Capacitor Failure Modes**

| Capacitor Type | Normal Use | Overvoltage |
|---|---|---|
| Aluminum electrolytic | Open | Short |
| Ceramic | Open | Short |
| Mica | Short | Short |
| Mylar | Short | Short |
| Metalized mylar | Leakage | Noisy |
| Solid tantalum | Short | Short |

**TABLE 5-2.   Self-Resonant Frequencies of Ceramic Capacitors**

| | Self-Resonant Frequency (MHz) | |
|---|---|---|
| Capacitance Value (pf) | 1/4-in Leads | 1/2-in Leads |
| 10,000 | 12 | — |
| 1000 | 35 | 32 |
| 500 | 70 | 65 |
| 100 | 150 | 120 |
| 50 | 220 | 200 |
| 10 | 500 | 350 |

resonance will be determined solely by the inductance. Under this condition, any value of capacitance will have the same high-frequency impedance, and larger capacitance values can be used to improve low-frequency performance. In this case, the only way to lower the capacitor's high-frequency impedance is by decreasing the inductance of the capacitor structure and its leads.

It should be noted that at the series resonant frequency, the impedance of a capacitor is actually lower (Fig. 5-2) than that of an ideal capacitor (one without inductance). Above resonance, however, the inductance will cause the impedance to increase with frequency.

The resonant frequency of a capacitor can be increased by using a feed-through capacitor designed to mount through, or on, a metal chassis. Figure 5-6 shows such a capacitor mounted in a chassis or shield, along with its schematic representation. Feed-through capacitors are three terminal devices. The capacitance is between the leads and the case of the capacitor, not between the two leads. Feed-through capacitors have very low inductance ground connections, because there is no lead present. Any lead inductance that does exist is in series with the signal lead and actually improves the capacitor's effectiveness, because it transforms the feed-through capacitor into a low-pass T-filter. Figure 5-7 shows the equivalent circuits, including lead inductance, both for a standard and a feed-through capacitor. As a result, feed-through capacitors have very good high-frequency performance. Figure 5-8 shows the impedance versus frequency characteristics of both a 0.05 µF feed-through capacitor and a standard 0.05 µF capacitor. The figure clearly shows the improved (lower) high-frequency impedance of the feed-through capacitor.

Feed-through capacitors are often used to feed power (ac or dc), as well as other low-frequency signals, to a circuit while at the same time shunting any high-frequency noise on the power or signal lead to ground. They are extremely effective, but are more expensive than standard capacitors.

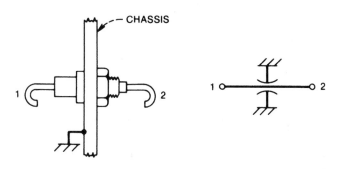

MOUNTED IN CHASSIS          SCHEMATIC REPRESENTATION

**FIGURE 5-6.** Typical feed-through capacitor.

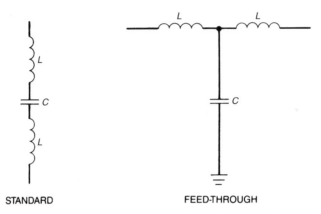

**FIGURE 5-7.** Lead inductance in standard and feed-through capacitors.

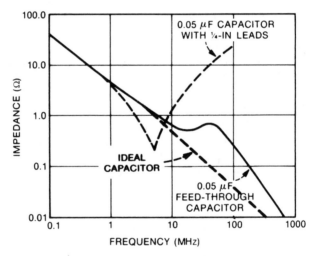

**FIGURE 5-8.** Impedance of 0.05-μF capacitors, showing improved performance of feed-through capacitor.

### 5.1.5 Paralleling Capacitors

No single capacitor will provide satisfactory performance over the entire frequency range from low to high frequencies. To provide filtering over this range of frequencies, two different capacitor types are often used in parallel. For example, an electrolytic could be used to provide the large capacitance necessary for low-frequency filtering, paralleled with a small low-inductance mica or ceramic capacitor to provide a low impedance at high frequencies.

When capacitors are paralleled, however, resonance problems can occur as a result of the parallel and series resonances produced by the capacitors and the inductance of the leads that interconnect them (Danker, 1985). This can result in large impedance peaks at certain frequencies; these are most severe when the paralleled capacitors have widely different values, or when there are long interconnections between them. See Sections 11.4.3 and 11.4.4.

## 5.2  INDUCTORS

Inductors may be categorized by the type of core on which they are wound. The two most general categories are air core (any nonmagnetic material fits into this group) and magnetic core. Magnetic core inductors can be subdivided depending on whether the core is open or closed. An ideal inductor would have only inductance, but an actual inductor also has series resistance, in the wire used to wind it, and distributed capacitance between the windings. This is shown in the equivalent circuit in Fig. 5-9. The capacitance is represented as a lumped shunt capacitor, so parallel resonance will occur at some frequency.

Another important characteristic of inductors is their susceptibility to, and generation of, stray magnetic fields. *Air core or open magnetic core inductors are most likely to cause interference*, because their flux extends a considerable distance from the inductor, as shown in Fig. 5-10A. Inductors wound on a closed magnetic core have much reduced external magnetic fields, because nearly all the magnetic flux remains inside the core, as shown in Fig. 5-10B.

As far as susceptibility to magnetic fields is concerned the magnetic core is more susceptible than the air core inductor. An open magnetic core inductor is the most susceptible, because the core—a low reluctance path—concentrates the external magnetic field and causes more of the flux to flow through the coil. As a matter of fact, open magnetic core inductors (rod cores) are often used as receive antennas for small AM radios. A closed magnetic core is less susceptible than an open core but more susceptible than an air core.

It is often necessary to shield inductors to confine their magnetic and electric fields within a limited space. Shields made of low-resistance material such as copper or aluminum confine the electric fields. At high frequencies, these shields also prevent magnetic flux passage, because of the eddy currents set up within

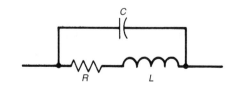

**FIGURE 5-9.** Equivalent circuit for an inductor.

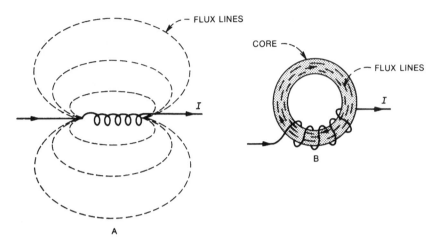

**FIGURE 5-10.** Magnetic fields from (*A*) air core and (*B*) closed magnetic core inductors.

the shield. At low frequencies, however, high-permeability magnetic material must be used to confine the magnetic field.*

For example, high-quality audio frequency transformers are often shielded with mumetal.

## 5.3 TRANSFORMERS

Two or more inductors intentionally coupled together, usually on a magnetic core, form a transformer. Transformers are often used to provide galvanic isolation between circuits. An example is the isolation transformer used to break a ground loop, as shown in Fig. 3-34. In these cases, the only desirable coupling is that which results from the magnetic field. Actual transformers, not being ideal, have capacitance between the primary and secondary windings, as shown in Fig. 5-11, this allows noise coupling from primary to secondary.

This coupling can be eliminated by providing an electrostatic, or Faraday, shield (a grounded conductor placed between the two windings), as shown in Fig. 5-12. If properly designed, this shield does not affect the magnetic coupling, but it eliminates the capacitive coupling provided the shield is grounded. The shield must be grounded at point *B* in Fig. 5-12. If it is grounded to point *A*, the shield is at a potential of $V_G$ and still couples noise through the capacitor $C_2$ to the load. Therefore, the transformer should be located near the load in order to simplify the connection between the

---

* See Chapter 6 for a detailed analysis of magnetic-field shielding

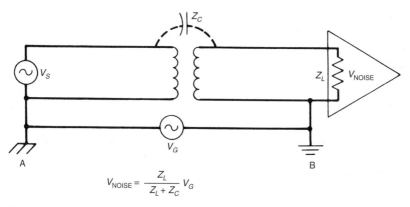

$$V_{NOISE} = \frac{Z_L}{Z_L + Z_C} V_G$$

**FIGURE 5-11.** An actual transformer has capacitive as well as magnetic coupling between primary and secondary windings.

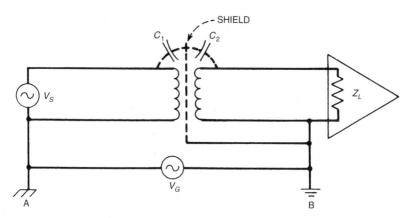

**FIGURE 5-12.** Grounded electrostatic shield between transformer windings breaks capacitive coupling.

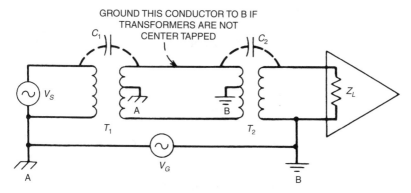

**FIGURE 5-13.** Two unshielded transformers can provide electrostatic shielding.

shield and point *B*. As a general rule, the shield should be connected to a point that is the other end of the noise source.

Electrostatic shielding may also be obtained with two unshielded transformers, as shown in Fig. 5-13. The primary circuit of $T_2$ must be grounded, preferably with a center tap. The secondary of $T_1$, if it has a center tap, may also be grounded to hold one end of $C_2$ near ground potential. As indicated in Fig. 5-13, if the transformers do not have center taps, one of the conductors between the transformers can be grounded. This configuration is less effective than a transformer with a properly designed electrostatic shield. The configuration of Fig. 5-13 is, however, useful in the laboratory to determine whether an electrostatically shielded transformer will effectively decrease the noise coupling in a circuit.

## 5.4 RESISTORS

Fixed resistors can be grouped into the following three basic classes: (1) wirewound, (2) film type, and (3) composition. The exact equivalent circuit for a resistor depends on the type of resistor and the manufacturing processes. The circuit of Fig. 5-14, however, is satisfactory in most cases. In a typical composition resistor, the shunt capacitance is in the order of 0.1–0.5 pF. The inductance is primarily lead inductance, except in the case of wirewound resistors, where the resistor body is the largest contributor. Except for wirewound resistors, or very low value resistors of other, types, the inductance can normally be neglected during circuit analysis. The inductance of a resistor does, however, make it susceptible to pickup from external magnetic fields. Inductance of the external lead can be approximated by using the data in Table 5-4.

The shunt capacitance can be important when high-value resistors are used. For example, consider a 22-M$\Omega$ resistor with 0.5 pF of shunt capacitance. At 145 kHz, the capacitive reactance will be 10% of the resistance. If this resistor is used above this frequency, then the capacitance may affect the circuit performance.

Table 5-3 shows measured impedance, magnitude, and phase angle, for a 1/2-W carbon resistor at various frequencies. The nominal resistance value is 1 M$\Omega$. Note that at 500 kHz the magnitude of the impedance has dropped to 560 k$\Omega$, and the phase angle has become −34°. Capacitive reactance has thus become significant.

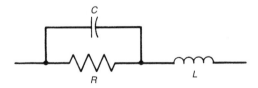

**FIGURE 5-14.** Equivalent circuit for a resistor.

**TABLE 5-3.  Impedance of a 1-MΩ, 1-W Carbon Resistor Measured at Various Frequencies**

| Frequency (kHz) | Impedance | |
| --- | --- | --- |
| | Magnitude (kΩ) | Phase Angle (degrees) |
| 1 | 1000 | 0 |
| 9 | 1000 | −3 |
| 10 | 990 | −3 |
| 50 | 920 | −11 |
| 100 | 860 | −16 |
| 200 | 750 | −23 |
| 300 | 670 | −28 |
| 400 | 610 | −32 |
| 500 | 560 | −34 |

### 5.4.1  Noise in Resistors

All resistors, regardless of their construction, generate a noise voltage. This voltage results from thermal noise and other noise sources, such as shot and contact noise. Thermal noise can never be eliminated, but the other sources can be minimized or eliminated. The total noise voltage therefore is equal to or greater than the thermal noise voltage. This is explained in Chapter 8.

Of the three basic resistor types, wirewound resistors are the quietest. The noise in a good quality wirewound resistor should be no greater than that resulting from thermal noise. At the other extreme is the composition resistor, which has the most noise. In addition to thermal noise, composition resistors also have contact noise, because they are made of many individual particles molded together. When no current flows in composition resistors, the noise approaches that of thermal noise. When current flows, additional noise is generated proportional to the current. Figure 5-15 shows the noise generated by a 10-kΩ carbon composition resistor at two current levels. At low frequencies, the noise is predominantly contact noise, which has an inverse frequency characteristic. The frequency at which the noise levels off, at a value equal to the thermal noise, varies widely between different type resistors and is also dependent on current level.

The noise produced by film-type resistors, is much less than that produced by composition resistors, but it is more than that produced by wirewound resistors. The additional noise is again contact noise, but because the material is more homogeneous, the amount of noise is considerably less than for composition resistors.

Another important factor that affects the noise in a resistor is its power rating. If two resistors of the same value and type both dissipate equal power,

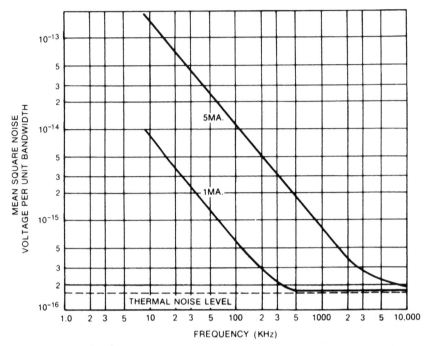

**FIGURE 5-15.** Effect of frequency and current on noise voltage for a 10-kΩ. carbon composition resistor.

the resistor with the higher power rating normally has the lower noise. Campbell and Chipman (1949) present data showing approximately a factor of 3 between the root mean square (rms) noise voltage of a 1/2-W composition resistor versus a 2-W composition resistor operating under the same conditions. This difference is caused by the factor $K$ in Eq. 8-19 (Chapter 8), which is a variable that depends on the geometry of the resistors.

Variable resistors generate all the inherent noises of fixed resistors, but in addition they generate noise from wiper contact. This additional noise is directly proportional to current through the resistor and the value of its resistance. To reduce the noise, the current through the resistor and the resistance itself should both be minimized.

## 5.5  CONDUCTORS

Conductors are not normally considered components; however, they do have characteristics that are very important to the noise and high-frequency performance of electronic circuits. In many cases, they are actually the most important component in the circuit. For conductors whose length is a small fraction of a wavelength, the two most important characteristics are resistance and inductance. Resistance should be obvious, but inductance is often over-looked, and in many cases, it is more important than resistance. Even at relatively low frequencies, a conductor usually has more inductive reactance than resistance.

### 5.5.1   Inductance of Round Conductors

The external loop inductance of a round, straight conductor of diameter $d$, whose center is located a distance $h$ above a ground plane, is

$$L = \frac{\mu}{2\pi} \ln\left(\frac{4h}{d}\right) \text{ H/m.} \tag{5-1}$$

This assumes that $h > 1.5\ d$. The permeability of free space $\mu$ is equal to $4\pi \times 10^{-7}$ H/m. Equation 5-1 therefore can be rewritten as

$$L = 200 \ln\left(\frac{4h}{d}\right) \text{ nH/m.} \tag{5-2a}$$

Changing units to nanohenries per inch gives

$$L = 5.08 \ln\left(\frac{4h}{d}\right) \text{ nH/in.} \tag{5-2b}$$

In Eqs. 5-2a and 5-2b, $h$ and $d$ can be in any units as long as they are both the same, because it is only the ratio of the two numbers that matters.

The preceding equations represent the external inductance of a conductor, because they do not include the effects of the magnetic field within the conductor itself. The total inductance is actually the sum of the internal plus the external inductances. The internal inductance of a straight wire of circular cross section carrying a current distributed uniformly over its cross section (a low-frequency current) is 1.27 nH/in., independent of wire size. The internal inductance is usually negligible compared with the external inductance except for closely spaced conductors. The internal inductance is reduced even more when high-frequency currents are considered because, as the result of the skin effect, the current is concentrated near the surface of the conductor. The external inductance therefore is normally the only inductance of significance.

Table 5-4 lists values of external loop inductance and resistance for various gauge solid conductors. The table shows that moving the conductor closer to

TABLE 5-4.  **Inductance and Resistance of Round Conductors**

| Wire size (AWG) | Diameter (in) | DC Resistance (mΩ/in) | Inductance (nH per in) | | |
|---|---|---|---|---|---|
| | | | 0.25 in Above Ground Plane | 0.5 in Above Ground Plane | 1 in Above Ground Plane |
| 26 | 0.016 | 3.38 | 21 | 25 | 28 |
| 24 | 0.020 | 2.16 | 20 | 23 | 27 |
| 22 | 0.025 | 1.38 | 19 | 22 | 26 |
| 20 | 0.032 | 0.84 | 17 | 21 | 25 |
| 18 | 0.040 | 0.54 | 16 | 20 | 23 |
| 14 | 0.064 | 0.21 | 14 | 17 | 21 |
| 10 | 0.102 | 0.08 | 12 | 15 | 19 |

the ground plane decreases the inductance, assuming the ground plane is the return current path. Raising the conductor higher above the ground plane increases the inductance.

Table 5-4 also shows that the larger the conductor diameter, the lower is the inductance. The inductance and the conductor diameter are logarithmically related as shown in Eq. 5-1. For this reason, low values of inductance are not easily obtained by increasing the conductor diameter.

For two parallel round conductors that carry uniform current in opposite directions, the *loop inductance*, neglecting effect of the magnetic flux in the wires themselves, is

$$L = 10 \ln\left(\frac{2D}{d}\right) \text{ nH/in.} \tag{5-3}$$

In Eq. 5-3, $D$ is the center-to-center spacing, and $d$ is the conductor diameter.

## 5.5.2  Inductance of Rectangular Conductors

The loop inductance of a rectangular conductor, such as a printed circuit board (PCB) trace, can be determined by starting with the well-known relationship that the characteristic impedance $Z_0$ of a transmission line is equal to $\sqrt{L/C}$ (Eq. 5-16). Therefore, the inductance will be

$$L = CZ_0^2. \tag{5-4}$$

IPC-D-317A (1995) gives equations for the characteristic impedance and the capacitance of a narrow rectangular trace located a distance $h$ above a ground plane (microstrip line). Substituting the IPC equations into Eq. 5-4 gives for the loop inductance of a rectangular PCB trace

$$L = 5.071 \ln\left[\frac{5.98h}{0.8w + t}\right] \text{ nH/in,} \qquad (5\text{-}5)$$

where $w$ is the trace width, $t$ is the trace thickness, and $h$ is the height of the trace above the ground plane. Equation 5-5 is only valid for the case where $h > w$. In Eq. 5-5, $h$, $w$, and $t$ can be in any units, because it is only their ratio that matters.

**Example 5-1.** A rectangular conductor with a width of 0.080 in. and a thickness of 0.0025 in. has the same cross-sectional area as a 26 Ga. round conductor. For the case where both conductors are located 0.5 in. above a ground plane, the inductance of the 26 Ga. round conductor (from Eq. 5-2b) is 25 nH/in, whereas the inductance of the rectangular conductor (from Eq. 5-5) is only 19 nH/in. This result demonstrates that a flat rectangular conductor has less inductance than a round conductor with the same cross-sectional area.

### 5.5.3 Resistance of Round Conductors

Resistance is the second important characteristic of a conductor. Selection of conductor size is generally determined by the maximum allowable dc voltage drop in the conductor. The dc voltage drop is a function of conductor resistance and the maximum current. The resistance per unit length of any conductor can be written as

$$R = \frac{\rho}{A}, \qquad (5\text{-}6)$$

where $\rho$ is the resistivity (the reciprocal of the conductivity $\sigma$) of the conductor material and $A$ is the cross-sectional area over which the current flows. For copper $\rho$ equals $1.724 \times 10^{-8}$ $\Omega$-m ($67.87 \times 10^{-8}$ $\Omega$-in). At dc, the current will be distributed uniformly across the cross section of the conductor and the dc resistance of a conductor, of circular cross section will be

$$R_{dc} = \frac{4\rho}{\pi d^2}, \qquad (5\text{-}7)$$

where $d$ is the diameter of the conductor. If the constant substituted for $\rho$ is in ohm-meters, then $d$ must be in meters and $R_{dc}$ will be in $\Omega$/m. If the constant substituted for $\rho$ is in ohm-inches, then $d$ must be in inches and $R_{dc}$ will be in $\Omega$/in. Table 5-4 lists the value of dc resistance for different size solid conductors.

At high frequency, the skin effect causes the resistance of a conductor to increase. The skin effect describes a condition where, because of the magnetic fields produced by current in a conductor, the current crowds toward the outer surface of the conductor. The skin effect is discussed in Section 6.4. As the frequency increases, the current is concentrated in a thinner and thinner annular ring at the surface of the conductor (see Fig. P5-7). This decreases the cross sectional area through which the current flows and increases the resistance. Therefore, *at high frequency all currents are surface currents*, and a hollow cylinder will have the same ac resistance as a solid conductor.

For solid round copper conductors, the ac and dc resistances are related by the following expression (Jordan, 1985).

$$R_{ac} = \left(96d\sqrt{f_{MHz}} + 0.26\right)R_{dc}, \tag{5-8}$$

where $d$ is the conductor diameter in inches and $f_{MHz}$ is the frequency in MHz. Equation 5-8 is accurate within 1% for $d\sqrt{f_{MHz}}$ greater than 0.01 ($d$ in inches), and it should not be used when $d\sqrt{f_{MHz}}$ is less than 0.08. For a 22-gauge wire, $d\sqrt{f_{MHz}}$ greater than 0.01 will occur at frequencies above 0.15 MHz. For $d\sqrt{f_{MHz}}$ less than 0.004, the ac resistance will be within 1% of the dc resistance. If the conductor material is other than copper, the first term of Eq. 5-8 must be multiplied by the factor

$$\sqrt{\frac{\mu_r}{\rho_r}},$$

where $\mu_r$ is the relative permeability of the conductor material and $\rho_r$ is the relativity resistivity of the material compared with copper. Relative permeability and conductivity (the reciprocal of resistivity) of various materials are listed in Table 6-1.

Substituting Eq. 5-7 into Eq. 5-8 and assuming that the frequency is high enough that the 0.26 term can be neglected, we get the following equation for the ac resistance of a round copper conductor

$$R_{ac} = \frac{8.28 \times 10^{-2}\sqrt{f_{MHz}}}{d} \quad m\Omega/in, \tag{5-9a}$$

where $d$ is in inches. For $d\sqrt{f_{MHz}}$ greater than 0.03 ($d$ in inches), Eq. 5-9a will be accurate within 10%. For a 22-gauge wire this will be true above 1.5 MHz. For $d\sqrt{f_{MHz}}$ greater than 0.08, Eq. 5-9 will be accurate within a few percent.

Changing units to milliohms/m produces

$$R_{ac} = \frac{82.8\sqrt{f_{MHz}}}{d} \ m\Omega/m, \qquad (5\text{-}9b)$$

where d is in millimeters.

Equation 5-9 shows that the ac resistance of a conductor is directly proportional to the square root of the frequency.

### 5.5.4  Resistance of Rectangular Conductors

The ac resistance of a conductor can be decreased by changing its shape. A rectangular conductor will have less ac resistance than a round conductor of the same cross-sectional area, because of its greater surface area (perimeter). Remember, high-frequency currents only flow on the surface of conductors. Because a rectangular conductor has less ac resistance and also less inductance than a round conductor with the same cross-sectional area, it is a better high-frequency conductor. Flat straps or braids are therefore commonly used as ground conductors.

For a rectangular conductor of width $w$ and thickness $t$, the dc current will be distributed uniformly over the cross section of the conductor, and the dc resistance, from Eq. 5-6, will be

$$R_{dc} = \frac{\rho}{wt}. \qquad (5\text{-}10)$$

The ac resistance of an isolated rectangular conductor can easily be calculated by realizing that most of the high-frequency current is concentrated in a thickness of approximately one skin depth at the surface of the conductor, as shown in Fig. 5-16. The cross-sectional area through which the current flows

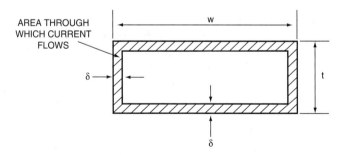

**FIGURE 5-16.** High-frequency current in a rectangular conductor is contained within a thickness of one skin depth of the surface.

will then be equal to $2\,(w+t)\,\delta$, where $w$ and $t$ are the width and thickness of the rectangular conductor, respectively, and $\delta$ is the skin depth of the conductor material. This assumes that $t > 2\,\delta$. Substituting $2\,(w+t)\,\delta$ for the area in Eq. 5-6 gives

$$R_{ac} = \frac{\rho}{2(w+t)\delta}. \tag{5-11}$$

The skin depth for copper (Eq. 6-11a) is

$$\delta_{copper} = \frac{66 \times 10^{-6}}{\sqrt{f_{MHz}}} \text{ m.} \tag{5-12}$$

Substituting Eq. 5-12 into Eq. 5-11 gives for the ac resistance of a rectangular copper conductor

$$R_{ac} = \frac{131\sqrt{f_{MHz}}}{(w+t)} \text{ m}\Omega/\text{m}, \tag{5-13a}$$

where $w$ and $t$ are in millimeters.

Changing units to milliohms/inch produces

$$R_{ac} = \frac{0.131\sqrt{f_{MHz}}}{(w+t)} \text{ m}\Omega/in, \tag{5-13b}$$

where $w$ and $t$ are now in inches.

The ac resistance of a rectangular conductor is proportional to the square root of the frequency and is inversely proportional to the width plus the thickness of the conductor. If $t \ll w$, which is often the case, then the ac resistance is inversely proportional to the width of the conductor.

All of the above ac resistance equations assume an isolated straight conductor. If the conductor is close to another current-carrying conductor, the ac resistance will be greater than predicted by these equations. This additional resistance results from the current crowding to one side of the conductor, as a result of the influence of the current in the other conductor. This current crowding decreases the area of copper through which the current flows, hence increasing the resistance. For a circular cross-section conductor, this effect will be negligible if the conductor is spaced at least 10 times its diameter from any adjacent current-carrying conductors.

## 5.6    TRANSMISSION LINES

When conductors become long, that is, they become a significant fraction of the wavelength of the signals on them, they can no longer be represented as a simple *lumped-parameter* series R–L network as was done in Section 5.5. Because of the phase shift that occurs as the signal travels down the conductor, the voltage and current will be different at different points along the conductor. At some points, the current (or voltage) will be a maximum, at other points the current (or voltage) will be a minimum, or possibly even zero. Therefore, the behavior of an impedance (resistance, inductance, or capacitance) will vary as a result of its location along the conductor. For example, a resistance located where the current is zero will have no voltage drop across it, whereas a resistance located where the current is a maximum will have a large voltage drop across it. Under these circumstances, the signal conductor and its return path must be considered together as a transmission line, and a *distributed-parameter* model of the line must be used.

A common rule, when working in the frequency domain, is that the conductor should be treated as a transmission line if its length is greater than 1/10th of a wavelength, or in the case of a digital signal, in the time domain, when the signal's rise time is less than twice the propagation delay (the reciprocal of the velocity of propagation) of the line.

What then is a transmission line? A transmission line is a series of conductors, often but not necessarily two, used to guide electromagnetic energy from one place to another. The important concept to understand is that we are moving an electromagnetic field, or energy, from one point to another, not a voltage or current. The voltage and current exist, but only as a consequence of the presence of the field. We can classify transmission lines by their geometry and the number of conductors that they have. Some of the more common types of transmission lines are as follows:

- Coaxial cable (2)
- Microstrip line (2)
- Stripline (3)
- Balanced line (2)
- Waveguide (1)

The numbers in parentheses represent the number of conductors in the transmission line. The geometry of all five cases listed above are depicted in Fig. 5-17.

Probably the most common transmission line is a coaxial cable (coax). In a coax, the electromagnetic energy is propagated through the dielectric between the center conductor and the inside surface of the outer conductor (shield).

On a printed circuit board, transmission lines usually consist of a flat, rectangular conductor adjacent to one or more planes (e.g., microstrip or stripline).

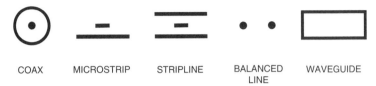

**FIGURE 5-17.** Some common transmission line geometries.

In the case of a stripline, the electromagnetic energy is propagated through the dielectric between the conductors. For the case of a microstrip, where the signal conductor is on a surface layer of the PCB, the field is propagated partially in air and partially in the dielectric of the PCB.

A balanced line consists of two conductors of the same size and shape, with equal impedances to ground and all other conductors (e.g., two parallel round conductors). In this case, the electromagnetic energy is propagated through the dielectric, often air, surrounding the conductors.

A waveguide consists of a single hollow conductor used to guide the electromagnetic energy. In a waveguide, energy is propagated through the hollow center of the conductor. In almost all cases, the propagation medium is air. Waveguides are mostly used in the gigahertz frequency range. A waveguide has an important characteristic different from all the other transmission lines described above, in that it cannot pass dc signals.

Note that the conductors of a transmission line are just the guides for the electromagnetic energy. The electromagnetic energy is propagated in the dielectric material. In a transmission line, the velocity of propagation $v$ of the electromagnetic energy is equal to

$$v = \frac{c}{\sqrt{\varepsilon_r}}, \tag{5-14}$$

where $c$ is the speed of light in a vacuum (free space) and $\varepsilon_r$ is the relative dielectric constant of the medium through which the wave is being propagated. The larger the dielectric constant, the slower the velocity of propagation will be. Table 4-3 listed the relative dielectric constants for various materials. The speed of light $c$ is approximately equal to $300 \times 10^6$ m/s (12 in/ns).* For most transmission lines, the velocity of propagation varies from approximately 1/3 of the speed of light to the speed of light, depending on the dielectric material. For many dielectrics used in transmission lines, the velocity of propagation is about one half the speed of light in a vacuum therefore the speed at which a signal propagates down a transmission line will be about 6 inches per nanosecond. This rate of propagation is a useful number to remember.

It is important to note that what travels at, or close to, the speed of light on a transmission line is the electromagnetic energy, which is in the dielectric

---

* The speed of light in a vacuum is actually 299,792,485 m/s (186,282.397 mi/s).

material not the electrons in the conductors. The speed of the electrons in the conductors is approximately 0.01 m/s (0.4 in./s) (Bogatin, 2004, p. 211) which is 30 billion times slower than the speed of light in free space. *In a transmission line, the most important material is therefore the dielectric through which the electromagnetic energy (field) is propagated, not the conductors* that are just the guides for the energy.

Instead of the simple series R–L network used to model a short conductor, a transmission line must be represented by a large number, ideally an infinite number, of R–L–C–G elements, as shown in Fig. 5-18. Remember, the elements cannot all be lumped together, because their actual location on the transmission line matters. The more sections used, the more accurate the model will be. In Fig. 5-18, R represents the resistance of the conductors in ohms per unit length. L represents the inductance of the conductors in henries per unit length. C represents the capacitance between the conductors in farads per unit length, and G represents the conductance (the reciprocal of resistance) of the dielectric material separating the two conductors in siemens per unit length.

Most transmission line analysis assumes that the propagation is solely by the transverse electromagnetic (TEM) mode. In the TEM mode, the electric and magnetic fields are perpendicular to each other, and the direction of propagation is transverse (perpendicular) to the plane that contains the electric and magnetic fields. To support the TEM mode of propagation, a transmission line must consist of two or more conductors. Therefore, a waveguide cannot support the TEM mode of propagation. Waveguides transmit energy in either the transverse electric ($TE_{m,n}$) or the transverse magnetic ($TM_{m,n}$) modes. Subscripts $m$ and $n$ represent the number of half wavelengths in the x and y directions, respectively, of the cross section of a rectangular waveguide.

The three most important properties of a transmission line are its characteristic impedance, its propagation constant, and its high-frequency *loss*.

### 5.6.1   Characteristic Impedance

When a signal is injected into a transmission line, the electromagnetic wave propagates down the line at the velocity of propagation $v$ of the dielectric material, using the conductors as guides. The electromagnetic wave will induce

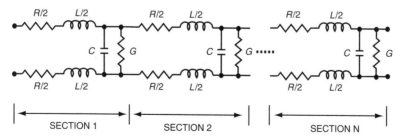

**FIGURE 5-18.** Distributed parameter model of a two-conductor transmission line.

current into the transmission line's conductors. This current flows down the signal conductor, through the capacitance between the conductors, and back to the source through the return conductor as shown in Fig. 5-19. The current flow through the capacitance between the transmission line conductors exists only at the rising edge of the propagating wave, because that is the only place on the line where the voltage is changing, and the current through a capacitor is equal to $I = C \, (dv/dt)$.

Because of the finite velocity of propagation, the injected signal does not initially know what termination is at the end of the line, or indeed where the end of the line is. Therefore, the voltage and current are related by the characteristic impedance of the line. Figure 5-19 clearly demonstrates the important principle, that it is possible to propagate both a voltage and a current down an open circuited transmission line.

In terms of the transmission line parameters shown in Fig. 5-18, the characteristic impedance $Z_0$ of a transmission line is equal to,

$$Z_0 = \sqrt{\frac{R + j\omega L}{G + j\omega C}}. \tag{5-15}$$

The analysis of a transmission line can be greatly simplified if the line is assumed to be lossless. Many practical transmission lines are low loss, and the equations for a lossless line are adequate to describe their performance. For the case of a lossless line, both $R$ and $G$ will be equal to zero. The model of a

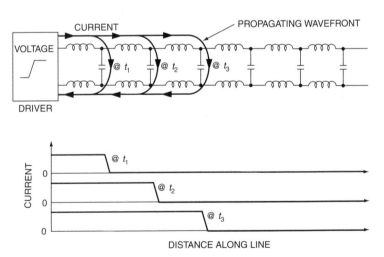

**FIGURE 5-19.** Signal and return currents both flow on the conductors of a transmission line as the rising edge of the signal propagates down the line. Note, $t_3 > t_2 > t_1$.

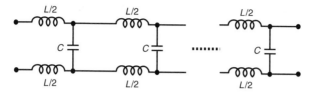

**FIGURE 5-20.** Distributed parameter model of a loseless transmission line.

lossless transmission line is shown in Fig. 5-20. Substituting $R = 0$ and $G = 0$ into Eq. 5-15 produces the well-known, and often quoted, equation for the characteristic impedance of a lossless transmission line,

$$Z_0 = \sqrt{\frac{L}{C}}. \tag{5-16}$$

It is important to note if any two parameters in Eq. 5-16 are known, then the third parameter can be calculated. Often, the properties quoted for a transmission line are just the characteristic impedance and the capacitance per unit length. This however, provides sufficient information to then calculate the inductance per unit length.

Except for three cases, all closed form expressions for the characteristic impedance of a transmission line, expressed in terms of the geometry of the line, are only approximate. The three exceptions are a coax, two identical parallel round conductors, and a round conductor over a plane. Variations of 10% or more are not uncommon between published formulas for characteristic impedance of transmission lines. Many published equations are only accurate over a limited range of characteristic impedance. The three exact equations are as follows:

The characteristic impedance of a coaxial line is given by

$$Z_0 = \frac{60}{\sqrt{\varepsilon_r}} \ln\left[\frac{r_2}{r_1}\right], \tag{5-17}$$

where $r_1$ is the radius of the inner conductor, $r_2$ is the radius of the outer conductor, and $\varepsilon_r$ is the relative dielectric constant of the material between the conductors.

The characteristic impedance for two identical parallel round conductors is given by

$$Z_0 = \frac{120}{\sqrt{\varepsilon_r}} \ln\left[\left(\frac{D}{2r}\right) + \sqrt{\left(\frac{D}{2r}\right)^2 - 1}\right], \tag{5-18a}$$

where $r$ is the radius of each conductor, $D$ is the distance or spacing between the conductors and, $\varepsilon_r$ is the relative dielectric constant of the material surrounding the conductors. This equation, however, is often approximated as

$$Z_0 = \frac{120}{\sqrt{\varepsilon_r}} \ln\left[\frac{D}{r}\right],$$    (5-18b)

for the case of $D \gg 2r$.

As a result of the symmetry of the problem, the characteristic impedance of a round conductor located a distance $h$ above a plane will be exactly one half that of two round conductors located a distance $2\,h$ apart. Therefore, the characteristic impedance of a round conductor over a plane is given by

$$Z_0 = \frac{60}{\sqrt{\varepsilon_r}} \ln\left[\left(\frac{h}{r}\right) + \sqrt{\left(\frac{h}{r}\right)^2 - 1}\right],$$    (5-19a)

where $r$ is the radius of the conductor and $h$ is the height of the conductor above the plane. For the case where $h \gg r$, Eq. 5-19a can be approximated as

$$Z_0 = \frac{60}{\sqrt{\varepsilon_r}} \ln\left[\frac{2h}{r}\right].$$    (5-19b)

The characteristic impedance of most practical transmission lines range from about 25 to 500$\Omega$, with the 50 to 150 $\Omega$ range being the most common.

### 5.6.2  Propagation Constant

The propagation constant describes the attenuation and phase shift of the signal as it propagates down the transmission line. In terms of the parameters shown in Fig. 5-18, the propagation constant $\gamma$ of a transmission line is equal to,

$$\gamma = \sqrt{(R + j\omega L)(G + j\omega C)}.$$    (5-20)

In the general case, the propagation constant will be a complex number with a real and imaginary part. If we define the real part as $\alpha$ and the imaginary part as $\beta$, then we can write the propagation constant as

$$\gamma = \alpha + j\beta.$$    (5-21)

The real part $\alpha$ is the attenuation constant and the imaginary part $\beta$ is the phase constant. For the case of a lossless line, the real and the imaginary parts of Eq. 5-20 are equal to

$$\alpha = 0, \tag{5-22a}$$

$$\beta = \omega\sqrt{LC}. \tag{5-22b}$$

From Eq. 5-22a, we observe that the attenuation of a lossless line is zero, as must be the case. Equation 5-22b represents the phase shift of the signal as it propagated down the line, in radians per unit length.

**Example 5-2.** A transmission line has a capacitance of 12 pF/ft and an inductance of 67.5 nH/ft. From Eq. 5-22b, a 100-MHz signal propagating down the line will have a phase shift of 0.565 radians/ft or 32.4°/ft. From Eq. 5-16, the characteristic impedance of the line will be 75 $\Omega$.

### 5.6.3   High-Frequency Loss

Although the lossless line model discussed above is a good representation of many actual transmission lines over a wide frequency range, in many cases from one to hundreds of megahertz, it does not account for the attenuation of the signal as it propagates down the line. To account for signal attenuation, we must factor in the loss of the line.

The two primary types of transmission line loss are (1) ohmic loss resulting from the resistance of the conductors and (2) dielectric loss resulting from the dielectric material absorbing energy from the propagating electric field, and heating the material. The first type affects the $R$ term in Eq. 5-20, and the second affects the $G$ term in Eq. 5-20.

The general equations for the loss of a transmission line are complex. To simplify the mathematics, a low-loss approximation is usually used. This approximation assumes that although $R$ and $G$ are not zero they are small, such that $R \ll \omega L$ and $G \ll \omega C$. This assumption is reasonable for most actual transmission lines at high frequency.

The derivation of the attenuation constant for a lossy line is beyond the scope of this book. However, if the loss is assumed to be small, the attenuation constant (the real part of Eq. 5-20) can be approximated by (Bogatin, 2004, p. 374)

$$\alpha = 4.34\left[\frac{R}{Z_0} + GZ_0\right] \text{ dB/unit length}, \tag{5-23}$$

where both the $R$ and $G$ are frequency dependent and increase with frequency. Equation 5-23 represents the loss per unit length of the line. The first term of

Eq. 5-23 is the attenuation caused by the ohmic loss of the conductors, and the second term is the attenuation caused by the loss in the dielectric material of the line.

**5.6.3.1 Ohmic Loss.**   Ohmic loss is the only transmission line parameter that is a function of the characteristics of the conductors used in the line. All the other parameters are only a function of the dielectric material and/or the line geometry. The attenuation that results from just the ohmic loss of the conductors is

$$\alpha_{ohmic} = 4.34\left[\frac{R}{Z_0}\right], \tag{5-24}$$

where $R$ is the ac resistance of the conductors, which was previously derived in Sections 5.5.3 and 5.5.4. If the two conductors that make up a transmission line are of significantly different dimensions, such as the case of a coax or a microstrip or stripline, then most of the resistance, and therefore loss, will be from the smaller conductor, and the resistance of the large conductor is often neglected. In these cases, the smaller conductor is usually the signal conductor, and the larger conductor is the return conductor.

In some cases, the ac resistance of the signal conductor is multiplied by a correction factor, possibly 1.35, to account for the additional resistance of the return conductor. For a microstrip signal line, the resistance will be larger than predicted by Eq. 5-13, because most of the current is just along the bottom of the conductor. In this case, a correction factor of 1.7 might be appropriate.

Converting Eq. 5-9a into ohms per inch and substituting this for $R$ in Eq. 5-24, gives for the attenuation constant of a circular cross section conductor

$$\alpha_{ohmic} = \frac{36000\sqrt{f_{MHz}}}{dZ_0} \quad \text{dB/in}, \tag{5-25}$$

where $d$ is the conductor diameter in inches.

Converting Eq. 5-13b to ohms per inch and substituting it for $R$ in Eq. 5-24 gives for the attenuation constant of a rectangular cross-section conductor

$$\alpha_{ohmic} = \frac{0.569 \times 10^{-3}\sqrt{f_{MHz}}}{(w+t)Z_0} \quad \text{dB/in}, \tag{5-26}$$

where $w$ is the conductor width and $t$ is the conductor thickness, both of which are in inches.

**5.6.3.2 Dielectric Loss.** The attenuation that results just from the dielectric absorption is

$$\alpha_{dielectric} = 4.34[GZ_0]. \qquad (5\text{-}27)$$

The loss in a dielectric material is determined by the dissipation factor of the material. The dissipation factor is defined as the ratio of the energy stored to the energy dissipated in the material per hertz, and it is usually listed as the tangent of the loss angle, $\tan(\delta)$. The larger $\tan(\delta)$ is for a material, the higher the loss. Table 5-5 lists the dissipation factor (loss tangent) for some common dielectric materials.

Using several transmission line identities that relate $G$, $C$, and $Z_0$ as well as substituting for the speed of light, Eq. 5-27 can be rewritten in the form (Bogatin, 2004, p. 378)

$$\alpha_{dielectric} = 2.3 \, f_{GHz} \tan(\delta) \sqrt{\varepsilon_r} \ \text{dB/in}, \qquad (5\text{-}28)$$

where $\tan(\delta)$ and $\varepsilon_r$ are the dissipation factor and the relative dielectric constant of the dielectric material respectively. Note that the dielectric loss is not a function of the geometry of the transmission line, it is only a function of the dielectric material.

As can be observed from Eqs. 5-25 and 5-26, the ohmic loss is proportional to the square root of the frequency, whereas from Eq. 5-28 the dielectric loss is directly proportional to frequency. Therefore, the dielectric loss will predominate at high frequency.

For short transmission lines (e.g., signal traces on a typical PCB), the transmission line losses can normally be neglected, until the frequency approaches 1 GHz or higher.

For the case of a sine wave signal, the loss or attenuation will decrease the amplitude of the transmitted wave. In the case of a square wave signal,

**TABLE 5-5. Dissipation Factor [tan(δ)] of Some Common Dielectric Materials.**

| Material | Tan (δ) |
| --- | --- |
| Vacuum/Free Space | 0 |
| Polyethylene | 0.0002 |
| Teflon[®][a] | 0.0002 |
| Ceramic | 0.0004 |
| Polypropylene | 0.0005 |
| Getek[®][b] | 0.01 |
| FR4 Epoxy Glass | 0.02 |

[a] Registered trademark of DuPont, Wilmington, DE.
[b] Registered trademark of General Electric, Fairfield, CT.

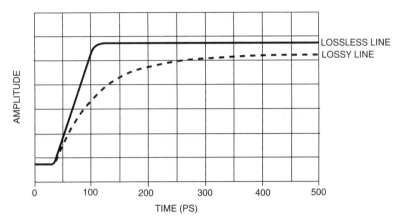

**FIGURE 5-21.** Time domain response of a square wave on a lossy transmission line, showing both amplitude and rise time degradation.

however, the high-frequency components will be attenuated more that the low-frequency components. Therefore, as the square wave propagates down the line its amplitude will decrease and its rise time will increase, as shown in Fig. 5-21. In most cases, the increase in the rise time is more detrimental to the signal integrity of the transmitted signal than is the loss in amplitude. As a rule of thumb, the rise time of a square wave propagating down a PCB transmission line, with FR4 epoxy-glass dielectric, will increase about 10 ps/in of travel (Bogatin, 2004, p. 389).

### 5.6.4 Relationship Among $C$, $L$ and $\varepsilon_r$.

Because the velocity of propagation is a function of the dielectric material, and the capacitance and inductance of the line are also related to the dielectric material, as well as the geometry of the line, the capacitance, inductance, and velocity are all interrelated. The velocity of propagation can be written as

$$v = \frac{c}{\sqrt{\varepsilon_r}} = \frac{1}{\sqrt{LC}}. \tag{5-29}$$

From the relationship between the characteristic impedance (Eq. 5-16) and the velocity of propagation (Eq. 5-29), the following equations for $L$ and $C$ of the transmission line, in terms of the characteristic impedance, can be derived:

$$L = \frac{\sqrt{\varepsilon_r}}{c} Z_0, \tag{5-30}$$

and

$$C = \frac{\sqrt{\varepsilon_r}}{c} \frac{1}{Z_0},$$    (5-31)

where $c$ is the speed of light in free space ($300 \times 10^6$ m/s).

Equations 5-30 and 5-31 relate the characteristic impedance, capacitance, inductance, and dielectric constant of the transmission line, and they can be very useful. If you know any two of the four parameters, then you can find the other two using Eqs. 5-30 and 5-31.

From Eq. 5-30, we determine that the inductance $L$ is only a function of the dielectric constant and the characteristic impedance of the line. Equation 5-31 shows a similar relationship for the capacitance $C$. Therefore, all transmission lines that have the same characteristic impedance and dielectric material, will have the same inductance and capacitance per unit length regardless of the size, geometry, or construction of the line. For example, all 70 Ω transmission lines with a dielectric constant of four will have a capacitance of 95 pF/m (2.4 pF/in), and an inductance of 467 nH/m (11.8 nH/in).

### 5.6.5  Final Thoughts

The question is often asked, when is a signal interconnection a transmission line and when is it not? The answer is simple: A signal path is always a transmission line. However, if the interconnection is short enough, that fact can be ignored and the answer obtained will be close enough to reality to predict the performance. Applying the criteria listed in the second paragraph of Section 5.6 to the case of a 1-ns rise time square wave, a signal interconnection of 3 in or more is a long line and should be analyzed as a transmission line.

On a transmission line, reflections will occur whenever the signal encounters a change in impedance, whether at the end of the line, or caused by a change in the geometry of the line. Vias and right angle bends all act as impedance discontinuities. The topic of transmission line reflections is not the subject of this book. The subject is covered adequately in any good transmission line text. An excellent, although dated, reference on classical transmission line theory is *Electric Transmission Lines* by Skilling (1951). Two excellent references on the subject of signal integrity and the applicability of transmission line theory to digital circuits are *High-Speed Digital Design* by Johnson and Graham (1993) and *High-Speed Digital System Design* by Hall et al. (2000).

## 5.7  FERRITES

Ferrite is a generic term for a class of nonconductive ceramics that consists of oxides of iron, cobalt, nickel, zinc, magnesium, and some rare earth metals.

The variety of ferrites available is large because each manufacturer has developed their own oxide composition. No two manufacturers use precisely the same combination; therefore, multiple sourcing of ferrites is difficult. Ferrites have one major advantage over ferromagnetic materials, which is high electrical resistivity that results in low eddy-current losses up into the gigahertz frequency range. In ferromagnetic materials, eddy-current losses increase with the square of the frequency. Because of this, in many high-frequency applications, ferrites are the materials of choice.

The material used in a ferrite determines the frequency range of applicability. Ferrites are available in many different configurations (see Fig. 5-22), such as beads, beads on leads, surface-mount beads (not shown in figure), round cable cores, flat cable cores, snap on cores, multiaperture cores, toroids, and so on.

Ferrites provide an inexpensive way of coupling high-frequency resistance into a circuit without introducing power loss at dc or affecting any low-frequency signals present. Basically, ferrites can be thought of as a high-frequency ac resistors with little or no resistance at low frequency or dc. Ferrite beads are small and can be installed simply by slipping them over a component lead or conductor. Surface-mount versions are also readily available. Ferrites are most effective in providing attenuation of unwanted signals above 10 MHz, although in some applications they can be effective as low 1 MHz. When properly used,

**FIGURE 5-22.** Some of the various ferrite configurations available.

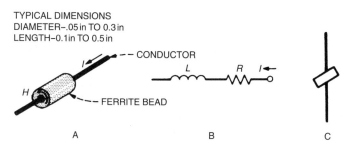

**FIGURE 5-23.** (*A*) Ferrite bead on conductor, (*B*) high-frequency equivalent circuit, and (*C*) typical schematic symbol.

ferrites can provide the suppression of high-frequency oscillations, common-and differential-mode filtering, and the reduction of conducted and radiated emissions from cables.

Figure 5-23A shows a small cylindrical ferrite bead installed on a conductor, and Fig. 5-23B shows the high-frequency equivalent circuit—an inductor in series with a resistor. The values of both the resistor and the inductor are dependent on frequency. The resistance is from the high-frequency hysteresis loss in the ferrite material. Figure 5-23C shows one schematic symbol often used for ferrite beads.

Most ferrite manufacturers characterize their components by specifying the magnitude of the impedance versus frequency. The magnitude of the impedance is given by

$$|Z| = \sqrt{R^2 + (2\pi f L)^2}, \tag{5-32}$$

where $R$ is the equivalent resistance of the bead and $L$ is the equivalent inductance—both values vary with frequency. Some manufacturers, however, only specify the impedance at one frequency, usually 100 MHz, or at a few frequencies.

Figure 5-24 shows the impedance data for a typical ferrite core (Fair-Rite, 2005, p. 147). When used in noise suppression, ferrites are usually used in the frequency range where their impedance is primarily resistive. The recommended frequency range for various ferrite materials when used in noise suppression applications is shown in Fig. 5-25 (Fair-Rite, 2005, p. 155). As can be observed, ferrites are available for use over the frequency range of 1 MHz to 2 GHz.

By using multiple turns, the ferrite impedance can be increased proportional to the number of turns squared. However, this also increases the interwinding capacitance and degrades the high-frequency impedance of the ferrite. If an improvement in the impedance of the ferrite is needed near its lower frequency range of applicability, the possibility of using multiple turns, however, should

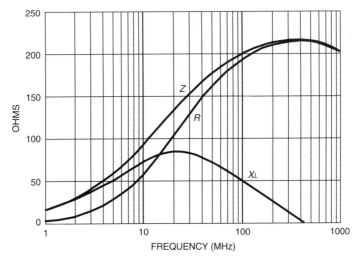

**FIGURE 5-24.** Impedance, resistance, and inductance of a Type 43 ferrite core. (© 2005 Fair-Rite Corp., reproduced with permission.)

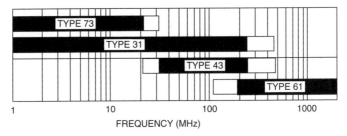

**FIGURE 5-25.** Recommended frequency range of various ferrite materials when used in noise suppression applications. (© 2005 Fair-Rite Corp., reproduced with permission.)

not be overlooked. From a practical point of view, seldom are more than two or three turns used. Most ferrites, however, when used in noise-reduction applications only have a single turn.

The most common ferrite geometry used in noise-suppression applications is the cylindrical core or bead. The greater the length of the cylinder, the higher the impedance. Increasing the length of the core is equivalent to using multiple ferrites.

The attenuation provided by a ferrite depends on the source and the load impedances of the circuit that contains the ferrite. To be effective, the ferrite must add an impedance greater than the sum of the source and load impedance, at the frequency of interest. Because most ferrites have impedances of a few hundred ohms or less, they are used most effectively in low-impedance circuits.

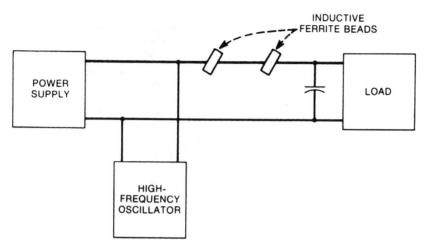

**FIGURE 5-26.** Ferrite bead used to form a L-filter to keep high-frequency oscillator noise from the load.

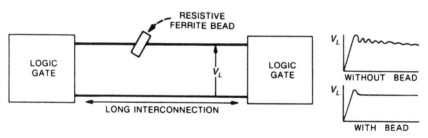

**FIGURE 5-27.** Resistive ferrite bead used to damp out ringing on long interconnection between fast logic gates.

If a single ferrite does not provide sufficient impedance, then multiple turns or multiple ferrites may be used.

Small ferrite beads are especially effective when used to damp out high-frequency oscillations generated by switching transients or parasitic resonances within a circuit. In addition, ferrite cores placed around multiconductor cables act as common-mode chokes and are useful in preventing high-frequency noise from being conducted out of, or into, a circuit.

Figures 5-26 through 5-29 show some applications of ferrite beads. In Fig. 5-26 two ferrite beads are used to form a low-pass R–C filter to keep the high-frequency oscillator signal out of the load, without reducing the dc voltage to the load. The ferrites used are resistive at the oscillator frequency. In Fig. 5-27, a resistive bead is used to damp out the ringing generated by a long interconnection between two fast logic gates.

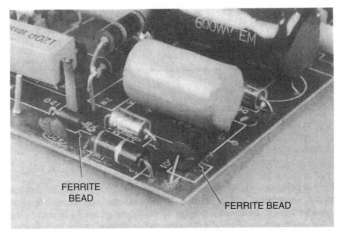

**FIGURE 5-28.** Ferrite beads installed in a color TV to suppress parasitic oscillations in horizontal output circuit.

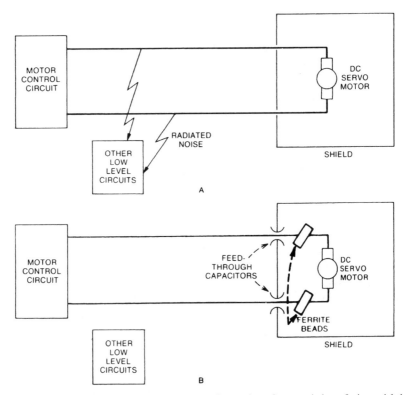

**FIGURE 5-29.** (*A*) High-frequency commutation noise of motor is interfering with low-level circuits, (*B*) ferrite bead used in conjunction with feed-through capacitors to eliminate interference.

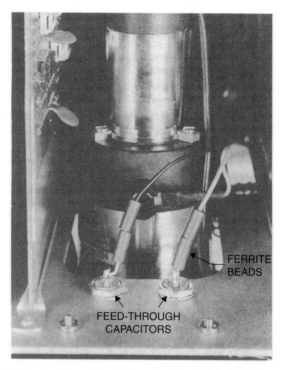

FIGURE 5-30. Ferrite beads and feed-through capacitors used to filter commutation noise on dc motor's power leads.

Figure 5-28 shows two ferrite beads mounted on a printed circuit board. The circuit is part of the horizontal output circuit for a color television set, and the beads are used to suppress parasitic oscillations.

Yet another application for ferrite beads is shown in Fig. 5-29. Figure 5-29A shows a dc servo motor connected to a motor control circuit. High-frequency commutation noise from the motor is being conducted out of the motor's shielded enclosure on the leads, and then it is radiated from the leads to interfere with other low-level circuits within the equipment. Because of acceleration requirements on the motor, resistors cannot be inserted in the leads. The solution in this case was to add two ferrite beads and two feed-through capacitors, as shown in Fig. 5-29B. A photograph of the motor with ferrite beads and feed-through capacitors is shown in Fig. 5-30. As can be observed in the figure, two ferrite beads were used on each of the motor leads to increase the series impedance.

When using ferrites as differential-mode filters in circuits with dc current, the effect of the dc current on the ferrite impedance must be addressed. The ferrite impedance will decrease with increasing current. Figure 5-31 shows the impedance of a small ferrite bead [0.545 in long, 0.138 in outside diameter (OD)] as a function of dc bias current (Fair-Rite, 2005). As can be observed, the

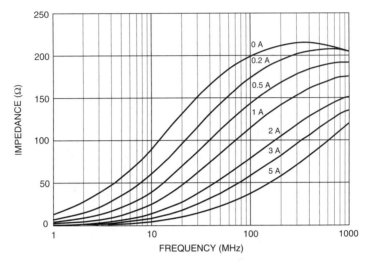

**FIGURE 5-31.** Impedance versus frequency plot of a ferrite bead as a function of the dc bias current. (© 2005, Fair-Rite Products Corp., reproduced with permission.)

**FIGURE 5-32.** Ferrite core used on a USB cable to suppress radiated emissions.

100 MHz impedance at zero current is 200 Ω and falls to 140 Ω with 0.5 A of current, and to 115 Ω with 1 A of current.

Ferrite cores are commonly used as common-mode chokes (see Section 3.5) on multiconductor cables. For example, most video cables used to connect personal computers to their video monitor have ferrite cores on them. The ferrite core acts as a one-turn transformer or common-mode choke, and can be effective in reducing the conducted and/or radiated emission from the cable, as well as suppressing high-frequency pickup in the cable. Figure 5-32 shows a ferrite core on a universel serial bus (USB) cable used to reduce the radiated emission from the cable. Snap-on cores (shown in Fig 5-22) can also be applied easily as an after-the-fact fix to cables, even if they have large connectors at the ends.

## SUMMARY

- Electrolytics are low-frequency capacitors.
- All capacitors become self-resonant at some frequency, which limits their high-frequency use.
- Mica and ceramic are good high-frequency capacitors.
- Air core inductors create more external magnetic fields than do closed core inductors, such as toroids.
- Magnetic core inductors are more likely to pick up interfering magnetic fields than are air core inductors.
- An electrostatic, or Faraday, shielded transformer can be used to reduce capacitive coupling between the windings.
- All resistors, regardless of type, generate the same amount of thermal noise.
- Variable resistors in low-level circuits should be placed so that no dc current flows through them.
- Above audio frequencies, a conductor normally has more inductive reactance than resistance.
- A flat rectangular conductor will have less ac resistance and inductance than a round conductor.
- The ac resistance of a conductor is proportional to the square root of the frequency.
- A transmission line is a series of conductors used to transmit electromagnetic energy from one place to another.
- When a conductor becomes longer than one tenth of a wavelength, it should be treated as a transmission line.
- A conductor should be treated as a transmission line when the rise time of a square wave signal is less than twice the propagation delay on the line.
- The characteristic impedance of a lossless transmission line is equal to $\sqrt{L/C}$.
- The velocity of propagation on a transmission line is $c/\sqrt{\varepsilon_r}$.
- The most important properties of a transmission line are as follows:
    - Characteristic impedance
    - Propagation constant
    - High-frequency loss

- It takes 1 ns to propagate a signal a distance of 6 in on a typical PCB.
- The rise time of a square wave propagating on a PCB will increase approximately 10 ps/in of travel.
- All transmission lines that have the same characteristic impedance and dielectric constant will have the same inductance and capacitance per unit length.
- The two primary types of loss on a transmission line are as follows:
  - Ohmic loss
  - Dielectric loss
- Ohmic loss is proportional to the square root of frequency, and dielectric loss is proportional to frequency.
- The dielectric loss will predominate at high frequency.
- The most important material in a transmission line is the dielectric, not the conductors.
- AC current can and will flow on an open-circuit transmission line.
- Only three transmission line topologies have exact closed form equations for the characteristic impedance. They are as follows:
  - Coax
  - Two round parallel conductors
  - A round conductor over a plane
- When used for noise suppression, ferrites are used in the frequency range where their impedance is resistive.
- Ferrite cores and beads act as ac resistors, coupling high-frequency resistance (loss) into a circuit with little or no low-frequency impedance.
- Ferrites are normally characterized by specifying their impedance versus frequency.
- A ferrite core placed on a cable acts as a common-mode choke, and it can be effective in reducing both conducted and radiated emission.

## PROBLEMS

5.1  a. Capacitors are usually characterized by what parameter?
   b. What is the most important consideration in choosing a type of capacitor?

5.2  a. Name two types of low-frequency capacitors?
   b. Name two types of medium-frequency capacitors?
   c. Name two types of high-frequency capacitors?

5.3  What would be an appropriate type of capacitor to use
   a. In a high-frequency, low voltage application?

b. In a high frequency, high voltage application?

c. For decoupling digital logic?

5.4 How is the inductance of a conductor related to its diameter?

5.5 Make a table of the ratio of the ac resistance to the dc resistance of a 22-gauge copper conductor at the following frequencies: 0.2, 0.5, 1, 2, 5, 10, and 50 MHz.

5.6 A copper conductor has a rectangular cross section of $0.5 \times 2$ cm.

a. What is the dc resistance per meter of the conductor?

b. What is the resistance per meter at 10 MHz?

5.7 a. Derive Eq. 5-9b, realizing that at high frequency most of the current will be confined to an annular ring located at the surface of a copper conductor that has a width equal to the skin depth $\delta$ of the conductor as shown in Fig. P5-7. Assume that $d \gg \delta$.

b. Assume that the criteria that $d \gg \delta$, in part a, is satisfied when $d \geq 10$ $\delta$. Under these conditions, what must $d\sqrt{f}$ be in order for the answer to part 'a' be applicable?

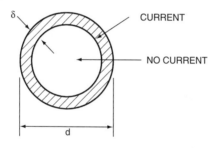

FIGURE P5-7.

5.8 How do the inductive reactance and ac resistance of a conductor vary with frequency?

5.9 Figure P5-9 shows a log-log plot of the ac and dc resistance of a rectangular conductor, of width $w$ and thickness $t$, versus frequency.

a. For a rectangular conductor the break frequency occurs when the skin depth of the conductor is equal to what?

b. Repeat part a assuming $t \ll w$.

c. Rationalize your answer to part b.

d. What is the slope of the ac resistance portion of the plot shown in Fig. P5-9?

5.10 Consider the following two conductors, a 0.25 in diameter round conductor and a 0.5 in wide by 0.1in thick rectangular conductor each located 1in above a ground plane.

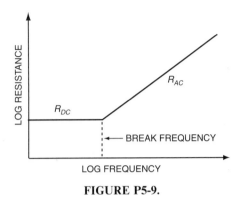

**FIGURE P5-9.**

a. What is the cross-sectional area of each conductor?

b. Calculate the dc resistance, the ac resistance at 10 MHz, and the inductance of the round conductor.

c. Calculate the dc resistance, the ac resistance at 10 MHz, and the inductance of the rectangular conductor.

d. Compare the results and draw your conclusions with respect to the characteristics of the two conductors.

5.11 A PCB trace is 0.008 in wide and 0.0014 in thick. The trace is located 0.020 in above a ground plane. What is the resistance and inductive reactance of the trace at 100 MHz?

5.12 Name two characteristics unique to waveguide?

5.13 In a typical transmission line, approximately how long does it take to propagate a signal a distance of 3ft?

5.14 A 75-$\Omega$ transmission line has a capacitance of 17 pF/ft. What is the inductance of the line?

5.15 What is the characteristic impedance of a coaxial cable that has an inner conductor diameter 0.108 in, an outer conductor diameter of 0.350 in, and a relative dielectric constant of 2?

5.16 The velocity of propagation and dielectric loss are both functions of what property of a transmission line?

5.17 What is the inductance/inch and capacitance/inch of a 50-$\Omega$ transmission line that has a relative dielectric constant of 2?

5.18 a. A transmission line has an inductance of 8.25 nH/in and a capacitance of 3.3 pF/in

b. What is the characteristic impedance of the line?

c. What will be the phase shift of a 10-MHz sine wave after it travels a distance of ten feet on the line?

5.19  What will be the approximate attenuation, at 3 GHz, of a 0.006-in wide by 0.0014-in thick, 50-$\Omega$ stripline on an FR4 epoxy-glass PCB?

5.20  Name two ways to increase the impedance of a ferrite core.

## REFERENCES

Bogatin, E. *Signal Integrity—Simplified*. Upper Saddle River, NJ, Prentice Hall, 2004.

Campbell, R. H. Jr. and Chipman, R. A. "Noise From Current-Carrying Resistors, 25–500 kHz," *Proceedings of the IRE*, vol. 37, August 1949, pp. 938–942.

Danker, B. "New Methods to Decrease Radiation from Printed Circuit Boards," *6th Symposium on Electromagnetic Compatibility*, Zurich, Switzerland, March 5-7, 1985.

Fair-Rite Products Corp. Fair-Rite Products Catalog, 15th ed. Wallkill, NY, 2005.

Hall, S. H. Hall, G. W. and McCall, J. A. *High-Speed Digital System Design*. New York, Wiley, 2000.

IPC-D-317A, *Design Guidelines for Electronic Packaging Utilizing High-Speed Techniques*. Northbrook, IL, IPC, 1995.

Johnson, H. W. and Graham, M. *High-Speed Digital Design*. Englewood, NJ, Prentice Hall 1993.

Jordan. E. C., ed. *Reference Data for Engineers: Radio, Electronics, Computer, and Communications*, 7th ed. Indianapolis, IN, Howard W. Sams, 1985, p. 6–7.

Skilling, H. H. *Electric Transmission Lines*. New York, McGraw Hill, 1951.

Whalen, J. J. and Paludi, C. "Computer Aided Analysis of Electronic Circuits—the Need to Include Parasitic Elements."*International Journal of Electronics*, vol. 43, no. 5, November 1977.

## FURTHER READING

Henney, K. and Walsh, C. *Electronic Components Handbook*, Vol. 1. New York, McGraw-Hill, 1957.

Rostek, P. M. "Avoid Wiring-Inductance Problems." *Electronic Design*, vol. 22, December 6, 1974.

# 6 Shielding

A shield is a metallic partition placed between two regions of space. It is used to control the propagation of electromagnetic fields from one region to the other. Shields may be used to contain electromagnetic fields, if the shield surrounds the noise source as shown in Fig. 6-1. This configuration provides protection for all susceptible equipment located outside the shield. A shield may also be used to keep electromagnetic radiation out of a region, as shown in Fig. 6-2. This technique provides protection only for the specific equipment contained within the shield. From an overall systems point of view, shielding the noise source is more efficient than shielding the receptor. However, in some cases, the source must be allowed to radiate (i.e., broadcast stations), and the shielding of individual receptors may be necessary.

It is of little value to make a shield, no matter how well designed, and then to allow electromagnetic energy to enter (or exit) the enclosure by an alternative path such as cable penetrations. Cables will pick up noise on one side of the shield and conduct it to the other side, where it will be reradiated. To maintain the integrity of the shielded enclosure, noise voltages should be filtered from all cables that penetrate the shield. This approach applies to power cables as well as signal cables. Cable shields that penetrate a shielded enclosure must be bonded to that enclosure to prevent noise coupling across the boundary.

This chapter is divided into two parts. The first covers the behavior of solid shields that contain no apertures. The second, which starts with Section 6.10, covers the effect of apertures on the shielding effectiveness.

## 6.1 NEAR FIELDS AND FAR FIELDS

The characteristics of a field are determined by the source (the antenna), the media surrounding the source, and the distance between the source and the point of observation. At a point close to the source, the field properties are determined primarily by the source characteristics. Far from the source, the properties of the field depend mainly on the medium through which the field

*Electromagnetic Compatibility Engineering*, by Henry W. Ott
Copyright © 2009 John Wiley & Sons, Inc.

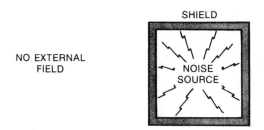

**FIGURE 6-1.** Shield application where the noise source is shielded, which prevents noise coupling to equipment outside the shield.

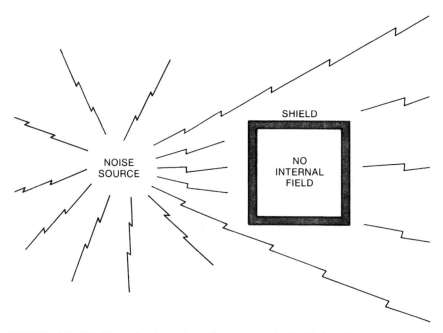

**FIGURE 6-2.** Shield application where the receptor is shielded, which prevents noise infiltration.

is propagating. Therefore, the space surrounding a source of radiation can be broken into two regions, as shown in Fig. 6-3. Close to the source is the near or induction field. At a distance greater than the wavelength ($\lambda$) divided by $2\pi$ (approximately one sixth of a wavelength) is the far or radiation field. The region around $\lambda/2\pi$ is the transition region between the near and far fields.

The ratio of the electric field ($E$) to the magnetic field ($H$) is the wave impedance. In the far field, this ratio equals the characteristic impedance of the medium (e.g., $E/H = Z_0 = 377\ \Omega$ for air or free space). In the near field, the

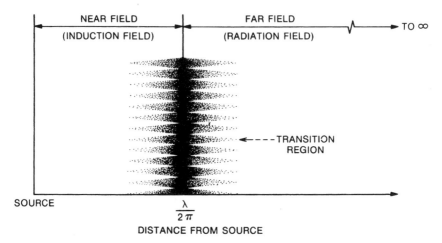

**FIGURE 6-3.** The space surrounding a source of radiation can be divided into two regions, the near field and the far field. The transition from near to far field occurs at a distance of $\lambda/2\pi$.

ratio is determined by the characteristics of the source and the distance from the source to where the field is observed. If the source has high current and low voltage ($E/H < 377$), the near field is predominantly magnetic. Conversely, if the source has low current and high voltage ($E/H > 377$), the near field is predominantly electric.

For a rod or straight wire antenna, the source impedance is high. The wave impedance near the antenna—predominantly an electric field—is also high. As distance is increased, the electric field loses some of its intensity as it generates a complementary magnetic field. In the near field, the electric field attenuates at a rate of $(1/r)^3$, whereas the magnetic field attenuates at a rate of $(1/r)^2$. Thus, the wave impedance from a straight wire antenna decreases with distance and asymptotically approaches the impedance of free space in the far field, as shown in Fig. 6-4.

For a predominantly magnetic field—such as produced by a loop antenna—the wave impedance near the antenna is low. As the distance from the source increases, the magnetic field attenuates at a rate of $(1/r)^3$ and the electric field attenuates at a rate of $(1/r)^2$. The wave impedance therefore increases with distance and approaches that of free space at a distance of $\lambda/2\pi$. In the far field, both the electric and magnetic fields attenuate at a rate of $1/r$.

In the near field the electric and magnetic fields must be considered separately, because the ratio of the two is not constant. In the far field, however, they combine to form a plane wave having an impedance of 377 $\Omega$. Therefore, when plane waves are discussed, they are assumed to be in the far field. When individual electric and magnetic fields are discussed, they are assumed to be in the near field.

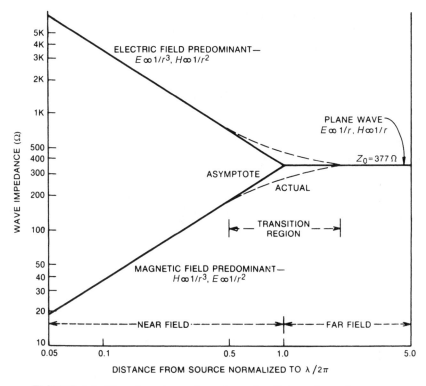

**FIGURE 6-4.** Wave impedance depends on the distance from the source.

## 6.2   CHARACTERISTIC AND WAVE IMPEDANCES

The following characteristic constants of a medium are used in this chapter:

| | |
|---|---|
| Permeability, | $\mu(4\pi \times 10^{-7}$ H/m for free space). |
| Dielectric constant, | $\varepsilon(8.85 \times 10^{-12}$ F/m for free space). |
| Conductivity, | $\sigma(5.82 \times 10^{7}$ siemens/m for copper). |

For any electromagnetic wave, the wave impedance is defined as

$$Z_w = \frac{E}{H}. \tag{6-1}$$

The characteristic impedance of a medium is defined (Hayt, 1974) by the following expression:

$$Z_0 = \sqrt{\frac{j\omega\mu}{\sigma + j\omega\varepsilon}}. \tag{6-2}$$

In the case of a plane wave in the far field, $Z_0$ is also equal to the wave impedance $Z_w$. For insulators ($\sigma \ll j\omega\varepsilon$) the characteristic impedance is independent of frequency and becomes

$$Z_0 = \sqrt{\frac{\mu}{\varepsilon}}. \tag{6-3}$$

For free space, $Z_0$ equals 377 Ω. In the case of conductors ($\sigma \gg j\omega\varepsilon$), the characteristic impedance is called the shield impedance $Z_s$ and it becomes

$$Z_s = \sqrt{\frac{j\omega\mu}{\sigma}} = \sqrt{\frac{\omega\mu}{2\sigma}}(1+j), \tag{6-4a}$$

$$|Z_s| = \sqrt{\frac{\omega\mu}{2\sigma}}. \tag{6-4b}$$

For copper at 1 MHz, $|Z_s|$ equals $3.68 \times 10^{-4}$ Ω. Substituting numerical values for the constants of Eq. 6-4b gives the following results:
For copper,

$$|Z_s| = 3.68 \times 10^{-7}\sqrt{f}. \tag{6-5a}$$

For aluminum,

$$|Z_s| = 4.71 \times 10^{-7}\sqrt{f}. \tag{6-5b}$$

For steel,

$$|Z_s| = 3.68 \times 10^{-5}\sqrt{f}. \tag{6-5c}$$

For any conductor, in general,

$$|Z_s| = 3.68 \times 10^{-7}\sqrt{\frac{\mu_r}{\sigma_r}}\sqrt{f}. \tag{6-5d}$$

Representative values of the relative permeability ($\mu_r$) and the relative conductivity ($\sigma_r$) are listed in Table 6-1.

**TABLE 6-1.  Relative Conductivity and Permeability of Various Materials**

| Material | Relative conductivity $\sigma_r$ | Relative permeability $\mu_r$ |
|---|---|---|
| Silver | 1.05 | 1 |
| Copper—annealed | 1.00 | 1 |
| Gold | 0.7 | 1 |
| Chromium | 0.664 | 1 |
| Aluminum (soft) | 0.61 | 1 |
| Aluminum (tempered) | 0.4 | 1 |
| Zinc | 0.32 | 1 |
| Beryllium | 0.28 | 1 |
| Brass | 0.26 | 1 |
| Cadmium | 0.23 | 1 |
| Nickel | 0.20 | 100 |
| Bronze | 0.18 | 1 |
| Platinum | 0.18 | 1 |
| Magnesium alloy | 0.17 | 1 |
| Tin | 0.15 | 1 |
| Steel (SAE 1045) | 0.10 | 1000 |
| Lead | 0.08 | 1 |
| Monel | 0.04 | 1 |
| Conetic (1 kHz) | 0.03 | 25,000 |
| Mumetal (1 (kHz) | 0.03 | 25,000 |
| Stainless steel (Type 304) | 0.02 | 500 |

## 6.3  SHIELDING EFFECTIVENESS

The following sections discuss shielding effectiveness in both the near and far fields. Shielding effectiveness can be analyzed in many different ways. One approach is to use circuit theory as shown in Fig. 6-5. In the circuit theory approach, the incident fields induce currents in the shield, and these currents in turn generate additional fields that cancel the original field in certain regions of space. We will take this approach when dealing with apertures (Section 6.10).

For most of this chapter, however, we will use the approach originally developed by S.A. Schelkunoff (1943, pp. 303–312). Schelkunoff's approach is to treat shielding as a transmission line problem with both loss and reflection components. The loss is the result of heat generated inside the shield, and the reflection is the result of the difference in impedance between the incident wave and the shield impedance.

Shielding can be specified in terms of the reduction in magnetic and/or electric field strength caused by the shield. It is convenient to express this shielding effectiveness in units of decibels (dB).* Use of decibels then permits the shielding produced by various effects to be added to obtain the total shielding. Shielding effectiveness ($S$) is defined for electric fields as

---

* See Appendix A for a discussion of the decibel.

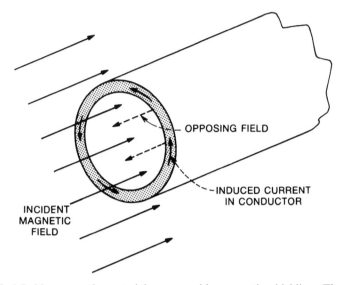

INCIDENT
MAGNETIC
FIELD

OPPOSING FIELD

INDUCED CURRENT
IN CONDUCTOR

**FIGURE 6-5.** Nonmagnetic material can provide magnetic shielding. The incident magnetic field induces currents in the conductor, which produces an opposing field to cancel the incident field in the region of space enclosed by the shield.

$$S = 20 \log \frac{E_0}{E_1} \text{ dB}, \qquad (6\text{-}6)$$

and for magnetic fields as

$$S = 20 \log \frac{H_0}{H_1} \text{ dB}. \qquad (6\text{-}7)$$

In the preceding equations, $E_0(H_0)$ is the incident field strength, and $E_1(H_1)$ is the field strength of the transmitted wave as it emerges from the shield.

In the design of a shielded enclosure, there are two prime considerations: (1) the shielding effectiveness of the shield material itself and (2) the shielding effectiveness resulting from discontinuities and apertures in the shield. These two items are considered separately in this chapter.

First, the shielding effectiveness of a solid shield with no seams or holes is determined, and then the effect of discontinuities and holes is considered. At high-frequencies, it is the shielding effectiveness of the apertures that determines the overall shielding effectiveness of a shield, not the intrinsic shielding effectiveness of the shield material.

Shielding effectiveness varies with frequency, geometry of shield, position within the shield where the field is measured, type of field being attenuated, angle of incidence, and polarization. This section will consider the shielding provided by a plane sheet of conducting material. This simple geometry serves to introduce general shielding concepts and shows which material properties determine shielding effectiveness, but it does not include those effects caused by the geometry of the shield. The results of the plane sheet calculations are useful for estimating the relative shielding effectiveness of various materials.

Two types of loss are encountered by an electromagnetic wave striking a metallic surface. The wave is partially reflected from the surface, and the transmitted (nonreflected) portion of the wave is attenuated as it passes through the shield. This latter effect, called absorption or penetration loss, is the same in either the near or the far field and for electric or magnetic fields. Reflection loss, however, is dependent on the type of field, and the wave impedance.

The total shielding effectiveness of a solid material with no apertures is equal to the sum of the absorption loss ($A$) plus the reflection loss ($R$) plus a correction factor ($B$) to account for multiple reflections in thin shields.* Total shielding effectiveness therefore can be written as

$$S = A + R + B \ \ \text{dB}. \tag{6-8}$$

All the terms in Eq. 6-8 must be expressed in decibels. The multiple reflection factor $B$ can be neglected if the absorption loss $A$ is greater than 9 dB. From a practical point of view, $B$ can also be neglected for electric fields and plane waves.

## 6.4  ABSORPTION LOSS

When an electromagnetic wave passes through a medium, its amplitude decreases exponentially (Hayt, 1974) as shown in Fig. 6-6. This decay occurs because currents induced in the shield produce ohmic losses and heating of the material. Therefore, we can write

$$E_1 = E_0 e^{-t/\delta} \tag{6-9}$$

and

$$H_1 = H_0 e^{-t/\delta}, \tag{6-10}$$

where $E_1(H_1)$ is the wave intensity at a distance $t$ within the shield as shown in Fig. 6-6. The distance required for the wave to be attenuated to $1/e$ or 37% of its original value is defined as the skin depth, which is equal to

$$\delta = \sqrt{\frac{2}{\omega\mu\sigma}} \ \ \text{m.}^\dagger \tag{6-11a}$$

---

* $R + B$ is actually the total reflection loss. For convenience, it is broken into two parts, the reflection loss neglecting multiple reflections $R$, and the connection factor for the neglected multiple reflections $B$.
$\dagger$ Skin depth calculated by Eq. 6-11a is in meters when the constants listed in Section 6.2 (MKS system) are used.

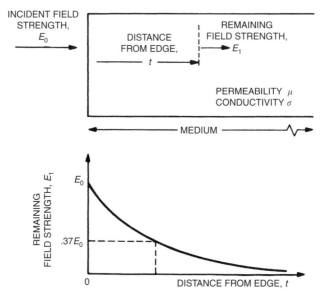

**FIGURE 6-6.** Electromagnetic wave passing through an absorbing material is attenuated exponentially.

Substituting numerical values for $\mu$ and $\sigma$ into Eq. 6.11a, and changing units so the skin depth is in inches gives

$$\delta = \frac{2.6}{\sqrt{f\mu_r\sigma_r}} \text{ in,} \tag{6-11b}$$

where $\mu_r$ and $\sigma_r$ are the relative permeability and relative conductivity of the shield material. Values for relative permeability and relative conductivity of various materials are listed in Table 6-1.

Some representative skin depths for copper, aluminum, steel, and mumetal are listed in Table 6-2.

The absorption loss through a shield can now be written as

$$A = 20\log\frac{E_0}{E_1} = 20\log e^{t/\delta} \tag{6-12a}$$

$$A = 20\left(\frac{t}{\delta}\right)log(e) \text{ dB,} \tag{6-12b}$$

$$A = 8.69\left(\frac{t}{\delta}\right) \text{ dB.} \tag{6-12c}$$

**TABLE 6-2.  Skin Depths of Various Materials**

| Frequency | Copper (in) | Aluminum (in) | Steel (in) | Mumetal (in) |
|-----------|-------------|---------------|------------|--------------|
| 60 Hz     | 0.335       | 0.429         | 0.034      | 0.014        |
| 100 Hz    | 0.260       | 0.333         | 0.026      | 0.011        |
| 1 kHz     | 0.082       | 0.105         | 0.008      | 0.003        |
| 10 kHz    | 0.026       | 0.033         | 0.003      | —            |
| 100 kHz   | 0.008       | 0.011         | 0.0008     | —            |
| 1 MHz     | 0.003       | 0.003         | 0.0003     | —            |
| 10 MHz    | 0.0008      | 0.001         | 0.0001     | —            |
| 100 MHz   | 0.00026     | 0.0003        | 0.00008    | —            |
| 1000 MHz  | 0.00008     | 0.0001        | 0.00004    | —            |

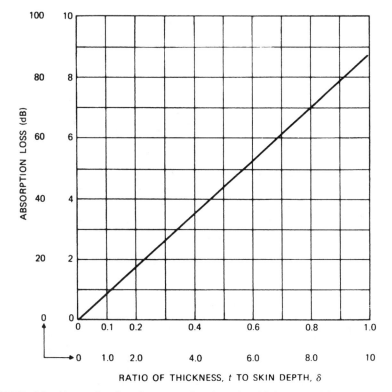

**FIGURE 6-7.** Absorption loss is proportional to the thickness and inversely proportional to the skin depth of the shield material. This plot can be used for electric fields, magnetic fields, or plane waves.

As can be observed from the preceding equation, the absorption loss in a shield one skin-depth thick is approximately 9 dB. Doubling the thickness of the shield doubles the loss in decibels.

Figure 6-7 is a plot of absorption loss in decibels versus the ratio $t/\delta$. This curve is applicable to plane waves, electric fields, or magnetic fields.

Substituting Eq. 6-11b into Eq. 6-12c gives the following general expression for absorption loss:

$$A = 3.34t\sqrt{f\mu_r\sigma_r} \text{ dB.} \tag{6-13}$$

In this equation, $t$ is equal to the thickness of the shield in inches. Equation 6-13 shows that the absorption loss (in dB) is proportional to the square root of the product of the permeability times the conductivity of the shield material.

Equation 6-13 is plotted in Fig. 6-8, which is a universal absorption loss curve. It is a plot of the absorption loss $A$ in decibels versus the parameter, $t\sqrt{f\mu_r\sigma_r}$, where $t$ is the shield thickness in inches, $f$ is the frequency in Hertz, and $\mu_r$ and $\sigma_r$ are the relative permeability and conductivity of the shield material, respectively.

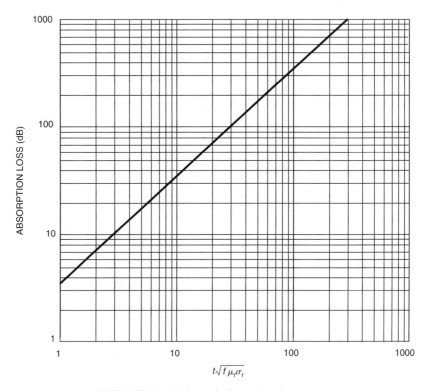

**FIGURE 6-8.** Universal absorption loss curve.

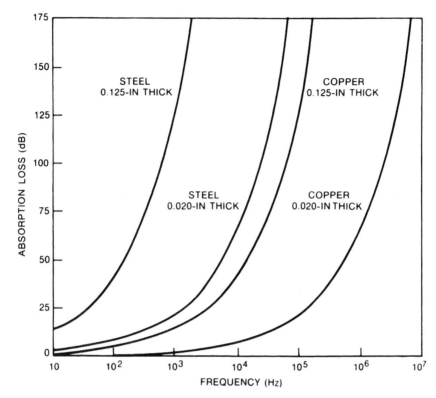

**FIGURE 6-9.** Absorption loss increases with frequency and with shield thickness; steel offers more absorption loss than copper of the same thickness.

Absorption loss versus frequency is plotted in Fig. 6-9 for two thickness of copper and steel. As can be observed, a thin (0.02 in) sheet of copper provides significant absorption loss (66 dB) at 1 MHz but virtually no loss at frequencies below 1000 Hz. Figure 6-9 clearly shows the advantage of steel over copper in providing absorption loss. Even when steel is used, however, a thick sheet must be used to provide appreciable absorption loss below 1000 Hz.

## 6.5 REFLECTION LOSS

The reflection loss at the interface between two media is related to the difference in characteristic impedances between the media as shown in Fig. 6-10. The

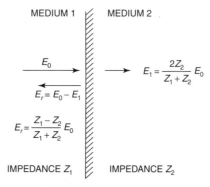

**FIGURE 6-10.** An incident wave is partially reflected from, and partially transmitted through, an interface between two media. The transmitted wave is $E_t$ and the reflected wave is $E_r$.

intensity of the transmitted wave from a medium with impedance $Z_1$ to a medium with impedance $Z_2$ (Hayt, 1974) is

$$E_1 = \frac{2Z_2}{Z_1 + Z_2} E_0, \tag{6-14}$$

and

$$H_1 = \frac{2Z_1}{Z_1 + Z_2} H_0, \tag{6-15}$$

$E_o$ ($H_o$) is the intensity of the incident wave, and $E_1$ ($H_1$) is the intensity of the transmitted wave.

When a wave passes through a shield, it encounters two boundaries, as shown in Fig. 6-11. The secondary boundary is between a medium with impedance $Z_2$ and a medium with impedance $Z_1$. The transmitted wave $E_t$ ($H_t$) through this boundary is given by

$$E_t = \frac{2Z_1}{Z_1 + Z_2} E_1, \tag{6-16}$$

and

$$H_t = \frac{2Z_2}{Z_1 + Z_2} H_1, \tag{6-17}$$

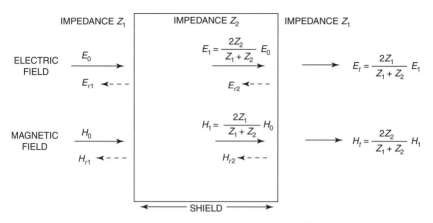

**FIGURE 6-11.** Partial reflection and transmission occur at both boundaries of a shield.

If the shield is thick* compared with the skin depth, then the total transmitted wave intensity is found by substituting Eqs. 6-14 and 6-15 into Eqs. 6-16 and 6-17, respectively. This neglects the absorption loss, which has been accounted for previously (Eq. 6-13). Therefore, for thick shields the total transmitted wave is

$$E_t = \frac{4Z_1 Z_2}{(Z_1 + Z_2)^2} E_0, \tag{6-18}$$

and

$$H_t = \frac{4Z_1 Z_2}{(Z_1 + Z_2)^2} H_0, \tag{6-19}$$

Note that even though the electric and magnetic fields are reflected differently at each boundary, the net effect across both boundaries is the same for both fields. If the shield is metallic and the surrounding area an insulator, then $Z_1 \gg Z_2$. Under these conditions, the largest reflection (smallest transmitted wave) occurs when the wave enters the shield (first boundary) for the case of electric fields, and when the wave leaves the shield (second boundary) for the case of magnetic fields. *Because the primary reflection occurs at the first surface in the case of electric fields, even very thin materials provide good reflection loss.* In the case of magnetic fields, however, the primary reflection occurs at the second surface, and as will be shown later, multiple

---

* If the shield is not thick, multiple reflections occur between the two boundaries because the absorption loss in the shield is small. (See the Section 6.5.6 "Multiple Reflections in Thin Shields").

reflections within the shield can significantly reduce the shielding effectiveness. When $Z_1 \gg Z_2$, Eqs. 6-18 and 6-19 reduce to

$$E_t = \frac{4Z_2}{Z_1} E_0, \qquad (6\text{-}20)$$

and

$$H_t = \frac{4Z_2}{Z_1} H_0, \qquad (6\text{-}21)$$

Substituting the wave impedance $Z_w$ for $Z_1$, and the shield impedance $Z_S$ for $Z_2$ the reflection loss, neglecting multiple reflection, for either the $E$ or $H$ field can be written as

$$R = 20 \log \frac{E_0}{E_1} = 20 \log \frac{Z_1}{4Z_2} = 20 \log \frac{|Z_w|}{4|Z_s|} \text{ dB}, \qquad (6\text{-}22)$$

where

$Z_w$ = impedance of wave prior to entering the shield (Eq. 6-1),
$Z_s$ = impedance of shield (Eq. 6-5d).

These reflection loss equations are for a plane wave approaching the shield at normal incidence. If the wave approaches at other than normal incidence, then the reflection loss increases with the angle of incidence. The results also apply to a curved interface, provided the radius of curvature is much greater than the skin depth.

### 6.5.1    Reflection Loss to Plane Waves

In the case of a plane wave (far field), the wave impedance $Z_w$ equals the characteristic impedance of free space $Z_0$ (377 $\Omega$). Therefore, Eq. 6-22 becomes

$$R = 20 \log \frac{94.25}{|Z_s|} \text{ dB}, \qquad (6\text{-}23\text{a})$$

Therefore, the lower the shield impedance, the greater is the reflection loss. Substituting Eq. 6-5d for $|Z_s|$ and rearranging Eq. 6-23a gives

$$R = 168 + 10 \log(\sigma_r/\mu_r f) \text{ dB} \qquad (6\text{-}23\text{b})$$

Figure 6-12 is a plot of the reflection loss versus frequency for three materials: copper, aluminum, and steel. Comparing this with Fig. 6-9 shows that although steel has more absorption loss than copper, it has less reflection loss.

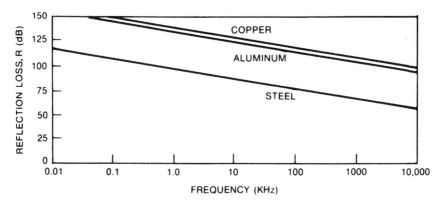

**FIGURE 6-12.** Reflection loss for plane waves is greater at low frequencies and for high conductivity material.

### 6.5.2  Reflection Loss in the Near Field

In the near field, the ratio of the electric field to the magnetic field is no longer determined by the characteristic impedance of the medium. Instead, the ratio of the electric field to the magnetic field depends more on the characteristics of the source (antenna). If the source has high voltage and low current, the wave impedance is greater than 377 $\Omega$, and the field will be a high-impedance, or electric-field. If the source has low voltage and high current, then the wave impedance will be less than 377 $\Omega$, and the field will be a low-impedance, or magnetic, field.

Because the reflection loss (Eq. 6-22) is a function of the ratio between the wave impedance and the shield impedance, the reflection loss varies with the wave impedance. A high-impedance (electric) field has higher reflection loss than a plane wave. Similarly, a low-impedance (magnetic) field has lower reflection loss than a plane wave. This is shown in Fig. 6-13 for a copper shield separated from the source by distances of 1 and 30 m. Also shown for comparison is the plane wave reflection loss.

For any specified distance between source and shield, the three curves (electric field, magnetic field, and plane wave) of Fig. 6-13 merge at the frequency that makes the separation between source and shield equal to $\lambda/2\pi$, (where $\lambda$ is the wavelength). When the spacing is 30 m, the electric and magnetic field curves come together at a frequency of 1.6 MHz.

The curves shown in Fig. 6-13 are for point sources that produce only an electric field or only a magnetic field. Most practical sources, however, are a combination of both electric and magnetic fields. The reflection loss for a practical source therefore lies somewhere between the electric field lines and the magnetic field lines shown in the figure.

Figure 6-13 shows that the reflection loss of an electric field decreases with frequency until the separation distance $\lambda/2\pi$. Beyond that, the reflection loss is the same as for a plane wave. The reflection loss of a magnetic field increases with frequency, again until the separation becomes $\lambda/2\pi$. Then, the loss begins to decrease at the same rate as that of a plane wave.

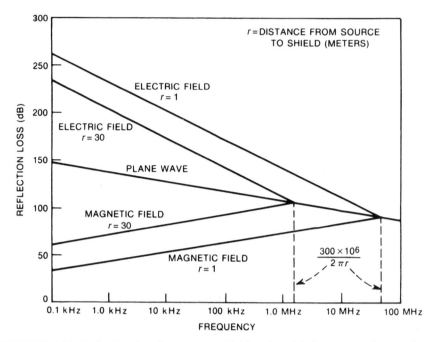

**FIGURE 6-13.** Reflection loss in a copper shield varies with frequency, distance from the source, and type of wave.

### 6.5.3  Electric Field Reflection Loss

The wave impedance from a point source of electric field can be approximated by the following equation when $r < \lambda/2\pi$:

$$|Z_w|_e = \frac{1}{2\pi f \varepsilon r}, \tag{6-24}$$

where $r$ is the distance from the source to the shield in meters and $\varepsilon$ is the dielectric constant. The reflection loss can be determined by substituting Eq. 6-24 into Eq. 6-22, giving

$$R_e = 20 \log \frac{1}{8\pi f \varepsilon r |Z_s|} \text{ dB}, \tag{6-25}$$

or substituting the free space value of $\varepsilon$

$$R_e = 20 \log \frac{4.5 \times 10^9}{f r |Z_s|} \text{ dB}, \tag{6-26a}$$

where $r$ is in meters. Substituting Eq. 6-5d for $|Z_s|$ and rearranging terms, Eq. 6-26a becomes

$$R_e = 322 + 10 \log \frac{\sigma_r}{\mu_r f^3 r^2} \text{ dB,} \qquad (6\text{-}26\text{b})$$

In Fig 6-13 the lines labeled "electric field" are plots of Eq. 6-26b for a copper shield with $r$ equal to 1 and 30 m. The equation and the plot represent the reflection loss at a specified distance from a point source producing only an electric field. An actual electric field source, however, has some small magnetic field component in addition to the electric field. It therefore has a reflection loss somewhere between the electric field line and the plane wave line of Fig. 6-13. Because, in general, we do not know where between these two lines the actual source may fall, the plane wave calculations (Eq. 6-23b) are normally used in determining the reflection loss for an electric field. The actual reflection loss is then equal to or greater than that calculated in Eq. 6-23b.

### 6.5.4  Magnetic Field Reflection Loss

The wave impedance from a point source of magnetic field can be approximated by the following equation, assuming $r < \lambda/2\pi$:

$$|Z_w|_m = 2\pi f \mu r, \qquad (6\text{-}27)$$

where $r$ is the distance from the source to the shield and $\mu$ is the permeability. The reflection loss can be determined by substituting Eq. 6-27 into Eq. 6-22, giving

$$R_m = 20 \log \frac{2\pi f \mu r}{4|Z_s|} \text{ dB,} \qquad (6\text{-}28)$$

or substituting the free space value of $\mu$

$$R_m = 20 \log \frac{1.97 \times 10^{-6} f r}{|Z_s|} \text{ dB,} \qquad (6\text{-}29\text{a})$$

where $r$ is in meters. Substituting Eq. 6-5d for $|Z_s|$ and rearranging Eq. 6-29a gives

$$R_m = 14.6 + 10 \log \left( \frac{f r^2 \sigma_r}{\mu_r} \right) \text{ dB,}^* \qquad (6\text{-}29\text{b})$$

with $r$ in meters.

---

* If a negative value is obtained in the solution for $R$, use $R = 0$ instead and neglect the multiple reflection factor $B$. If a solution for $R$ is positive and near zero, Eq. 6-29b is slightly in error. The error occurs because the assumption $Z_1 \gg Z_2$, made during the derivation of the equation, is not satisfied in this case. The error is 3.8 dB when $R$ equals zero, and it decreases as $R$ gets larger. From a practical point of view, however, this error can be neglected.

In Fig. 6-13 the curves labeled "magnetic field" are plots of Eq. 6-29b for a copper shield with $r$ equal to 1 and 30 m. Equation 6-29b and the plot in Fig. 6-13 represent the reflection loss at the specified distance from a point source producing only a magnetic field. Most real magnetic field sources have a small electric field component in addition to the magnetic field, and the reflection loss lies somewhere between the magnetic field line and the plane wave line of Fig. 6-13. Because we do not generally know where between these two lines the actual source may fall, Eq. 6-29b should be used to determine the reflection loss for a magnetic field. The actual reflection loss will then be equal to or greater than that calculated in Eq. 6-29b.

Where the distance to the source is not known, the near field magnetic reflection loss can usually be assumed to be zero at low frequencies.

### 6.5.5   General Equations for Reflection Loss

Neglecting multiple reflections a generalized equation for reflection loss can be written as

$$R = C + 10 \log \left( \frac{\sigma_r}{\mu_r} \right) \left( \frac{1}{f^n r^m} \right), \tag{6-30}$$

where the constants $C$, $n$ and $m$ are listed in Table 6-3 for plane waves, electric fields, and magnetic fields, respectively.

Equation 6-30 is equivalent to Eq. 6-23b for plane waves, Eq. 6-26b for electric fields, and Eq. 6-29b for magnetic fields. Equation 6-30 shows that the reflection loss is a function of the shield material's conductivity divided by its permeability.

### 6.5.6   Multiple Reflections in Thin Shields

If the shield is thin, the reflected wave from the second boundary is rereflected off the first boundary, and then it returns to the second boundary to be reflected again, as shown in Fig. 6-14. This can be neglected in the case of a thick shield, because the absorption loss is high. By the time the wave reaches the second boundary for the second time, it is of negligible amplitude, because by then it has passed through the thickness of the shield three times.

TABLE 6-3.   Constants to be Used in Eq. 6-30

| Type of Field | $C$ | $n$ | $m$ |
|---|---|---|---|
| Electric field | 322 | 3 | 2 |
| Plane wave | 168 | 1 | 0 |
| Magnetic field | 14.6 | −1 | −2 |

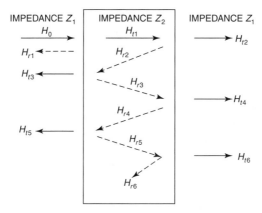

**FIGURE 6-14.** Multiple reflections occur in a thin shield; part of the wave is transmitted through the second boundary at each reflection.

For electric fields, most of the incident wave is reflected at the first boundary, and only a small percentage enters the shield. This can be observed from Eq. 6-14 and the fact that $Z_2 \ll Z_1$. Therefore, multiple reflections within the shield can be neglected for electric fields.

For magnetic fields most of the incident wave passes into the shield at the first boundary, as shown in Eq. 6-15 when $Z_2 \ll Z_1$. The magnitude of the transmitted wave is actually double that of the incident wave. With a magnetic field of such large magnitude within the shield, the effect of multiple reflections inside the shield must be considered.

The correction factor for the multiple reflection of magnetic fields in a shield of thickness $t$ and skin depth $\delta$ is

$$B = 20 \log \left( 1 - e^{-2t/\delta} \right) \text{ dB,}^* \tag{6-31}$$

Figure 6-15 is plot of the correction factor $B$ as a function of $t/\delta$. Note that the correction factor is a negative number, indicating that less shielding (than predicated by Eq. 6-30) is obtained from a thin shield as the result of multiple reflections.

## 6.6 COMPOSITE ABSORPTION AND REFLECTION LOSS

### 6.6.1 Plane Waves

The total loss for plane waves in the far field is a combination of the absorption and reflection losses, as indicated in Eq. 6-8. The multiple reflection correction

---

* See Appendix C for this calculation.

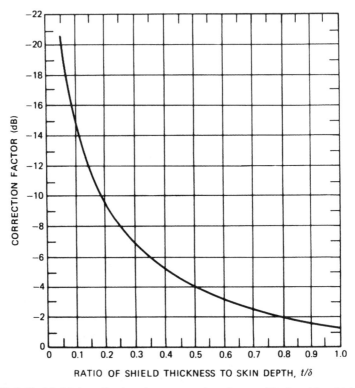

**FIGURE 6-15.** Multiple reflection loss correction factor (*B*) for thin shields, with magnetic fields. See Table C-1 for values of *B* for very small values of $t/\delta$.

term *B* is normally neglected for plane waves, because the reflection loss is so high and the correction term is small. If the absorption loss is greater than 1 dB, the correction term is less than 11 dB; if the absorption loss is greater than 4 dB, then the correction is less than 2 dB.

Figure 6-16 shows the overall attenuation or shielding effectiveness of a 0.020-in thick solid copper shield. As can be observed, the *reflection loss decreases with increasing frequency,* because the shield impedance $Z_s$ increases with frequency. The *absorption loss, however, increases with frequency,* because of the decreasing skin depth. The minimum shielding effectiveness occurs at some intermediate frequency, in this case at 10 kHz. From Fig. 6-16 it is apparent that for low-frequency plane waves, reflection loss accounts for most of the attenuation, whereas most of the attenuation at high frequencies comes from absorption loss.

### 6.6.2 Electric Fields

The total loss for an electric field is obtained by combining the absorption (Eq. 6-13) and reflection losses (Eq. 6-26), as indicated in Eq. 6-8. The multiple

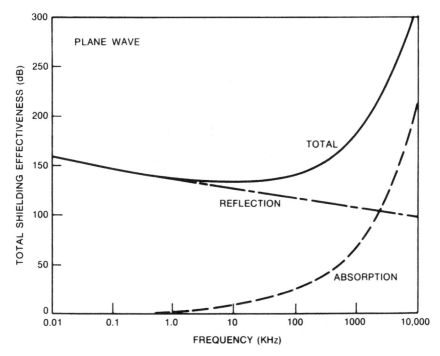

**FIGURE 6-16.** Shielding effectiveness of a 0.02-in thick copper shield in the far field.

reflection correction factor $B$ is normally neglected in the case of an electric field, because the reflection loss is so great and the correction term is small. At *low-frequency*, reflection loss in the primary shielding mechanism for electric fields. At *high-frequency*, absorption loss is the primary shielding mechanism.

### 6.6.3  Magnetic Fields

The total loss for a magnetic field is obtained by combining the absorption loss (Eq. 6-13) and the reflection loss (Eq. 6-29), as indicated in Eq. 6-8. If the shield is thick (absorption loss > 9 dB), the multiple reflection correction factor $B$ can be neglected. If the shield is thin, then the correction factor from Eq. 6-31 or Fig. 6-15 must be included.

In the near field, the reflection loss to a low-frequency magnetic field is small. Because of multiple reflections, this effect is even more pronounced in a thin shield. *The primary loss for magnetic fields is absorption loss.* Because both the absorption and reflection loss are small at low frequencies, the total shielding effectiveness is low. It is therefore difficult to shield low-frequency magnetic fields. Additional protection against low-frequency magnetic fields can be achieved only by providing a low-reluctance magnetic shunt path to divert the field around the circuit being protected. This approach is shown in Fig. 6-17.

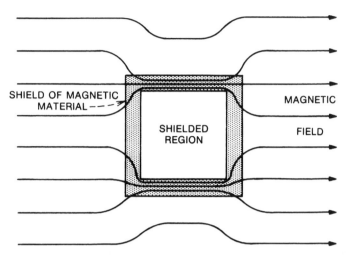

**FIGURE 6-17.** Magnetic material used as a shield by providing a low-reluctance path for the magnetic field, diverting it around the shielded region.

## 6.7 SUMMARY OF SHIELDING EQUATIONS

Figure 6-18 shows the composite shielding effectiveness of a 0.02-in thick solid aluminum shield for an electric field, plane wave, and a magnetic field. As can be observed in the figure, considerable shielding exists in all cases exept for low-frequency magnetic fields.

At high frequencies (above 1MHz), absorption loss predominates in all cases, and any *solid* shield thick enough to be practical provides more than adequate shielding for most applications.

Figure 6-19 is a summary that shows which equations are used to determine shielding effectiveness under various conditions. A qualitative summary of the shielding provided by solid shields under various conditions is given in the summary at the end of this chapter (Table 6-9).

## 6.8 SHIELDING WITH MAGNETIC MATERIALS

If a magnetic material is used as a shield in place of a good conductor there will be an increase in the permeability $\mu$ and a decrease in the conductivity $\sigma$. This has the following effects:

1. It increases the absorption loss, because the permeability increases more than the conductivity decreases for most magnetic materials. (See Eq. 6-13.)
2. It decreases the reflection loss. (See Eq. 6-30)

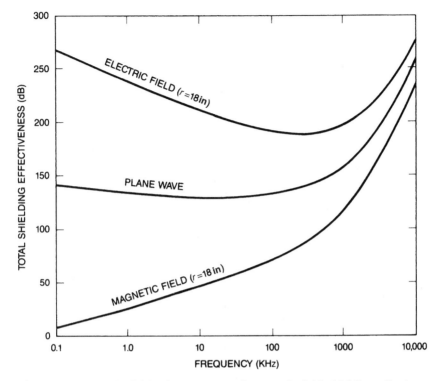

**FIGURE 6-18.** Electric field, plane wave, and magnetic field shielding effectiveness of a 0.02-in-thick solid aluminum shield.

The total loss through a shield is the sum of that from absorption and that from reflection. *In the case of low-frequency magnetic field, very little reflection loss occurs, and absorption loss is the primary shielding mechanism. Under these conditions, it is often advantageous to use a magnetic material to increase the absorption loss.* In the case of low-frequency electric fields or plane waves, the primary shielding mechanism is reflection. Thus, using a magnetic material would decrease the shielding.

When magnetic materials are used as a shield, three often overlooked properties must be taken into account. These properties are as follows:

1. Permeability decreases with frequency.
2. Permeability depends on field strength.
3. Machining or working high permeability magnetic materials, such as mumetal, may change their magnetic properties.

Most permeability values given for magnetic materials are static, or direct current (dc), permeabilities. As frequency increases, the permeability decreases.

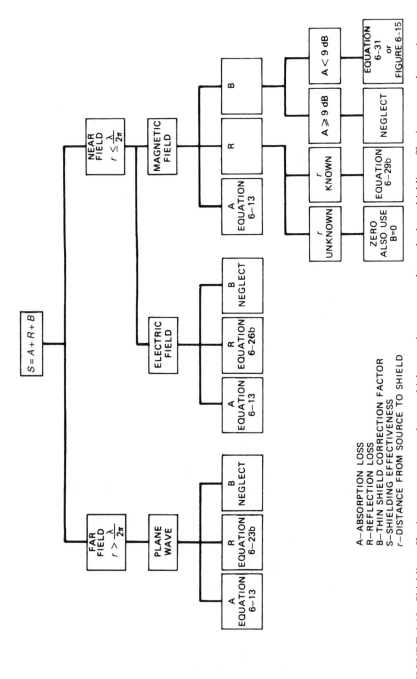

FIGURE 6-19. Shielding effectiveness summary shows which equations are used to calculate shielding effectiveness under various conditions.

A—ABSORPTION LOSS
R—REFLECTION LOSS
B—THIN SHIELD CORRECTION FACTOR
S—SHIELDING EFFECTIVENESS
r—DISTANCE FROM SOURCE TO SHIELD

Usually, the larger the dc permeability, the greater will be the decrease with frequency. Figure 6-20 plots permeability against frequency for a variety of magnetic materials. As can be observed, mumetal (an alloy of nickel, iron, copper, and molybdenum) is no better than cold rolled steel at 100 kHz, even though the dc permeability is 13 times that of cold rolled steel. High-permeability materials are most useful as magnetic field shields at frequencies below 10 kHz.

Above 100 kHz, steel gradually starts to lose its permeability. Table 6-4 lists representative values for the permeability of steel versus frequency.

Absorption loss predominates at high frequency, and as Eq. 6-13 shows, the absorption loss is a function of the square root of the product of the permeability times the conductivity. For copper, the product of the relative conductivity times the relative permeability equals one. Because the conductivity of steel is about $1/10$ that of copper (Table 6-1), the product of the relative conductivity and the relativity permeability will equal 1 when the relative permeability drops to 10. For steel, this occurs at a frequency of 1.5 GHz (Table 6-4). Therefore, above that frequency, steel will actually provide less absorption loss than copper, because the permeability is not large enough to overcome the reduced conductivity of the steel.

The usefulness of magnetic materials as a shield varies with the field strength, $H$. A typical magnetization curve is shown in Fig. 6-21. The static permeability is the ratio of $B$ to $H$. As can be observed, maximum permeability, and

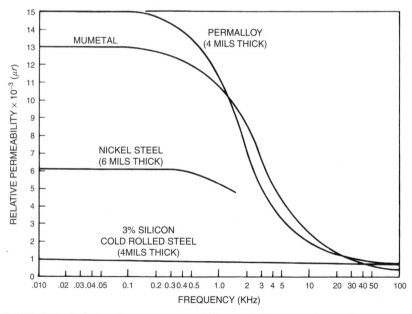

**FIGURE 6-20.** Relation between permeability and frequency for various magnetic materials.

**TABLE 6-4.   Relative Permeability of Steel versus Frequency**

| Frequency | Relative permeability, $\mu_r$ |
|---|---|
| 100 Hz | 1000 |
| 1 kHz | 1000 |
| 10 kHz | 1000 |
| 100 kHz | 1000 |
| 1 MHz | 700 |
| 10 MHz | 500 |
| 100 MHz | 100 |
| 1 GHz | 50 |
| 1.5 GHz | 10 |
| 10 GHz | 1 |

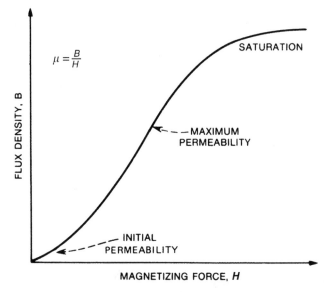

**FIGURE 6-21.** Typical magnetization curve. Permeability is equal to the slope of the curve.

therefore shielding, occurs at a medium level of field strength. At both higher and lower field strengths, the permeability, and hence the shielding, is lower. The effect at high field strengths is caused by saturation, which varies, depending on the type of material and its thickness. At field strengths well above saturation, the permeability falls off rapidly. In general, the higher the permeability, the lower is the field strength that causes saturation. Most magnetic material specifications give the best permeability, namely that at optimum frequency and field strength. Such specifications can be misleading.

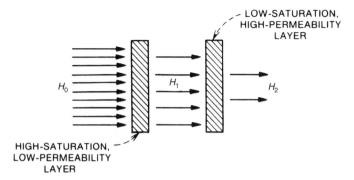

**FIGURE 6-22.** Multilayer magnetic shields can be used to overcome the saturation phenomenon.

To overcome the saturation phenomenon, multilayer magnetic shields can be used. An example is shown is Fig. 6-22. There, the first shield (a low-permeability material such as steel) saturates at a high level, and the second shield (a high-permeability material such as mumetal) saturates at a low level. The first shield reduces the magnitude of the magnetic field to a point that does not saturate the second shield; the second shield then provides most of the magnetic field shielding. These shields can also be constructed using a conductor, such as copper, for the first shield, and a magnetic material for the second. The low-permeability, high-saturation material is always placed on the side of the shield closest to the source of the magnetic field. In some difficult cases, additional shield layers may be required to obtain the desired magnetic field attenuation. Another advantange of multilayer shields is that increased reflection loss occurs from the additional reflecting surfaces.

Machining or working of some high-permeability materials, such as mumetal or permalloy, may degrade their magnetic properties. This can also happen if the material is dropped or subjected to shock. The material must then be properly reannealed after machining or forming to restore its original magnetic properties.

## 6.9 EXPERIMENTAL DATA

Results of tests performed to measure the low-frequency magnetic field shielding effectiveness of various types of metallic sheets are shown in Fig. 6-23 and 6-24. The measurements were made in the near field with the source and receptor 0.1 in apart. The shields were from 3 to 60 mils (in $\times$ $10^{-3}$) thick, and the test frequency ranged from 1 to 100 kHz. Figure 6-23 clearly shows the superiority of steel over copper for shielding magnetic fields at 1 kHz. But at 100 kHz, steel is only slightly better than copper. Somewhere between 100 kHz and 1 MHz, however, a point is reached where copper becomes a better shield than steel.

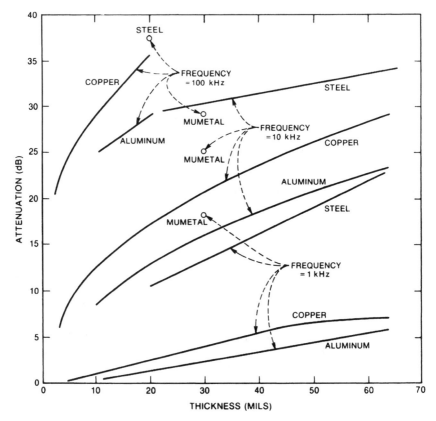

**FIGURE 6-23.** Experimental data on magnetic attenuation of metallic sheets in the near field.

Figure 6-23 also demonstrates the effect of frequency on mumetal as a magnetic shield. At 1 kHz, mumetal is more effective than steel, but at 10 kHz, steel is more effective than mumetal. At 100 kHz, steel, copper, and aluminum are all better than mumetal.

In Fig. 6-24, some data from Fig. 6-23 are replotted to show the magnetic field attenuation provided by thin copper and aluminum shields at various frequencies from 1kHz to 1MHz.

In summary, a magnetic material such as steel or mumetal makes a better magnetic field shield at low frequencies than does a good conductor such as aluminum or copper. At high frequencies, however, the good conductors provide the better magnetic shielding.

The magnetic shielding effectiveness of solid nonmagnetic shields increases with frequency. Therefore, measurements of shielding effectiveness should be made at the lowest frequency of interest. The shielding effectiveness of magnetic materials may decrease with increasing frequency as a result of the decreasing

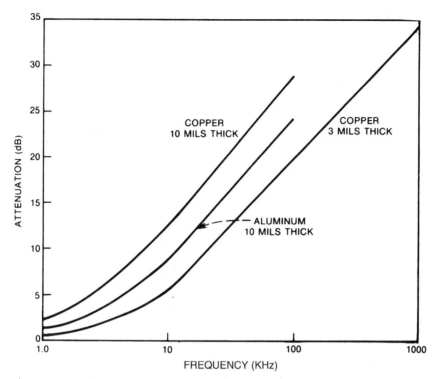

**FIGURE 6-24.** Experimental results of tests to determine magnetic field attenuation of conducting sheets in the near field.

permeability. The effectiveness of nonsolid shields will also decrease with frequency because of the increased leakage through the apertures.

## 6.10   APERTURES

The previous sections of this chapter have assumed a solid shield with no apertures. It has been shown that, with the exception of low-frequency magnetic fields, it is easy to obtain more than 100 dB of shielding effectiveness. In practice, however, most shields are not solid. There must be access covers, doors, holes for cables, ventilation, switches, displays, and joints and seams. All of these apertures considerably reduce the effectiveness of the shield. *As a practical matter, at high frequency, the intrinsic shielding effectiveness of the shield material is of less concern than the leakage through the apertures.*

Apertures have more effect on the magnetic field leakage than on the electric field leakage. Accordingly, greater emphasis is given to methods of minimizing the magnetic field leakage. In almost all cases, these same methods are more than adequate for minimizing the electric field leakage.

The amount of leakage from an aperture depends mainly on the following three items:

1. The maximum linear dimension, not area, of the aperture.
2. The wave impedance of the electromagnetic field.
3. The frequency of the field.

The fact that maximum linear dimension, not area, determines the amount of leakage can be visualized best by using the circuit theory approach to shielding. In this approach, the incident electromagnetic field induces current into the shield, and this current then generates an additional field. The new field cancels the original field in some regions of space, specifically the region on the opposite side of the shield from the incident field. For this cancellation to occur, the induced shield current must be allowed to flow undisturbed in the manner in which it is induced. If an aperture forces the induced current to flow in a different path, then the generated field will not completely cancel the original field, and the shielding effectiveness will be reduced. The more the current is forced to detour, the greater will be the decrease in the shielding effectiveness.

Figure 6-25 shows how apertures affect the induced shield current. Figure 6-25A shows a section of shield that contains no apertures. Also shown is the

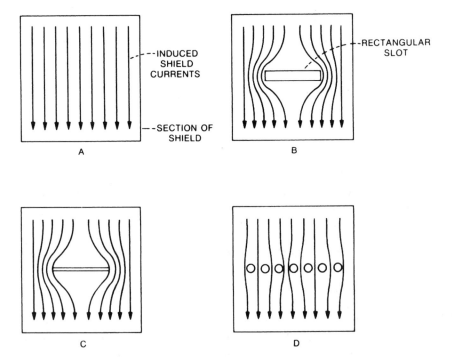

**FIGURE 6-25.** Effect of shield discontinuity on magnetically induced shield currents.

induced shield current. Figure 6-25B shows how a rectangular slot detours the induced current and hence produces leakage. Figure 6-25C shows a much narrower slot of the same length. This narrower slot has almost the same effect on the induced current as the wider slot of Fig. 6-25B and therefore produces the same amount of leakage even though the area of the opening is considerably reduced. Figure 6-25D shows that a group of small holes has much less detouring effect on the current than the slot of Fig. 6-25B, and therefore it produces less leakage even if the total area of the two apertures are the same. From this, it should be apparent that a large number of small holes produce less leakage than a large hole of the same total area.

The rectangular slots shown in Figs. 6-25B and 6-25C form slot antennas (Kraus and Marhefka, 2002). Such a slot, even if very narrow, can cause considerable leakage if its length is greater than 1/10 wavelength. Seams and joints often form efficient slot antennas. Maximum radiation from a slot antenna occurs when the maximum linear dimension is equal to 1/2 wavelength.

Based on slot antenna theory, we can determine the shielding effectiveness of a single aperture. Because a slot antenna is a most efficient radiator when its maximum linear dimension is equal to 1/2 wavelength, we can define the shielding for this dimension to be 0 dB. As the aperture becomes shorter, the radiation efficiency will decrease at a rate of 20 dB per decade, hence the shielding effectiveness will increase at the same rate. Therefore, for an aperture with a maximum linear dimension *equal to or less than 1/2 wavelength*, the shielding effectiveness in dB is equal to

$$S = 20 \log\left(\frac{\lambda}{2l}\right), \tag{6-32a}$$

where $\lambda$ is the wavelength and $l$ is the maximum linear dimension of the aperture. Equation 6-32a can be rewritten as

$$S = 20 \log\left[\frac{150}{f_{MHz} l_{meters}}\right]. \tag{6-32b}$$

Solving Eq. 6-32b for the length $l$ of the aperture gives

$$l_{meters} = \frac{150}{10^{\frac{S}{20}} f_{MHz}}, \tag{6-33}$$

where $S$ is the shielding effectiveness in decibels.

Slot antenna theory states that the magnitude and pattern of the radiation from a slot antenna will be identical to that of its complementary antenna, with the exception that the $E$ and $H$ fields will be interchanged, that is the polarization is rotated 90 degrees (Kraus and Marhefka, 2002 p. 307).

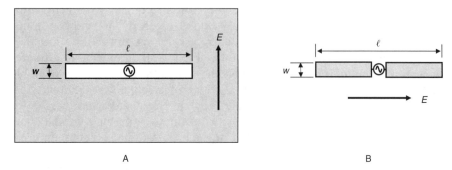

**FIGURE 6-26.** (A) A slot antenna, and (B) its complementary dipole antenna.

A complementary antenna is one in which the parts are interchanged. There-fore, the air (slot) is replaced by metal, and the metal into which the slot was cut is replaced by air as shown in Fig. 6-26. The complement of a slot antenna is a dipole antenna.

Figure 6-27 shows the shielding effectiveness versus frequency for various slot lengths. Both Eq. 6-32 and Fig. 6-27 represent the shielding effectiveness of only one aperture. In controlling slot lengths for commercial products, it is best to avoid apertures greater than 1/20 of a wavelength (this provides a shielding effectiveness of 20 dB). Table 6-5 gives the maximum slot lengths equivalent to 1/20 wavelength at various frequencies.

### 6.10.1   Multiple Apertures

More than one aperture will reduce the shielding effectiveness of an enclosure. The amount of reduction depends on (1) the number of apertures, (2) the frequency, and (3) the spacing between the apertures.

For a linear array of closely spaced apertures, the reduction in shielding effectiveness is proportional to the square root of the number of apertures $(n)$. Therefore, the shielding effectiveness, in decibels, from multiple apertures is

$$S = -20 \log \sqrt{n}, \tag{6-34a}$$

or

$$S = -10 \log n. \tag{6-34b}$$

Equation. 6-34 applies to a linear array of, equal sized and closely spaced apertures, where the total length of the array is less than 1/2 wavelength. Note that $S$ in Eq. 6-34 is a negative number.

The *net shielding effectiveness of a linear array of equal size holes* will be the shielding effectiveness of one of the holes (Eq. 6-32b) plus the shielding

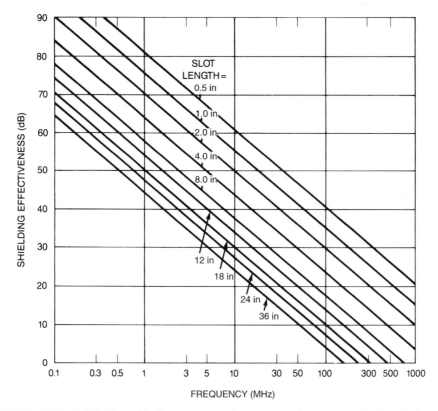

**FIGURE 6-27.** Shielding effectiveness versus frequency and maximum slot length for a single aperture.

**TABLE 6-5. Maximum Slot Length versus Frequency for 20-dB Shielding Effectiveness.**

| Frequency (MHz) | Maximum Slot Length (in) |
|---|---|
| 30 | 18 |
| 50 | 12 |
| 100 | 6 |
| 300 | 2 |
| 500 | 1.2 |
| 1000 | 0.6 |
| 3000 | 0.2 |
| 5000 | 0.1 |

effectiveness from multiple apertures (Eq. 6-34a), or

$$S = 20 \log \left[ \frac{150}{f_{MHz} l_{meters} \sqrt{n}} \right]. \tag{6-35}$$

Apertures located on different surfaces do not decrease the overall shielding effectiveness, because they radiate in different directions. Therefore, it is advantageous to distribute apertures around the surfaces of a product to minimize the radiation in any one direction.

Table 6-6 shows the reduction in shielding effectiveness from multiple apertures based on Eq. 6-34.

Equation 6-34 is applicable to a linear array of closely spaced apertures as shown in Fig. 6-28A. It cannot be directly applied to a two-dimensional array of holes as shown in Fig. 6-28B. However, if only the first row of holes is considered, then Eq. 6-34 would apply because that is a linear array of closely spaced apertures. The first row of holes cause the field induced shield current to be diverted around the holes as was shown in Fig. 6-25D, hence reducing the shielding effectiveness.

If a second row of holes is now added, it will not produce any significant additional current diversion; hence, the second row of holes will not have any additional detrimental affect on the shielding effectiveness. The same is true of the third through sixth row of holes. Therefore, only the holes in the first row of Fig. 6-28B, in this case six, need be considered. This approach is only an approximation, but experience has shown that it is a reasonable design

**TABLE 6-6.  Reduction in Shielding Effectiveness Versus the Number of Apertures.**

| Number of Apertures | S (dB) | Number of Apertures | S (dB) |
|---|---|---|---|
| 2 | −3 | 20 | −13 |
| 4 | −6 | 30 | −15 |
| 6 | −8 | 40 | −16 |
| 8 | −9 | 50 | −17 |
| 10 | −10 | 100 | −20 |

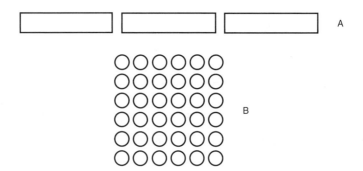

**FIGURE 6-28.** (A) Linear array of three closely spaced rectangular apertures; (B) two dimensional array of 36 round holes.

approach to an otherwise complex problem, and it allows one to apply Eq. 6-34 to a two dimensional array of holes. For a more detailed treatment of multiple apertures, see the article by Quine (1957).

As a general rule, when applying Eq. 6-34 to a two dimensional array of holes, determine the maximum number of holes lined up in a straight line, horizontal, vertical, or diagonal, and only use that number for $n$. For example, in the case of a $6 \times 12$ array of 72 holes, $n$ would be equal to 12 and the shielding would be reduced by 11 dB because of the multiplicity of holes.

## 6.10.2 Seams

A seam is a long narrow slot that may or may not make electrical contact at various points along its length.

It is sometimes difficult to visualize a seam as a radiating structure because one of the dimensions is so small, possibly only a few thousands of an inch. This can often best be visualized by considering the fact, as previously stated, that the magnitude of the radiation from a slot will be identical to that of its complementary antenna.

The complement of a long narrow seam is a long skinny wire that clearly looks like an efficient antenna—a dipole. This concept clearly illustrates why the length of the slot is more important than the area in determining the radiated emission. The length of the slot represents the length of the equivalent complimentary dipole. If the seam length happens to be in the order of 1/2 wavelength, it will become an efficient antenna. Therefore, it is necessary to guarantee electrical contact points at frequent intervals along a seam in order to reduce the length of the resulting antenna. A seam with periodic contact points can then be considered a linear array of closely spaced apertures.

Along the length of the seam, there should be firm electrical contact at intervals small enough to provide the desired shielding effectiveness. Contact can be obtained by using (1) multiple fasteners, (2) contact buttons, (3) contact fingers, or (4) conductive gaskets. Continuous electrical contact along the seam, although desirable, is often not required except when very large amounts of shielding are required or at high frequencies.

A properly designed seam will provide good electrical continuity between the mating parts. The transfer impedance across the seam should be in the neighborhood of 5 m$\Omega$ over the frequency range of interest, and it should not increase significantly with time or aging. Society of automotive engineers (SAE) standard ARP 1481 suggests 2.5 m$\Omega$ for the seam impedance. For a discussion of the meaning of transfer impedance, see Section 6.10.3.

Low contact resistance between mating surfaces is primarily a function of the following two items:

1. Having a conductive surface or finish
2. Providing adequate pressure

The material on both sides of a seam should be conductive. Although most bare metals initially will provide a good low-impedance surface, when left uncoated they will oxidize, anodize, or corrode, which increases the surface impedance dramatically. In addition, if two dissimilar metals are in contact, they will form a galvanic cell, or battery, in the presence of any atmospheric moisture, and galvanic corrosion will take place (see Section 1.12.1). One way to avoid this result is to use the same metal on both contact surfaces of the seam.

Most metals require a conductive finish to provide a low joint impedance throughout the life of the product. Some acceptable conductive finishes are listed in Table 6-7. The better performing finishes are at the top of the list, with shielding performance decreasing as you go down the list. The table assumes that the two mating surfaces are the same. Note that most materials require in the neighborhood of 100 to 200 psi of pressure for adequate performance. Gold and tin, both being soft and malleable, require much less pressure than the other finishes to produce a low impedance joint.

Gold being a noble metal provides a low contact resistance and is stable when exposed to most environments. Gold being malleable requires little contact pressure to provide a low impedance joint. The only negative consideration with respect to gold is its cost. In high-reliability designs, gold is often used and is usually compatible with all environments. Gold coatings as thin as 50 $\mu$in are often adequate. Gold is usually plated over 100 $\mu$in of nickel to provide a barrier to prevent diffusion of the gold into the base metal. Gold is galvanically compatible in a seam with nickel, silver, and stainless steel. Compatible means the materials can be used together in a seam and will maintain a low impedance joint, initially and with aging, without any other treatment when adequate pressure is applied. Gold, however, is not compatible with tin except in a controlled humidity environment.

Tin also provides a low contact resistance and is stable in most environments. Tin is so malleable that pressures above about 5 psi do not decrease the

**TABLE 6-7.   Conductive Finishes For Metals, Listed in Order of Effectiveness.**

| Finish | Pressure (psi) | Aging Performance |
|---|---|---|
| Gold | < 5 | Excellent |
| Tin | < 5 | Excellent |
| Electro-galvanize (Zinc) | 50–60 | Very Good |
| Nickel | > 100 | Very Good |
| Hot-dipped (Aluminum-Zinc) | > 100 | Good |
| Zinc plated | > 100 | Good |
| Stainless steel (Passivated) | > 100 | Very Good |
| Aluminum (Untreated) | > 200 | Moderate |
| Clear chromate on aluminum | > 200 | Moderate |
| Yellow chromate on aluminum | > 200 | Moderate/poor |

impedance of the joint. Tin performs almost as well as gold, costs significantly less, and is compatible with all environments. Tin is galvanically compatible in a seam with, nickel, silver, stainless steel, and aluminum, but not gold.

The primary negative with respect to tin is the possibility of metal whisker growth. Whiskers are small microscopic crystals that grow from the surface of some metals. The whiskers are extremely small, a few micrometers in diameter and from 50 to 100 $\mu$m long. They can break off from the surface and cause short circuits in electronic equipment.

The growth of metal whiskers on tin can be eliminated or minimized by the following techniques (MIL-HDBK-1250, 1995):

1. Use a heavy, rather than thin plating.
2. Use hot dip tin rather than electrodeposited tin.
3. Reflow tin plating to relive stresses.
4. Minimize the humidity in the environment.
5. Use tin-plating with 2% to 5% co–deposited lead.
6. Avoid using organic brighteners.

Nickel is also stable in most environments, and with adequate pressure will provide low contact resistance. Because nickel is hard, a pressure of 100 psi or more is required. Nickel is galvanically compatible in a seam with tin, stainless steel, silver, and gold.

Stainless steel is becoming popular for use in many low-cost commercial products as well as in medical electronics. It is often a lower cost alternative to tin or nickel plated steel, because it avoids the necessity of a secondary plating operation. Although stainless steel is much less conductive than aluminum, or steel, its inherent surface stability in most environments more than offsets its diminished conductivity when used in seams, as long as sufficient pressure is applied. In most applications, it requires no additional treatment or coating. Stainless steel is galvanically compatible in a seam with tin, nickel, and gold.

Untreated aluminum on exposure to air will form a thin initial oxide that is stable, limiting in thickness, and has excellent corrosion resistance. Although this film is nonconductive, it can easily be penetrated, to provide electrical conductivity, if sufficient pressure is applied. Untreated aluminum is compatible in a seam with tin.

Chromate conversion coatings are used to protect metals, usually aluminum or magnesium alloys, from corrosion. The process consists of first acid etching the aluminum and then applying a strong oxidizer, which produces a chemical conversion of the surface (oxidizes it). The result is a microscopically thin film on the surface (typically only several micro-inches thick). These coatings, however, are *nonconductive*. For two mating surfaces to make good electrical contact, the nonconductive chromate surface finish must be penetrated. The ease at which this penetration occurs is a function of the thickness of the chromate coating, which is neither easily controlled nor easily measurable other

TABLE 6-8.   **Classifications of Chromate Coatings.**

| Class | Appearance | Corrosion Protection |
|-------|------------|---------------------|
| 1 | Yellow to brown | Maximum |
| 2 | Yellow | Moderate |
| 3 | Clear | Minimum |
| 4 | Light green to green | Moderate |

than by its color. As the thickness increases, the color of the coating darkens from clear, to yellow, to green, to brown. The thicker coatings provide increased corrosion resistance as well as increased electrical resistance.

Table 6-8 lists the classifications of chromate coatings (ASTM B 449, 1994).

Only clear and in some cases yellow chromate coatings should be considered for joints that require electrical contact. It is best to specify a clear chromate finish per MIL-STD-C-5541E, Class 3 coating. MIL-STD-C-5541E, Class 3 requires an initial surface resistance no greater than 5 mΩ per square inch, and 10 mΩ per square inch after 168 hours of salt spray exposure, when measured with an electrode pressure of 200 psi. In contrast, highly colored brown and green coatings can have surface resistances well in excess of 1000 Ω.

Alodine® and Iridite™ are trade names for two commercially available chromate conversion processes.*

Other variables that are also important to the contact resistance of chromate finishes, in addition to the coating thickness, are as follows:

1. Pressure
2. Surface roughness
3. Flatness of mating surfaces

Contact resistance will decrease as the pressure is increased, as the surface roughness increases, and as the panel flatness increases.

Aluminum with a clear chromate finish is galvanically compatible in a seam with tin and nickel.

Magnesium alloy castings are becoming popular in many small portable electronic devices, such as digital cameras. The advantages of magnesium in these applications are (1) it is light weight, and (2) it has excellent fluidity, which allows for thin wall casting with minimum draft and maximum dimensional accuracy. The problem with magnesium when a seam is involved is that it is very susceptible to corrosion. It is the least noble of all structural metals, and appears at the anodic end of the galvanic series (the opposite end from gold) as was shown in Table 1-5. If gold is the best metal for a joint, then magnesium is the worst. Magnesium is so corrodible that the only surface finish that has been

---

*Alodine is a registered trade mark of Henkel Surface Technologies, Madison Heights, MI; and Iridite is a trade mark of MacDermid Industrial Solutions, Waterbury, CT.

found to be satisfactory for a majority of environments is a minimum of 0.001 in thick tin plate (ARP 1481, 1978).

Just having two conductive metals in contact is not sufficient to provide a low impedance joint. Sufficient pressure between the surfaces must be provided as part of the mechanical design. This can be accomplished by the means of fasteners or by somehow incorporating spring pressure into the joint design. The design must then guarantee the pressure even under the worst-case dimensional tolerances and with aging of the product. Most surface finishes require a pressure in the vicinity of 100 psi or more.

If a conductive gasket is to be used, then it can provide the required pressure as well as take up any gaps in the joint design resulting from mechanical tolerances. However, sufficient pressure must still be provided, as specified by the gasket manufacturer.

### 6.10.3 Transfer Impedance

The shielding effectiveness of a seam or joint is a difficult parameter to measure accurately because it involves a complicated radiated emission test set up with many variables. A better, more reliable, and more repeatable method of measuring the quality of electrical contact between mating parts of a shield is by measuring its transfer impedance. The concept of transfer impedance originated over 75 years ago as a means of measuring the effectiveness of shielded cables; see Section 2.11 (Schelkunoff, 1934). The concept has since been extended to include the behavior of joints in shielded enclosures (Faught, 1982).

Basically, the transfer impedance test measures the voltage across the joint with a known high-frequency current flowing across it. The ratio of the voltage to the current is the transfer impedance. Using a spectrum analyzer with a tracking generator and an appropriate coaxial test fixture such as shown in Fig. 6-29, the transfer impedance of test samples can very easily be measured over a wide frequency spectrum.*

The shielding effectiveness will be proportional to the inverse of the measured transfer impedance. The shielding quality (which is a good approximation to the shielding effectiveness) is defined in SAE ARP 1705 as the ratio of the incident wave impedance $Z_W$ divided by the transfer impedance $Z_T$, thereby giving as the relationship between transfer impedance and shielding effectiveness

$$S = 20 \log \frac{Z_W}{Z_T}. \tag{6-36a}$$

---

* The measurement is simple; however, the test setup is not. The complete setup, which includes the test fixture itself, must maintain 50-$\Omega$ impedance over the frequency range of the test, usually up to 1 GHz. This is necessary to prevent reflections produced in the test system from inducing errors into the measurement.

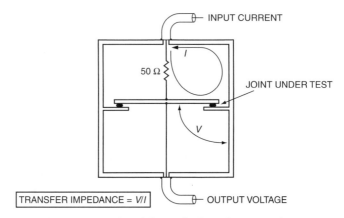

**FIGURE 6-29.** Coaxial transfer impedance test fixture.

For the case of a plane wave, Eq. 6-36a becomes

$$S = 20 \log \frac{377}{Z_T}. \tag{6-36b}$$

Figure 6-30 shows measured transfer impedances data for various combinations of dissimilar metals both before and after accelerated aging (Archambeault and Thibeau, 1989). The aging was intended to simulate 5 to 8 years' exposure in a typical commercial product environment. The samples were aged for 14 days in an environmental chamber that contained 10 ppb of chlorine, 200 ppb of nitric oxide, and 10 ppb of hydrogen sulfide with controlled temperature and humidity.

Each graph is for a coated metal, mated with three different conductive gaskets. The gaskets were tin plated, nickel plated, and tin-lead plated beryllium copper spring fingers. The black bars show the initial transfer impedance, and the white bars show the impedance after aging. No data were available after aging for the tin-plated gasket when used with the aluminum-zinc coated steel or with the stainless steel.

Although the test measured the transfer impedance of the test samples across the frequency range of 1 MHz to 1 GHz, only the worst-case impedance is plotted in Fig. 6-30. Note that the vertical axis represents scale numbers rather than the actual transfer impedance. The range of transfer impedance corresponding to each scale number is listed in the figure.

Using the criteria of maintaining a transfer impedance of 5 mΩ or less would be equivalent to readings on the vertical axis of four or less on the bar graphs of Fig. 6-30. Notice that the tin-plated steel was the only material tested that showed no increase in impedance with aging.

The impedance across a seam consists of a resistive and a capacitive component in parallel, as shown in Fig. 6-31A. The resistance is a function of

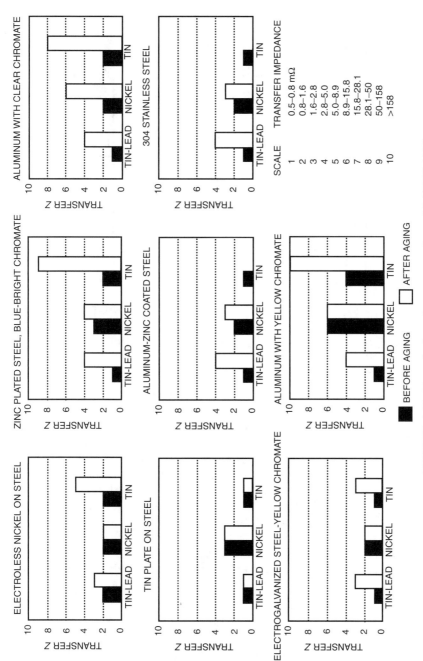

**FIGURE 6-30.** Transfer impedance comparisons (© IEEE 1989).

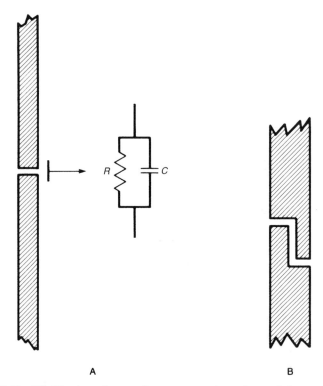

**FIGURE 6-31.** (A) The impedance of a seam consists of a resistive and capacitive component; (B) seam overlap increases the capacitance across the joint.

actual electrical contact between the mating parts, and it depends primarily on the surface finish and pressure. The capacitance does not require actual electrical contact, rather it depends on the spacing and the surface area of the two half's of the seam. Designing a seam to have a flange or overlap (Fig. 6-31B), as opposed to a butt joint, as shown in Fig. 6-31A is helpful in increasing the capacitance. Increasing the capacitance alone, however, is not sufficient to produce a low-impedance seam without also providing direct electrical contact with sufficient pressure between the mating parts. As previously indicated, for good shielding effectiveness the transfer impedance across the seam should not be more than a few milliohms. A capacitance of 100 pf, a large amount of capacitance for a seam, at a frequency of 1 GHz has an impedance of 1.6 Ω. This result is almost three orders of magnitude greater than a few milliohms; hence the increased capacitance is not very effective in reducing the seam impedance because it has a small effect on the total joint impedance.

## 6.11 WAVEGUIDE BELOW CUTOFF

Additional attenuation can be obtained from an aperture if the hole has depth, that is, if it is shaped to form a waveguide, as shown in Fig. 6-32. A waveguide

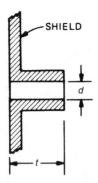

**FIGURE 6-32.** Cross section of a hole formed into a waveguide with diameter $d$ and depth $t$.

has a cutoff frequency below which it becomes an attenuator. For a round waveguide the cutoff frequency is

$$f_c = \frac{6.9 \times 10^9}{d} \text{ Hz,} \tag{6-37}$$

where $d$ is the diameter in inches. For a rectangular waveguide the cutoff frequency is

$$f_c = \frac{5.9 \times 10^9}{l} \text{ Hz,} \tag{6-38}$$

where $l$ is the largest dimension of the waveguide's cross section in inches.

As long as the operating frequency is much less than the cutoff frequency, the magnetic field shielding effectiveness of a round waveguide (Quine, 1957) is

$$S = 32\frac{t}{d} \text{ dB,} \tag{6-39}$$

where $d$ is the diameter and $t$ is the depth of the hole, as shown in Fig. 6-32. For a rectangular waveguide (Quine, 1957), the shielding effectiveness is

$$S = 27.2\frac{t}{l} \text{ dB,} \tag{6-40}$$

where $l$ is the largest linear dimension of the hole's cross-section and $t$ is the length or depth of the hole.

The shielding determined from Eqs. 6-39 or 6-40 is in *addition* to that resulting from the size of the aperture (Eq. 6-32b). A waveguide having a

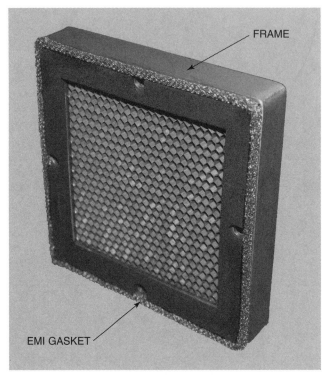

**FIGURE 6-33.** Honeycomb ventilation panel, rear view showing EMI gasket used to make electrical contact to chasis that it is mounted on (courtesy of MAJR Products Corp.).

length three times its diameter will have close to 100 dB of additional shielding.

The classic example of the application of this principal is a honeycomb ventilation panel as shown in Fig. 6-33. When mounted, the entire perimeter of the panel must make electrical contact to the chassis. The maximum dimension of the holes is usually 1/8-in, and the panels are usually 1/2-in., thick, giving a $t/d$ ratio of 4 and 128 dB additional shielding from the waveguide effect.

## 6.12 CONDUCTIVE GASKETS

The ideal shield is a continuous conductive enclosure with no apertures, a Faraday cage. Joints made with continuous welding, brazing, or soldering provides the maximum shielding. Rivets and screws make less desirable, but often more practical, joints. If screws are used, they should be spaced as close together as practical. Spacing of contact points should be such that the maximum dimension of the slot formed is limited to an appropriate length for the degree of shielding required. If the required screw spacing becomes too

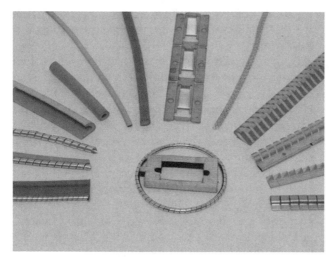

**FIGURE 6-34.** Various types of EMI gaskets.

small to be practical, then electromagnetic interference (EMI) gaskets should be considered.

The primary function of an EMI gasket is to provide a conductive path between the two mating parts of a seam. An EMI gasket combined with the proper surface finish on the enclosure will provide good electrical continuity between the mating parts, thus minimizing the impedance of the joint and increasing the shielding effectiveness of the enclosure. It is important to remember that EMI gaskets work by providing a low-impedance conductive path across the seam, not by just filling the gap created by the seam.

Some of the more common types of EMI gaskets include conductive elastomers, metal spring fingers, wire mesh, spiral ribbons, and fabric over foam. Newer gasket technologies include wire knit, form in place, and die cut. Each type has its advantages and disadvantages. The gaskets are available in a variety of cross-section designs as well as in various materials and surface finishes, which include tin, nickel, beryllium-copper, silver, stainless steel, and so on. Conductive adhesives, caulks, and sealants are also available. A sampling of some various gaskets types are shown in Fig. 6-34.

### 6.12.1   Joints of Dissimilar Metals

When two dissimilar metals are joined, a galvanic couple is formed as was discussed in Section 1.12.1. The gasket material must therefore be selected to be galvanically compatible with the mating surface to minimize corrosion.

Tin, nickel, and stainless steel are all compatible with each other and should cause no problem when mated together in a joint or seam. All of these materials will retain close to their original electrical conductivity after aging when mated with 100 psi or more of pressure.

Aluminum presents more of a problem. Aluminum is compatible with itself and also with tin. Aluminum with a clear or yellow chromate finish is also compatible with nickel.

Silver is not compatible with aluminum. Silver, however, is often used as the conductive material on fabric over foam and conductive elastomer gaskets. Experience has shown that silver-filled conductive elastomers and silver cloth-over-foam gaskets do not behave galvaniclly the same as pure silver. Rather, they show much less galvanic corrosion than predicted, especially when mated with aluminum. The gasket manufacturer's data should be consulted to determine the compatibility of these gaskets with different metals.

Spring finger EMI gaskets are often made of beryllium copper, because beryllium copper is the most conductive of the spring materials. Beryllium copper, however, is not very compatible with other materials. Therefore, when beryllium-copper spring fingers are used, they should be plated with tin or nickel to form a compatible galvanic couple with the enclosure. If the spring fingers are to mate against beryllium-copper, however, they can be left unplated.

Figure 6-30 showed measured transfer impedance data for various galvanic couples, before and after aging. The Society of Automotive Engineers Aerospace Recommended Practice ARP 1481, Corrosion Control and Electrical Conductivity in Enclosure Design, contains a more extensive matrix of the compatibility of various combinations of galvanic couples for almost fifty metals, alloys, and finishes, developed through experience, testing, and actual use.

### 6.12.2   Mounting of Conductive Gaskets

Figure 6-35 shows the correct and incorrect way to install an EMI gasket between an enclosure and its cover. The gasket should be in a slot and on the inside of the screws to protect against leakage around the screw holes. For electrical continuity across the joint or seam, the metal should be free of paints, oxides, and insulating films. The mating surfaces should be protected from corrosion with a conductive finish.

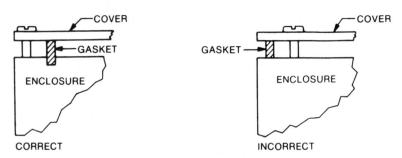

**FIGURE 6-35.** EMI gaskets, correct and incorrect installation.

If both EMI protection and environmental protection are required, then two separate gaskets, or a combination of EMI and environmental gasket, may be used. The combination gasket usually has a conductive EMI gasket combined with a silicon rubber environmental gasket. If both environmental and EMI gaskets are installed, either as a combination unit or two separate gaskets, the EMI gasket should be on the inside of the environmental gasket.

With a sheet metal enclosure, the EMI gasket may be mounted by one of the methods shown in Fig. 6-36.

In the seam designs shown in both Figs. 6-35 and 6-36, the EMI gaskets are in compression and require screws or some other means to provide the necessary pressure to properly compress the gasket. With the seam designs shown in Fig. 6-37, however, the gaskets are mounted in sheer, and no fasteners are required to provide the necessary compression. When the gasket is in sheer, the design of the seam itself provides the pressure, when assembled. Also, with the designs shown Fig. 6-37 the performance of the seam is not a function of the exact positioning of the cover or front panel.

In addition to surface finish and mounting methods, the following should also be considered when choosing an EMI gasket:

- The degree of shielding required
- The environment (air conditioned, high humidity, salt spray, etc.)
- Compression set
- Compression pressure required
- Fastener spacing requirements
- Gasket attachment method
- Maintainability, open and close life cycles, and so on

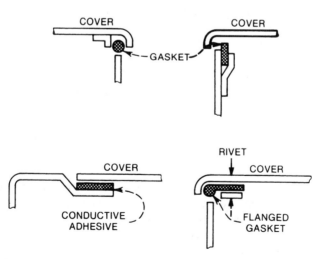

**FIGURE 6-36.** Suitable ways to install EMI gaskets in sheet metal enclosures.

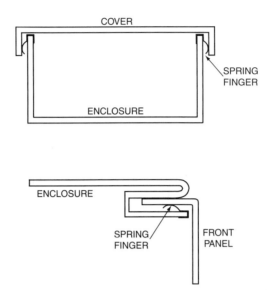

**FIGURE 6-37.** EMI gasket mounting methods that require no fasteners. Gasket is mounted in shear not compression.

The seam design must provide sufficient pressure to compress the gasket material properly to guarantee a low impedance joint. If, however, the gasket is compressed too much, it will be permanently deformed and lose its resiliency. Subsequent opening and closing will not then provide sufficient compression for the gasket to perform properly. Compression set must therefore be avoided.

To avoid compression set, mechanical stops should be designed into the seam to prevent exceeding the maximum compression value for the gasket, which is typically 90% of the gasket diameter or height. Another approach is to mount the gasket in a groove. When a groove is used, not only is its depth important but also its width. The cross section of the grove must be designed with sufficient room to hold the fully compressed gasket. Dovetail grooves, although more expensive, often can be used very effectively in this application. A properly designed dovetail groove has the added advantage of preventing the gasket from falling out of the groove when the seam is disassembled.

Most of the major gasket manufacturers have excellent application engineering departments that should be consulted in choosing the proper gasket and surface finish for the intended application.

If perforated sheet stock or screening is used to cover a large aperture, then the material should have electrical continuity between the strands. In addition, the entire perimeter of the screening must make electrical contact to the enclosure. Remember, the purpose of the screen is to allow current to flow across the aperture, not just to cover up the hole.

Conductive gaskets can also be used around switches and controls mounted in the shield, which can be mounted as shown in Fig. 6-38.

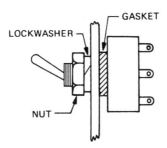

**FIGURE 6-38.** Switch mounted in a panel using an EMI gasket.

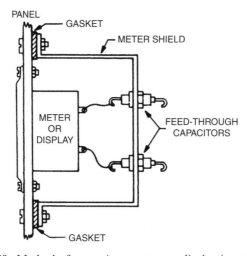

**FIGURE 6-39.** Method of mounting a meter or display in a shielded panel

Larger holes cut in panels for meters or liquid crystal display (LCD) displays can completely destroy the effectiveness of an otherwise good shield. If meters or LCD displays are used in a shield panel, then they should be mounted as shown in Fig. 6-39 to provide shielding of the meter or display hole. Effectively, this puts the shield behind the display. All wires that enter or leave the display must be shielded or filtered, as discussed in Chapter 3 and 4, to avoid compromising the shielding effectiveness.

For optimum shielding, the enclosure should be thought of as "electrically watertight," with EMI gaskets used in place of normal environmental gaskets.

## 6.13   THE "IDEAL" SHIELD

Let us say that you want to design the "ideal" shield, with no concern for the cost or the complexity of the design. You would probably make the shield from a magnetic material to maximize the absorption loss—steel would be a good choice. To provide protection against rusting and to provide good electrical continuity across any seams, you would probably plate the steel with a material that is stable in most environments and has low-contact resistance, such as tin. To avoid electromagnetic energy leakage at the seams, you would weld or solder all the seams except one; remember, cost is not a concern. The one seam, the cover, would not be welded or soldered so that you could install and remove the electronics from the enclosure.

The one seam, between the cover and the enclosure, would have to have a conductive surface finish. The tin plate already applied to the steel should provide that, without any additional cost or complexity. For maximum shielding effectiveness, the design of the seam should be such that the two mating parts makes good electrical contact all along their length. The seam design must also provide high contact pressure, 100 to 200 psi or more between the mating parts. Although cost is not a concern, it would be nice if no gasket or fasteners were required to guarantee the electrical continuity, a pressure fit between the cover and the enclosure might be a viable, although unusual, design approach.

Congratulations, you just designed the "ideal" shield, also known as a paint can! The last time I checked (2008) you could buy, at retail, a gallon-size paint can in small quantities for under $3.00 US. In volume, the cost would be less than half of this. Conclusion, a shield if properly designed does not have to be expensive to be effective.

## 6.14   CONDUCTIVE WINDOWS

Large viewing windows are difficult to shield because they require a high degree of optical transparency. As a result, special conductive windows are made for these applications. Two primary approaches to making a window conductive, are as follows: (1) transparent conductive coatings and (2) wire mesh screens.

### 6.14.1   Transparent Conductive Coatings

Extremely thin, transparent conductive coatings can be vacuum deposited onto optical substrates, such as plastic or glass. This approach results in good shielding properties with moderate optical transparency. Because the deposited film thickness is in the micro inches, little absorption loss occurs, and the primary shielding comes from reflection loss. Hence, material with good conductivity is used. Because the shielding effectiveness is a function of the resistivity of the coating (which depends on the thickness), and the optical transparency depends on the thinness of the coating, trade offs must be made. Typical surface resistivity ranges from 10 to 20 $\Omega$ per square, with optical transmission of 70% to 85%.

Typically, gold, indium, silver, or zinc oxide are used as the deposited materials, because of their high stability and good conductivity. Conductive coatings can also be deposited onto optical quality polyester film, which is then laminated between two pieces of glass or plastic to form the window.

### 6.14.2  Wire Mesh Screens

A conductive wire mesh, or perforated metal, screen can be laminated between two clear plastic or glass sheets to form a shielding window. Another approach is to cast the wire mesh screen within a clear plastic. The primary advantage of wire mesh screen windows over coated windows is optical transparency. Wire mesh screens can have optical transparencies as high as 98%. The door on a microwave oven is a good example of the use of a perforated metal screen as a viewing window.

Wire mesh screens are often made from copper or stainless steel mesh with 10 to 50 conductors per inch. These screens provide the highest optical transparency, typically 80% to 98%. Screen made from mesh with 50 to 150 conductors per inch provides the highest shielding effectiveness but reduced optical transparency. In all wire mesh screens; bonding of the wires at their crossover is required for good shielding effectiveness.

Generally, wire mesh or perforated metal screens provide greater shielding effectiveness than conductive-coated windows. Their primary disadvantage is that they can sometimes inhibit viewing because of diffraction.

### 6.14.3  Mounting of Windows

The method used for mounting a shielded window (whether wire mesh or conductive coated) is as important as the material of the window itself in determining the overall performance. The mounting of a window must be such that there is good electrical contact between the wire mesh screen (or conductive coating) and the surface of the mounting enclosure, along the entire perimeter of the window. The window design will be such that the conductive coating or wire mesh is accessible at the edge around the periphery of the assembly. Figure 6-40 shows one possible method of mounting a conductive window.

## 6.15  CONDUCTIVE COATINGS

Plastics are popular for packaging electronic products. To provide shielding, these plastics must be made conductive, and the following two basic ways can be used to accomplish this: (1) by coating the plastic with a conductive material or (2) by using a conductive filler molded into the plastic. When conductive plastics are used, the important considerations are the required shielding effectiveness, the cost, and the aesthetic appeal of the final product.

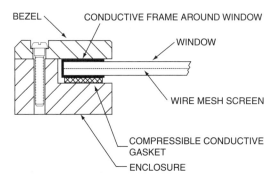

**FIGURE 6-40.** Mounting of a conductive window.

As discussed, previously shielding effectiveness depends not only on the material used but also on the control of the leakage through the seams and holes of the enclosure. Everything previously said in Section 6.10 about controlling apertures is applicable to conductive-coated plastic shields. As is the case for solid metal shields, the apertures are usually the limiting factor in high-frequency shielding effectiveness. The most difficult, and therefore expensive, part of using conductive plastics is often the controlling of the leakages through the apertures, in particular the seams. The maximum length of an aperture or slot was given in Eq. 6-33, and the shielding effectiveness of a fixed length slot by Eq.6-32b.

The conductive coatings used on plastics are usually thin; therefore, absorption loss does not become significant until high frequencies. As a result, reflection loss is often the primary shielding mechanism.

To be effective as shields, conductive plastics must have a surface resistivity of a few ohms per square or less. Therefore, high conductivity materials such as silver, copper, zinc, nickel, or aluminum should be used. The reflection loss of a conductive coating can be approximated by Eq. 6-22, where $Z_w$ is the wave impedance and $Z_s$ is the surface resistance of the coating in ohms per square. For example, a conductive coating with a surface resistivity of 1 $\Omega$ per square, will have a plane wave ($Z_w = 377$) reflection loss of approximately 39.5 dB.

To provide protection only against electrostatic discharge (ESD), however, considerably higher surface resistivity may be used, up to a few hundred ohms per square. Therefore, carbon or graphite materials can often be used if ESD protection is the only requirement.

Another consideration when using conductive-coated plastic is that IEC 60950-1 and other electrical safety requirements requires that any exposed metal can withstand a 25-A surge current. Some thin conductive coatings will not be able to pass this test.

Some methods of producing a conductive plastic enclosure are as follows:

1. Conductive paints
2. Flame/arc spray

3. Vacuum Metalizing
4. Electroless plating
5. Metal foil linings
6. Metallic fillers molded into the plastic

Conductive paints, and electroless plating are the two methods most commonly used for EMI shielding.

### 6.15.1  Conductive Paints

Many electronic products used today are coated with conductive paints. The coating consists of a binder (usually urethane or acrylic) and a conductive pigment (silver, copper, nickel, or graphite). A typical mixture can contain as much as 80% conductive filler and only 20% organic binder. Nickel and copper are the most common fillers used.

Conductive paints can provide good surface conductivity (less than 1 $\Omega$ per square), although not as good as the pure metal, because electrical conductivity is achieved only through contact between the conductive particles in the paint. The surface conductivity of conductive paint is typically one to two orders of magnitude less than that of the pure metal.

Conductive paints can be applied easily with standard spray equipment, and parts of the enclosure can be masked off so that they are not coated. This process is inexpensive; however, to obtain maximum effectiveness, the enclosure must be designed to provide adequate pressure and electrical continuity across the seams—thereby complicating the mold for the plastic part. If the plastic enclosure is designed with no consideration for the possibly of being conductively coated, then the seam design most likely will not be adequate to provide good shielding effectiveness.

Conductive paints are commonly used and, with the exception of those that contain silver, are the least expensive of the common coating methods.

### 6.15.2  Flame/Arc Spray

In the flame/arc spray coating method, a low-melting-point metal wire, or powder (usually zinc), is melted in a special spray gun and sprayed onto the plastic material. This method produces a hard, dense coating of metal with excellent conductivity. Its disadvantage is that the application process requires special equipment and skill. Therefore this method is more expensive than spray painting, but it provides good results because a pure metal is deposited onto the plastic, not just metal particles in a nonconductive binder as is the case for conductive paints.

### 6.15.3  Vacuum Metalizing

In vacuum metalizing, a pure metal, usually aluminum, is evaporated in a vacuum chamber and then condenses on and bonds to the surface of the plastic parts in the chamber, forming a fairly uniform coating of pure aluminum on the surface. The coating is not as uniform, however, as in the case of electroless plating, but it is usually more than adequate. Areas can easily be masked to control where the metal is deposited and where it is not. This procedure produces a surface with excellent adhesion and conductivity. The disadvantage of this method is that expensive special equipment is required.

Vacuum metalizing is often used for small production runs, or when fast turn around is required. This process is also used to produce a chrome look decorative trim on plastic, for use in automotive trim, toys, models, decorative fixtures, and so on.

### 6.15.4    Electroless Plating

Electroless plating—or chemical deposition, which is a more accurate name—consists of depositing a metallic coating (usually nickel or copper) by a controlled chemical reaction that is catalyzed by the metal being deposited. It is accomplished by submerging the parts to be plated into a chemical bath. The process produces a uniform coating of pure metal with very good conductivity, and it can be applied to simple or complex shapes. The conductivity approaches that of the pure metal. The process, although more costly than the previous coating methods, is commonly used because of its excellent performance.

This process is similar to that used to produce plated-through holes (vias) on printed circuit boards

For electromagnetic compatibility (EMC) shielding, a two-step process is usually used, with a thin layer of nickel being deposited over a thicker layer of copper. Most of the shielding comes from the copper layer, however, the nickel overcomes two disadvantages of copper, (1) that copper is soft and can rub off or be worn through easily with multiple assembly and disassembly of the joint, and (2) that copper oxidizes and will form a nonconductive surface over time. Nickel being hard and environmentally stable overcomes these two problems. Although more complicated, selective plating only to specific areas of the plastic part can also be done. This, however, requires a few additional steps, and therefore it increases the cost of the process.

In many instances, electroless plating is the preferred method of obtaining high levels of shielding effectiveness in plastic enclosures. It is considered to be the standard by which all other coating methods are judged. When combined with proper seam design and aperture control, electroless plating can provide high levels of shielding effectiveness.

### 6.15.5    Metal Foil Linings

A pressure-sensitive metallic foil (usually copper or aluminum), with an adhesive backing, is applied to the interior of the plastic part. The metal foil lining provides very good conductivity. This method of coating parts is most often used in experimental work. It is not usually desirable for production because it is slow and requires a lot of manual labor. Complex shaped parts are also difficult to cover in this manner.

### 6.15.6 Filled Plastics

Conductive plastic can be produced by mixing a conductive agent with the plastic resin prior to injection molding. The result is an injection-moldable composite. The conductive material may be in the form of fibers, flakes, or powders. A wide range of conductivity can be achieved by this approach, without the necessity of a second coating operation. Typical conductive fillers are aluminum flakes, nickel or silver coated carbon fibers, or stainless steel fibers. The resulting conductivity is usually limited, because conductivity is the result of contact between the conductive particles in the plastic.

Loading levels of conductive fillers can vary from 10% to 40% to achieve the desired electrical properties. High loading levels, however, often alter the mechanical properties, colorability, and aesthetics of the base material to the point where the altered mechanical properties may no longer fit the application.

The primary advantage of filled plastics is that the secondary step of coating the material to achieve conductivity can be eliminated. However, because the conductive material is inside the plastic, the surface may not be conductive. This makes the controlling of the conductivity across a seam or joint difficult. A secondary machining operation may be necessary on the edges of the material to expose the conductive particles in the seam, which defeats the advantage of not requiring a secondary processing step.

The idea of using conductive filled plastic for EMC shielding was introduced about 35 years ago with much hype, and it has not lived up to its expectations. Conductive filled plastic have been successfully used in some instances for ESD control, where only a limited amount of conductivity is required, but the process has not lived up to its promise as a simple, inexpensive, universal solution for EMC shielding of plastics.

## 6.16 INTERNAL SHIELDS

Shielding does not always have to be done at the enclosure level. It can also be done within the enclosure at the subassembly level, by shielding individual modules, card cages, and so on. Shielding can also be brought down to the printed circuit board (PCB) level with board level shields around individual groups of components. At the board level, a five-sided shield can be mounted on the PCB and connected to the ground plane with either through-hole or surface-mount technology. The ground plane in the board then becomes the sixth side of the shield.

Figure 6-41 shows a cell phone PCB and its board mounted shield. The shield provides not only overall shielding for the board, but also shielding between different sections of the board itself. In this example, the shield is

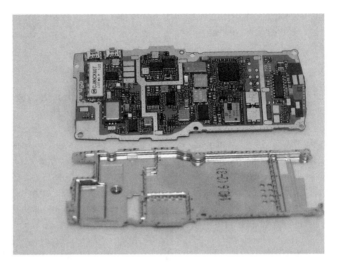

**FIGURE 6-41.** A PCB (top) with its board level shield (bottom).

stainless steel and the mating surfaces on the PCB are gold plated. The six screws that normally hold the two halves of the cell phone clamshell together are also used to attach the shield to the PCB. Closely spaced bent spring tabs on the shield make contact to the board at periodic intervals between the screw holes. Even though the screw holes are 1 3/4 in apart, the spring tabs limit the maximum linear dimension of the aperture to 0.1 in. This dimension is equivalent to 1/20 wavelength at 5.9 GHz.

This chapter has so far only considered shields that completely enclose a product, and it has emphasized the importance of controlling apertures. In some cases, however, partial shields can be effective for internal shielding. For example, putting a fence on a PCB around sensitive components can be effective in reducing near-field capacitive coupling as shown in Fig. 6-42. One could argue, rightfully so, that this is not truly a shield but just a method of breaking the capacitive coupling between adjacent circuits. The fence must be connected to the PCB ground plane, as shown in Fig. 6-42b, to provide a return current path for the intercepted capacitive coupled current. A fence can often be effectively used between the digital logic circuits and the input/output (I/O) circuitry to minimize the coupling of the digital noise to the I/O.

Shielding can even be applied within a PCB by using power and/or ground planes to shield layers located between them. An example of this is shown in Fig. 16-17. The high-speed signals are routed on layers 3 and 4, and they are shielded by the planes on layers 2 and 5. A 1-oz copper plane in a PCB will be greater than three skin depths thick at all frequencies above 30 MHz. For a 2 oz copper plane, this will be true above 10 MHz.

Some radiation, however, may still occur from the edges of the board. With larger layer count boards, multiple ground planes can be used as the shields and connected together around the periphery of the board with closely spaced

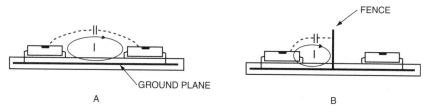

**FIGURE 6-42.** (*A*) Parasitic capacitance coupling between integrated circuits (ICs), (*B*) parasitic coupling blocked by the use of a PCB fence. Current *I* represents noise coupling through parasitic capacitance.

ground vias to reduce or eliminate any edge radiation from the board. On some high-frequency boards, with gigahertz plus signals, the edges of the board are plated with copper to complete the shield.

## 6.17 CAVITY RESONANCE

Energy enclosed in a metal box will bounce off the box's interior walls. If electromagnetic energy gets into the box (from a source inside the box or from leakage of external energy into the box), the box will act as a cavity resonator. As the energy bounces (reflects) from one surface to another, it produces standing waves within the box, depending on the size and shape of the enclosure. At frequencies that correspond to the box resonance, these standing waves will produce high field strengths at some points within the enclosure and nulls at other points. For a *cubical box*, the *lowest resonant frequency* is (White, 1975)

$$f_{MHz} = \frac{212}{l}, \qquad (6\text{-}41)$$

where $f_{MHz}$ is the resonant frequency in megahertz and $l$ is the length of one side of the box in meters. Additional resonances will occur at frequencies above this.

In general, a rectangular box will have multiple resonant frequencies located at (White, 1975)

$$f_{MHz} = 150\sqrt{\left(\frac{k}{l}\right)^2 + \left(\frac{m}{h}\right)^2 + \left(\frac{n}{w}\right)^2}, \qquad (6\text{-}42)$$

where $l$, $h$, and $w$ are the length, height, and width (in meters) of the box respectively, and $k$, $m$, and $n$ are positive integers (0, 1, 2, 3,.........etc.) that represent the various possible modes of propagation. No more than one of the integers can be zero at any one time. For example, k $= 1$, $m = 0$, and $n = 1$

represents the lowest resonant frequency for the TE101 mode of propagation. The resonant frequency that corresponds to the lowest order of propagation will have two of the integers ($k$, $m$, or $n$) equal to one and the other integer equal to zero.

Equation 6-42 can be applied to a cubical box by letting $w = h = l$. For the case of the lowest resonant frequency the integers $k$, $m$, and $n$ will be equal to 1, 1, and 0 in any sequence. If the above conditions are substituted into Eq. 6-42, then it reduces to Eq. 6-41.

## 6.18   GROUNDING OF SHIELDS

A solid shield that completely surrounds a product (a Faraday cage) can be at any potential and still provide effective shielding. That is, it will prevent outside electromagnetic fields from affecting circuits inside the shield, and vice versa. Thus, the shield does not need to be grounded, nor must it have its potential defined in any way to behave as a shield. In most cases, however, the shield should be connected to the circuit common, to prevent any potential difference between the shield and the circuits inside the shielded enclosure.

In many practical instances, however, the shield should be grounded for other reasons. If the product is powered off the alternating current (ac) power line, grounding is a safety requirement. The grounded shield provides a fault current return path to trip a circuit breaker and remove power from the equipment, which then renders the product safe. Grounding can also prevent the buildup of static charge on the shield.

All metallic parts of a system should be electrically bonded together, whether the shield is grounded or not. Each part should have a low-impedance connection, preferably in at least two places, to the other metal parts. Screws can be used to bond the parts if they have a conductive finish. Alternately, star washers or thread cutting screws can be used to cut through nonconductive paint and finishes.

## SUMMARY

- All cables that enter or leave a shielded enclosure should be shielded or filtered.
- Shielded cables that enter a shielded enclosure should have their shields bonded to the enclosure.
- Reflection loss is large for electric fields and plane waves.
- Reflection loss is normally small for low-frequency magnetic fields.
- A shield one-skin depth thick provides approximately 9 dB of absorption loss.
- Reflection loss decreases with frequency.

- Absorption loss increases with frequency.
- Magnetic fields are harder to shield against than electric fields.
- Use a material with a high relative permeability to shield against low-frequency magnetic fields.
- Use a highly conductive material to shield against electric fields, plane waves, and high-frequency magnetic fields.
- Absorption loss is a function of the square root of the permeability times the conductivity.
- Reflection loss is a function of the conductivity divided by the permeability.
- Increasing the permeability of a shield material increases the absorption loss and decreases the reflection loss.
- Aperture control is the key to high-frequency shielding.
- The maximum linear dimension, not the area, of an aperture determines the amount of leakage.
- In order to minimize leakage, electrical contact must exist across the seams of shielded enclosures.
- Shielding effectiveness decreases proportional to the square root of the number of apertures.
- For most shield materials the absorption loss predominates above 1 MHz.
- Low-frequency shielding effectiveness depends primarily on the shield material.
- Aperture control is just as important in conductive coated plastic enclosures as in metal shields.
- Shielding not only can be done at the enclosure level, but also at the module level and at the PCB level.
- Shields do not have to be grounded to be effective.
- Table 6-9 is a qualitative summary of shielding applicable to a solid shield with no apertures.

## PROBLEMS

6.1 What is the magnitude of the characteristic impedance of silver, brass, and stainless steel at 10 kHz?

6.2 Calculate the skin depth and absorption loss for a brass shield 0.062 in thick at the following frequencies: a. 0.1 kHz, b. 1 kHz, c. 10 kHz, and d. 100 kHz?

6.3 Considering absorption loss only, discuss the design of a shield to provide 30 dB of attenuation against a 60-Hz field.

6.4 a. What is the reflection loss of a 0.001-in thick copper shield to a 1000-Hz plane wave?
   b. If the thickness is increased to 0.01 in, what is the reflection loss?

**TABLE 6-9. Qualitative Summary of Shielding Effectiveness.**

| Material | Frequency (kHz) | Absorption Loss[a] All Fields | Reflection Loss | | |
|---|---|---|---|---|---|
| | | | Magnetic Field[b] | Electric Field | Plane Wave |
| Magnetic ($\mu_r = 1000$, $\sigma_r = 0.1$) | <1 | Bad–poor | Bad | Excellent | Excellent |
| | 1–10 | Average–good | Bad–poor | Excellent | Excellent |
| | 10–100 | Excellent | Poor | Excellent | Good |
| | >100 | Excellent | Poor–average | Good | Average–good |
| Nonmagnetic ($\mu_r = 1$, $\sigma_r = 1$) | <1 | Bad | Poor | Excellent | Excellent |
| | 1–10 | Bad | Average | Excellent | Excellent |
| | 10–100 | Poor | Average | Excellent | Excellent |
| | >100 | Average–good | Good | Excellent | Excellent |

*Key*

| | Attenuation |
|---|---|
| Bad | 0–10 dB |
| Poor | 10–30 dB |
| Average | 30–60 dB |
| Good | 60–90 dB |
| Excellent | >90 dB |

[a] Absorption loss for 1/32-in thick shield.

[b] Magnetic field reflection loss for a source distance of 1 m (Shielding is less if distance is less than 1 m and more if distance is greater than 1 m)

6.5  Calculate the shielding effectiveness of a 0.015-in-thick copper shield located 1 in from the source of a 10-kHz magnetic field.

6.6  What would be the shielding effectiveness of the shield of the previous problem if it were located in the far field?

6.7  What is the shielding effectiveness of a 0.032-in thick, soft aluminum shield located 1 ft away from the source of a 10-kHz electric field?

6.8  A shield is located 6 in from the source of an electric or magnetic field. Above what frequency should the far field equations be used?

6.9  Calculate the absorption loss of three different copper shields, 0.020 in, 0.040 in, and 0.060 in thick, to a 1-kHz magnetic field.

6.10  A shield that contains 10 identical holes in a linear array is required to have 30 dB of shielding effectiveness at 100 MHz. What is the maximum linear dimension of one hole?

6.11  A shield has a 5 by 12 array of 1/8 in diameter cooling holes. What will be the shielding effectiveness at 250 MHz?

6.12  A shielded ventilation panel consists of a 20 by 20 array of 400, 1/8 in round holes. The panel is 1/2 in thick (hence the holes have a depth of 1/2 in). What is the approximate shielding effectiveness of the panel at 250 MHz?

6.13  A product requires 20 dB of shielding at 200 MHz. It is planned to use 100 small round cooling holes (all the same size) arranged in a 10 by 10 array. What is the maximum diameter for one of the holes?

6.14  What two items are required to provide good electrical continuity across a seam?

6.15  A gasketed joint has a measured transfer impedance of 10 mΩ. What will the shielding effectiveness of the joint be to an incident plane wave?

6.16  What is the lowest resonant frequency of a 2-ft cubical enclosure?

6.17  What is the lowest resonant frequency of a 2-ft by 3-ft by 6-ft cabinet? (The solution to this problem may require some extra thought.)

# REFERENCES

Archambeault, B. and Thibeau, R. "Effects of Corrosion on the Electrical Properties of Conducted Finishes for EMI Shielding," *IEEE National Symposium on Electromagnetic Compatibility,* Denver, CO, 1989.

ARP 1481. *Corrosion Control and Electrical Conductivity in Enclosure Design*, Society of Automotive Engineers, May 1978.

ARP 1705. *Coaxial Test Procedure to Measure the RF Shielding Characteristics of EMI Gasket Materials*, Society of Automotive Engineers, 1994.

ASTM Standard B 449. *Standard Specifications for Chromates on Aluminum,* April 1994.

Faught, A. N. "An Introduction to Shield Joint Evaluation Using EMI Gasket Transfer Impedance Data," *IEEE International Symposium on EMC*, August 1982, pp. 38–44.

Hayt, W. H. Jr. *Engineering Electromagnetics*, 3rd. ed. New York, McGraw Hill, 1974.

IEC 60950-1, *Information Technology Equipment—Safety—Part 1: General Requirements*, International Electrotechnical Commission, 2006.

Kraus, J. D., and Marhefka, R. J. *Antennas*. 3rd ed., New York, McGraw-Hill, 2002.

MIL-HDBK-1250, *Handbook For Corrosion Prevention and Deterioration Control in Electronic Components and Assemblies*, August 1995.

MIL-STD-C-5541E, *Chemical Conversion Coatings on Aluminum and Aluminum Alloys*, November 1990.

Quine, J. P. "Theoretical Formulas for Calculating the Shielding Effectiveness of Perforated Sheets and Wire Mesh Screens," *Proceedings of the Third Conference on Radio Interference Reduction*, Armour Research Foundation, February, 1957, pp. 315–329.

Schelkunoff, S. A. "The Electromagnetic Theory of Coaxial Transmission Lines and Cylindrical Shields," *Bell Sys Tech J*, Vol.13, October 1934. pp. 532–579.

Schelkunoff, S. A. *Electromagnetic Waves*. Van Nostrand, New York, 1943.

White, R. J. *A Handbook of Electromagnetic Shielding Materials and Performance*, Don White Consultants, Germantown, M.D., 1975.

## FURTHER READING

Carter, D. "RFI Shielding Windows," Tecknit Europe, May 2003, Available at www.tecknit.co.uk/catalog/windowsguide.pdf. Accesed July 2008.

Cowdell, R. B. "Nomographs Simplify Calculations of Magnetic Shielding Effectiveness," *EDN*, vol. 17, September 1, 1972. Available at www.lustrecal.com/enginfo.html. Accessed July 2008.

Cowdell, R. B. Nomograms Solve Tough Problems of Shielding," *EDN*, April 17, 1967. Available at www.lustrecal.com/enginfo.html. Accessed July 2008.

Chomerics. *EMI Shielding Engineering Handbook*. Woburn, MA, 2000.

*IEEE Transactions on Electromagnetic Compatibility, Special Issue on Electromagnetic Shielding*, vol 30-3, August 1988.

Kimmel, W. D. and Gerke, D. D. "Choosing the Right Conductive Coating," *Conformity*, August, 2006.

Miller, D. A. and Bridges, J. E. "Review of Circuit Approach to Calculate Shielding Effectiveness," *IEEE Transactions on Electromagnetic Compatibility*, vol. EMC-10, March 1968.

Molyneux-Child, J. W. *EMC Shielding Materials, A Designers Guide*, Butterworth-Heinemann, London, U.K. 1997.

Paul, C. R. *Introduction to Electromagnetic Compatibility*, Chapter 11, Shielding, 2nd ed., Wiley, New York, 2006.

Schultz, R. B. Plantz, V. C. and Bush, D. R. "Shielding Theory and Practice," *IEEE Transactions on Electromagnetic Compatibility*, vol. 30-3, August 1988.

Schultz, R. B. *Handbook of Electromagnetic Compatibility*, (Chapter 9, Electromagnetic Shielding), Perez, R., ed., New York, Academic Press, 1995.

Vitek, C. "Predicting the Shielding Effectiveness of Rectangular Apertures," *IEEE National Symposium on Electromagnetic Compatibility*, 1989.

White, R. J. and Mardiguian, M. *Electromagnetic Shielding*, Interference Control Technologies, Gainesville, V.A., 1988.

Young, F. J. "Ferromagnetic Shielding Related to the Physical Properties of Iron," *IEEE Electromagnetic Compatibility Symposium Record,* IEEE, New York, 1968.

# 7 Contact Protection

Whenever contacts open or close a current-carrying circuit, electrical break-down may develop between the contacts. This breakdown begins while the contacts are close together, but not quite touching. In the case of contacts that are closing, the breakdown continues until the contacts are closed. In the case of contacts that are opening, the breakdown continues until conditions can no longer support the breakdown. Whenever breakdown occurs, some physical damage is done to the contacts, which decreases their useful life. In addition, breakdown can also lead to high-frequency radiation and to voltage and current surges in the wiring. These surges may be the source of interference affecting other circuits.

The techniques used to minimize physical damage to the contacts are similar to those used to eliminate radiated and conducted interference. All of the contact protection networks discussed in this chapter greatly reduce the amount of noise generated by the contacts and the load, as well as extend the life of the contacts. Two types of breakdown are important in switching contacts. They are the gas or glow discharge and the metal-vapor or arc discharge.

## 7.1 GLOW DISCHARGES

A regenerative, self-supporting, glow discharge or corona can occur between two contacts when the gas between the contacts becomes ionized.* Such a breakdown is also called a Townsend discharge. The voltage necessary to initiate a glow discharge is a function of the gas, the spacing of the contacts, and the gas pressure. If the gas is air at standard temperature and pressure, 320 V is required at a gap length of 0.0003 in to initiate a glow discharge. If the gap is shorter or longer, more voltage is required. Figure 7-1 shows the required breakdown voltage ($V_B$) for starting the glow discharge versus the separation

---

* In a gas, a few free electrons and ions are present, from radioactive decay, cosmic radiation, or light. When an electric field is applied, these free electrons and ions produce a small current flow. If the electric field is large enough, the electrons will attain sufficient velocity (energy) to break other electrons loose from neutral atoms or molecules with which they collide, thus ionizing the gas. This process becomes self-sustaining if the externally applied voltage can maintain the necessary electric field strength.

---

*Electromagnetic Compatibility Engineering,* by Henry W. Ott
Copyright © 2009 John Wiley & Sons, Inc.

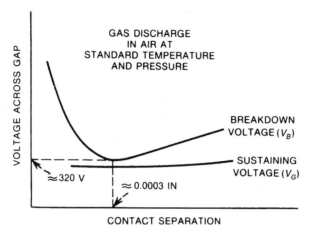

**FIGURE 7-1.** Voltage versus distance for glow discharge.

distance of the contacts. After the breakdown occurs, a somewhat smaller voltage ($V_G$) is sufficient to keep the gas ionized. In air $V_G$ is approximately 300 V. As can be observed in Fig.7-1, this sustaining voltage is nearly constant regardless of the contact spacing. A minimum current is also necessary to sustain the glow; typically, this is a few milliamperes.

*To avoid a glow discharge, the voltage across the contacts should be kept below 300 V.* If this is done, the only concern is contact damage from an arc discharge.

## 7.2  METAL-VAPOR OR ARC DISCHARGES

An arc discharge can occur at contact spacings and voltages much below those required for a glow discharge. It can even occur in a vacuum, because it does not require the presence of a gas. An arc discharge is started by field-induced electron emission,* which requires a voltage gradient of approximately 0.5 MV/cm (5 V at $4 \times 10^{-6}$ in).

An arc is formed whenever an energized, but unprotected, contact is opened or closed, because the voltage gradient usually exceeds the required value when the contact spacing is small. When the arc discharge forms, the electrons emanate from a small area of the cathode—where the electric field is strongest.

---

* Electrons in a metal are free to move around within the metal. Some of these electrons will have sufficient velocity to escape from the surface of the material. However, when they leave, they produce an electric field, which pulls them back to the surface. If an external electric field is present with sufficient voltage gradient, it can overcome the force that normally returns the electrons to the surface. The electrons are, therefore, removed from the surface and set free.

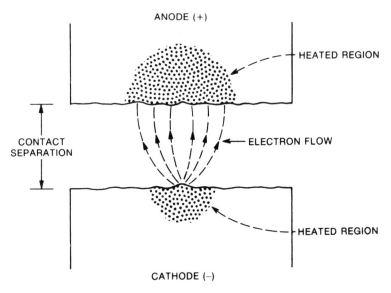

**FIGURE 7-2.** Initiation of an arc discharge.

Because, on a microscopic scale, all surfaces are rough, the highest and sharpest point on the cathode has the largest voltage gradient and becomes the source of electrons for the field emission, as shown in Fig.7-2. The electron stream fans out as it crosses the gap and finally bombards the anode. The localized current has a high density, and it heats the contact material (as the result of $I^2R$ losses) to a few thousand degrees Kelvin. This temperature may be sufficient to vaporize the contact metal. In general, either the anode or cathode may vaporize first, depending on the rates at which heat is delivered to and removed from the two contacts. This in turn depends on the size, material, and spacing of the contacts.

The appearance of molten metal marks the transition from field emission (electron flow) to a metal-vapor arc. This transition typically takes place in times less than a nanosecond. The molten metal, once present, forms a conductive "bridge" between the contacts, thus maintaining the arc even though the voltage gradient may have decreased below the value necessary to initiate the discharge. This metal-vapor bridge draws a current limited only by the supply voltage and the impedance of the circuit. After the arc has started, it persists as long as the external circuit provides enough voltage to overcome the cathode contact potential and enough current to vaporize the anode or the cathode material. As the contacts continue to separate, the molten metal "bridge" stretches and eventually ruptures. The minimum voltage and current required to sustain the arc are called the minimum arcing voltage ($V_A$) and the minimum arcing current ($I_A$). Typical values of minimum arcing voltage and current are shown in Table 7-1 (Relay and Switch Industry Associations, 2006). If either the voltage or curent falls below these values, the arc is extinguished.

**TABLE 7-1.   Contact Arcing Characteristics**

| Material | Minimum Arcing Voltage ($V_A$) | Minimum Arcing Current ($I_A$) |
|---|---|---|
| Silver | 12 | 400 mA |
| Gold | 15 | 400 mA |
| Gold alloy[a] | 9 | 400 mA |
| Palladium | 16 | 800 mA |
| Platinum | 17.5 | 700 mA |

[a] 69% gold, 25% silver, 6% platinum.

For arcs between contacts of different materials, $V_A$ is determined by the cathode (negative contact) material, and $I_A$ is assumed to be whichever contact material (anode or cathode) has the lowest arcing current. Note, however, that the minimum arcing currents listed in Table 7-1 are for clean, undamaged contacts. After the contacts have become damaged from some arcing, the minimum arcing current may decrease to as low as one tenth of the value listed in the table.

In summary, an arc discharge is a function of the contact material, and it is characterized by a relatively low voltage and a high current. In contrast, a glow discharge is a function of the gas, usually air, between the contacts, and it is characterized by a relatively high voltage and low current. As will be demonstrated in a Section 7.8, it is difficult to prevent an arc discharge from forming, because only a low voltage is required. If the arc does form, however, it should be prevented from sustaining itself by keeping the available current below the minimum arcing current.

## 7.3   AC VERSUS DC CIRCUITS

If the contact is to survive, the arc once formed must be broken rapidly to minimize damage to the contact material. If it is not broken rapidly enough, some metal transfers from one contact to the other. The damage done by an arc is proportional to the energy in it—namely (voltage) × (current) × (time).

The higher the voltage across the contacts, the more difficult it is to interrupt the arc. Under arcing conditions, a set of contacts can usually handle their rated number of volt-amperes at a voltage equal to or less than the rated voltage, but not necessarily at a higher voltage.

A set of contacts can normally handle a much higher ac than dc voltage, for the following reasons:

1. The average value of an ac voltage is less than the rms value.
2. During the time that the voltage is less than 10–15 V, an arc is very unlikely to start.

3. As the result of polarity reversal, each contact is an anode and a cathode an equal number of times.

4. The arc will be extinguished when the voltage goes through zero.

A contact rated at 30 V dc can therefore typically handle 115 V ac. One disadvantage of switching ac, however, is that it is much harder to provide adequate contact protection networks when they are required.

## 7.4   CONTACT MATERIAL

Various load levels (currents) require different types of contact materials. No one material is useful from zero current (dry circuit) up to high current. Palladium is good for high-current loads under eroding contact conditions. Silver and silver cadmium operate well at high current but may fail under conditions of no arcing. Gold and gold alloys work well under low-level or dry-circuit conditions but erode excessively at high currents.

Many so-called "general purpose relays" are on the market, rated from dry circuit to 2 A. These relays are usually made by plating hard gold over a heavy load contact material such as silver or palladium. When used for low current, the contact resistance remains low because of the gold plating. When used for high load current, the gold is burned off during the first few operations, and the high current contact material remains. For this reason, once a general purpose relay is used with high currents, it is no longer usable in a low-current application.

A problem sometimes occurs when soft gold is plated over silver. The silver migrates through the gold and forms a high-resistance coating (silver-sulfide) on the contact. This may then cause the contact to fail because of the high-resistance surface coating.

## 7.5   CONTACT RATING

Contacts are normally rated by the maximum values of voltage and current they can handle feeding a resistive load. When a contact is operated at its rated conditions, there is some momentary arcing on "make" and "break."* When operated under these conditions, a contact operates for a time equal to its rated electrical life. Ratings of mechanical life are for dry circuits (drawing no current).

Some contacts are also rated for an inductive load in addition to their resistive load rating. A third common rating is a motor or lamp rating for loads that draw much higher initial current than the normal steady-state current.

---

* A small amount of arcing may actually be useful in burning off any thin insulating film that has formed on the contacts.

All of these ratings assume that no contact protection is used. If proper contact protection is used, the rated voltage and/or current can be handled for a greater number of operations, or a higher voltage and/or current can be handled for the rated number of operations.

## 7.6  LOADS WITH HIGH INRUSH CURRENTS

If the load is not resistive, the contacts must be appropriately derated or protected. Lamps, motors, and capacitive loads all draw much higher current when the contacts are initially closed than their steady-state current. The initial current in a lamp filament, for example, can be 10–15 times the normal rated current, as shown in Fig.7-3. Typically, a contact is rated at only 20% of its normal resistive load capacity for lamp loads.

Capacitive loads also can draw extremely high initial currents. The charging current of a capacitor is limited only by the series resistance of the external circuit.

Motors typically draw initial currents that are 5–10 times their normal rated currents. In addition, the motor inductance causes a high voltage to be generated when the current is interrupted (an inductive kick). This also causes arcing. Motors therefore are difficult to switch because they cause contact damage on both "make" and "break."

To protect a contact used in a circuit with high inrush current, the initial current must be limited. Using a resistor in series with the contact to limit the initial current is not always feasible, because it also limits the steady-state current. If a resistor is not satisfactory, then a low dc resistance inductor can be used to limit the initial current. In some light duty applications, ferrite beads placed on the contact lead may provide sufficient initial current limiting, without affecting the steady-state current.

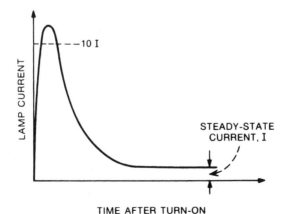

**TIME AFTER TURN-ON**

**FIGURE 7-3.** Lamp current versus time.

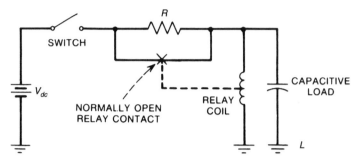

**FIGURE 7-4.** Use of a switchable current-limiting resistor to protect a closing contact.

In severe cases, a switchable current-limiting resistor, as shown in Fig.7-4, may be required. Here a relay is placed across the capacitive load with its normally open contact placed across the current-limiting resistor.* When the switch is closed, the capacitor charging current is limited by resistor $R$. When the voltage across the capacitor becomes large enough to operate the relay, the normally open contact closes, shorting out the current limiting resistor.

Another problem associated with closing contacts is chatter, or bounce. After the contacts initially touch, they may bounce open again and break the circuit. In some contacts, this may continue for 10 or more times, and each time the contacts must make and break the current. Not only can the repeated arcing cause operational problems in the circuit, but it produces considerably more contact damage and high-frequency radiation.

## 7.7  INDUCTIVE LOADS

The voltage across an inductance ($L$) is given by

$$V = L\left(\frac{di}{dt}\right). \tag{7-1}$$

This expression explains the large voltage transient encountered when the current through an inductor is suddenly interrupted. The rate of change, $di/dt$, becomes large and negative, which results in the large reverse voltage transient or inductive "kick." Theoretically, if the current goes from some finite value to zero instantaneously, the induced voltage would be infinite. But in reality, contact arcing and circuit capacitance will never let this happen. Nevertheless,

---

* In place of the relay, a power field effect transistor (FET) could also be used.

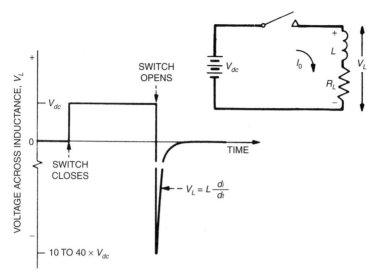

**FIGURE 7-5.** Voltage across inductive load when switch closes and opens.

very large induced voltages do occur. Suppression of high voltage inductive transients, consists of minimizing the $di/dt$ term.

It is not at all uncommon for an inductance operating from a 12–V dc power supply to generate voltages of 50–500 V when the current through the inductor is suddenly interrupted. Figure 7-5 shows the voltage waveshape across an inductor as the current is interrupted. The high voltage generated when a contact breaks the current to an inductive load causes severe contact damage. It is also the source of radiated and conducted noise, unless appropriate contact protection circuits are used. Under such conditions, most of the energy stored in the inductance is dissipated in the arc, which causes excessive damage.

The circuit shown in Fig. 7-6 can be used to illustrate the damage done to a set of contacts by an inductive load. In this figure, a battery is connected to an inductive load through a switch contact. The load is assumed to have negligible resistance. In practice, this condition can be approximated by a low-resistance dc motor. The steady-state current becomes limited by the back EMF of the motor rather than the circuit resistance. Then, let the switch be opened while a current $I_0$ is flowing through the inductance. The energy stored in the magnetic field of the inductance is equal to $(1/2) \, LI_0^2$.

When the switch is opened, what happens to the energy stored in the magnetic field of the inductance? If the circuit resistance is negligible, than all the energy must be dissipated in the arc that forms across the contacts or be radiated. Without some type of protective circuitry, a switch used in this application does not last very long.

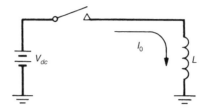

**FIGURE 7-6.** Inductive load controlled by a switch. When the switch opens, most of the energy stored in the inductor is dissipated in the arc formed across the switch contacts.

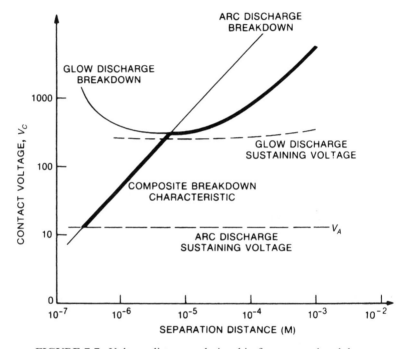

**FIGURE 7-7.** Voltage-distance relationship for contact breakdown.

## 7.8   CONTACT PROTECTION FUNDAMENTALS

Figure 7-7 summarizes the conditions for contact breakdown in terms of the required voltage-distance relationships. The required breakdown voltage for starting a glow discharge is shown, as is the minimum voltage required to sustain the glow discharge. Also shown is a voltage gradient of 0.5 MV/cm, which is that required to produce an arc discharge. The minimum voltage required to maintain the arc discharge is also shown in this figure. The heavy line therefore represents the composite requirements for producing contact breakdown. To the right and below this curve, no breakdown occurs, whereas above and to the left of this curve, contact breakdown occurs.

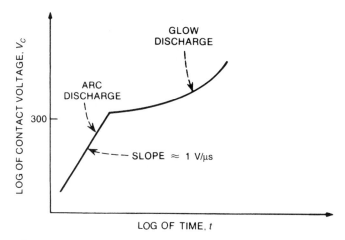

**FIGURE 7-8.** Contact breakdown characteristics versus time.

A more useful presentation of the breakdown information contained in Fig. 7-7 is to plot the breakdown voltage versus time, instead of distance. This conversion can be accomplished using the separation velocity of the contacts. A typical composite breakdown characteristic as a function of time is shown in Fig. 7-8. As can be observed, the two requirements for avoiding contact breakdown are as follows:

1. Keep the contact voltage below 300 V to prevent a glow discharge.
2. Keep the initial rate of rise of contact voltage below the value necessary to produce an arc discharge. (A value of 1 V/$\mu$s is satisfactory for most contacts.)

If it is not possible to avoid contact breakdown in a specific application, then the breakdown should be kept from being self-sustaining. This usually means arranging the circuit so that the current available is always below that necessary to sustain the breakdown.

To determine whether or not breakdown can occur in a specific case, it is necessary to know what voltage is produced across the contacts as they open. This voltage is then compared with the breakdown characteristics in Fig. 7-8. If the contact voltage is above the breakdown characteristics, contact breakdown occurs.

Figure 7-9 shows an inductive load connected to a battery through a switch $S$. The voltage that would be produced across the contacts of the opening switch, if no breakdown occurred, is called the "available circuit voltage." This is shown in Fig. 7-10 for the circuit in Fig. 7-9. $I_0$ is the current flowing through the inductor the instant the switch is opened, and $C$ is the stray

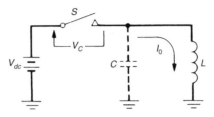

**FIGURE 7-9.** Contact controlling an inductive load. Capacitor $C$ represents the stray capacitance of the wiring.

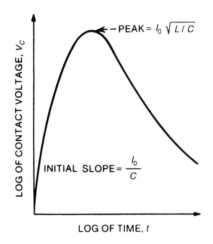

**FIGURE 7-10.** Available circuit voltage across opening contact for circuit in Fig. 7-9, assuming no contact breakdown.

capacitance of the wiring. Figure 7-11 compares the available circuit voltage (Fig. 7-10) with the contact breakdown characteristics (Fig. 7-8). The voltage exceeds the breakdown characteristics from $t_1$ to $t_2$, and therefore contact breakdown occurs in this region.

Knowing that breakdown occurs, let us consider in more detail exactly what happens as the contacts in Fig. 7-9 are opened. When the switch is opened, the magnetic field of the inductance tends to keep the current $I_0$ flowing. Because the current-cannot flow through the switch, it flows through the stray capacitance $C$ instead. This then charges the capacitor and the voltage across the capacitor rises at an initial rate of $I_0/C$, as shown in Fig. 7-12, As soon as this voltage exceeds the breakdown curve, an arc occurs across the contacts. If the available current at this point is less than the minimum arcing current $I_A$, the arc lasts only long enough to discharge the capacitance $C$ to a voltage below the sustaining voltage $V_A$. After the capacitor is discharged, the current again charges $C$ and the

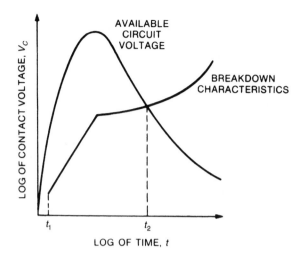

**FIGURE 7-11.** Comparison of available circuit voltage and contact breakdown characteristics for circuit in Fig. 7-9.

process is repeated until the voltage exceeds the glow discharge voltage (point $A$ in Fig. 7-12). At this point, a glow discharge occurs. If the smaller sustaining current necessary to maintain the glow discharge is still not available, then the glow lasts only until the voltage drops below the minimum glow voltage $V_G$* This process is repeated until time $t_2$, after which sufficient voltage is not available to produce any additional breakdowns.

If at any time the available current exceeds the minimum arcing current $I_A$, a steady arc occurs and continues until the available voltage or current falls below the minimum glow voltage or current. Figure 7-13 shows the waveshape when sufficient current is available to maintain a glow discharge, but not enough for an arc discharge.

If the stray capacitance $C$ is increased sufficiently, or if a discrete capacitor is placed in parallel with it, then the peak voltage and the initial rate of rise of contact voltage can be reduced to the point where no arcing occurs. This waveshape is shown in Fig. 7-14. Using a capacitor this way, however, may causes contact damage on closure because of the large capacitor charging current.

The electrical oscillations that occur in the resonant circuit of Fig. 7-9, when the switch is opened, can become the source of high-frequency interference to nearby equipment. These oscillations can be avoided if sufficient resistance and capacitance are provided to guarantee that the circuit is overdamped. The required condition for nonoscillation is given by Eq. 7-6 in Section 7.10.2.

---

* If sufficient current is now available, then the glow discharge may transfer to an arc, and the voltage will fall to $V_A$ instead of $V_G$ . Sufficient current, however, is usually not available at the low voltage $V_A$ to maintain the arc, so it is extinguished at this point.

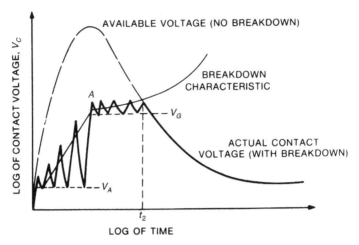

**FIGURE 7-12.** Actual contact voltage for circuit in Fig. 7-9 When current is insufficient to maintain a continuous glow discharge.

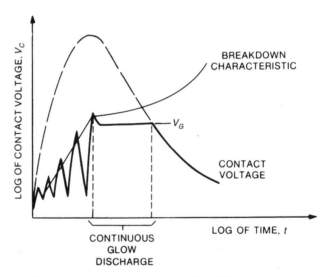

**FIGURE 7-13.** Contact voltage for circuit in Fig. 7-9 when current is sufficient to maintain a continuous glow discharge.

## 7.9   TRANSIENT SUPPRESSION FOR INDUCTIVE LOADS

To protect contacts that control inductive loads and to minimize radiated and conducted noise, some type of contact protection network must normally be placed across the inductance, the contacts, or both. In some cases, the protection network can be connected across either the load or the contact

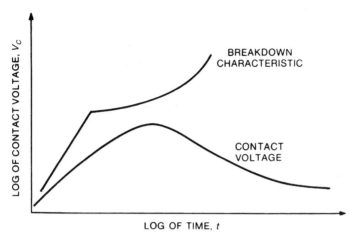

**FIGURE 7-14.** Contact voltage for circuit in Fig. 7-9 when capacitance is large enough to prevent breakdown.

with equal effectiveness. In large systems, a load may be controlled by more than one contact, and it may be more economic to provide protection at the load rather than at each individual contact.

In severe cases, protection networks may have to be applied across both the inductance and the contacts to eliminate interference and protect the contacts adequately. In other cases, the amount of protection that can be used is limited by operational requirements. For example, protection networks across the coil of a relay increase the release time. In this case, the protection network has to be a compromise between meeting operational requirements and providing adequate protection to the contacts controlling the relay.

From a noise reduction point of view, it is usually preferable to provide as much transient suppression as possible across the noise source—in this case, the inductor. In most cases this provides sufficient protection for the contacts. When it does not, additional protection can be used across the contacts.

Precise calculations for the component values of a contact protection network are difficult. It involves parameters, the values of which are normally unknown by the circuit designer, such as the inductance and capacitance of the interconnecting wiring and the contact separation velocity. The simplifed design equations that follow are a starting point, and in many cases provide an acceptable contact protection network. Tests should be used, however, to verify the effectiveness of the network in the intended application.

Protection networks can be divided into the following two categories: those usually applied across the inductor and those usually applied across the contacts. Some of these networks, however, can be applied in either place.

Figure 7-15 shows six networks commonly placed across a relay coil or other inductance to minimize the transient voltage generated when current is interrupted. In Fig. 7-15A, a resistor is connected across the inductor. When the switch opens, the inductor drives whatever current was flowing before the opening of the contact through the resistor. The transient voltage peak therefore increases with increasing resistance but is limited to the steady-state current times the resistance. If $R$ is made equal to the load resistance $R_L$, then the voltage transient is limited to a magnitude equal to the supply voltage. In this case, the voltage across the contact is the supply voltage plus the induced coil voltage, or twice the supply voltage. This circuit is wasteful of power because the resistor draws current whenever the load is energized. If $R$ should equal the load resistance, then the resistor dissipates as much steady-state power as the load.

Another arrangement is shown in Fig. 7-15B, where a varistor (a voltage variable resistor) is connected across the inductor. When the voltage across the varistor is low, its resistance is high, but when the voltage across it is high, its resistance is low. This device works the same as the resistor in Fig. 7-15A, except that the power dissipated by the varistor while the circuit is energized is reduced.

A better arrangement is shown in Fig. 7-15C. Here, a resistor and capacitor are connected in series and placed across the inductor. This circuit dissipates no power when the inductor is energized. When the contact is opened, the capacitor initially acts as a short circuit, and the inductor drives its current through the resistor. The values for the resistor and the capacitor can be determined by the method described in Section 7.10.2 for the $R$–$C$ network.

In Fig. 7-15D, a semiconductor diode is connected across the inductor. The diode is poled so that when the circuit is energized, no current flows through the diode. However, when the contact opens, the voltage across the inductor is of opposite polarity than the supply voltage. This voltage forward-biases the diode, which then limits the transient voltage across the inductor to a very low value (the forward voltage drop of the diode plus any $IR$ drop in the diode). The voltage across the opening contact is therefore approximately equal to the supply voltage. This circuit is very effective in suppressing the voltage transient. However, the time required for the inductor current to decay is more than for any of the previous circuits and may cause operational problems.

For example, if the inductor is a relay, its release time is increased. A small resistor can be added in series with the diode in Fig. 7-15D to decrease the release time of the relay, but at the expense of generating a higher transient voltage. The diode must have a voltage rating greater than the maximum supply voltage. The current rating of the diode must be greater than the maximum load current. If the contacts only operate occasionally, then the peak current rating of the diode can be used. If the contacts operate more than a few times per minute, then the continuous current rating of the diode should be used.

Adding a zener diode in series with a rectifier diode, as shown in Fig. 7-15E, allows the inductor current to decay faster. This protection, however, is not as good as that for the previously mentioned diode and uses an extra component. In this case, the voltage across the opening contact is equal to the zener voltage plus the supply voltage.

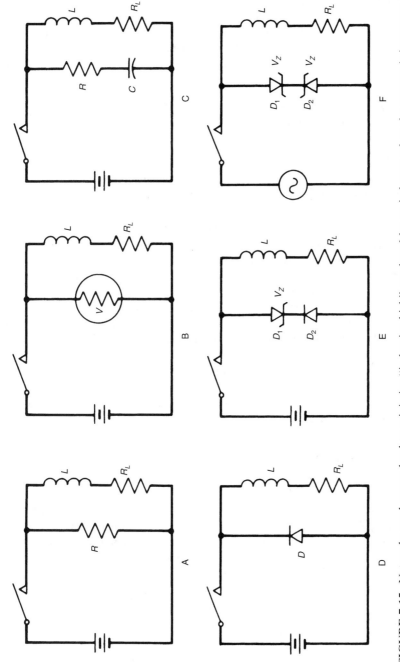

**FIGURE 7-15.** Networks used across load to minimize "inductive kick" produced by an inductor when the current is interrupted.

Neither of the diode circuits (Fig. 7-15D or 7-15E) can be used with ac circuits. Circuits that operate from ac sources or circuits that must operate from two dc polarities can be protected using the networks in Fig. 7-15 A through 7-15C, or by two zener diodes connected back to back, as shown in Fig. 7-15F. Each zener must have a voltage breakdown rating greater than the peak value of the ac supply voltage and a current rating equal to the maximum load current.

## 7.10   CONTACT PROTECTION NETWORKS FOR INDUCTIVE LOADS

### 7.10.1   *C* Network

Figure 7-16 shows three contact protection networks commonly used across contacts that control inductive loads. One of the simplest methods of suppressing arcs caused by interrupting dc current is to place a capacitor across the contact, as shown in Fig. 7-16A. If the capacitor is large enough, the load current is momentarily diverted through it as the contact is opened, and arcing does not occur. However, when the contact is open the capacitor charges up to the supply voltage $V_{dc}$. When the contact is then closed, the capacitor discharges through the contact with the initial discharge current limited only by the parasitic resistance of the wiring and the contacts.

The larger the value of the capacitor and the higher the supply voltage, the more damage the arc on "make" does, because of the increased energy stored in the capacitor. If the contacts bounce or chatter on "make" additional damage is done because of multiple making and breaking of the current. Because of these reasons, using a capacitor alone across a set of contacts is not generally recommended. If used, the value of capacitance is determined as explained in the following section.

### 7.10.2   *R–C* Network

Figure 7-16B shows a circuit that overcomes the disadvantage of the circuit in Fig. 7-16A by limiting the capacitor discharge current when the contact is closed. This is done by placing a resistor $R$ in series with the capacitor. For contact closing, it is desirable to have the resistance as large as possible to limit the discharge current. However, when the contact is opened, it is desirable to have the resistance as small as possible, because the resistor decreases the effectiveness of the capacitor in preventing arcing. The actual value of $R$ must therefore be a compromise between the two conflicting requirements.

The minimum value of $R$ is determined by closing conditions. It can be set by limiting the capacitor discharge current to the minimum arcing current $I_A$* for

---

* Limiting the discharge current to 0.1 $I_A$ is preferable. However, because the value of the resistor $R$ is a compromise between two conflicting requirements, this usually cannot be done in the case of the *R–C* network.

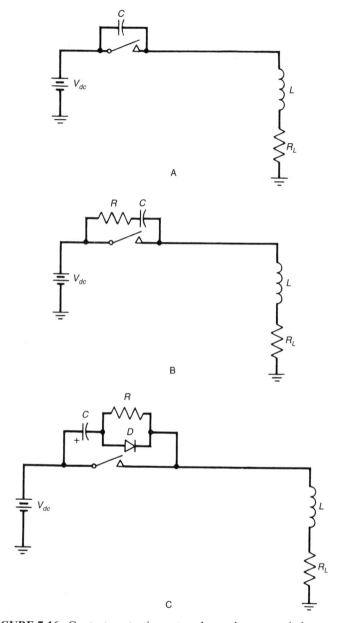

**FIGURE 7-16.** Contact protection networks used across switch contacts.

the contact. The maximum value is determined by opening conditions. The initial voltage across the opening contact is equal to $I_0 R$. If $R$ is equal to the load resistance, the instantaneous voltage across the contact equals the supply voltage. The maximum value of $R$ is usually taken equal to the load resistance to limit the initial voltage developed across the opening contacts to the supply voltage. The limits on $R$ can therefore be stated as

$$\frac{V_{dc}}{I_a} < R < R_L, \tag{7-2}$$

where $R_L$ is equal to the load resistance.

The value of $C$ is chosen to meet the following two requirements: (1) the peak voltage across the contacts should not exceed 300 V (to avoid a glow discharge) and (2) the initial rate of rise of contact voltage should not exceed 1 V per $\mu$s (to avoid an arc discharge). The latter requirement is satisfied if $C$ is at least 1 $\mu$F/A of load current.

The peak voltage across the capacitor is usually calculated by neglecting the circuit resistance and assuming all the energy stored in the inductive load is transferred to the capacitor. Under these conditions,

$$V_{C(peak)} = I_0 \sqrt{\frac{L}{C}}, \tag{7-3}$$

where $I_0$ is the current through the load inductance when the contact is opened. The value of the capacitor $C$ should always be chosen so that $V_c$ (peak) does not exceed 300 V. Therefore,

$$C \geq \left(\frac{I_0}{300}\right)^2 L. \tag{7-4}$$

In addition, to limit the initial rate of rise of contact voltage to 1 V/$\mu$s,

$$C \geq I_0 \times 10^{-6} \tag{7-5}$$

In some cases, it is preferable that the resonant circuit formed by the inductor and capacitor be nonoscillating (overdamped). The condition for nonoscillation is

$$C \geq \frac{4L}{R_1^2}, \tag{7-6}$$

where $R_1$ is the total resistance in series with the $L$–$C$ circuit. In the case of Fig. 7-16B, this would be $R_1 = R_L + R$. The requirement for nonoscillation, however, is not usually adhered to because it requires a large value capacitor.

The $R$–$C$ protection network (Fig. 7-16B) is the most widely used because of its low cost and small size. In addition, it only has a small effect on the release time of the load. The $R$–$C$ network is not, however, 100% effective. The presence of the resistor causes an instantaneous voltage (equal to $I_0R$) to develop across the opening contact, and therefore some early arcing is present. Figure 7-17 shows the voltage developed across the contact, with a properly designed $R$–$C$ network, superimposed on the contact breakdown characteristic. This figure shows the early arcing from the instantaneous voltage increase across the contact.

### 7.10.3    R–C–D Network

Figure 7-16C shows a more expensive circuit that overcomes the disadvantages of the circuits in Fig. 7-16A and 7-16B. When the contact is open, capacitor $C$ charges up to the supply voltage with the polarity shown in the figure. When the contact closes, the capacitor discharges through resistor $R$, which limits the current. When the contact opens, however, diode $D$ shorts out the resistor, thus allowing the load current momentarily to flow through the capacitor while the contact opens. The diode must have a breakdown voltage greater than the supply voltage with an adequate surge current rating (greater than the maximum load current). The capacitor value is chosen the same as for the $R$-$C$

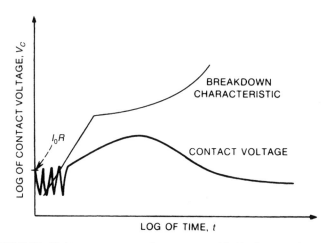

**FIGURE 7-17.** Voltage across opening contact with $R$–$C$ protection network.

network. Because the diode shorts out the resistor when the contacts open, a compromise resistance value is no longer required. The resistance can now be chosen to limit the current on closure to less than one tenth the arcing current,

$$R \geq \frac{10V_{dc}}{I_A}. \tag{7-7}$$

*The R–C–D network provides optimum contact protection,* but it is more expensive than other methods and cannot be used in an ac circuit.

## 7.11   INDUCTIVE LOADS CONTROLLED BY A TRANSISTOR SWITCH

If an inductive load is controlled by a transistor switch, then care must be taken to guarantee that the transient voltage generated by the inductor when the current is interrupted does not exceed the breakdown voltage of the transistor. One of the most effective, and common, ways to do this is to place a diode across the inductor, as shown in Fig. 7-18. In this circuit, the diode clamps the transistor collector to $+V$ when the transistor interrupts the current through the inductor, which limits the voltage across the transistor to $+V$. Any of the networks of Fig. 7-15 may also be used. A zener diode connected across the transistor is another common method. In any case the network should be designed to limit the voltage across the transistor to less than its breakdown voltage rating.

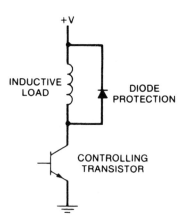

**FIGURE 7-18.** Diode used to protect transistor-controlling inductive loads.

Very large transient currents flow through the path between the inductive load and the protection diode. Therefore, this loop area should be minimized to limit the radiation that occurs from the transient current. The diode should be located as close as possible to the inductive load. This consideration is especially important when the protection diode is contained in a relay driver *IC*; otherwise, a large loop will occur.

## 7.12   RESISTIVE LOAD CONTACT PROTECTION

In the case of resistive loads operating with a source voltage of less than 300 V, a glow discharge cannot be started and therefore is of no concern. If the supply voltage is greater than the minimum arcing voltage $V_A$ (about 12 V), then an arc discharge occurs when the contacts are either opened or closed. Whether the arc, once started, sustains itself, depends on the magnitude of the load current.

If the load current is below the minimum arcing current $I_A$, the arc is extinguished quickly after initially forming. In this case, only a minimal amount of contact damage occurs, and in general, no contact protection networks are needed. Because of parasitic circuit capacitance or contact bounce, the arc starts, stops, and reignites many times. This type of arcing may be the source of high-frequency radiation and may require some protection to control interference.

If the load current is greater than the minimum arcing current $I_A$, then a steady arc forms. This steady arc does considerable damage to the contacts. If the current is less than the resistive circuit current rating of the contact, however, and the rated number of operations is satisfactory, then contact protection may not be required.

If contact protection is required for a resistive load, what type of network should be used? In a resistive circuit, the maximum voltage across an opening or closing contact is the supply voltage. Therefore, provided the supply voltage is under 300 V, the contact protection network does not have to provide high voltage protection. This function is already provided by the circuit. The required function of the contact protection network, in this case, is to limit the initial rate of rise of contact voltage to prevent initiating an arc discharge. This can best be accomplished by using the *R-C-D* network in Fig. 7-16C across the contact.

## 7.13   CONTACT PROTECTION SELECTION GUIDE

The following guide can be used to determine the type of contact protection for various loads:

1. Noninductive loads drawing less than the arcing current, in general, require no contact protection.

2. Inductive loads drawing less than the arcing current should have an *R–C* network or a diode for protection.

3. Inductive loads drawing greater than the arcing current should have an *R–C–D* network or a diode for protection.

4. Noninductive loads drawing greater than the arcing current should use the *R–C–D* network. Equation 7-4 does not have to be satisfied in this case provided the supply voltage is less than 300 V.

## 7.14  EXAMPLES

Proper selection of contact protection may be better understood with some numerical examples.

**Example 7-1.** A 150 Ω, 0.2 H relay coil is operated from a 12-V dc power source through a silver switch contact. The problem is to design a contact protection network for use across the relay.

The steady-state load current is 80 mA, which is less than the arcing current for silver contacts; therefore an *R-C* network or diode is appropriate. To keep the voltage gradient across the contact below 1 V/$\mu$s, the capacitance of the protection network must be greater than 0.08 $\mu$F (from Eq. 7-5). To keep the maximum voltage across the opening contact below 300 V, the capacitance must be greater than 0.014 $\mu$F (from Eq. 7-4). From Eq. 7-2, the value of the resistor should be between 30 and 150 Ω. Therefore, an appropriate contact protection network is 0.1 $\mu$F in series with 100 Ω placed either across the contact or load.

**Example 7-2.** A magnetic brake that has 1 H inductance and 53 Ω resistance is operated from a 48-V dc source through a switch with silver contacts. If an *R–C* contact protection network is used, then the resistor should have a value (from

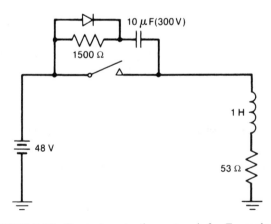

**FIGURE 7-19.** Contact protection network for Example 7-2.

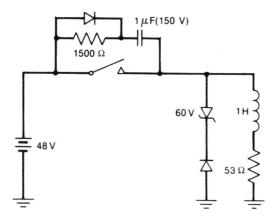

**FIGURE 7-20.** Alternate contact protection network for Example 7-2. This circuit allows the use of a physically smaller capacitor.

Eq. 7-2) of $120 < R < 53$. Because this is impossible, a more complicated protection network must be used, such as the *R-C-D* network. For the *R-C-D* network, the resistor should have a value greater than 1200 Ω (from Eq. 7-7). The steady-state dc current in the brake is 0.9 A. Therefore, from Eq. 7-5, the capacitor must be greater than 0.9 $\mu$F to limit the voltage gradient across the contacts on opening. From Eq. 7-4, the capacitor must also be greater than 9 $\mu$F. A 10-$\mu$F capacitor with a 300-V rating could be used, with a 1500 Ω resistor and a diode, as shown in Fig. 7-19.

The 10-$\mu$F, 300-V capacitor must of necessity be relatively large physically. To avoid using such a large capacitor, the following alternate solution could be used. If a series combination of a 60-V zener diode and a rectifier diode is placed across the load, the maximum transient voltage across the load would be limited to 60 V. The maximum voltage across the contact during opening would then be the zener voltage plus the supply voltage, or 108 V. Therefore, the capacitor in the protection network does not have to be chosen to limit the maximum voltage across the contacts to 300 V, because this voltage is already limited by the diode to 108 V. The only requirement now on the capacitor is that it satisfy Eq. 7-5. Therefore, a 1-$\mu$F, 150-V capacitor can be used as shown in Fig. 7-20, which avoids the need for a physically large size, 10-$\mu$F, 300-V capacitor.

## SUMMARY

- Two types of breakdown are important in switching contacts: the glow, or gas, discharge, and the arc, or metal-vapor, discharge.
- An arc discharge is a function of the contact material.
- A glow discharge is a function of the gas between the contacts.

- An arc discharge is characterized by a low voltage and a high current.
- A glow discharge is characterized by a high voltage and a low current.
- To prevent a glow discharge, keep contact voltage below 300 V.
- To prevent an arc discharge, keep the initial rate of rise of contact voltage to less than 1 V/$\mu$s.
- Lamps, motors and capacitor loads cause contact damage on closure because of the high inrush currents.
- Inductive loads are most damaging because of the high voltages they generate when current is interrupted.
- The R–C network is the most widely used contact protection network.
- The R–C–D network or just a diode are the most effective protection networks.
- The effect of the contact protection network on the release time of a relay must be considered.
- A diode connected across an inductor is a very effective transient suppression network; however, it may cause operational problems since it prevents the rapid decay of the inductor current.

## PROBLEMS

7.1   Which type of breakdown is the hardest to prevent, glow discharge or arc discharge, and why?

7.2   A 1H, 400-$\Omega$ relay coil is to be operated from a 30-V dc supply. The switch controlling the relay has platinum contacts. Design an R–C contact-protection network for this circuit.

7.3   For the zener diode protection circuit of Fig. 7-15E, plot the following three waveshapes when the contact closes and then opens. Assume no contact breakdown.
   a. The voltage across the load ($V_L$).
   b. The current through the load ($I_L$).
   c. The voltage across the contact ($V_c$).

7.4   a. A mechanical switch, having silver contacts, is used to control a 24-V dc relay with a winding resistance of 240 $\Omega$ and inductance of 10 mH. If the switch contacts are protected by an R–C network, what value resistor and capacitor should be used?
   b. If a 100-$\Omega$ resistor is used in the contact protection network, then what value capacitor would be required for the protection network to be nonoscillating (overdamped)?

7.5   If the R–C–D network of Fig. 7-16C is used to protect the contact of Problem 7.4a, determine the appropriate valued for R, C, and the characteristics of diode D.

# REFERENCES

Relay and Switch Industry Association (RSIA), *Engineers' Relay Handbook*. 6th ed. Arlington, VA 2006.

# FURTHER READING

Auger, R. W. and Puerschner, K. *The Relay Guide*, New York, Reinhold, 1960.

Bell Laboratories. *Physical Design of Electronic Systems*. Vol. 3: Integrated Device and Connection Technology, Chapter 9 (Performance Principles of Switching Contacts), Englewood Cliffs, N.J, Prentice-Hall, 1971.

Dewey, R. "Everyone Knows that Inductive Loads Can Greatly Shorten Contact Life." *EDN*, April 5, 1973.

Duell, J. P. Jr. "Get Better Price/Performance from Electrical Contacts." *EDN*, June 5, 1973.

Holm, R. *Electrical Contacts*. 4th ed. Berlin, Germany, Springer-Verlag, 1967.

Howell, K. E. "How Switches Produce Electrical Noise." *IEEE Transactions on Electromagnetic Compatibility*, vol. EMC-21, No. 3, August, 1979.

Oliver, F. J. *Practical Relay Circuits*. New York, Hayden Book Co., 1971.

Penning, F. M. *Electrical Discharge in Gases. New York,* Phillips Technical Library, 1957.

# 8 Intrinsic Noise Sources

Even if all external noise coupling could be eliminated from a circuit, a theoretical minimum noise level would still exist because of certain intrinsic or internal noise sources. Although the root mean square (rms) value of these noise sources can be well defined, the instantaneous amplitude can only be predicted in terms of probability. Intrinsic noise is present in almost all electronic components.

This chapter covers the three most important intrinsic noise sources: thermal noise, shot noise, and contact noise. In addition popcorn noise and methods of measuring random noise are discussed.

## 8.1  THERMAL NOISE

Thermal noise comes from thermal agitation of electrons within a resistance, and it sets a lower limit on the noise present in a circuit. Thermal noise is also referred to as resistance noise or "Johnson noise" (for J. B. Johnson, its discoverer). Johnson (1928) found that a nonperiodic voltage exists in all conductors and its magnitude is related to temperature. Nyquist (1928) subsequently described the noise voltage mathematically, using thermodynamic reasoning. He showed that the open-circuit rms noise voltage produced by a resistance is

$$V_t = \sqrt{4KTBR}, \qquad (8\text{-}1)$$

where

$K$ = Boltzmann's constant ($1.38 \times 10^{-23}$ joules/K)
$T$ = absolute temperature (K)
$B$ = noise bandwidth (Hz)
$R$ = resistance ($\Omega$)

At room temperature (290K), $4kT$ equals $1.6 \times 10^{-20}$ W/Hz. The bandwidth $B$ in Eq. 8-1 is the equivalent noise bandwidth of the system being considered. The calculation of equivalent noise bandwidth is covered in Section 8.3.

*Electromagnetic Compatibility Engineering,* by Henry W. Ott
Copyright © 2009 John Wiley & Sons, Inc.

Thermal noise is present in all elements containing resistance. A plot of the thermal noise voltage at a temperature of 17°C (290 K) is shown in Fig. 8-1. Normal temperature variations have a small effect on the value of the thermal noise voltage. For example, at 117°C the noise voltage is only 16% greater than that given in Fig. 8-1 for 17°C.

Equation 8-1 shows that the thermal noise voltage is proportional to the square root of the bandwidth and the square root of resistance. It would therefore be advantageous to minimize the resistance and bandwidth of a system to reduce the thermal noise voltage. If thermal noise is still a problem, considerable reduction is possible by operating the circuit at extremely low temperatures (close to absolute zero), or by using a parametric amplifier. Because the gain of a parametric amplifier comes from a reactance varied at a rapid rate, it does not have thermal noise.

The thermal noise in a resistor can be represented by adding a thermal noise voltage source $V_t$ in series with the resistor, as shown in Fig. 8-2. The magnitude of $V_t$ is determined from Eq. 8-1. In some cases, it is preferable to represent the thermal noise by an equivalent rms noise current generator of magnitude

$$I_t = \sqrt{\frac{4kTB}{R}} \qquad (8\text{-}2)$$

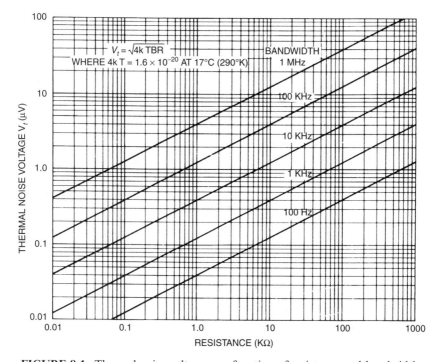

FIGURE 8-1. Thermal noise voltage as a function of resistance and bandwidth.

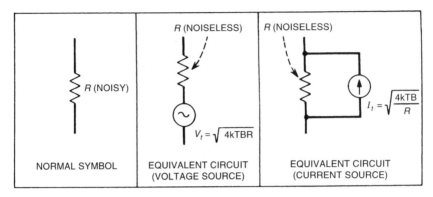

**FIGURE 8-2.** Thermal noise in a resistor (left) can be represented in an equivalent circuit as a voltage source (center) or a current source (right).

in parallel with the resistor. This is also shown in Fig. 8-2.

Thermal noise is a universal function, independent of the composition of the resistance. For example, a 1000-$\Omega$ carbon resistor has the same amount of thermal noise as a 1000-$\Omega$ tantalum thin-film resistor. An actual resistor may have more noise than that due to thermal noise, but never less. This additional, or excess, noise is because of the presence of other noise sources. A discussion of noise in actual resistors was given in Section 5.4.1.

Electric circuit elements can produce thermal noise only if they can dissipate energy. Therefore, a reactance cannot produce thermal noise. This can be demonstrated by considering the example of a resistor and capacitor connected, as shown in Fig. 8-3. Here, we make the erroneous assumption that the capacitor generates a thermal noise voltage $V_{tc}$. The power that generator $V_{tc}$ delivers to the resistor is $P_{cr} = N(f)V_{tc}^2$, where $N(f)$ is some nonzero network function.* The power that generator $V_{tr}$ delivers to the capacitor is zero, because the capacitor cannot dissipate power. For thermodynamic equilibrium the power that the resistor delivers to the capacitor must equal the power that the capacitor delivers to the resistor. Otherwise, the temperature of one component increases and the temperature of the other component decreases. Therefore,

$$P_{cr} = N(f)V_{tc}^2 = 0. \qquad (8\text{-}3)$$

The function $N(f)$ cannot be zero at all frequencies, because it is a function of the network. Voltage $V_{tc}$ must therefore be zero, which demonstrates that a capacitor cannot generate thermal noise.

Let us now connect two unequal resistors (at the same temperature) together, as shown in Fig. 8-4, and check for thermodynamic equilibrium.

---

* In this example $N(f) = [jw/(jw + 1/RC)]^2/R$.

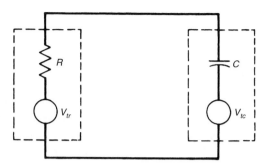

**FIGURE 8-3.** An R–C circuit can only be in thermodynamic equilibrium if $V_{tc}$ equals zero.

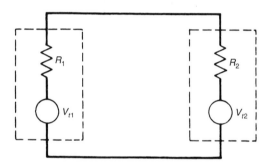

**FIGURE 8-4.** Two resistors connected in parallel are in thermodynamic equilibrium.

The power that generator $V_{t1}$ delivers to resistor $R_2$ is

$$P_{12} = \frac{R_2}{(R_1 + R_2)^2} V_{t1}^2. \tag{8-4}$$

Substituting Eq 8-1 for $V_{t1}$ gives

$$P_{12} = \frac{4kTBR_1R_2}{(R_1 + R_2)^2}. \tag{8-5}$$

The power that generator $V_{t2}$ delivers to $R_1$ is

$$P_{21} = \frac{R_1}{(R_1 + R_2)^2} V_{t2}^2. \tag{8-6}$$

Substituting Eq. 8-1 for $V_{t2}$ gives

$$P_{21} = \frac{4kTBR_1R_2}{(R_1 + R_2)^2}. \tag{8-7}$$

Comparing Eq. 8-5 to Eq. 8-7, we conclude that

$$P_{12} = P_{21}, \tag{8-8}$$

which shows that the two resistors are in thermodynamic equilibrium.

The power that generator $V_{t1}$ delivers to resistor $R_1$ does not have to be considered in the preceding calculation. This power comes from and is dissipated in resistor $R_1$. Thus, it produces no net effect on the temperature of resistor $R_1$. Similarly, the power that generator $V_{t2}$ delivers to resistor $R_2$ need not be considered.

Let us now consider the case when the two resistors in Fig. 8-4 are equal in value, and maximum power transfer occurs between the resistors. We can then write

$$P_{12} = P_{21} = P_n = \frac{V_t^2}{4R}. \tag{8-9}$$

Substituting Eq. 8-1 for $V_t$ gives

$$P_n = kTB. \tag{8-10}$$

The quantity $kTB$ is referred to as the "available noise power." At room temperature (17°C), this noise power per hertz of bandwidth is $4 \times 10^{-21}$ W, and it is independent of the value of the resistance.

It can be shown (van der Ziel, 1954, p. 17) that the thermal noise generated by any arbitrary connection of passive elements is equal to the thermal noise that would be generated by a resistance equal to the real part of the equivalent network impedance. This fact is useful for calculating the thermal noise of a complex passive network.

## 8.2  CHARACTERISTICS OF THERMAL NOISE

The frequency distribution of thermal noise power is uniform. For a specified bandwidth anywhere in the spectrum, the available noise power is constant and independent of the resistance value. For example, the noise power in a 100-Hz band between 100 and 200 Hz is equal to the noise power in a 100-Hz band between 1,000,000 and 1,000,100 Hz. When viewed on a wideband oscilloscope, thermal noise appears as shown in Fig. 8-5. Such noise—with a uniform power distribution with respect to frequency—is called "white noise," which implies that it is made up of many frequency components. Many noise sources other than thermal noise share this characteristic and are similarly referred to as white noise.

Although the rms value for thermal noise is well defined, the instantaneous value can only be defined in terms of probability. The instantaneous amplitude of

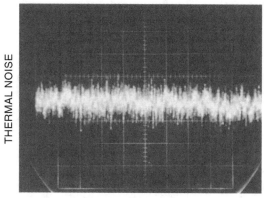

TIME, 200 μs PER DIVISION

**FIGURE 8-5.** Thermal noise as observed on a wideband oscilloscope (horizontal sweep 200 μs per division).

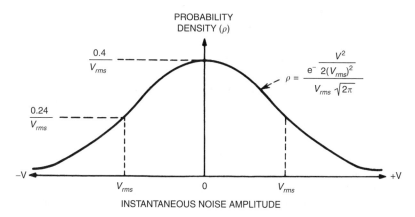

**FIGURE 8-6.** Probability density function for thermal noise (Gaussian distribution).

thermal noise has a Gaussian, or normal, distribution. The average value is zero, and the rms value is given by Eq. 8-1. Figure 8-6 shows the probability density function for thermal noise. The probability of obtaining an instantaneous voltage between any two values is equal to the integral of the probability density function between the two values. The probability density function is greatest at zero magnitude, which indicates that values near zero are most common.

The crest factor of a waveform is defined as the ratio of the peak to the rms value. For thermal noise the probability density function, shown in Fig. 8-6, asymptotically approaches zero for both large positive and large negative amplitudes. Because the curve never reaches zero, the magnitude of the instantaneous noise voltage has no finite limit. On this basis, the crest factor

TABLE 8-1.    Crest Factors for Thermal Noise

| Percent of Time Peak Exceeded | Crest Factor (peak/rms) |
| --- | --- |
| 1.0 | 2.6 |
| 0.1 | 3.3 |
| 0.01 | 3.9 |
| 0.001 | 4.4 |
| 0.0001 | 4.9 |

would be infinite, which is not a very useful result. A more useful result is obtained if we calculate the crest factor for peaks that occur at least a specified percentage of the time. Table 8-1 shows the results. Normally, only peaks that occur at least 0.01% of the time are considered, and a crest factor of approximately 4 is used for thermal noise.

## 8.3   EQUIVALENT NOISE BANDWIDTH

The noise bandwidth $B$ is the voltage-gain-squared bandwidth of the system or circuit being considered. The noise bandwidth is defined for a system with uniform gain throughout the passband and zero gain outside the passband. Figure 8-7 shows this ideal response for a low-pass circuit and a bandpass circuit.

   Practical circuits do not have these ideal characteristics but have responses similar to those shown in Fig. 8-8. The problem then is to find an equivalent noise bandwidth that can be used in equations to give the same results as the actual nonideal bandwidth does in practice. In the case of a white noise source (equal noise power for a specified bandwidth anywhere in the spectrum), the objective is met if the area under the equivalent noise bandwidth curve is made equal to the area under the actual curve. This is shown in Fig. 8-9 for a low pass circuit.

   For any network transfer function, $A(f)$ (expressed as a voltage or current ratio), an equivalent noise bandwidth exists with constant magnitude of transmission $A_0$ and bandwidth of

$$B = \frac{1}{|A_0|^2} \int_0^\infty |A(f)|^2 df. \tag{8-11}$$

   A typical bandpass function is shown in Fig. 8-10. $A_0$ is usually taken as the maximum absolute value of $A(f)$.

**Example 8-1.** Calculate the equivalent noise bandwidth for the simple $R$–$C$ circuit of Fig. 8-11. The voltage gain of this single pole (time constant) circuit versus frequency is

$$A(f) = \frac{f_0}{jf + f_0}, \tag{8-12}$$

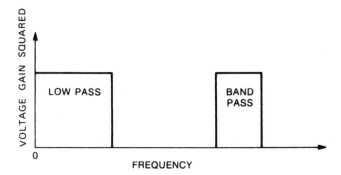

**FIGURE 8-7.**  Ideal bandwidth of low-pass and band-pass circuit elements.

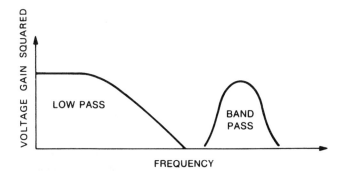

**FIGURE 8-8.**  Actual bandwidth of low-pass and band-pass circuit elements.

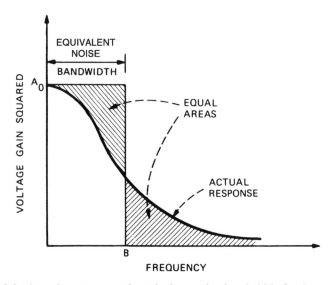

**FIGURE 8-9.**  Actual response and equivalent noise bandwidth for low-pass circuit. This curve is drawn with a linear scale.

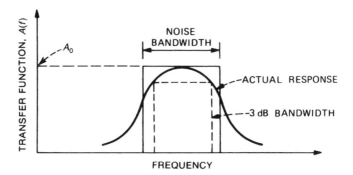

**FIGURE 8-10.** Any network transfer function can be expressed as an equivalent bandwidth with constant transmission ratio.

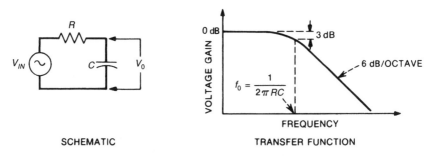

**FIGURE 8-11.** R–C circuit.

where

$$f_0 = \frac{1}{2\pi RC}. \tag{8-13}$$

Frequency $f_0$ is where the voltage gain is down 3 dB, as shown in Fig. 8-11. At $f = 0$, $A(f) = A_0 = 1$. Substituting Eq. 8-12 into Eq. 8-11 gives

$$B = \int_0^\infty \left( \frac{f_0}{\sqrt{f_0^2 + f^2}} \right)^2 df, \tag{8-14a}$$

$$B = f_0^2 \int_0^\infty (f_0^2 + f^2)^{-1} df. \tag{8-14b}$$

This can be integrated by letting $f = f_0 \tan \theta$; therefore, $df = f_0 \sec^2 \theta \, d\theta$. Making this substitution into Eq. 8-14b gives

$$B = f_0 \int_0^{\pi/2} d\theta. \tag{8-15}$$

Integrating gives

$$B = \frac{\pi}{2} f_0 \tag{8-16}$$

Therefore, the equivalent noise bandwidth for this circuit is $\pi/2$ or 1.57 times the 3-dB voltage bandwidth $f_0$. This result can be applied to any circuit that can be represented as a single-pole, low-pass filter. This result is also applicable to certain active devices, such as transistors, which can be modeled as single-pole, low-pass circuits.

Table 8-2 gives the ratio of the noise bandwidth to the 3-dB bandwidth for circuits with various numbers of identical poles. As can be observed, when the number of poles increases, the noise bandwidth approaches the 3-dB bandwidth. In the case of three or more poles the 3-dB bandwidth can be used in place of the noise bandwidth with only a small error.

A second method of determining noise bandwidth is to perform the integration graphically. This is done by plotting the voltage-gain-squared versus frequency on linear graph paper. A noise bandwidth rectangle is then drawn such that the area under the noise bandwidth curve is equal to the area under the actual curve, as shown in Fig. 8-9.

## 8.4   SHOT NOISE

Shot noise is associated with current flow across a potential barrier. It is caused by the fluctuation of current around an average value that results from the random emission of electrons (or holes). This noise is present in both vacuum tubes and semiconductors. In vacuum tubes, shot noise comes from the random

**TABLE 8-2.   Ratio of the Noise Bandwidth B to the 3-db Bandwidth $f_0$**

| Number of Poles | $B/f_0$ | High-Frequency Rolloff (dB per octave) |
|---|---|---|
| 1 | 1.57 | 6 |
| 2 | 1.22 | 12 |
| 3 | 1.15 | 18 |
| 4 | 1.13 | 24 |
| 5 | 1.11 | 30 |

emission of electrons from the cathode. In semiconductors, shot noise is caused by random diffusion of carriers through the base of a transistor and the random generation and recombination of hole electron pairs.

The shot effect was analyzed theoretically by W. Schottky in 1918. He showed that the rms noise current was equal to (van der Ziel, 1954, p. 91)

$$I_{sh} = \sqrt{2qI_{dc}B},\qquad(8\text{-}17)$$

where

$q$ = electron charge ($1.6 \times 10^{-19}$ coulombs)
$I_{dc}$ = average dc current (A)
$B$ = noise bandwidth (Hz)

Equation 8-17 is similar in form to Eq. 8-2. The power density for shot noise is constant with frequency and the amplitude has a Gaussian distribution. The noise is white noise and has the same characteristic as previously described for thermal noise. Dividing Eq. 8-17 by the square root of the bandwidth gives

$$\frac{I_{sh}}{\sqrt{B}} = \sqrt{2qI_{dc}} = 5.66 \times 10^{-10}\sqrt{I_{dc}}.\qquad(8\text{-}18)$$

In Eq. 8-18, the noise current per square root of bandwidth is only a function of the direct current (dc)-current flowing through the device. Therefore, by measuring the dc current through the device, the amount of noise can be accurately determined.

In making amplifier noise figure measurements (as discussed in Section 9.2, the availability of a variable source of white noise can considerably simplify the measurement. A diode can be used as a white noise source. If shot noise is the predominant noise source in the diode, then the rms value of the noise current can be determined simply by measuring the dc current through the diode.

## 8.5  CONTACT NOISE

Contact noise is caused by fluctuating conductivity from an imperfect contact between two materials. It occurs anywhere when two conductors are joined together, such as in switches and relay contacts. It also occurs in transistors and diodes, because of imperfect contacts, and in composition resistors and carbon microphones that are composed of many small particles molded together.

Contact noise is also referred to by many other names. When found in resistors, it is referred to as "excess noise." When observed in vacuum tubes, it is usually referred to as "flicker noise." Because of its unique frequency characteristic it is often called "$1/f$ noise," or "low-frequency noise."

Contact noise is directly proportional to the value of direct current flowing through the device. The power density varies as the reciprocal of frequency ($1/f$) and the magnitude is Gaussian. The noise current $I_f$ per square root of bandwidth can be expressed approximately (van der Ziel, 1954, p. 209) as

$$\frac{I_f}{\sqrt{B}} \approx \frac{KI_{dc}}{\sqrt{f}}, \tag{8-19}$$

where

$I_{dc}$ = average value of dc current (A)
$f$ = frequency (Hz)
$K$ = a constant that depends on the type of material and its geometry
$B$ = bandwidth in hertz centered about the frequency ($f$)

It should be noted that the magnitude of contact noise can become very large at low frequencies because of its $1/f$ characteristic. Most of the theories advanced to account for contact noise predict that at some low frequencies, the amplitude becomes constant. However, measurements of contact noise at frequencies as low as a few cycles per day still show the $1/f$ characteristic. Because of its frequency characteristics, contact noise is usually the most important noise source in low-frequency circuits.

$1/f$ or contact noise is also referred to as "pink noise." Pink noise is band-limited white noise. Its characteristics are similar to that of white noise that has been passed through a filter with a 3 dB per octave rolloff. White noise has equal noise power in each unit of bandwidth, giving it a 3 dB per octave rise in power when plotted on a logarithmic frequency scale. Pink noise, however, has equal noise power per octave (or per decade) of bandwidth, and therefore, it has a flat response when plotted on a logarithmic frequency scale. For example, the pink noise power in the octave between 2 kHz and 4 kHz will be equal to the noise power in the octave from 20 kHz to 40 kHz. White noise, however, would have 10 times (3 dB) more noise power in the 20-to-40 kHz frequency band as in the 2-to-4 kHz band.

The characteristics of pink noise, low-frequency emphasis and equal power per octave, closely match the response of human hearing. In addition, pink noise somewhat resembles speech in the frequency spectrum, and in not having a specific frequency or amplitude. As a result, pink noise sources are often used in the testing of audio systems. When used for audio testing, pink noise is usually generated by the filtering of a white noise source such as a diode.

## 8.6 POPCORN NOISE

Popcorn noise, which is also called burst noise, was first discovered in semiconductor diodes and has also appeared in integrated circuits(ICs). If

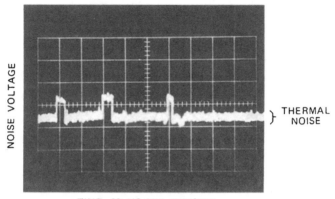

NOISE VOLTAGE

TIME, 20 MS PER DIVISION

**FIGURE 8-12.** Output waveform of an IC operatic-amplifier with popcorn noise. The random noise on the baseline and at the top of the burst is thermal noise.

burst noise is amplified and fed into a loudspeaker, it sounds like corn popping, with thermal noise providing a background frying sound—thus the name popcorn noise.

Unlike the other noise sources discussed in this chapter, popcorn noise is caused by a manufacturing defect, and it can be eliminated by improved manufacturing processes. This noise is caused by a defect in the junction, usually a metallic impurity, of a semiconductor device. Popcorn noise occurs in bursts and causes a discrete change in level, as shown in Fig. 8-12. The width of the noise bursts varies from microseconds to seconds. The repetition rate, which is not periodic, varies from several hundred pulses per second to less than one pulse per minute. For any particular sample of a device, however, the amplitude is fixed because it is a function of the characteristics of the junction defect. Typically, the amplitude is from 2–100 times the thermal noise.

The power density of popcorn noise has a $1/f^n$ characteristic, where $n$ is typically two. Because the noise is a current-related phenomenon, popcorn noise voltage is greatest in a high-impedance circuit, for example, the input circuit of an operational amplifier.

## 8.7  ADDITION OF NOISE VOLTAGES

Noise voltages, or currents, produced independently with no relationships between each other are uncorrelated. When uncorrelated noise sources are added together, the total power is equal to the sum of the individual powers. Adding two noise voltage generators, $V_1$ and $V_2$, together on a power basis, gives

$$V_{\text{total}}^2 = V_1^2 + V_2^2. \tag{8-20}$$

The total noise voltage can then be written as

$$V_{\text{total}} = \sqrt{V_1^2 + V_2^2}. \tag{8-21}$$

Therefore, uncorrelated noise voltages can be added by taking the square root of the sum of the squares of the individual noise voltages.

Two correlated noise voltages can be added by using

$$V_{\text{total}} = \sqrt{V_1^2 + V_2^2 + 2\gamma V_1 V_2}, \tag{8-22}$$

where $\gamma$ is a correlation coefficient that can have any value from $+1$ to $-1$. When $\gamma$ equals 0, the voltages are uncorrelated; when $|\gamma|$ equals 1, the voltages are totally correlated. For values of $\gamma$ between 0 and $+1$ or 0 and $-1$ the voltages are partially correlated.

## 8.8   MEASURING RANDOM NOISE

Noise measurements are usually made at the output of a circuit or amplifier. This is done for the following two reasons: (1) the output noise is larger and therefore easier to measure and (2) it avoids the possibility of the noise meter upsetting the shielding, grounding, or balancing of the input circuit of the device being measured. If a value of equivalent input noise is required, then the output noise is measured and divided by the circuit gain to obtain the equivalent input noise.

Because most voltmeters were intended to measure sinusoidal voltages, their response to a random noise source must be investigated. Three general requirements for a noise meter are (1) it should respond to noise power, (2) it should have a crest factor of four or greater, and (3) its bandwidth should be at least 10 times the noise bandwidth of the circuit being measured. We will now consider the response of various types of meters when used to measure white noise.

A true rms meter is obviously the best choice, provided its bandwidth and crest factor are sufficient. A crest factor of three provides less than 1.5% error, whereas a crest factor of four gives an error of less than 0.5% No correction to the meter indication is required.

The most common alternating current (ac) voltmeter responds to the average value of the waveform but has a scale calibrated to read rms. This meter uses a rectifier and a dc meter movement to respond to the average value of the waveform being measured. For a sine wave, the rms value is 1.11 times the average value. Therefore, the meter scale is calibrated to read 1.11 times the measured value. For white noise, however, the rms value is 1.25 times the average value. Therefore, when used to measure white noise, an average-responding meter reads too low. If the bandwidth and crest factor are sufficient, such a meter may be

**TABLE 8-3.  Charateristics of Meters Used to Measure White Noise**

| Type of Meter | Correction Factor | Remarks |
|---|---|---|
| True rms | None | Meter bandwidth greater than ten times noise bandwidth and meter crest factor 3 or grater. |
| RMS calibrated average responding | Multiply reading by 1.13 or add 1.1 dB | Meter bandwidth grater than ten times noise bandwidth, and meter crest factor 3 or grater. Read below one-half sclae to avoid clipping peaks. |
| RMS calibrated peak responding | Do not use | |
| Oscilloscope | RMS $\approx$ 1/8 peak-to-peak value | Waveshape can be observed to be sure it is random noise and not pickup. Ignore occasional, emtreme peaks. |

used to measure white noise by multiplying the meter reading by 1.13 or by adding 1.1 dB. Measurements should be made on the lower half of the meter scale to avoid clipping the peaks of the noise waveform.

Peak-responding voltmeters should not be used to measure noise since their response depends on the charge and discharge time constants of the individual meter used.

An oscilloscope is an often overlooked, but excellent, device for measuring white noise. One advantage it has over all other indicators is that the waveshape being measured can be seen. In this way, you can be sure that you are measuring the desired random noise, not pickup or 60-Hz hum. The rms value of white noise is approximately equal to the peak-to-peak value taken from the oscilloscope, divided by eight.* When determining the peak-to-peak value on the oscilloscope, one or two peaks that are considerably greater than the rest of the waveform should be ignored. With a little experience, rms values can be accurately determined by this method. With an oscilloscope, random noise can be measured even when 60-Hz hum or other nonrandom noise sources are present, because the waveforms can be distinguished and measured separately on the display.

Table 8-3 summarizes the characteristics of various types of meters when used to measure white noise.

## SUMMARY

- Thermal noise is present in all elements that contains resistance.
- A reactance does not generate thermal noise.

---

* This assumes a crest factor of 4 for white noise.

- The thermal noise in any connection of passive elements is equal to the thermal noise that would be generated in a resistance equal to the real part of the equivalent network impedance.
- Shot noise is produced by current flow across a potential barrier.
- Contact noise ($1/f$ noise) is present whenever current flows through a nonhomogeneous material.
- Contact noise is only a problem at low frequencies.
- Popcorn noise can be eliminated by improved manufacturing processes.
- The noise bandwidth is greater than the 3-dB bandwidth.
- As the number of poles (time constants) increase, the noise bandwidth approaches the 3-dB bandwidth.
- The crest factor for thermal noise is normally assumed to be four.
- Noise having equal power in each unit of bandwidth (such as thermal and shot noise) is refered to as white noise.
- Noise having equal power per active (or decade) of bandwidth (such as $1/f$ or contact noise) is refered to as pink noise.
- The characteristics of pink noise are similiar to white noise that has been passed through a 3 dB per octave roll off low-pass filter.
- Uncorrelated noise voltages add on a power basis; therefore

$$V_{\text{total}} = \sqrt{V_1^2 + V_2^2 + \cdots V_m^2}.$$

## PROBLEMS

8.1 At room temperature (290 K), what is the minimum noise voltage possible in a 50-$\Omega$ measuring system that has a bandwidth of 100 kHz?

8.2 Calculate the noise voltage produced by a 5000-$\Omega$ resistor in a system with a 10-kHz bandwidth at a temperature of:
   a.  27°C (300 K).
   b.  100°C (373 K).

8.3 Calculate the thermal noise voltage per square root of bandwidth for the circuit shown in Fig. P8-3.

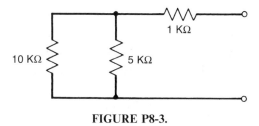

10 KΩ        5 KΩ        1 KΩ

**FIGURE P8-3.**

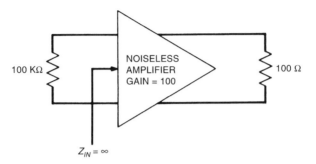

**FIGURE P8-4.**

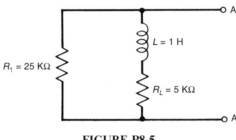

**FIGURE P8-5.**

8.4 Determine the noise voltage at the amplifier output for the circuit of Fig. P8-4. Assume the amplifier has a frequency response equivalent to:

a. An ideal low-pass filter with a cutoff frequency of 2 kHz.

b. An ideal bandpass filter with cutoff frequencies of 99 and 101 kHz.

8.5 Determine the voltage per square root of bandwidth generated across terminals A-A for the circuit of Fig. P8-5. at room temperature and at a frequency of 1590 Hz.

8.6 Determine the equivalent noise current generator for a 100k-$\Omega$ resistor. Assume $T = 290$ K, and $B = 1$ MHz.

8.7 Determine the noise bandwidth of a low-pass filter having a 3 dB bandwidth of 3.3 kH and a 12 dB per octave roll off.

8.8 A diode is used as a white-noise generator. The current through the diode is equal to 10 mA. What will be the noise current per $\sqrt{B}$?

8.9 A noise source is known to have a $1/f$ spectral density. The noise voltage measured in the frequency band from 100 to 200 Hz is 2 $\mu$V.

a. What will be the noise voltage in the frequency band of 200 to 800 Hz?

b. What will be the noise voltage in the frequency band of 2000 to 4000 Hz?

8.10  Three uncorrelated noise voltage sources exist in a system. The magnitudes of the three noise voltages are 10 µV, 20 µV, and 32 µV. What will the total noise voltage be?

## REFERENCES

Johnson, J. B. "Thermal Agitation of Electricity in Conductors." *Physical Review,* vol. 32, July 1928, pp. 97–109.

Nyquist, H. "Thermal Agitation of Electric Charge in Conductors." *Physical Review,* vol. 32, July 1928, pp. 110–113.

van der Ziel, A. *Noise.* Englewood Cliffs, N.J. Prentice-Hall, 1954.

## FURTHER READING

Baxandall, P. J. "Noise in Transistor Circuits, Part 1." *Wireless World*, vol. 74, November 1968.

Bennett, W. R. "Characteristics and Origins of Noise—Part I." *Electronics*, vol. 29, March 1956, pp. 154–160.

Campbell, R. H., Jr. and Chipman, R. A. "Noise from Current-Carrying Resistors 20 to 500 Kc." *Proceedings of I.R. E.,* vol. 37, August 1949, pp. 938–942.

Dummer, G. W. A. *Fixed Resistors.* London, U.K. Sir Isaac Pitman, 1956

Lathi, B. P. *Signals, Systems and Communications.* Chapter 13. New York, Wiley, 1965.

Mumford, W. W.and Scheibe, E. H. *Noise Performance Factors in Communication Systems.* Horizon House, Dedham, MA, 1969.

van der Ziel, A. *Fluctuation Phenomena in Semi-Conductors.* New York, Academic, 1959.

# 9 Active Device Noise

Bipolar transistors, field effect transistors (FETs), and operational amplifiers (op-amps) have inherent noise-generation mechanisms. This chapter discusses these internal noise sources and shows the conditions necessary to optimize noise performance.

Before covering active device noise, the general topics of how noise is specified and measured are presented. This general analysis provides a standard set of noise parameters that can then be used to analyze noise in various devices. The common methods of specifying device noise are (1) noise factor and (2) the use of a noise voltage and current model.

## 9.1 NOISE FACTOR

The concept of noise factor was developed in the 1940s as a method of evaluating noise in vacuum tubes. Despite several serious limitations, the concept is still widely used today.

The noise factor ($F$) is a quantity that compares the noise performance of a device to that of an ideal (noiseless) device. It can be defined as

$$F = \frac{\text{Noise power output of actual device } (P_{no})}{\text{Noise power output of ideal device}}. \qquad (9\text{-}1)$$

The noise power output of an ideal device is due to the thermal noise power of the source resistance. The standard temperature for measuring the source noise power is 290K. Therefore, the noise factor can be written as

$$F = \frac{\text{Noise power output of actual device } (P_{no})}{\text{Power output due to source noise}}. \qquad (9\text{-}2)$$

---

*Electromagnetic Compatibility Engineering,* by Henry W. Ott
Copyright © 2009 John Wiley & Sons, Inc.

An equivalent definition of noise factor is the input signal-to-noise ($S/N$) ratio divided by the output signal-to-noise ratio

$$F = \frac{S_i/N_i}{S_o/N_o}. \tag{9-3}$$

These signal-to-noise ratios must be power ratios unless the input impedance is equal to the load impedance, in which case they can be voltage squared, current squared, or power ratios.

All noise factor measurements must be taken with a resistive source, as shown in Fig. 9-1. The open circuit input noise voltage is therefore just the thermal noise of the source resistance $R_s$, or

$$V_t = \sqrt{4kTBR_s}. \tag{9-4}$$

At 290K, this is

$$V_t = \sqrt{1.6 \times 10^{-20} BR_s}. \tag{9-5}$$

If the device has a voltage gain $A$, defined as the ratio of the output voltage measured across $R_L$ to the open-circuit source voltage, then the component of output voltage due to the thermal noise in $R_s$ is $AV_t$. Using $V_{no}$ for the total output noise voltage measured across $R_L$, the noise factor can be written as

$$F = \frac{(V_{no})^2/R_L}{(AV_t)^2/R_L}, \tag{9-6}$$

or

$$F = \frac{(V_{no})^2}{(AV_t)^2}. \tag{9-7}$$

$V_{no}$ includes the effects of both the source noise and the device noise. Substituting Eq. 9-4 into Eq. 9-7 gives

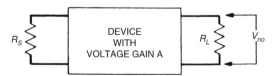

**FIGURE 9-1.** Resistive source is used for noise factor measurements.

$$F = \frac{(V_{no})^2}{4kTBR_sA^2}. \tag{9-8}$$

The following three characteristics of noise factor can be observed by examining Eq. 9-8:

1. It is independent of load resistance $R_L$.
2. It does depend on source resistance $R_s$.
3. If a device were completely noiseless, the noise factor would equal one.

Noise factor expressed in decibels is called noise figure *(NF)* and is equal to

$$NF = 10 \log F. \tag{9-9}$$

In a qualitative sense, noise figure and noise factor are the same, and in casual conversation they are often interchanged.

Because of the bandwidth term in the denominator of Eq. 9-8, the noise factor can be specified in the following two ways: (1) a spot noise, measured at a specified frequency over a 1-Hz bandwidth, or (2) an integrated, or average noise, measured over a specified bandwidth. If the device noise is "white" and is generated prior to the bandwidth-limiting portion of the circuit both the spot and integrated noise factors are equal. This is because, as the bandwidth is increased, both the total noise and the source noise increase by the same factor.

The concept of noise factor has three major limitations:

1. Increasing the source resistance may decrease the noise factor while increasing the total noise in the circuit.
2. If a purely reactive source is used, noise factor is meaningless because the source noise is zero, which makes the noise factor infinite.
3. When the device noise is only a small percentage of the source thermal noise (as with some low noise FETs), the noise factor requires taking the ratio of two almost equal numbers. This approach can produce inaccurate results.

A direct comparison of two noise factors is only meaningful if both are measured at the same source resistance. Noise factor varies with the bias conditions, frequency, and temperature as well as source resistance, and all of these should be defined when specifying noise factor.

Knowing the noise factor for one value of source resistance does not allow the calculation of the noise factor at other values of source resistance. This is because both the source noise and device noise vary as the source resistance is changed.

## 9.2   MEASUREMENT OF NOISE FACTOR

A better understanding of noise factor can be obtained by describing the methods used to measure it. Two methods follow: (1) the single-frequency method, and (2) the noise-diode, or white-noise, method.

### 9.2.1   Single-Frequency Method

The test set up for the single-frequency method is shown in Fig. 9-2. $V_s$ is an oscillator set to the frequency of the measurement, and $R_s$ is the source resistance. With the source $V_s$ turned off, the output root mean square (rms) noise voltage $V_{no}$ is measured. This voltage consists of two parts: the first from the thermal noise voltage $V_t$ of the source resistor, and the second from the noise in the device.

$$V_{no} = \sqrt{(AV_t)^2 + (\text{Device noise})^2}. \tag{9-10}$$

Next, the generator $V_s$ is turned on, and an input signal is applied until the output power doubles (output rms voltage increases by 3 dB over that previously measured). Under these conditions the following equation is satisfied

$$(AV_s)^2 + (V_{no})^2 = 2V_{no}^2; \tag{9-11}$$

therefore

$$AV_s = V_{no}. \tag{9-12}$$

Substituting Eq. 9-12 into Eq. 9-7 gives

$$F = \left(\frac{V_s}{V_t}\right)^2. \tag{9-13}$$

**FIGURE 9-2.** Single-frequency method for measuring noise factor.

Substituting from Eq. 9-5 for $V_t$ produces

$$F = \frac{V_s^2}{1.6 \times 10^{-20} BR_s}.$$

(9-14)

Because the noise factor is not a function of $R_L$, any value of load resistor can be used for the measurment.

The disadvantage of this method is that the noise bandwidth of the device* must be known.

### 9.2.2  Noise Diode Method

A better method of measuring noise factor is to use a noise diode as a white noise source. The measuring circuit is shown in Fig. 9-3. $I_{dc}$ is the direct current through the noise diode, and $R_s$ is the source resistance. The shot noise in the diode is

$$I_{sh} = \sqrt{3.2 \times 10^{-19} I_{dc} B}.$$

(9-15)

Using Thevenin's theorem, the shot-noise current generator can be replaced by a voltage generator $V_{sh}$ in series with $R_s$, where

$$V_{sh} = I_{sh} R_s.$$

(9-16)

The rms noise voltage output $V_{no}$ is first measured with the diode current equal to zero. This voltage consists of two parts: that from the thermal noise of the source resistor, and that from the noise in the device. Therefore

$$V_{no} = \sqrt{(AV_t)^2 + (\text{Device noise})^2}.$$

(9-17)

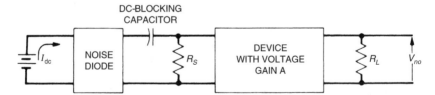

**FIGURE 9-3.** Noise-diode method of measuring noise factor.

---

* It should be remembered that the noise bandwidth is usually not equal to the 3-dB bandwidth (see Section 8.3).

The diode current is then allowed to flow and is increased until the output noise power doubles (output rms voltage increases by 3 dB). Under these conditions, the following equation is satisfied:

$$(AV_{sh})^2 + (V_{no})^2 = 2(V_{no})^2; \tag{9-18}$$

therefore

$$V_{no} = AV_{sh} = AI_{sh}R_s. \tag{9-19}$$

Substituting $V_{no}$ from Eq. 9-19 into Eq. 9-7, gives

$$F = \frac{(I_{sh}R_s)^2}{V_t^2}. \tag{9-20}$$

Substituting Eqs. 9-15 and 9-5 for $I_{sh}$ and $V_t$, respectively, gives

$$F = 20I_{dc}R_s. \tag{9-21}$$

The noise factor is now a function of only the direct current through the diode, and the value of the source resistance. Both of these quantities are easily measured. Neither the gain nor the noise bandwidth of the device need be known.

## 9.3   CALCULATING $S/N$ RATIO AND INPUT NOISE VOLTAGE FROM NOISE FACTOR

Once noise factor is known, it can be used to calculate the signal-to-noise ratio and the input noise voltage. For these calculations, it is important that the source resistance used in the circuit be the same as that used to make the noise factor measurement. Rearranging Eq. 9-8 gives

$$V_{no} = A\sqrt{4kTBR_sF}. \tag{9-22}$$

If the input signal is $V_s$, the output signal voltage is $V_o = AV_s$. Therefore the output signal-to-noise power ratio is

$$\frac{S_o}{N_o} = \frac{P_{signal}}{P_{noise}}, \tag{9-23}$$

or

$$\frac{S_o}{N_o} = \left(\frac{AV_s}{V_{no}}\right)^2. \tag{9-24}$$

Using Eq. 9-22 to substitute for $V_{no}$,

$$\frac{S_o}{N_o} = \frac{(V_s)^2}{4kTBR_sF}. \tag{9-25}$$

The signal-to-noise ratio, as used in Eqs. 9-23, 9-24, and 9-25, refers to a power ratio. However, signal-to-noise is sometimes expressed as a voltage ratio. Care should be taken as to whether a specified signal-to-noise ratio is a power or voltage ratio, because the two are not numerically equal. When expressed in decibels, the power signal-to-noise ratio is 10 log $(S_o/N_o)$.

Another useful quantity is the total equivalent input noise voltage $(V_{nt})$, which is the output noise voltage (Eq. 9-22) divided by the gain

$$V_{nt} = \frac{V_{no}}{A} = \sqrt{4kTBR_sF}. \tag{9-26}$$

The total equivalent input noise voltage is a single noise source that represents the total noise in the circuit. *For optimum noise performance, $V_{nt}$ should be minimized.* Minimizing $V_{nt}$ is equivalent to maximizing the signal-to-noise ratio, provided the signal voltage remains constant. This is discussed further in Section 9.7 on optimum source resistance.

The equivalent input noise voltage consists of two parts, one from the thermal noise of the source and the other from the device noise.

Representing the device noise by $V_{nd}$, we can write the total equivalent input noise voltage as

$$V_{nt} = \sqrt{(V_t)^2 + (V_{nd})^2}, \tag{9-27}$$

where $V_t$ is the open-circuit thermal noise voltage of the source resistance. Solving Eq. 9-27 for $V_{nd}$ gives

$$V_{nd} = \sqrt{(V_{nt})^2 - (V_t)^2}. \tag{9-28}$$

Substituting Eqs. 9-4 and 9-26 into Eq. 9-28 gives

$$V_{nd} = \sqrt{4kTBR_s(F-1)}. \tag{9-29}$$

## 9.4  NOISE VOLTAGE AND CURRENT MODEL

A better approach, and one that overcomes the limitations of noise factor, is to model the noise in terms of an equivalent noise voltage and current. The actual network can be modeled as a noise-free device with two noise generators, $V_n$ and $I_n$, connected to the input side of a network, as shown in Fig. 9-4. $V_n$ represents the device noise that exists when $R_s$ equals zero, and $I_n$ represents the additional device noise that occurs when $R_s$ does not equal zero. The use of these two noise generators plus a complex correlation coefficient (not shown) completely characterizes the noise performance of the device (Rothe and Dahlke, 1956). Although $V_n$ and $I_n$ are normally correlated to some degree, values for the correlation coefficient are seldom given on manufacturers data sheets. In addition, the typical spread of values of $V_n$ and $I_n$ for a device normally overshadows the effect of the correlation coefficient. Therefore, it is common practice to assume the correlation coefficient is equal to zero. This will be done in the remainder of this chapter.

Figure 9-5 shows representative curves of noise voltage and noise current. As can be seen in Fig, 9-5, the data normally consist of a plot of $V_n/\sqrt{B}$ and $I_n/\sqrt{B}$ versus frequency. The noise voltage or current over a band of frequencies can be found by integrating $[V_n/\sqrt{B}]^2$ or $[I_n/\sqrt{B}]^2$ versus frequency and then taking the square root of the result. In the case when $V_n/\sqrt{B}$ or $I_n/\sqrt{B}$ is constant over the desired bandwidth, the total noise voltage or current can be found simply by multiplying $V_n/\sqrt{B}$ or $I_n/\sqrt{B}$ by the square root of the bandwidth.

Using these curves and the equivalent circuit of Fig. 9-4, the total equivalent input noise voltage, signal-to-noise ratio, or noise factor for any circuit can be determined. This can be done for any source impedance, resistive or reactive, and across any frequency spectrum. The device must, however, be operated at or near the bias conditions for which the curves are specified. Often, additional curves are given showing the variation of these noise generators with bias points. With a set of these curves the noise performance of the device is completely specified under all operating conditions.

The representation of noise data in terms of the equivalent parameters $V_n$ and $I_n$ can be used for any device. Field-effect transistors and op-amps are

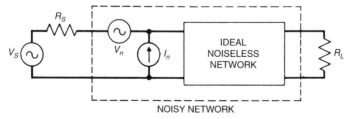

NOISY NETWORK

**FIGURE 9-4.**  A noisy network modeled as an ideal noisaless network with the addition of an input noise voltage and input noise current source.

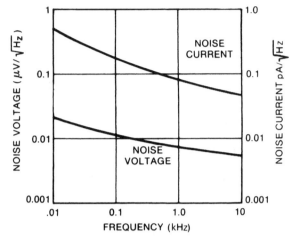

**FIGURE 9-5.** Typical noise voltage $V_n/\sqrt{B}$ and noise current $I_n/\sqrt{B}$ curves.

usually specified in this manner. Some bipolar transistor manufacturers are also beginning to use the $V_n$–$I_n$ parameters instead of noise factor.

The total equivalent input noise voltage of a device is an important parameter. Assuming no correlation between noise sources, this voltage, which combines the affect of $V_n$, $I_n$, and the thermal noise of the source, can be written as

$$V_{nt} = \sqrt{4kTBR_s + V_n^2 + (I_nR_s)^2},$$ (9-30)

where $V_n$ and $I_n$ are the noise voltage and noise current over the bandwidth $B$. For optimum noise performance, the total noise voltage represented by Eq. 9-30 should be minimized. This is discussed in more detail in Section 9.7 on Optimum Source Resistance.

The total equivalent input noise voltage per square root of bandwidth can be written as

$$\frac{V_{nt}}{\sqrt{B}} = \sqrt{4kTR_s + \left(\frac{V_n}{\sqrt{B}}\right)^2 + \left(\frac{I_nR_s}{\sqrt{B}}\right)^2}.$$ (9-31)

The equivalent input noise voltage from the device noise only can be calculated by subtracting the thermal noise component from Eq. 9-30. The equivalent input device noise then becomes

$$V_{nd} = \sqrt{V_n^2 + (I_nR_s)^2}.$$ (9-32)

Figure 9-6 is a plot of the total equivalent noise voltage per square root of the bandwidth for a typical low-noise bipolar transistor, junction field effect transistor, and op-amp. The thermal noise voltage generated by the source resistance is also shown. The thermal noise curve places a lower limit on the total input noise voltage. As can be observed from this figure, when the source resistance is between 10,000 $\Omega$ and 1 M$\Omega$, this FET has a total noise, voltage only slightly greater than the thermal noise in the source resistance. On the basis of noise, this FET approaches an ideal device when the source resistance is in this range. With low source resistance, however, a bipolar transistor generally has less noise than an FET. In most cases, the op-amp has more noise than either of the other devices. The reasons for this are discussed in Section 9.12 on Noise in Operational Amplifiers.

## 9.5  MEASURMENT OF $V_n$ AND $I_n$

It is a relatively simple matter to measure the parameters $V_n$ and $I_n$ for a device. The method can best be described by referring to Fig. 9-4 and recalling from Eq. 9-30 that the total equivalent noise voltage $V_{nt}$ is

$$V_{nt} = \sqrt{4kTBR_s + V_n^2 + (I_n R_s)^2}. \qquad (9-33)$$

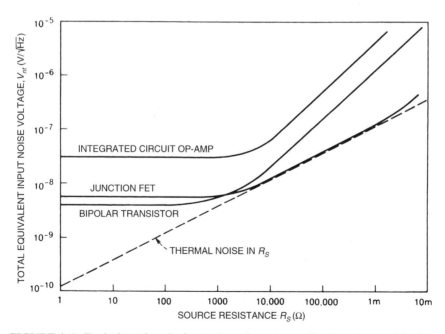

**FIGURE 9-6.** Typical total equivalent noise voltage curves for three types of devices.

To determine $V_n$, the source resistance is set equal to zero, which causes the first and last terms in Eq. 9-33 to equal zero, and the output noise voltage $V_{no}$ is measured. If the voltage gain of the circuit is $A$,

$$V_{no} = AV_{nt} = AV_n, \qquad \text{for } R_s = 0. \qquad (9\text{-}34)$$

The equivalent input noise voltage is

$$V_n = \frac{V_{no}}{A}, \qquad \text{for } R_s = 0. \qquad (9\text{-}35)$$

To measure $I_n$, a second measurement is made with a large source resistance. The source resistance should be large enough so that the first two terms in Eq. 9-33 are negligible. This will be true if the measured output noise voltage $V_{no}$ is

$$V_{no} \gg A\sqrt{4kTBR_s + V_n^2}.$$

Under these conditions the equivalent input noise current is

$$I_n = \frac{V_{no}}{AR_s}, \qquad \text{for } R_s \text{ large.} \qquad (9\text{-}36)$$

## 9.6   CALCULATING NOISE FACTOR AND S/N RATIO FROM $V_n$-$I_n$

Knowing the equivalent input noise voltage $V_n$, the current $I_n$, and the source resistance $R_s$, the noise factor can be calculated by referring to Fig. 9-4. This derivation is left for the reader, see Problem 9.10. The result is

$$F = 1 + \frac{1}{4kTB}\left(\frac{V_n^2}{R_s} + I_n^2 R_s\right), \qquad (9\text{-}37)$$

where $V_n$ and $I_n$ are the equivalent input noise voltage and current over the bandwidth $B$ of interest.

The value of $R_s$ producing the minimum noise factor can be determined from Eq. 9-37 by differentiating it with respect to $R_s$. The resulting $R_s$ for minimum noise factor is

$$R_{so} = \frac{V_n}{I_n}. \qquad (9\text{-}38)$$

If Eq. 9-38 is substituted back into Eq. 9-37, the minimum noise factor can be determined and is

$$F_{\min} = 1 + \frac{V_n I_n}{2kTB}.$$ (9-39)

The output power signal-to-noise ratio can also be calculated from the circuit of Fig. 9-4. This derivation is left for the reader, see problem 9.11. The result is

$$\frac{S_o}{N_o} = \frac{(V_s)^2}{(V_n)^2 + (I_n R_s)^2 + 4kTBR_s},$$ (9-40)

where $V_s$ is the input signal voltage.

For constant $V_s$ maximum signal-to-noise ratio occurs when $R_s = 0$, and is

$$\left.\frac{S_o}{N_o}\right|_{\max} = \left(\frac{V_s}{V_n}\right)^2.$$ (9-41)

It should be noted that when $V_s$ is constant and $R_s$ is variable, minimum noise factor occurs when $R_s = V_n/I_n$, but maximum signal-to-noise ratio occurs at $R_s = 0$. Minimum noise factor therefore does not necessarily represent maximum signal-to-noise ratio or minimum noise. This can best be understood by referring to Fig. 9-7, which is a plot of the total equivalent input noise voltage $V_{nt}$ for a typical device. When $R_s = V_n/I_n$, the ratio of the device noise to the thermal noise is a minimum. However, the device noise and the thermal noise are both minimum when $R_s = 0$. Although minimum equivalent input noise voltage (and maximum signal-to-noise ratio) occurs mathematically at $R_s = 0$, there is actually a range of values of $R_s$ over which it is almost constant, as shown in Fig. 9-7. In this range, $V_n$ of the device is the predominant noise source. For large values of source resistance, $I_n$ is the predominant noise source.

## 9.7  OPTIMUM SOURCE RESISTANCE

Because the maximum signal-to-noise ratio occurs at $R_s = 0$ and minimum noise factor occurs at $R_s = V_n/I_n$, the question of what is the optimum source resistance for the best noise performance arises. The requirement of a zero resistance source is impractical because all actual sources have a finite source resistance. However, as was shown in Fig. 9-7, as long as $R_s$ is small there is a range of values over which the total noise voltage is almost constant.

In practice, the circuit designer does not always have control over the source resistance. A source of fixed resistance is used for one reason or another. The

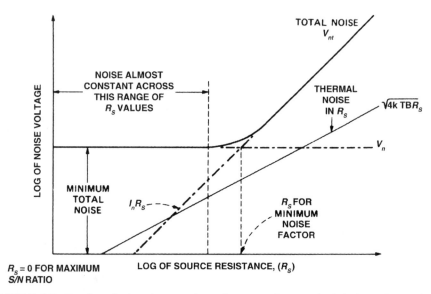

**FIGURE 9-7.** Total equivalent input noise voltage $V_{nt}$ for a typical device. The total noise voltage is made up of three components (thermal noise, $V_{nt}$ and $I_n R_s$) as was shown in Eq. 9-30.

question then arises as to whether this source resistance should be transformed to the value that produces minimum noise factor. The answer to this question depends on how the transformation is made.

If the actual source resistance is less than $R_s = V_n/I_n$, a physical resistor should not be inserted in series with $R_s$ to increase the resistance. To do this would produce three detrimental effects:

1. It increases the thermal noise due to the larger source resistance. (This increase is proportional to $\sqrt{R}$.)
2. It increases the noise from the input noise current generator flowing through the larger resistor. (This increase is proportional to $R$.)
3. It decreases the amount of the signal getting to the amplifier.

The noise performance can, however, be improved by using a transformer to effectively raise the value of $R_s$ to a value closer to $R_s = V_n/I_n$, thus minimizing the noise produced by the device. At the same time, the signal voltage is stepped up by the turns ratio of the transformer. This effect is cancelled by the fact that the thermal noise voltage of the source resistance is also stepped up by the same factor. There is, however, a net increase in signal-to-noise ratio when this is done.

If the actual source resistance is greater than that required for minimum noise factor, noise performance can still be improved by transforming the higher value of $R_s$ to a value closer to $R_s = V_n/I_n$. The noise will, however, be greater than if a lower-impedance source were used.

*For optimum noise peformance, the lowest possible source impedance should be used.* Once this is decided, noise performance can be improved by transformer coupling this source to match the impedance $R_s = V_n/I_n$.

The improvement in signal-to-noise ratio that is possible by using a transformer can best be seen by rewriting Eq. 9-3 as

$$\frac{S_o}{N_o} = \frac{1}{F}\left(\frac{S_i}{N_i}\right). \tag{9-42}$$

Assuming a fixed source resistance, adding an ideal transformer of any turns ratio does not change the input signal-to-noise ratio. With the input signal-to-noise ratio fixed, the output signal-to-noise ratio will be maximized when the noise factor $F$ is a minimum. $F$ is a minimum when the device sees a source resistance $R_s = V_n/I_n$. Therefore, transformer coupling the actual source resistance minimizes $F$ and maximizes the output signal-to-noise ratio. If the value of the source resistance is not fixed, then choosing $R_s$ to minimize $F$ does not necessarily produce optimum noise performance. However, for a given source resistance $R_s$, the least noisy circuit is the one with the smallest $F$.

When using transformer coupling, the thermal noise in the transformer winding must be accounted for. This can be done by adding to the source resistance the primary winding resistance, plus the secondary winding resistance divided by the square of the turns ratio. The turns ratio is defined as the number of turns of the secondary divided by the number of turns of the primary. Despite this additional noise introduced by the transformer, the signal-to-noise ratio is normally increased sufficiently to justify using the transformer if the actual source resistance is more than an order of magnitude different than the optimum source resistance.

Another source of noise to consider when using a transformer is its sensitivity to pickup from magnetic fields. Shielding the transformer is often necessary to reduce this pickup to an acceptable level.

The improvement in signal-to-noise ratio resulting from transformer coupling can be expressed in terms of the signal-to-noise improvement (SNI) factor defined as

$$\text{SNI} = \frac{(S/N) \text{ using transformer}}{(S/N) \text{ without transformer}}. \tag{9-43}$$

It can be shown that the signal-to-noise improvement factor can also be expressed in a more useful form as

$$\text{SNI} = \frac{(F) \text{ without transformer}}{(F) \text{ with transformer}}. \tag{9-44}$$

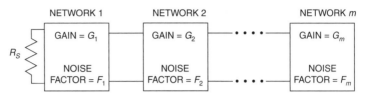

**FIGURE 9-8.** Networks in cascade.

## 9.8   NOISE FACTOR OF CASCADED STAGES

The signal-to-noise ratio and total equivalent input noise voltage should be used in designing the components of a system for optimum noise performance. Once the components of a system have been designed, it is usually advantageous to express the noise performance of the individual components in terms of noise factor. The noise factor of the various components can then be combined as follows.

The overall noise factor of a series of networks connected in cascade (see Fig. 9-8) was shown by Friis (1944) to be

$$F = F_1 + \frac{F_2 - 1}{G_1} + \frac{F_3 - 1}{G_1 G_2} + \cdots + \frac{F_m - 1}{G_1 G_2 \cdots G_{m-1}}, \qquad (9\text{-}45)$$

where $F_1$ and $G_1$ are the noise factor and available power gain* of the first stage, $F_2$, $G_2$ are those of the second stage, and so on.

Equation 9-45 clearly shows the important fact that *with sufficient gain $G_1$ in the first stage of a system, the total noise factor is primarily determined by the noise factor $F_1$ of the first stage.*

**Example 9-1.** Figure 9-9 shows several identical amplifiers operating in cascade on a transmission line. Each amplifier has an available power gain $G$, and the amplifiers are spaced so the loss in the section of cable between amplifiers is also $G$. This type of arrangement can be used in a telephone trunk circuit or a CATV distribution system. The amplifiers have an available power gain equal to $G$ and a noise factor $F$. The cable sections have an insertion gain $1/G$ and a noise factor $G$.[†] Equation 9-45 then becomes

$$F_t = F + \frac{G - 1}{G} + \frac{F - 1}{1} + \frac{G - 1}{G} + \frac{F - 1}{1} + \cdots + \frac{F - 1}{1}, \qquad (9\text{-}46)$$

---

* $G = A^2 R_s / R_o$, where $A$ is the open-circuit voltage gain (open-circuit output voltage divided by source voltage). $R_s$ is the source resistance, and $R_o$ is the network output impedance.
[†] This can be derived by applying the basic noise factor definition (Eq. 9-1) to the cable section. The cable is considered a matched transmission line operating at its characteristic impedance.

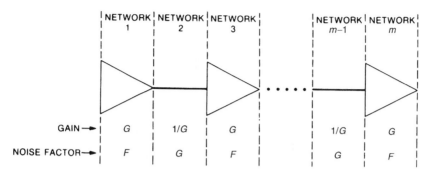

**FIGURE 9-9.** Identical amplifiers equally spaced on a transmission line.

$$F_t = F + 1 - \frac{1}{G} + F - 1 + 1 - \frac{1}{G} + F - 1 + \cdots + F - 1. \qquad (9\text{-}47)$$

For $K$ amplifiers and $K$–1 cable sections,

$$F_t = KF - \frac{K-1}{G}. \qquad (9\text{-}48)$$

If $FG \gg 1$,

$$F_t = KF. \qquad (9\text{-}49)$$

The overall noise figure equals

$$(NF)_t = 10 \log F + 10 \log K. \qquad (9\text{-}50)$$

The overall noise figure therefore equals the noise figure of the first amplifier plus ten times the logarithm of the number of stages. Another way of looking at this is that every time the number of stages is doubled, the noise figure increases by 3 dB. This limits the maximum number of amplifiers that can be cascaded.

**Example 9-2.** Figure 9-10 shows an antenna connected to a TV set by a section of 300-$\Omega$ matched transmission line. If the transmission line has 6 dB of insertion loss and the TV set has a noise figure of 14 dB, what signal voltage is required at the antenna terminal for a 40-dB signal-to-noise ratio at the terminals of the TV set? To solve this problem, all the noise sources in the system are converted to equivalent noise voltages at one point, in this case the input to the TV set. The noise voltages can then be combined, and the appropriate signal level needed to produce the required signal-to-noise ratio can be calculated.

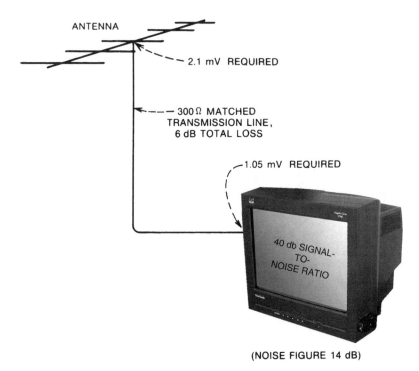

**FIGURE 9-10.** TV set connected to antenna.

The thermal noise at the input of the TV set from a 300-$\Omega$ input impedance with a 4-MHz bandwidth is –53.2 dBmV (2.2 $\mu$V).* Because the TV set adds 14 dB of noise to the input thermal noise, the total input noise level is –39.2 dBmV (thermal noise voltage in dB + noise figure). Because a signal-to-noise ratio of 40 dB is required, the signal voltage at the amplifier input must be +0.5 dBmV (total input noise in dB + signal-to-noise ratio in dB). The transmission line has 6 dB of loss, so the signal voltage at the antenna terminal must be +6.5 dBmV or 2.1 mV. In order to add terms directly, as in this example, all the quantities must be referenced to the same impedance level, in this case 300 $\Omega$.

## 9.9  NOISE TEMPERATURE

Another method of specifying noise performance of a circuit or device is by the concept of equivalent input noise temperature $(T_e)$.

---

*The open circuit noise voltage at room temperature (290 K) for a 300-$\Omega$ resistor and a 4-MHz bandwidth is 4.4 $\mu$V. When this source is connected to a 300-$\Omega$ load, it delivers one-half of this voltage, or 2.2 $\mu$V, to the load.

The equivalent input noise temperature of a circuit can be defined as the increase in source resistance temperature necessary to produce the observed noise power at the output of the circuit. The standard reference temperature $T_0$ for noise temperature measurements is 290K.

Figure 9-11 shows a noisy amplifier with a source resistance $R_s$ at temperature $T_0$. The total measured output noise is $V_{no}$. Figure 9-12 shows an ideal noiseless amplifier having the same gain as the amplifier in Fig. 9-11 and also a source resistance $R_s$. The temperature of the source resistance is now increased by $T_e$, so the total measured output noise $V_{no}$ is the same as in Fig. 9-11. $T_e$ is then the equivalent noise temperature of the amplifier.

The equivalent input noise temperature is related to the noise factor $F$ by

$$T_e = 290(F - 1) \qquad (9\text{-}51)$$

and to noise figure $NF$ by

$$T_e = 290(10^{NF/10} - 1). \qquad (9\text{-}52)$$

In terms of the equivalent input noise voltage and current $(V_n - I_n)$ the noise temperature can be written as

$$T_e = \frac{V_n^2 + (I_n R_s)^2}{4kBR_s}. \qquad (9\text{-}53)$$

The equivalent input noise temperature of a several amplifiers in cascade can be shown to be

$$T_{e(\text{total})} = T_{e1} + \frac{T_{e2}}{G_1} + \frac{T_{e3}}{G_1 G_2} + \cdots, \qquad (9\text{-}54)$$

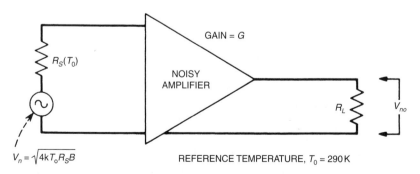

FIGURE 9-11. Amplifier with noise.

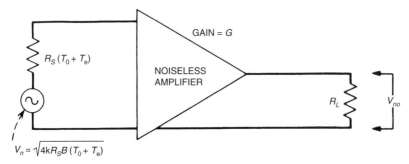

**FIGURE 9-12.** Source resistance temperature increased to account for amplifier noise.

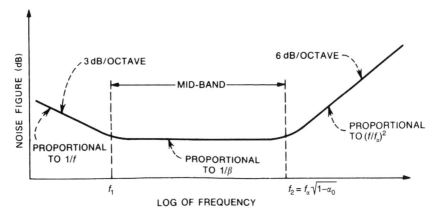

**FIGURE 9-13.** Noise figure versus frequency for bipolar transistor.

where $T_{e1}$ and $G_1$ are the equivalent input noise temperature and available power gain of the first stage, $T_{e2}$ and $G_2$ are the same for the second stage, and so on.

## 9.10  BIPOLAR TRANSISTOR NOISE

The noise figure versus frequency for a typical bipolar transistor is shown in Fig. 9-13. It can be observed that the noise figure is constant across some middle range of frequencies and rises on both sides. The low-frequency increase in noise figure is because of "$1/f$" or contact noise (see Section 8.5). The $1/f$ noise and the frequency $f_1$ increase with increasing collector current.

Above frequency $f_1$, the noise is caused by white noise sources that consist of thermal noise in the base resistance and shot noise in the emitter and collector junctions. The white noise sources can be minimized by choosing a transistor with small base resistance, large current gain, and high alpha cutoff frequency.

The increase in noise figure at frequencies above $f_2$ is caused by (1) the decrease in transistor gain at these frequencies and (2) the transistor noise produced in the output (collector) junction, which therefore is not affected by transistor gain.

For a typical audio transistor, the frequency $f_1$, below which the noise begins to increase, may be between 1 and 50 kHz. The frequency $f_2$, above which the noise increases, is usually greater than 10 MHz. In transistors designed for radio frequency (rf) use, $f_2$ may be much higher.

### 9.10.1   Transistor Noise Factor

The theoretical expression for bipolar transistor noise factor can be derived by starting with the T-equivalent circuit of a transistor, as shown in Fig. 9-14, neglecting the leakage term $I_{CBO}$. By neglecting $r_c (r_c \gg R_L)$ and adding the following noise sources—(1) thermal noise of the base resistance, (2) shot noise in emitter diode, (3) shot noise in collector, and (4) thermal noise in the source resistance—the circuit can be revised to form the equivalent circuit shown in Fig. 9-15.

The noise factor can be obtained from the circuit in Fig. 9-15 and the relationships

$$I_c = \alpha_o I_e, \tag{9-55}$$

$$r_e = \frac{kT}{qI_e} \approx \frac{26}{I_e(\text{ma})}, \tag{9-56}$$

$$|\alpha| = \frac{|\alpha_o|}{\sqrt{1 + (f/f_\alpha)^2}}, \tag{9-57}$$

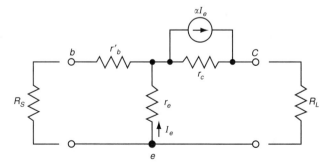

**FIGURE 9-14.**  T-equivalent circuit for a bipolar transistor.

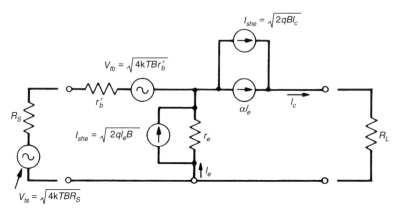

**FIGURE 9-15.** Noise equivalent circuit of a bipolar transistor.

where $\alpha_o$ is the direct current (dc) value of the transistor common-base current gain $\alpha$, $k$ is Boltzmann's constant, $q$ is the electron charge, $f_\alpha$ is the transistor alpha cutoff frequency, and $f$ is the frequency variable. Using this equivalent circuit, Nielsen (1957) has shown the noise factor of the transistor to be

$$F = 1 + \frac{r_b'}{R_s} + \frac{r_e}{2R_s} + \frac{(r_e + r_b' + R_s)^2}{2r_e R_s \beta_o}\left[1 + \left(\frac{f}{f_\alpha}\right)^2 (1 + \beta_o)\right], \tag{9-58}$$

where $\beta_o$ is the dc value of the common-emitter current gain $\beta$,

$$\beta_o = \frac{\alpha_o}{1 - \alpha_o}. \tag{9-59}$$

This equation does not include the effect of the $1/f$ noise and is valid at all frequencies above $f_1$ in Fig. 9-13. The $1/f$ noise can be represented as an additional noise current source in parallel with $\alpha I_e$ in the collector circuit.

The second term in Eq. 9-58 represents the thermal noise in the base, the third term represents shot noise in the emitter, and the fourth term represents shot noise in the collector. This equation is applicable to both the common-emitter and the common-base configurations.

The value of source resistance $R_{so}$ for the minimum noise factor can be determined by differentiating Eq. 9-58 with respect to $R_s$ and setting the result equal to zero. This source resistance is found to be

$$R_{so} = \left[(r_b' + r_e)^2 + \frac{(2r_b' + r_e)\beta_o r_e}{1 + (f/f_\alpha)^2(1 + \beta_o)}\right]^{1/2}. \tag{9-60}$$

For most bipolar transistors, the value of source resistance for minimum noise factor is close to the value that produces maximum power gain. Most transistor applications operate the transistor at a frequency considerably below the alpha cutoff frequency. Under this condition ($f \ll f_\alpha$), assuming $\beta_o \gg 1$, Eq. 9-60 reduces to

$$R_{so} = \sqrt{(2r_b' + r_e)\beta_o r_e}. \tag{9-61}$$

If in addition the base resistance $r_b'$ is negligible (not always the case), Eq. 9-61 becomes

$$R_{so} \approx r_e\sqrt{\beta_o}. \tag{9-62}$$

This equation is also useful for making quick approximations of the source resistance that produces minimum noise factor. Equation 9-62 shows that the higher the common-emitter current gain $\beta_o$ of the transistor, the higher will be the value of $R_{so}$.

### 9.10.2   $V_n$–$I_n$ for Transistors

To determine the parameters for the equivalent input noise voltage and current model, we must first determine the total equivalent input noise voltage $V_{nt}$. Substituting Eq. 9-58 into Eq. 9-26, and squaring the result, gives

$$V_{nt}^2 = 2kTB(r_e + 2r_b' + 2R_s) + \frac{2kTB(r_e + r_b' + R_s)^2}{r_e\beta_o}$$

$$\times \left[1 + \left(\frac{f}{f_\alpha}\right)^2 (1 + \beta_o)\right]. \tag{9-63}$$

The equivalent input noise voltage squared $V_n^2$ is obtained by making $R_s = 0$ in Eq. 9-63 (see Eqs. 9-34 and 9-35), giving

$$V_n^2 = 2kTB(r_e + 2r_b') + \frac{2kTB(r_e + r_b')^2}{r_e\beta_o}\left[1 + \left(\frac{f}{f_\alpha}\right)^2 (1 + \beta_o)\right]. \tag{9-64}$$

To determine $I_n^2$, we must divide Eq. 9-63 by $R_s^2$ and then make $R_s$ large (see Eqs. 9-34 and 9-36), giving

$$I_n^2 = \frac{2kTB}{r_e\beta_o}\left[1 + \left(\frac{f}{f_\alpha}\right)^2 (1 + \beta_o)\right]. \tag{9-65}$$

## 9.11    FIELD-EFFECT TRANSISTOR NOISE

The three important noise mechanisms in a junction FET are as follows: (1) the shot noise produced in the reverse biased gate, (2) the thermal noise generated in the channel between source and drain, and (3) the $1/f$ noise generated in the space charge region between gate and channel.

Figure 9-16 is the noise equivalent circuit for a junction FET. The noise generator $I_{sh}$ represents the shot noise in the gate circuit, and generator $I_{tc}$ represents the thermal noise in the channel. $I_{ts}$ is the thermal noise of the source admittance $G_s$. The FET has an input admittance $g_{11}$, and a forward transconductance $g_{fs}$.

### 9.11.1    FET Noise Factor

Assuming no correlation between $I_{sh}$ and $I_{tc}$* in Fig. 9-16, the total output noise current can be written as

$$I_{\text{out}} = \left[ \frac{4kTBG_s g_{fs}^2}{(G_s + g_{11})^2} + \frac{I_{sh}^2 g_{fs}^2}{(G_s + g_{11})^2} + I_{tc}^2 \right]^{1/2}. \tag{9-66}$$

The output noise current from the thermal noise of the source only is

$$I_{\text{out}}(\text{source}) = \left( \frac{\sqrt{4kTBG_s}}{G_s + g_{11}} \right) g_{fs}. \tag{9-67}$$

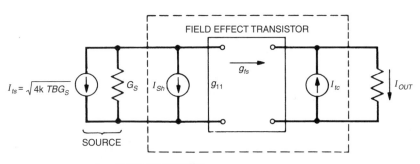

$g_{11}$ = INPUT ADMITTANCE
$g_{fs}$ = FORWARD TRANSCONDUCTACE (A/V)
$G_s$ = SOURCE ADMITTANCE

**FIGURE 9-16.**  Noise equivalent circuit of junction field effect transistor.

---

*At high frequencies, noise generators $I_{sh}$ and $I_{tc}$ show some correlation. As a practical matter, however, this is normally neglected.

The noise factor $F$ is Eq. 9-66 squared, divided by Eq. 9-67 squared, or

$$F = 1 + \frac{I_{sh}^2}{4kTBG_s} + \frac{I_{tc}^2}{4kTBG_s(g_{fs})^2}(G_s + g_{11})^2. \tag{9-68}$$

$I_{sh}$ is the input shot noise and equals

$$I_{sh} = \sqrt{2qI_{gss}B}, \tag{9-69}$$

where $I_{gss}$ is the total gate leakage current. $I_{tc}$ is the thermal noise of the channel and equals

$$I_{tc} = \sqrt{4kTBg_{fs}}. \tag{9-70}$$

Substituting Eqs. 9-69 and 9-70 into 9-68, and recognizing that

$$\frac{2q}{4kT}I_{gss} = g_{11}, \tag{9-71}$$

gives for the noise factor

$$F = 1 + \frac{g_{11}}{G_s} + \frac{1}{G_s g_{fs}}(G_s + g_{11})^2. \tag{9-72}$$

Rewriting Eq. 9-72 in terms of the resistances instead of admittances gives

$$F = 1 + \frac{R_s}{r_{11}} + \frac{R_s}{g_{fs}}\left(\frac{1}{R_s} + \frac{1}{r_{11}}\right)^2. \tag{9-73}$$

Neither Eq. 9-72 nor Eq. 9-73 include the effect of the $1/f$ noise. The second term in the equations represent the contribution from the shot noise in the gate junction. The third term represents the contribution of the thermal noise in the channel.

For low noise operation, an FET should have high gain (large $g_{fs}$) and a high input resistance $r_{11}$ (small gate leakage).

Normally, at low frequencies, the source resistance $R_s$ is less than the gate leakage resistance $r_{11}$. Under these conditions Eq. 9-73 becomes

$$F \approx 1 + \frac{1}{g_{fs}R_s}. \tag{9-74}$$

In the case of an insulated gate FET (IGFET) or metal oxide FET (MOSFET) there is no p–n gate junction and therefore no shot noise, so Eq. 9-74 applies. However, in the cases of IGFETs or MOSFETs the $1/f$ noise is often greater than in the case of junction field effect transistors (JFETs).

### 9.11.2    $V_n$–$I_n$ Representation of FET Noise

The total equivalent input noise voltage can be obtained by substituting Eq. 9-73 into Eq. 9-26, giving

$$V_{nt}^2 = 4kTBR_s \left[ 1 + \frac{R_s}{r_{11}} + \frac{R_s}{g_{fs}} \left( \frac{1}{R_s} + \frac{1}{r_{11}} \right)^2 \right]. \tag{9-75}$$

Making $R_s = 0$ in Eq. 9-75 gives the equivalent input noise voltage squared (see Eqs. 9-34 and 9-35) as

$$V_n^2 = \frac{4kTB}{g_{fs}}. \tag{9-76}$$

To determine $I_n^2$, we must divide Eq. 9-75 by $R_s^2$ and then make $R_s$ large (see Eqs. 9-34 and 9-36), giving

$$I_n^2 = \frac{4kTB(1 + g_{fs}r_{11})}{g_{fs}r_{11}^2}. \tag{9-77}$$

For the case when $g_{fs}r_{11} \gg 1$, Eq. 9-77 becomes

$$I_n^2 = \frac{4kTB}{r_{11}}. \tag{9-78}$$

## 9.12   NOISE IN OPERATIONAL AMPLIFIERS

The input stage of an operational amplifier is of primary concern in determining the noise performance of the device. Most monolithic op-amps use a differential input configuration that uses two and sometimes four input transistors. Figure 9-17 shows a simplified schematic of a typical two-transistor input circuit used in an operational amplifier. Because two input transistors are used, the noise voltage is approximately $\sqrt{2}$ times that for a single-transistor input stage. In addition, some monolithic transistors have lower current gains ($\beta$) than discrete transistors, and that also increases the noise.

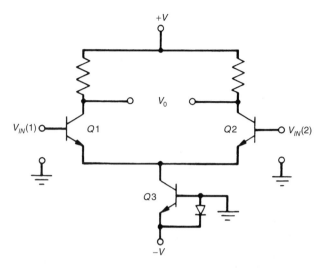

**FIGURE 9-17.** Typical input circuit schematic of an integrated circuit (IC) operational amplifier. Transistor $Q_3$ acts as a constant current source to provide dc bias for the input transistors $Q_1$ and $Q_2$.

Therefore, in general, operational amplifiers are inherently higher noise devices than discrete transistor amplifiers. This can be observed in the typical equivalent input noise voltage curves shown in Fig. 9-6. A discrete bipolar transistor stage preceding an op-amp can often provide lower noise perfor- mance along with the other advantages of the operational amplifier. Op-amps do have the advantage of a balanced input with low temperature drift and low- input offset currents.

The noise characteristics of an operational amplifier can best be modeled by using the equivalent input noise voltage and current $V_n$–$I_n$. Figure 9-18A shows a typical operational amplifier circuit. Figure 9-18B shows this same cicuit with the equivalent noise voltage and current sources included.

The equivalent circuit in Fig. 9-18B can be used to calculate the total equivalent input noise voltage, which is

$$V_{nt} = \left[4kTB(R_{s1} + R_{s2}) + V_{n1}^2 + V_{n2}^2 + (I_{n1}R_{s1})^2 + (I_{n2}R_{s2})^2\right]^{1/2}. \qquad (9\text{-}79)$$

It should be noted that $V_{n1}$, $V_{n2}$, $I_{n1}$, and $I_{n2}$ are also functions of the bandwidth $B$.

The two noise voltage sources of Eq. 9-79 can be combined by defining

$$(V_n')^2 = V_{n1}^2 + V_{n2}^2. \qquad (9\text{-}80)$$

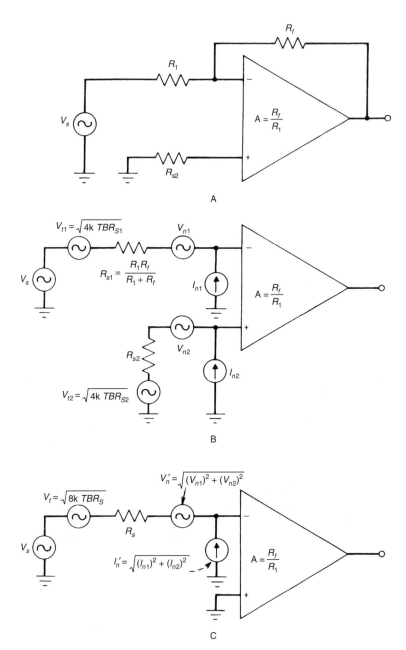

**FIGURE 9-18.** (*A*) Typical op-amp circuit; (*B*) Noise equivalent circuit of typical op-amp; (*C*) Circuit of B with noise sources combined at one input terminal for the case $R_{s1} = R_{s2} = R_{s3}$.

Equation 9-79 can then be rewritten as

$$V_{nt} = \left[4kTB(R_{s1} + R_{s2}) + (V'_n)^2 + (I_{n1}R_{s1})^2 + (I_{n2}R_{s2})^2\right]^{1/2}. \qquad (9\text{-}81)$$

Although the voltage sources have been combined, the two noise current sources are still required in Eq. 9-81. If, however, $R_{s1} = R_{s2}$, which is usually the case because this minimizes the dc output offset voltage resulting from input bias current, then the two noise current generators can also be combined by defining

$$(I'_n)^2 = I_{n1}^2 + I_{n2}^2. \qquad (9\text{-}82)$$

For $R_{s1} = R_{s2} = R_s$, Eq. 9-81 reduces to

$$V_{nt} = \left[8kTBR_s + (V'_n)^2 + (I'_nR_s)^2\right]^{1/2}. \qquad (9\text{-}83)$$

The equivalent circuit for this case is shown in Fig. 9-18C. To obtain optimum noise performance (maximum signal-to-noise ratio) from an op-amp, the total equivalent input noise voltage $V_{nt}$ should be minimized.

## 9.12.1    Methods of Specifying Op-Amp Noise

Various methods are used by op-amp manufacturers to specify noise for their devices. Sometimes they provide values for $V_n$ and $I_n$ at each input terminal, as represented by the equivalent circuit in Fig. 9-19A. Because of the symmetry of the input circuit, the noise voltage and noise current at each input are equal. A second method is to provide combined values, $V'_n$ and $I'_n$, which are then applied to one input only, as shown in Fig. 9-19B. To combine the two noise current generators, it must be assumed that equal source resistors are connected to the two input terminals. The magnitudes of the combined noise voltage generators in Fig. 9-19B, with respect to the individual generators in Fig. 9-19A, are

$$V'_n = \sqrt{2}V_n, \qquad (9\text{-}84)$$

$$I'_n = \sqrt{2}I_n. \qquad (9\text{-}85)$$

In still other cases, the noise voltage given by the manufacturer is the combined value $V'_n$, whereas the noise current is the value that applies to each input separately $I_n$. The equivalent circuit representing this arrangement is shown in Fig. 9-19C. The user, therefore, must be sure he or she understands which equivalent circuit is applicable to the data given by the manufacturer of the device before using the information. To date, there is no standard as to which of these three methods should be used for specifying op-amp noise.

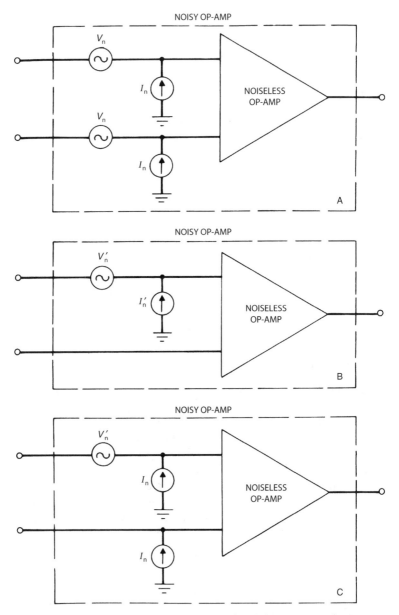

**FIGURE 9-19.** Methods of modeling op-amp noise: (*A*) separate noise generators at each input; (*B*) noise generators combined at one input; (*C*) separate noise current generators with combined noise voltage generator.

### 9.12.2  Op-Amp Noise Factor

Normally, the noise factor is not used in connection with op-amps. However, the noise factor can be determined by substituting Eq. 9-83 into Eq. 9-26, and solving for $F$. This gives

$$F = 2 + \frac{(V_n')^2 + (I_n' R_s)^2}{4kTBR_s}. \qquad (9\text{-}86)$$

Equation 9-86 assumes the source noise is from the thermal noise in just one of the source resistors $R_s$, not both. This is a valid assumption when the op-amp is used as a single-ended amplifier. The thermal noise in the resistor $R_s$ on the unused input is considered part of the amplifier noise and is a penalty paid for using this configuration.

In the case of the inverting op-amp configuration, the noise from $R_s$ at the unused input may be bypassed with a capacitor. This is not possible, however, in the noninverting configuration, because the feedback is connected to this point.

A second method of defining the noise factor for an op-amp is to assume the source noise is from the thermal noise of both source resistors ($2R_s$ in this case). The noise factor then can be written as

$$F = 1 + \frac{(V_n')^2 + (I_n' R_s)^2}{8kTBR_s}. \qquad (9\text{-}87)$$

Equation 9-87 is applicable if the op-amp is used as a differential amplifier with both inputs driven.

## SUMMARY

- If the source resistance is a variable and the source voltage a constant in the design of a circuit, minimizing noise factor does not necessarily produce optimum noise performance.
- For a given source resistance, the least noisy circuit is the one with the lowest noise factor.
- For the best noise performance the output signal-to-noise ratio should be maximized, this is equivalent to minimizing the total input noise voltage $(V_{nt})$.
- The concept of noise factor is meaningless when the source is a pure reactance.
- For best noise performance a low-source resistance should be used (assuming the source voltage remains constant).
- Noise peformance may be improved by transformer coupling the source resistance to a value equal to $R_s = V_n/I_n$.

- If the gain of the first stage of a system is high, the total system noise is determined by the noise of the first stage.
- Active device noise can be specified in a number of different ways as follows:
  - As 2 noise factors $F$.
  - As an equivalent input noise voltage $V_n$ and input noise current $I_n$.
  - As an equivalent input noise temperture $T_e$.

## PROBLEMS

9.1 Derive Eq. 9-3 from Eq. 9-1.

9.2 Which device produces the least equivalent input device noise $(V_{nd}/\sqrt{B})$?
  a. A bipolar transistor with a noise figure of 10 dB measured at $R_s = 10^4\,\Omega$.
  b. An FET with a noise figure of 6 dB measured at $R_s = 10^5\,\Omega$.

9.3 A transistor has a noise figure of 3 dB measured with a source resistance of 1.0 M$\Omega$. What is the output power–signal-to-noise ratio if this transistor is used in a circuit with an input signal of 0.1 mV and a source resistance of 1.0 M$\Omega$? Assume the system has an equivalent noise bandwidth of 10 kHz.

9.4 The noise of an FET is specified as follows. Equivalent input noise voltage is $0.06 \times 10^{-6}$ V/$\sqrt{\text{Hz}}$, and the equivalent input noise current is $0.2 \times 10^{-12}$ A/$\sqrt{\text{Hz}}$.
  a. If the FET is used in a circuit with a source resistance of 100 k$\Omega$ and an equivalent noise bandwidth of 10 kHz, what is the noise figure?
  b. What value of $R_s$ will produce the lowest noise figure, and what is the noise figure with this value of $R_s$?

9.5 A low-noise preamplifier is to be driven from a 10-$\Omega$ source. Data supplied by the manufacturer specify $V_n$ and $I_n$ at the operating frequency as

$$\frac{V_n}{\sqrt{B}} = 10^{-8} \quad \text{V}/\sqrt{\text{Hz}},$$

$$\frac{I_n}{\sqrt{B}} = 10^{-13} \quad \text{A}/\sqrt{\text{Hz}}.$$

  a. Determine the input-transformer turns ratio to provide optimum noise performance.
  b. Calculate the noise figure for the circuit using the transformer of part "a."
  c. What would be the noise figure with the preamplifier directly coupled to the 10-$\Omega$ source?
  d. What would be the SNI factor for this circuit?

9.6 Figure P9-6 shows an FM antenna connected to an FM receiver by a section of 75-$\Omega$ matched coaxial cable. The required signal-to-noise ratio at the input terminals to the set for good quality reception is 18 dB, and the noise figure of the receiver is 8 dB.

a. If the cable that connects the receiver to the antenna has 6 dB of insertion loss, what signal voltage is required at the point where the antenna connects to the cable, to provide good quality reception? The noise bandwidth of the receiver is 50 kHz.

b. Why is the voltage considerably less than that required in the case of a TV set as worked out in Example 9-2 in section 9.8 of the text?

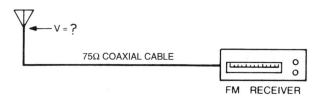

$75\Omega$ COAXIAL CABLE

FM  RECEIVER

**FIGURE P9-6.**

9.7 Find the noise figure for a system with an equivalent input noise temperature $(T_e)$ equal to 290K.

9.8 A transistor is operated at a frequency $f \ll f_\alpha$. The transistor parameters are $r'_b = 50\ \Omega$ and $\beta_0 = 100$. Calculate the minimum noise factor and the value of source resistance for which it occurs when the collector current is

a. 10 $\mu$A

b. 1.0 mA

*Note:* $r_e \approx 26/I_C(\text{mA})$

9.9 A junction FET has the following parameters measured at 100 MHz: $g_{fs} = 1500 \times 10^{-6}$ mhos and $g_{11} = 800 \times 10^{-6}$ mhos. If the transistor is to be used in a circuit with a source resistance of 1000 $\Omega$, what is the noise figure?

9.10 Derive Eq. 9-37. Start with the equivalent circuit of Fig. 9-4 and Eq. 9-1.

9.11 Derive Eq. 9-40. Start with the equivalent circuit of Fig. 9-4.

## REFERENCES

Friis, H. T. "Noise Figures of Radio Receivers." *Proceedings of the IRE*, Vol. 32, July 1944.

Mumford, W. W. and Scheibe, E. H. *Noise Performance Factors in Communication Systems*. Horizon House, Dedham, MA, 1968.

Nielsen, E. G. "Behavior of Noise Figure in Junction Transistors." *Proceedings of the IRE*, vol. 45, July 1957, pp. 957–963.

Rothe, H. and Dahlke, W. "Theory of Noisy Fourpoles." *Proceedings of IRE*, vol. 44, June 1956."

## FURTHER READING

Baxandall, P. J. "Noise in Transistor Circuits." *Wireless World*, Vol. 74, November-December 1968.

Cooke, H. F. "Transistor Noise Figure." *Solid State Design*, February 1963, pp. 37–42.

Gfeller, J. "FET Noise." *EEE*, June 1965, pp. 60–63.

Graeme, J. "Don't Minimize Noise Figure." *Electronic Design*, January 21, 1971.

Haus, H. A., et al. "Representation of Noise in Linear Twoports." *Proceedings of IRE*, vol. 48, January 1960.

Letzter, S. and Webster, N. "Noise in Amplifiers." *IEEE Spectrum*, vol. 7, no. 8, August 1970, pp. 67–75.

Motchenbacher, C. D. and Connelly, J. A., *Low Noise Electronic System Design*. New York, Wiley, 1993.

Robe, T. "Taming Noise in IC Op-Amps." *Electronic Design*, vol. 22, July 19, 1974.

Robinson, F. N. H. "Noise in Transistors." *Wireless World*, July 1970.

Schiek, B., Rolfes, I., and Siwevis, H. *Noise in High-Frequency Circuits and Oscillators*, New York, Wiley, 2006.

Trinogga, L. A. and Oxford, D. F. "J.F.E.T. Noise Figure Measurement." *Electronic Engineering*, April 1974.

van der Ziel, A. "Noise in Solid State Devices and Lasers." *Proceedings of the IEEE*, vol. 58, August 1970.

van der Ziel, A. *Noise in Measurements*. New York, Wiley, 1976.

Watson, F. B. "Find the Quietest JFETs." *Electronic Design*, November 8, 1974.

# 10 Digital Circuit Grounding

A digital system is also a radio-frequency (rf) system with significant noise and interference potential. Although most digital designers are knowledgeable about the subject of digital design, they are not always well equipped to handle the design and analysis of rf systems, which is exactly what they are designing. It often turns out that digital engineers are excellent antenna designers, but they do not know it!

In addition, many analog-circuit designers are now designing digital circuits, and they may not realize that different techniques are required for grounding, power distribution, and interconnection. For example, although a single-point ground may be desirable in some low-frequency analog circuits, it may be the primary source of noise coupling and emission in a digital circuit.

Small digital logic gates, which draw only a few milliamperes of direct current (dc), do not at first seem to be a serious source of noise. However, their high switching speed, combined with the inductance of the conductors that interconnect them, makes them a major source of noise. The magnitude of the voltage generated when current changes through an inductor is

$$V = L\frac{di}{dt},$$ (10-1)

where $L$ is the inductance and $di/dt$ is the rate of change of current. For example, consider the case where the power-supply wiring has an inductance of 50 nH. If the transient current, when a logic gate switches, is 50 mA and the gate switches in 1 ns, the noise voltage generated across the power-supply wiring when this one gate changes state will, by Eq. 10-1, be 2.5 V. Multiplying this result by the many gates in a typical system, and realizing that the typical supply voltage for such a system may be only 3.3 V, shows that this can be a major noise source. Noise voltages occur on the ground, power, and signal conductors of the system.

Chapters 10, 11, 12, and 14 cover the theory and design techniques for (1) grounding, (2) power distribution, (3) radiation control, and (4) susceptibility of digital circuits, respectively.

*Electromagnetic Compatibility Engineering,* by Henry W. Ott
Copyright © 2009 John Wiley & Sons, Inc.

## 10.1   FREQUENCY VERSUS TIME DOMAIN

Digital-circuit designers think in terms of the time domain. Considering electromagnetic capability (EMC), however, it is better to think in terms of the frequency domain. Legal requirements on the emission from systems are specified in the frequency domain, as are the characteristics of interference-control components, such as capacitors, ferrites, filters, and shields. The frequency domain and the time domain are related by the Fourier transform (Paul, 2006), which will be discussed in Section 12.1.3.

The harmonic content of a square wave extends out to infinity. However, there is a point beyond which the energy content in the harmonic is low and can be ignored. This point is considered to be the bandwidth of the logic and occurs at the break point where the Fourier coefficients start to decay at 40 dB/decade, instead of 20 dB/decade (as discussed in Section 12.1.3). Therefore, the bandwidth of a digital signal can be related to the rise time $t_r$ by the following equation:

$$BW = \frac{1}{\pi t_r}. \tag{10-2}$$

For example, a rise time of 1 ns is equivalent to a bandwidth of 318 MHz. With new integrated circuit (IC) technology, sub-nanosecond rise times are becoming common. For example, low-voltage differential signaling (LVDS) has a 300-ps rise time, which is equivalent to a 1-GHz bandwidth.

## 10.2   ANALOG VERSUS DIGITAL CIRCUITS

Analog circuits often contain amplifiers, and a small amount of external noise coupled into the circuit can cause interference. This occurs in circuits operating at very low signal levels (millivolts or microvolts) and/or those containing high-gain amplifiers.

In contrast, digital circuits do not contain amplifiers, and they operate at relatively large signal levels compared with many analog circuits. For complementary metal–oxide semiconductor (CMOS) circuits, the noise margin is about 0.3 times the $V_{cc}$ voltage, or 1.5 V for a 5-V supply. Therefore, digital circuits have an inherent immunity to low-level noise pickup. However, as a consequence of the trend toward lower and lower supply voltages for digital logic (e.g., 3.3 and 1.75 V), this inherent noise immunity is steadily decreasing.

## 10.3   DIGITAL LOGIC NOISE

In analog circuits, external noise sources are usually the primary concern. Whereas in digital circuits, the internal noise sources are usually the major concerns. Internal noise in digital circuits is the result of the following: (1) ground bus noise (often referred to as "ground bounce"), (2) power bus noise, (3) transmission line reflections, and (4) crosstalk. The most important of these, ground and power bus noise, are covered in this chapter and Chapter 11, respectively. Crosstalk was covered in Chapter 2, and reflections are covered in most good books on digital logic design, such as Blakeslee (1979), Barna (1980), or Johnson and Graham (2003), and these topics will not be covered here.

## 10.4 INTERNAL NOISE SOURCES

Figure 10-1 shows a simplified digital system that consists of four logic gates. Consider what happens when the output of gate 1 switches from high-to-low. Before gate 1 switches, its output is high, and the stray capacitance of the wiring between gates 1 and 2 is charged to the supply voltage. When gate 1 switches, the stray capacitance must be discharged before the low can be transmitted to gate 3. Therefore, a large transient current flows through the ground system to discharge this stray capacitance. As a result of ground inductance, this current produces a noise voltage pulse at the ground terminals of gates 1 and 2. If the output of gate 2 is low, then this noise pulse will be coupled to the input of gate 4, as shown is Fig. 10-1, which may cause gate 4 to switch producing a signal integrity problem.

Figure 10-1 also shows another problem associated with this ground noise, that is, cable radiation. Cables that leave or enter the system are also referenced to the circuit ground as shown in the figure. Therefore, some percentage of the ground noise voltage will excite the cable as an antenna, causing it to radiate

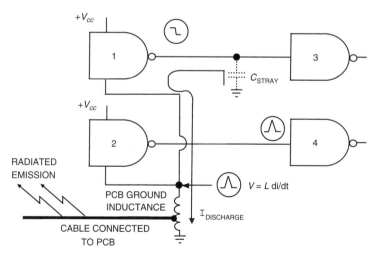

**FIGURE 10-1.** Ground noise is created when output of gate 1 switches from high to low.

and producing an EMC problem. The noise voltage necessary to cause an EMC problem is about three orders of magnitude less than the noise voltage required to create a signal integrity problem. The most practical way to decrease the magnitude of the ground noise voltage is to decrease the inductance of the ground system.

The discharge path from the stray capacitance through the output of gate 1 and the ground conductors contains little resistance. It forms a high-Q series-resonant circuit that is likely to oscillate, causing the output voltage of gate 1 to go negative as shown in Fig. 10-2.

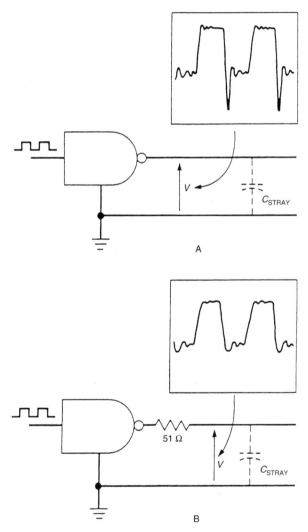

**FIGURE 10-2.** TTL output voltage waveshape with (A) ringing caused by stray capacitance and ground inductance, and (B) ringing damped by the addition of an output resistor.

The Q (or gain at resonance) of a series-resonant circuit is equal to

$$Q = \frac{1}{R}\sqrt{\frac{L}{C}}. \tag{10-3}$$

Additional damping with a resistor (see Fig. 10-2B), or ferrite bead, in the output of the gate will decrease this ringing.

In the case of a transistor–transistor logic (TTL) gate, most of the ringing will occur on the negative transition as shown in Fig. 10-2 because TTL has an internal resistor in series with the positive supply. For a CMOS logic gate, the ringing usually occurs equally on both the negative and the positive transition, because it does not contain an internal resistor, see Fig. 10-3.

A second source of noise in digital logic can be understood by referring to Fig. 10-3 that shows a typical schematic of a CMOS inverter logic gate with a *totem-pole* output circuit (a pull-up transistor on top of a pull-down transistor). When the output is high, the *p*-channel transistor (top) is on and the *n*-channel transistor (bottom) is off. Conversely, when the output is low, the *n*-channel transistor is on and the *p*-channel transistor is off. However, during switching there is a short period of time during which both transistors are on simultaneously. This conduction overlap results in a low-impedance connection between the power supply and ground, which produces a large power-supply transient current spike

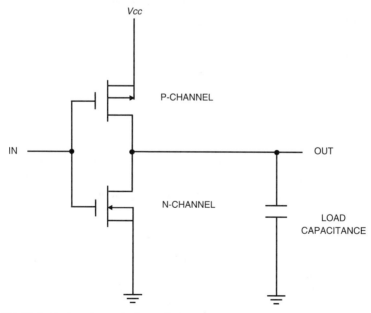

**FIGURE 10-3.** Basic schematic for a CMOS logic gate that has a totem pole output circuit.

of 50 to 100 mA per logic gate. Large integrated circuits, such as microprocessors, can have transient power supply currents in excess of 10 A. This current is referred to by different names such as overlap current, contention current, or shoot-through current. A similar effect occurs with TTL, as well as most other logic families, however the peak transient currents are lower, because TTL has a current-limiting resistor in series with the totem pole output circuit. Note that, although CMOS has smaller steady state, or average current, than TTL, it has larger transient currents. Emitter-coupled logic (ECL) however, does not use the totem pole output topology and therefore does not draw as large a transient current when it switches.

Therefore, whenever a digital logic gate switches, a large transient current is drawn from the power supply. This current charges the load capacitance and provides the short-circuit current for the totem pole output circuit. This current flows through the inductance of the power and ground conductors causing a large transient voltage drop in the supply voltage. This large transient power supply current is a major source of noise in CMOS logic circuits and contributes to the radiated emission from such circuits. The solution is (see Chapter 11) to provide a source of charge, a decoupling capacitor or capacitors, near each integrated circuit (IC) to supply the transient current without having to draw the current through all the inductance of the power and ground conductors.

To minimize the noise generated by these two internal noise sources, all digital logic systems must be designed with the following:

1. A low impedance (inductance) ground system
2. A source of charge (decoupling capacitors) near each logic IC

This chapter will cover the subject of achieving a low inductance ground, and Chapter 11 will cover the decoupling, or power distribution, issue.

## 10.5  DIGITAL CIRCUIT GROUND NOISE

Transient ground currents are the primary source of both intrasystem noise voltages, and conducted and radiated emissions. To minimize the noise from transient ground currents, the impedance of the ground must be minimized. A typical printed circuit board (PCB) trace (a 1-oz copper conductor 0.006 in. wide, and 0.02 in. from a return conductor) has a resistance of 82 m$\Omega$/in (Eq. 5-10) and a loop inductance of 15 nH/in (Eq. 10-5). The impedance of the 15-nH inductance versus frequency, which is related to the logic rise/fall time by Eq. 10-2, is given in Table 10-1.

As can be observed, at all the frequencies above 1 MHz the impedance of the 15-nH inductance is greater than the 82-m$\Omega$ resistance. At frequencies above 10 MHz, the inductance is many orders of magnitude greater than the resistance. For a digital signal with a 1-ns rise time (318-MHz bandwidth, from Eq. 10-2),

**TABLE 10-1.   Impedance of a 1-in Long Printed Circuit Board Trace (15 nH of inductance).**

| Frequency (MHz) | Rise Time (ns) | Impedance ($\Omega$) |
|---|---|---|
| 1 | 318 | 0.1 |
| 10 | 32 | 1.0 |
| 30 | 11 | 2.8 |
| 50 | 6.4 | 4.7 |
| 100 | 3.2 | 9.4 |
| 300 | 1.1 | 28 |
| 500 | 0.64 | 47 |
| 1000 | 0.32 | 94 |

the ground conductor will have an inductive reactance of about 30 $\Omega$ /in. It is, therefore, the inductance that is of most concern when laying out a digital printed wiring board. If the ground-circuit impedance is to be minimized, the inductance must be reduced by an order of magnitude or more.

## 10.5.1   Minimizing Inductance

To control inductance, it is helpful to understand how it depends on the physical properties of the circuit.

Inductance is directly proportional to the length of a conductor. This fact can be used to advantage by minimizing the lengths of high-frequency leads that carry large transient currents, such as clock leads and line or bus drivers. This is not a universal solution, however, because in a large system some leads must be long. This points out one of the advantages of large-scale integration (LSI); by putting a large amount of circuitry on a single IC chip the length, and therefore the inductance, of the interconnecting leads are significantly reduced.

Inductance is inversely proportional to the log of the conductor diameter, or the width of a flat conductor. For a single round conductor of diameter $d$ located a distance $h$ above a current-return plane, the loop inductance is equal to

$$L = 0.005 \ln\left(\frac{4h}{d}\right) \mu\text{H}/\text{in.} \tag{10-4}$$

for the case of $h > 1.5\, d$.

For a flat conductor, such as on a printed circuit board, the loop inductance is

$$L = 0.005 \ln\left(\frac{2\pi h}{w}\right) \mu\text{H}/\text{in.} \tag{10-5}$$

where $w$ is the width of the conductor. This equation is only applicable for a long narrow trace, where $h \geq w$.

If Eq. 10-4 is set equal to Eq. 10-5, the width $w$ needed for a flat conductor to have the same inductance as a round conductor of diameter $d$ can be determined. The result is

$$2w = \pi d \tag{10-6a}$$

or

$$w = 1.57d. \tag{10-6b}$$

Assuming that the trace width is much greater than the trace thickness, Eq. 10-6b indicates that a flat conductor will have the same inductance as a round conductor if it has the same surface area.

Because of the log relationship in Eqs. 10-4 and 10-5, it is difficult to achieve a large decrease in inductance by increasing the conductor diameter or width. In a typical case, doubling the diameter or width (an increase of 100%) will only decrease the inductance by 20%. The size would have to be increased by 1200% for a 50% decrease in inductance. If a large decrease in inductance is needed, some other method must be found to achieve it.

Another method for decreasing the inductance of a circuit is to provide alternative paths for current flow. These paths must be electrically, but not necessarily physically, parallel. If two equal inductances are paralleled, then the equivalent inductance will be one half that of one inductor, neglecting mutual inductance. If four paths are parallel, then the inductance will be one quarter. Because inductance is inversely proportional to the number of parallel paths, this method is effective in decreasing inductance, provided the mutual inductance can be controlled (minimized).

## 10.5.2   Mutual Inductance

When two conductors are paralleled, the effect of the mutual inductance must be considered in calculating the total inductance. The net partial-inductance ($L_t$) of two parallel conductors that carry current in the *same* direction can be written as

$$L_t = \frac{L_1 L_2 - M^2}{L_1 + L_2 - 2M}, \tag{10-7}$$

where $L_1$ and $L_2$ are the partial self-inductances of the two conductors and $M$ is the partial mutual-inductance between them. If the two conductors are identical ($L_1$ equal to $L_2$), Eq. 10-7 reduces to

$$L_t = \frac{L_1 + M}{2}. \tag{10-8}$$

Equation 10-8 shows that the mutual inductance limits the overall reduction in inductance from parallel inductors. If the inductors are close together (tightly coupled), the mutual inductance approaches the self-inductance ($L_1 \approx M$), and the overall inductance is approximately equal to the inductance of a single conductor.

If the conductors are spaced far apart (loosely coupled), then the mutual inductance becomes small, and the total inductance approaches one-half the original inductance of a single conductor. The effect of conductor spacing on mutual inductance must therefore be determined.

Using the theory of partial inductance (see Appendix E), the effect of spacing on the net partial inductance of parallel conductors that carry current in the same direction can be determined. Substituting Eq. E-14 for the partial self-inductance $L_1$, and Eq. E-17 for the partial mutual-inductance $M$ in Eq. 10-8 gives for the normalized net partial-inductance of two conductors in parallel

$$\frac{L_t}{L_1} = \frac{\ln\left(\frac{2l}{r}\right) + \ln\left(\frac{2l}{D}\right) - 2}{2\left[\ln\left(\frac{2l}{r}\right) - 1\right]}, \tag{10-9}$$

where $D$ is the spacing of the two conductors, $l$ is the length of the conductors ($l \gg D$), and $r$ is the radius of each conductor.

A plot of Eq. 10-9 is shown in Fig. 10-4, for two 3-in. long, 24-gauge conductors. As can be observed, most of the inductance decrease occurs within the first 0.5 in. of separation. Spacing greater than 0.5 in. does not produce significant additional decreases in inductance.

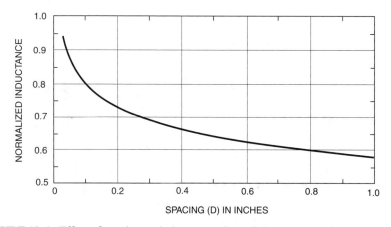

**FIGURE 10-4.** Effect of spacing on inductance of parallel conductors that carry current in the same direction. Inductance is normalized to that of a single conductor.

### 10.5.3  Practical Digital Circuit Ground Systems

A practical high-speed digital circuit ground system must provide a low-impedance (low-inductance) connection between all possible combinations of ICs that communicate with each other. The most practical way to accomplish this is to provide as many alternative (parallel) ground paths as possible. This result can most easily be achieved with a grid.

The impedance of an inductance (inductive reactance) is directly proportional to frequency. Therefore, to maintain the same ground impedance, the inductance of the ground must decrease proportional to frequency. This means that as digital logic frequencies increase the ground grid must be made finer and finer to provide more parallel paths. If this concept is taken to its limit, then the result is an infinite number of parallel paths, or a plane. Although a ground plane will provide optimum performance, it is important to remember that a grid is the basic topology that is desired.

A grid can be implemented on a PCB by printing horizontal and vertical ground traces on the board as shown in Fig. 10-5. On a double-sided PCB, the horizontal traces are routed on one side of the board, and the vertical traces on the other side. The ground traces on the two sides are then connected together with plated through holes (vias) where they cross. This arrangement leaves ample room for all the necessary signal interconnections.

A satisfactory grid can be implemented even on a crowded board with a little extra effort when the board is originally laid out. If this approach is used, it is important to put the ground grid on the board first, before routing the signal paths. Although not impossible, it may be difficult to place the grid on the board once the signal conductors are routed. A ground grid adds no per-unit cost to the product and is therefore a cost-effective noise reduction technique.

Although the primary ground-distribution conductors should be made with wide traces, which are necessary to handle the dc current (with minimal voltage drop), the ground grid can be closed with narrow traces because each added

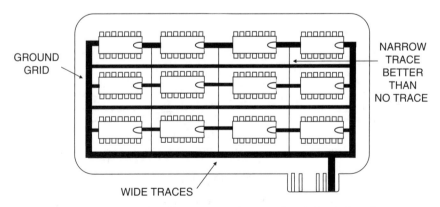

**FIGURE 10-5.** A grid-type ground on a printed circuit board.

trace provides many more additional parallel paths to the ground structure. This fact is important to understand, for designers may be reluctant to use narrow conductors as part of the ground system.

It is important to realize that the width of a trace is a dc or low-frequency consideration, which is used to decrease the resistance. Whereas, the gridding of the traces is a high-frequency consideration, which is used to reduce the inductance. The two effects are independent of each other.

A grid, even a fairly coarse grid, can reduce ground-noise by an order of magnitude or more over that of a single-point ground. For example, the data in Table 10-2 from German (1985) shows the ground-noise voltage measured between ground pins of various combinations of ICs on identical double-sided printed circuit boards, with identical component placement with and without a ground grid. In this case, the maximum ground-voltage differential decreased from 1000 to 250 mV, and the voltage between the ground pins on IC15 and IC16 decreased from 1000 to 100 mV when the ground grid was implemented, which is an order of magnitude improvement. The maximum radiated emission also decreased from 42.9 dB $\mu$V/m with the single-point ground structure to 35.8 dB $\mu$V/m with the ground grid, a 7.1-dB improvement.

Using the theory of partial inductance, see Appendix E, Smith and Paul (1991) studied the effect of grid spacing on inductance. They concluded, that a grid spacing of 0.5 in. or less should be used in order to obtain the most significant reduction of ground noise.

Ground grids have been used successfully on double-sided boards at frequencies up to a few tens of megahertz. Above about 5 or 10 MHz, however, a ground plane should be seriously considered.

**TABLE 10-2.  Peak Differential Ground Noise Voltage (mV).**

| Location | Single Point Ground | Ground Grid |
|---|---|---|
| IC1-IC2 | 150 | 100 |
| IC1-IC3 | 425 | 150 |
| IC1-IC4 | 425 | 150 |
| IC1-IC5 | 450 | 150 |
| IC1-IC6 | 450 | 150 |
| IC1-IC7 | 450 | 150 |
| IC1-IC8 | 425 | 225 |
| IC1-IC9 | 400 | 175 |
| IC1-IC10 | 400 | 150 |
| IC1-IC11 | 625 | 200 |
| IC1-IC12 | 400 | 150 |
| IC1-IC13 | 425 | 250 |
| IC14-IC11 | 900 | 200 |
| IC15-IC7 | 850 | 125 |
| IC15-IC10 | 900 | 125 |
| IC15-IC16 | 1000 | 100 |

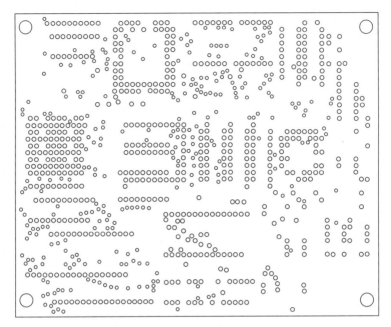

**FIGURE 10-6.** Typical printed circuit board ground plane.

Figure 10-6 shows a typical printed circuit board ground plane. Notice that the plane is not solid but is full of holes (more like Swiss cheese). The holes are required wherever a via, or through-hole component's lead, goes through the board, to not short it out to ground. Actually, the ground structure in Fig. 10-6 looks more like a fine grid than a ground plane, which is fine, because a grid was what was desired in the first place.

The ground system is the foundation of a digital logic printed wiring board. If the ground system is poor, then it is difficult to remedy the problem, short of starting over and implementing the ground properly. Therefore, all digital printed wiring boards should be designed with either a ground plane or a ground grid.

### 10.5.4 Loop Area

Another important method of reducing inductance is to minimize the area of the loop enclosed by the current flow. Two conductors with current in opposite directions (e.g., a signal and its ground return trace) have a total loop inductance $L_t$, equal to

$$L_t = L_1 + L_2 - 2M, \tag{10-10}$$

where $L_1$ and $L_2$ are the partial self-inductances of the individual conductors and $M$ is the partial mutual-inductance between them. If the two conductors are identical, then Eq. 10-10 reduces to

$$L_t = 2(L_1 - M).$$  (10-11)

To minimize the total loop inductance, the partial mutual-inductance between the conductors should be maximized. Therefore, the two conductors should be placed as close together as possible to minimize the area between them.

If the coefficient of magnetic coupling $k$ between the two conductors were unity, then the mutual inductance would be equal to the self-inductance, because

$$M = k\sqrt{L_1 L_2},$$  (10-12)

and the total inductance of the closed loop would be zero. At high frequencies, a coaxial cable approaches this ideal condition.

Placing signal and return current paths close together is therefore an effective way of reducing inductance. This can be done with a tightly twisted pair or a coaxial cable. With this configuration, inductances of less than 1 nH/in are possible.

What is the loop area for a system that contains multiple-ground return paths? The area of interest is the total area enclosed by the *actual* current flow (Ott, 1979). An important consideration, therefore, is the ground path taken by the current in returning to the source. Often, this is not the path intended by the designer.

By comparing Eqs. 10-8 and 10-11, a very important conclusion can be drawn. *To minimize the total inductance, two conductors that carry current in the same direction (such as two ground conductors) should be separated. However, two conductors that carry current in the opposite direction (such as power and ground, or signal and ground conductors) should be placed as close together as possible.*

Appendix E contains a more detailed discussion of inductance and it discusses the important distinction between loop and partial inductances.

## 10.6   GROUND PLANE CURRENT DISTRIBUTION AND IMPEDANCE

Because the ground topology is so important to the performance of a high-frequency printed circuit board, and because ground planes are used on most high-frequency printed circuit boards, it is important to understand the characteristics of ground planes. For example, what is the actual distribution of the return current underneath a signal trace, and what is the impedance of the ground plane?

Any voltage drop across a PCB ground plane will excite cables terminating on the board, which causes them to radiate as dipole or monopole antennas as was shown in Fig. 10-1. The amount of current needed in a 1-m long antenna to cause the radiation to exceed the Federal Communications Commission (FCC) emission requirements is very small, in the vicinity of a few microamps (see Table 12-1). Therefore, even small ground noise voltages are important, because it only takes a few millivolts of potential to produce currents of this magnitude.

Although it is true that a ground plane has significantly less inductance than a trace, ground plane inductance is not negligible. The mechanism by which a ground plane reduces the inductance is by allowing the current to spread out, thus providing many parallel paths. To calculate the impedance of a ground plane, the current distribution in the plane must first be determined.

Many papers and articles (for example, Leferink and van Doorn, 1993) have been written analyzing the inductance of ground planes, but the authors usually assume that the current flows uniformly across the full cross section of the plane, which is an unlikely occurrence.

### 10.6.1    Reference Plane Current Distribution

***10.6.1.1 Microstrip Line.*** A microstrip line consists of a trace above a reference plane. A representation of the fields surrounding a microstrip line is shown in Fig. 10-7. Because of skin effect (see Section 6.4) the high-frequency fields cannot penetrate the plane.[*] The reference plane currents (return currents) will exist where the electric field lines terminate on the adjacent plane.[†] It can be observed from Fig. 10-7 that the reference plane current under a microstrip trace spreads out beyond the width of the trace, thus providing many parallel paths for the return current flow. But what exactly is the distribution of the current in the reference plane?

Holloway and Kuester (1995) derived an expression for the reference plane current density of a microstrip trace. The configuration evaluated was that of a trace of width $w$ at a height $h$ above a plane as shown in Fig. 10-8. The reference plane current density $J(x)$ at a distance $x$ from the center of the trace was shown to be

$$J(x) = \frac{I}{w\pi} \left[ \tan^{-1} \left( \frac{2x - w}{2h} \right) - \tan^{-1} \left( \frac{2x + w}{2h} \right) \right]. \qquad (10\text{-}13)$$

---

[*] For a 1-oz copper plane, this is true above 30 MHz where the copper plane is more than three skin depths thick. For 1/2-oz copper, this is true above 120 MHz, and for 2-oz copper above 8 MHz.
[†] This is true regardless of what the adjacent plane is, a ground or a power plane. Therefore, in this section the adjacent plane will be referred to as simply a reference plane, not a ground or a power plane.

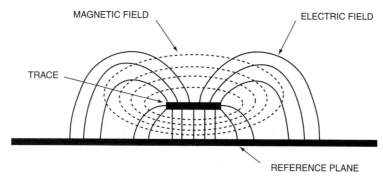

**FIGURE 10-7.** Electric and magnetic fields surrounding a microstrip line.

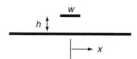

**FIGURE 10-8.** Configuration of a microstrip line.

$J(x)$ is the current density, and $I$ is the total current in the loop. The current density of Eq. 10-13 is the distribution necessary to produce minimum inductance. The current density will be the same regardless of frequency; the only restriction is that the frequency is high enough that the resistance of the plane is negligible compared with the inductive reactance. Typically, this will occur at frequencies above a few hundred kilohertz.

Figure 10-9 is a normalized plot $[J(x)/J(0)]$ of Eq. 10-13 as a function of $x/h$. The vertical axis of the plot is normalized to the current density in the plane directly under the center of the trace ($x = 0$), and the horizontal axis is normalized to the height of the trace ($x/h$). As can be observed, most of the current remains close to the trace. Notice that at a distance from the center of the trace of five times the trace height, the current density is small and the slope of the curve is very shallow. Because the slope is so shallow, there still will be some current far out from the center of the trace.

Integrating Eq. 10-13 between $+x$ and $-x$, gives the percentage of the microstrip return current contained in the portion of the plane between $\pm x/h$ of the centerline of the trace. The result is shown in Fig. 10-10 where the horizontal axis is normalized to the height of the trace ($x/h$). As can be observed, 50% of the current is contained within a distance of $\pm$ the height of the trace, 80% of the current is contained within a distance of $\pm$ 3 times the trace height, and 97% of the current will be present within a distance of $\pm$ 20 times the trace height.

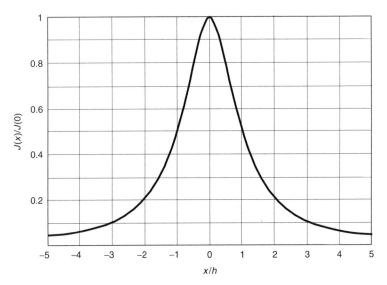

**FIGURE 10-9.** Normalized reference plane current density for a microstrip line.

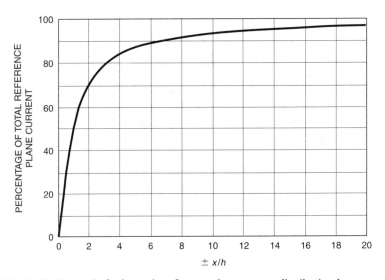

**FIGURE 10-10.** Integral of microstrip reference plane current distribution between $\pm x/h$.

If the width of the trace is less than the height ($w \leq h$), then Eq. 10-13 can be approximated by

$$\frac{J(x)}{J(0)} = \frac{1}{1 + \left(\frac{x}{h}\right)^2}. \tag{10-14}$$

Furthermore, if we are only concerned with the reference plane current at distances $x \gg h$ from the centerline of the trace, then Eq. 10-14 reduces to

$$\frac{J(x)}{J(0)} = \frac{h^2}{x^2}.$$

(10-15)

Crosstalk between adjacent traces is the result of the interaction of the fields produced by the traces, and the fields terminate where the current exists. Equation 10-15, therefore, indicates that the crosstalk between adjacent micro-strip traces is proportional to the square of the trace height divided by the square of the separation distance. Equation 10-15 demonstrates an important point, that *placing the trace closer to the reference plane, even if the spacing between the traces remains the same, will reduce the crosstalk.* This provides an effective way to reduce crosstalk without using up valuable real estate on the printed circuit board, as would be the case if the traces were moved farther apart.

***10.6.1.2 Stripline.***  A stripline consists of a trace symmetrically located between two planes as shown in Fig. 10-11. In the appendix of a 2000 paper, Chris Holloway derived the reference plane current density for a stripline configuration. The reference plane current density at a distance $x$ from the center of the trace was shown to be

$$J(x) = \frac{I}{w\pi}\left\{\tan^{-1}\left[e^{\left(\frac{\pi(x-w/2)}{2h}\right)}\right] - \tan^{-1}\left[e^{\left(\frac{\pi(x+w/2)}{2h}\right)}\right]\right\}.$$

(10-16)

Equation 10-16 represents the current density in just one of the two planes. The total reference plane current density will therefore be twice that of Eq. 10-16. Figure 10-12 compares the stripline current density to the microstrip current density. It plots the normalized current density $J(x)/J(0)$ from Eq. 10-13 and twice the normalized current density from Eq. 10-16 (which represents the total stripline current in both planes) versus $x/h$. As can be observed, for the case of a stripline, the current does not spread out nearly as far as in the case of a micro-strip line. Notice also that at a distance from the center of the trace of four times the trace height, the stripline current density is almost zero.

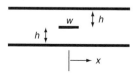

**FIGURE 10-11.** Configuration of a stripline.

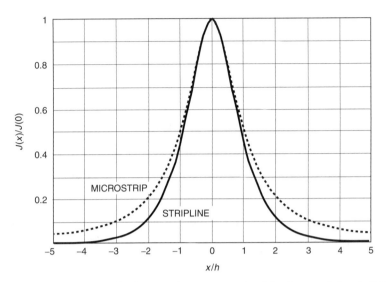

**FIGURE 10-12.** Normalized reference plane current densities for a stripline (solid) and microstrip (dotted).

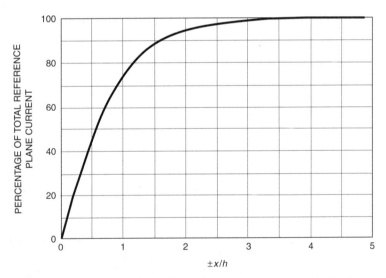

**FIGURE 10-13.** Integral of total stripline reference plane current distribution between $\pm x/h$. Half of this current is in each of the two planes.

Figure 10-13 shows twice the integral of Eq. 10-16 from $+x$ to $-x$, as a function of $x/h$. The curve represents the total current in both planes; half of the current is in each plane. The curve shows the percentage of the stripline return current contained in the portion of the plane between $\pm x/h$ of the centerline of the trace. Seventy four percent of the stripline current is contained within a distance of $\pm$ the height of the trace, and 99% of the current is contained within a distance of $\pm 3$ times the height. Therefore, the stripline reference plane current does not spread out nearly as much as the microstrip line's reference plane current.

### 10.6.1.3 Asymmetric Stripline.
An asymmetric stripline consists of a trace asymmetrically located between two planes as shown in Fig. 10-14.

Asymmetric stripline is used commonly on digital logic circuit boards, where two orthogonally routed signal layers are located between the two planes. Because the two signal layers are orthogonal, minimum interaction occurs between them. The reason for using this configuration is that only two planes are required for two stripline circuits, instead of three planes for two stripline circuits. For any one stripline circuit, however, the planes are asymmetrically spaced. With this configuration, $h_2$ would be equal to $2\,h_1$.

Holloway and Kuester (2007) have also derived an expression for the reference plane current density of an asymmetric stripline. For the asymmetric stripline, they show the reference plane current density in the plane closest to the trace, at a distance $x$ from the center of the trace to be

$$J_{close}(x) = \frac{I}{w\pi}\left[\tan^{-1}\left(\frac{e^{\frac{\pi(x-w/2)}{h_1+h_2}} - \cos\left(\frac{\pi h_1}{h_1+h_2}\right)}{\sin\left(\frac{\pi h_1}{h_1+h_2}\right)}\right) - \tan^{-1}\left(\frac{e^{\frac{\pi(x+w/2)}{h_1+h_2}} - \cos\left(\frac{\pi h_1}{h_1+h_2}\right)}{\sin\left(\frac{\pi h_1}{h_1+h_2}\right)}\right)\right], \quad (10\text{-}17a)$$

where $h_1$ is the distance between the trace and the closest plane and $h_2$ is the distance between the trace and the farthest plane (see Fig. 10-14).

The reference plane current density in the plane farthest from the trace is

$$J_{far}(x) = \frac{I}{w\pi}\left[\tan^{-1}\left(\frac{e^{\frac{\pi(x-w/2)}{h_1+2}} - \cos\left(\frac{\pi h_2}{h_1+h_2}\right)}{\sin\left(\frac{\pi h_2}{h_1+h_2}\right)}\right) - \tan^{-1}\left(\frac{e^{\frac{\pi(x+w/2)}{h_1+h_2}} - \cos\left(\frac{\pi h_2}{h_1+h_2}\right)}{\sin\left(\frac{\pi h_2}{h_1+h_2}\right)}\right)\right], \quad (10\text{-}17b)$$

Figure 10-15 is a normalized plot of Eqs. 10-17a and 10-17b as a function of $x/h_1$, for the case $h_2 = 2\,h_1$. The plot is normalized to the sum of the current

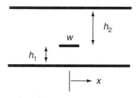

**FIGURE 10-14.** Configuration of an asymmetric stripline.

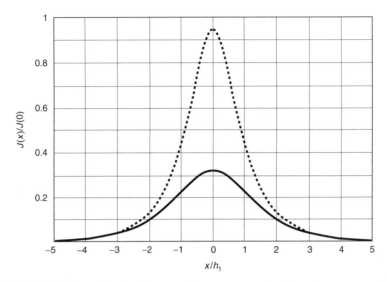

**FIGURE 10-15.** Normalized reference plane current densities for an asymmetric stripline, for the case where $h_2 = 2\,h_1$. The dotted line represents the current on the close plane; the solid line represents the current on the far plane.

densities in both planes directly underneath the center of the trace ($x = 0$). Directly under the trace, approximately 75% of the current is on the close plane and 25% of the current is on the far plane. However, at distances greater than approximately three times the trace height (for the case of a $h_2/h_1$ ratio of 2), the current in both the close and far planes become the same.

A plot of the current density for all three cases (asymmetric stripline with $h_2 = 2\,h_1$, stripline, and microstrip line) versus $x/h_1$ is shown in Fig. 10-16. For the stripline and the asymmetric stripline cases, the current plotted is the sum of the currents on the two planes. As can be observed from the figure, for $x/h_1$ less than 2, the asymmetric stripline current density is very similar to the microstrip current density. However, for $x/h_1$ greater than about four, the asymmetric stripline current density is closer to that of stripline than to that of that of microstrip.

Integrating Eqs. 10-17a and 10-17b from x equals − infinity to + infinity gives the total current on each plane. Doing the integration produces

$$I_{close} = \left(1 - \frac{h_1}{h_1 + h_2}\right)I \qquad (10\text{-}18a)$$

and

$$I_{far} = \left(\frac{h_1}{h_1 + h_2}\right)I, \qquad (10\text{-}18b)$$

where $h_1$ is the distance to the closest reference plane, and $h_2$ is the distance to the farthest reference plane,

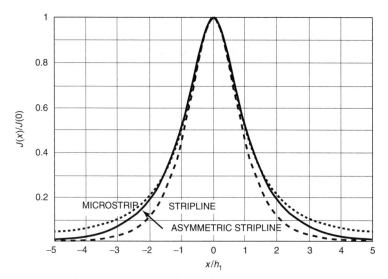

**FIGURE 10-16.** Normalized reference plane current densities for a microstrip (dotted), a stripline (dashed), and an asymmetric stripline (solid). The stripline and the asymmetric stripline plots are for the sum of the currents on the two planes.

**TABLE 10-3.   Percent of Total Current in Each Plane for Asymmetric Stripline.**

| $h_2/h_1$ | Close Plane | Far Plane |
|-----------|-------------|-----------|
| 1 | 50% | 50% |
| 2 | 67% | 33% |
| 3 | 75% | 25% |
| 4 | 80% | 20% |
| 5 | 83% | 17% |

Table 10-3 lists the percentage of the total current on each plane as a function of the $h_2/h_1$ ratio.

It is interesting to note that if $h_2 = h_1$ is substituted into Eqs. 10-17a and 10-17b then both equations reduce to Eq. 10-16, which is the equation for the current density for stripline. Similarly, if $h_2$ is allowed to approach infinity, then Eq. 10-17a reduces to Eq. 10-13, which is the equation for the current density of a microstrip line. Therefore, Eqs. 10-17a and 10-17b are universal equations that give the reference plane current distribution for any of the three common PCB transmission line topologies.

An additional discussion of reference plane current distribution is contained in Chapter 17 on Mixed-Signal PCB Layout, see Sections 17.2, 17.6.1, and 17.6.2.

### 10.6.2 Ground Plane Impedance

Although the calculation of the inductance of a wire or a PCB trace is fairly straightforward, the calculation of the inductance of a plane is complex. In 1994, I therefore made measurements to determine the net partial inductance of a microstrip ground plane.

*10.6.2.1 Measured Inductance.* The voltage drop across a ground plane is equal to the current times the impedance, and it can be expressed in the frequency domain as:

$$V_g = I_g [R_g + j\omega L_g].$$ (10-19)

Where $R_g$ is the resistance of the ground plane, and $L_g$ is the net partial inductance of the ground plane.

Inductance relates a voltage to a rate of change of current. Equation. 10-1 expressed this relationship in the time domain. The voltage drop across a section of ground plane can therefore be expressed in the time domain as

$$V_g = L_g \left(\frac{di}{dt}\right) + IR_g.$$ (10-20)

If we assume that at high frequencies, the inductance term predominates, then the resistive term can be neglected and we can write

$$L_g = \frac{V_g}{di/dt}.$$ (10-21)

The validity of this assumption will be discussed in more detail in section 10.6.2.3. From Eq. 10-21, we conclude that if the ground plane voltage can be measured and if $di/dt$ of the signal is known, then the ground plane inductance can be determined.*

The voltage drop across a section of the ground plane under a microstrip line being excited by a 10-MHz square wave was measured. The rate of change of the current was determined by measuring the signal across the load resistor. Knowing the ground voltage drop and $di/dt$, the ground inductance was calculated using Eq. 10-21.

The test PCB is shown in Fig. 10-17. It consisted of a double-sided board with a single trace on the top of the board and a ground plane on the bottom of the board, which thereby formed a microstrip line. Various thickness laminate boards were used, which allowed me to vary the spacing between the trace and

---

* This is difficult to do correctly. The method used here is described in Appendix E, Section E.4.

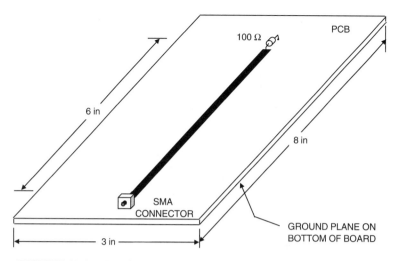

**FIGURE 10-17.** Test board for ground plane inductance measurements.

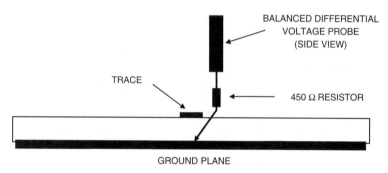

**FIGURE 10-18.** Ground voltage measurement test setup.

the ground plane. The line was 6-in. long and was terminated with a 100-$\Omega$ resistor. Test points for measuring the ground voltage drop were located every inch along the ground plane. The ground voltage was measured from the top of the ground plane (trace side) using a high-frequency 50-$\Omega$ differential probe as shown in Fig. 10-18. More details on the test setup are contained in Appendix E, Section E.4.

Figure 10-19 shows the results of the ground plane inductance measurements. *The inductance was measured over a 1-in. interval of the ground plane near the center of the trace length. Figure 10-19 clearly demonstrates the fact that the ground plane inductance decreases with trace height, as a result of the

---

*The 7-mil data point was left off of this plot. The problem associated with this data point is discussed in section 10.6.2.3

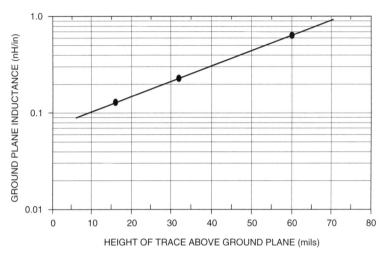

**FIGURE 10-19.** Measured ground plane inductance (nH/in) versus trace height in mils (1-mil = 0.001 in). Data point for 7-mil trace height omitted from plot because it is believed to be in error, see Section 10.6.2.3.

increasing mutual inductance. The inductance varies from about 0.1 to 0.7 nH/in. as the trace height varied from 0.010-in to 0.060-in. In comparison, the inductance of a typical PCB trace is 15 nH/in. *Decreasing trace height not only reduces ground plane inductance, but more importantly ground plane voltage,* that in turn will reduce the emissions from the board.

The inductance is, for all practical purposes, independent of the trace width because the inductance is related to the log of the width. For a trace height of 0.020 in., the measured ground plane inductance is approximately 0.15 nH/in. or *two orders of magnitude less than the inductance of a typical trace.* Although the ground plane inductances are low, they are not negligible. With 40 mA of digital logic current flowing, the voltage drop in the ground plane was measured as 15 mV/in.

***10.6.2.2 Calculated Ground Plane Inductance.*** Using their previously derived expression for the current distribution of a microstrip ground plane (Eq. 10-13), Holloway and Kuester in 1998 were able to calculate the inductance of a ground plane. However, because of the complexity of the integrals involved, they could not develop a closed form expression for the inductance. However, they, do present curves for the calculated inductance for various microstrip configurations. One curve presented is for the geometry that I measured to obtain the results presented in Fig. 10-19. Figure 10-20 compares Holloway and Kuester's 1998 calculated results to my 1994 measured results, for a 0.050-in. wide trace. The calculated results correlate well with the measured results for trace heights ≥10 mils (1 mil = 0.001 in.).

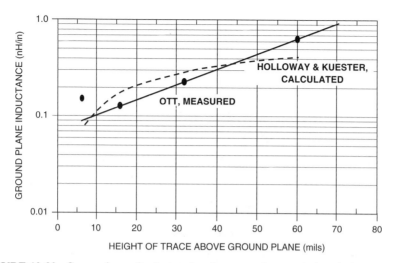

**FIGURE 10-20.** Comparison of calculated and measured ground plane inductance as a function of trace height.

### 10.6.2.3 Discrepancies For Small Trace Heights.

For trace heights < 10 mils, the measured and theoretical results seem to be diverging. The theoretical inductance is decreasing, whereas the measured inductance at a 7-mil trace height, actually increased slightly compare to the 16-mil data point. The first thought to come to mind is that the measured data are in error at this point. This could occur possibly because some other effect, not accounted for in the measurement, is coming into play at low trace heights.

From Fig. 10-20, we observe that for small values of trace height the ground plane inductance has been reduced to 1/150th of that of a trace. Also from Fig 10-9, we can conclude that the degree to which the ground plane current spreads out is a function of trace height, because the current spread is a function of $x/h$. For small heights, the ground plane current only spreads out a small amount, and therefore the ground plane resistance will increase, because the current flows through less copper. The above two effects (inductance decreasing and resistance increasing as trace height is reduced) causes us to refer back to Eq. 10-21 and ask the question, is it possible that the neglected ground plane resistance is now beginning to effect the measured data?

### 10.6.2.4 Ground Plane Resistance.

Holloway and Hufford (1997) provide the information necessary to calculate the alternating current (ac) resistance of a microstrip ground plane. The resistance is based on the ground plane current distribution of Eq. 10-13 and is:

$$R_g = \frac{2R_s}{w\pi}\left\{\tan^{-1}\left(\frac{w}{2h}\right) - \frac{h}{w}\left[\ln\left\langle 1 + \left(\frac{w}{2h}\right)^2\right\rangle\right]\right\}, \qquad (10\text{-}22)$$

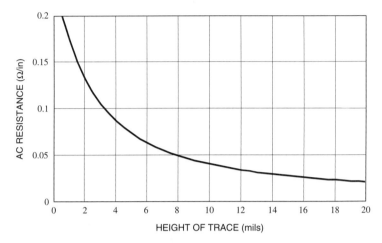

**FIGURE 10-21.** AC ground plane resistance ($\Omega$/in) for the case of a 0.010-in -wide trace at 100 MHz as a function of trace height in mils.

where $R_s$ is the Leontovich surface impedance and is equal to

$$R_s = \frac{1}{\sigma\delta},\tag{10-23}$$

where $\sigma$ is the conductivity (for copper $5.85 \times 10^7$ siemens) and $\delta$ is the skin depth of the material. Equation 6-11 defines the skin depth. Substituting Eq. 6-11 into Eq. 10-23 gives

$$R_S = \sqrt{\frac{\pi f \mu}{\sigma}}.\tag{10-24}$$

As can be observed, $R_s$ is a function of the square root of the frequency. The ground resistance $R_g$ is therefore a function of frequency, trace width, and trace height. The ground plane resistance will increase as the frequency is increased.

A plot of Equation 10-22 versus trace height for a 0.010-in-wide trace at 100 MHz is shown in Fig. 10-21. At trace heights $>$ 10 mils, the resistance is small. However, as the trace height becomes smaller, the resistance increases dramatically, which lends credibility to the idea that the ground plane resistance is affecting the inductance measurement for small trace heights.

***10.6.2.5 Comparison Between Ground Plane Inductance and Resistance.*** As mentioned above, we do not have a closed form equation for the ground plane inductance. Curve matching the data from Fig. 10.19 produced the following empirical equation for ground plane inductance in nH/in,

$$L_g = 0.073 \times 10^{15.62h},\tag{10-25}$$

where $h$ is the trace height in inches.

Although the above equation is for a 0.050-in.-wide trace, Holloway and Kuester (1998) have shown that the inductance is insensitive to trace width, which has been born out with my measurements. An order of magnitude increase in trace width decreased the inductance by less than 5%. Therefore, Eq. 10-25 will be used as a reasonable approximation to the ground plane inductance regardless of the trace width.

The inductive reactance of the ground plane is, therefore,

$$X_{Lg} = 2\pi f L_g = 4.59 \times 10^{-10} f 10^{15.62h}. \qquad (10\text{-}26)$$

Figure 10-22 is a plot of the ground plane inductive reactance $X_{Lg}$ (Eq. 10-26) and the ground plane resistance $R_g$ (Eq. 10-22) versus trace height for the case of a 0.010-in.-wide trace at 100 MHz.

Referring to Fig. 10-22, we can observe that as the trace height above the ground plane decreases, so does the inductive reactance, however, the ground plane resistance increases dramatically for small trace heights. Figure 10-22 also shows that for a 0.010-in-wide trace at 100 MHz, the magnitude of the ground plane resistance is equal to the ground plane inductive reactance at a height of about 6.5 mils. At trace heights less than this, the magnitude of the ground plane resistance is greater than the magnitude of the inductive reactance of the ground plane.

Figure 10-23 is a similar plot at a frequency of 200 MHz. Comparing Figs. 10-22 and 10-23 we observe that the trace height below which the ground plane

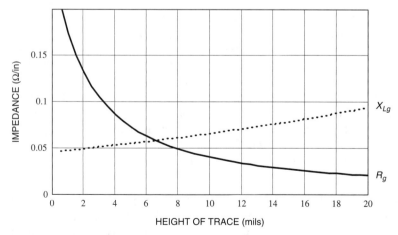

**FIGURE. 10-22.** Ground plane inductive reactance $X_{Lg}$ and resistance $R_g$ for a 0.010-in-wide trace, as a function of trace height in mils, at a frequency of 100 MHz.

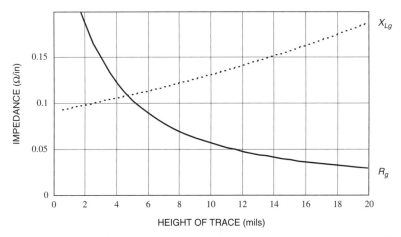

**FIGURE 10-23.** Ground plane inductive reactance $X_{Lg}$ and resistance $R_g$ for a 0.010-in-wide trace, as a function of trace height in mils at a frequency of 200 MHz.

resistance becomes larger than the inductive reactance decreases as the frequency increases. Figures 10-22 and 10-23 clearly demonstrate that *the ground plane resistance becomes the predominate effect for small trace heights*, typically less than 9 mils, which explains the higher then expected measured value of ground plane inductance for small trace heights. The significance of the above is that *there is a limit as to how low the ground plane impedance can be driven*, by placing the trace closer to ground.

The ground plane resistance (Eq. 10-22) is based on the current distribution of Eq. 10-13 and shown in Fig. 10-9. In calculating this distribution, Holloway and Kuester (1997) made the assumption that the ground plane inductance was the predominant impedance. In other words, the current distribution was calculated based on that required for minimum inductance, and then the ground plane resistance was calculated based on that current distribution. The current distribution, therefore, will be accurate only as long as the resistance is negligible compared with the inductive reactance. As demonstrated above, this will only be true for trace heights greater than the value where the magnitude of the resistance is equal to the magnitude of the inductive reactance. *Therefore, both the measured inductance and the calculated inductance plotted in Fig. 10-20 are both in error for the case of small trace heights ( < 10 mils).*

***10.6.2.6 Critical Height.*** Let us define the value of trace height where the ac resistance is equal to the inductive reactance as the *critical height* and designate it as $h_c$. The above discussion then leads to the question, what is the ground plane current distribution for trace heights less than $h_c$? We presently do not know the answer to this question, but can speculate on a likely possibility. It seems unlikely that the resistance continues to rise, as shown in Fig. 10-21, for small trace heights, because Eq. 10-22 is based on the current distribution of Fig. 10-9, which we just concluded was incorrect for trace heights less than the critical height $h_c$.

One, however, could make a heuristic argument as follows:

> For trace heights less than $h_c$ the resistance is the predominate effect. Let us now assume that we lower the trace a small incremental value. The ground plane current will want to spread out more than shown in Fig. 10-9 to decrease the resistance (which is now the dominant effect). However, if the current spreads out more, the inductance will increase and the resistance will no longer be the dominant effect, so the current will not do this. However, from an inductance point of view, the current will want to spread out less (as indicated by Fig. 10-9), but if it does the resistance will increase even more, so it will not do that. Continuing this line of reasoning, as the trace height is lowered another small increment brings us to the only logical conclusion, that is, *the current distribution will not change significantly as the trace is lowered below this critical value.*

Based on the above heuristic argument, we can conclude the following: For trace heights less than $h_c$ the current distribution will remain constant, and the ground impedance will therefore also be constant. In other words, *there is a limit to how much we can lower the ground plane impedance by varying the trace height.* For trace heights less than $h_c$, the ground impedance will no longer decrease but will remain constant.

Because the ground resistance and ground inductance are equal in magnitude at the critical height, and are 90° out of phase, the magnitude of the ground impedance at this height will be 1.41 $X_{Lg}$ (or 1.41 $R_g$). The value of the critical height is a function of both trace width and frequency. However, the variation with trace width is small and can be ignored.

Figure 10-24 shows the value of critical trace height ($h_c$) versus frequency for a 0.010-in-wide trace. If trace height is lowered below this value, then the ground impedance will no longer decrease but will remain constant.

Figure 10-25 is a plot of the magnitude of $R_g$ (or $X_{Lg}$), versus frequency at the critical trace height (where $R_g = X_{Lg}$) for a 0.010-in-wide trace. The magnitude of the ground plane impedance at this point will be equal to 1.41 $R_g$ (which is also equal to 1.41 $X_{Lg}$). Notice that this impedance increases linearly with frequency.

Summarizing the above, at large trace heights the ground plane inductance is the predominate impedance. As the trace is moved closer to the ground plane, the inductance decreases but the ground plane resistance increases. A point is finally reached (at a trace height of $h_c$) where the resistance becomes equal to the inductive reactance, and the impedance cannot be lowered any more by decreasing the trace height.

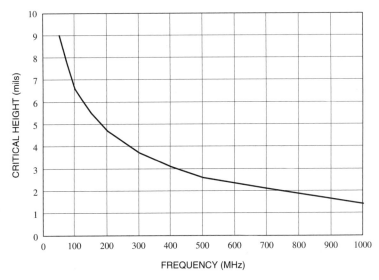

**FIGURE 10-24.** Critical height as a function of frequency, for a 0.010-in-wide trace.

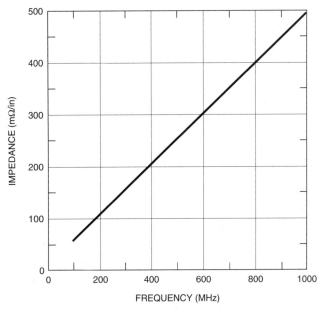

**FIGURE 10-25.** Ground plane impedance ($R_g$ or $X_{Lg}$) in milliohms/in for a 0.010-in-wide trace versus frequency when the trace height is equal to the critical height.

### 10.6.3 Ground Plane Voltage

From an EMC perspective, what is most important is not the ground plane impedance but the ground plane voltage drop that is produced by the

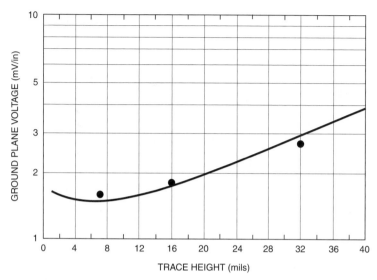

**FIGURE 10-26.** Calculated ground plane voltage (mV/in) at 50 MHz, for a 0.050-in-wide trace. Measured values are indicated by the dots.

impedance. This ground plane voltage then excites any cables connected to the PCB and causes common-mode radiation (see Fig. 12-2 and Section 12.3). We now have enough information to calculate the ground plane voltage. Substituting Eq. 10-22 and 10-25 into Eq. 10-19 gives the following expression for the ground plane voltage caused by the combined effect of ground plane resistance and ground plane inductance:

$$V_g = I_g \left\langle \frac{2\sqrt{\frac{\pi f \mu}{\sigma}}}{w\pi} \left\{ \tan^{-1}\left(\frac{w}{2h}\right) - \frac{h}{w}\left[\ln\left\langle 1 + \left(\frac{w}{2h}\right)^2\right\rangle\right] \right\} + j\omega 0.073 \cdot 10^{15.62h} \right\rangle. \quad (10\text{-}27)$$

Figure 10-26 is a plot of Eq. 10-27 for a 0.05-in -wide trace at 50 MHz with 40 mA of ground current. The ground plane voltage was also measured for the same configuration. The measured voltages, which are indicated by the dots in the Fig. 10-26, correlate well with the calculated result.

### 10.6.4   End Effects

The previous discussion of ground plane current distribution and ground plane impedance, as well as that in virtually all other published papers, analyzes the simple cross-sectional geometry shown in Fig. 10-8 and does not consider what happens at the beginning and at the end of the trace where the current is fed into or out of the ground plane through a via.

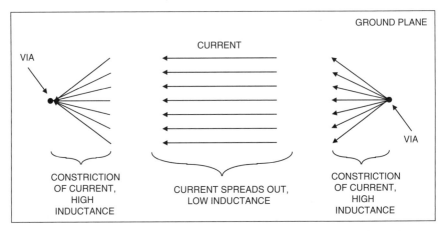

**FIGURE 10-27.** Representation of the ground plane current under a microstrip trace.

Figure 10-27 is a simple representation of the ground plane current distribution under a microstrip trace. Far away from the ends of the trace, the current has spread out to the current distribution defined by Eq. 10-13 and shown in Fig. 10-9. Because it is the spreading out of the current that decreases the inductance of the ground plane, what happens at each end where the current must constrict to flow into or out of the plane on a via? If the spreading out of the current decreases the inductance, then the constriction of the current at the ends must increase the inductance. Therefore, *the total ground plane inductance along the total length of the trace will be primarily determined by the large inductance at each end, which is caused by the current constricting to the size of the via.*

To confirm the hypothesis that the ground plane inductance increases in the vicinity of the via, measurements were made of the inductance over 1-in. intervals of the ground plane, at various distances from the via. Close to the via, inductance measurements were made at one-half-inch and at one-quarter-inch intervals. The results are plotted in Fig. 10-28. The distance listed on the horizontal axis is the distance to the center of the measured interval. As can be observed, far from the via, the inductance is small and constant. However as the distance from the via is decreased the inductance increases. In this case, the increase in the ground plane inductance resulting from the current constriction all occurs within one inch of the via. It is also interesting to note, that the curve approaches 15 nH/in, the inductance of a trace, at zero distance from the via.

The use of multiple vias can reduce this effect, because the current does not have to constrict as much. Figure 10-29 shows the measured inductance versus distance from the via for the case of a single via and for the case of three vias. The multiple-via case consisted of three vias 0.1 in apart along a line perpendicular to the trace. The ground plane inductance at zero distance from the via for the three via case is about one half that for the case of one via.

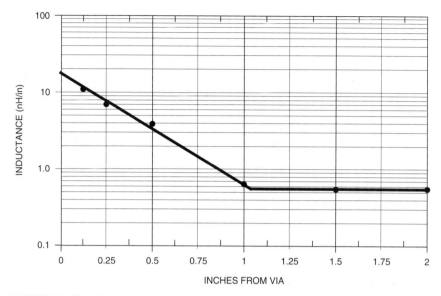

**FIGURE 10-28.** Ground plane inductance as a function of the distance from the via.

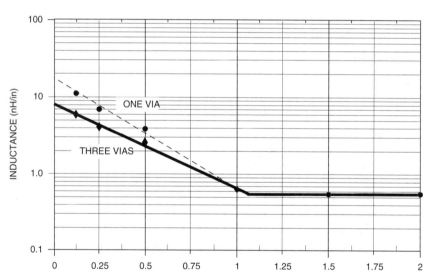

**FIGURE 10-29.** Ground plane inductance versus distance from the via for the cases of a single via and for the case of three vias. The dashed line is for one via and the solid line is for three vias.

In many cases, however, the use of multiple vias is not practical on a crowded, high-density PCB. However, in the case of decoupling capacitors, if room on the board allows, then multiple vias are very desirable because they will decrease the inductance in series with the decoupling capacitor and that can be advantageous, as discussed in Section 11.7.

## 10.7 DIGITAL LOGIC CURRENT FLOW

As discussed in the previous section, the lowest impedance (inductance) signal return path for high-frequency current is in a plane directly adjacent the signal trace. In the case of the four-layer printed circuit board stackup shown in Fig. 10-30, the return current path for the signal on the top layer will be the power plane. As discussed previously, the electric field lines that result from the signal on a microstrip trace terminate on the adjacent plane, as was shown in Fig. 10-7, regardless of what the purpose of the plane is. Because at high frequencies the fields cannot penetrate the plane, due to skin effect, the signal does not know that there is a ground plane below the power plane, and therefore the return current will be on the power plane. Does this create a problem? Is it not better to have the return current on the ground plane? To answer these questions, we must first analyze how the digital logic signal currents actually flow.

Many engineers and designers are confused about how and where digital return currents flow and what is the source of the digital logic current. First, let me state that *the driver IC is not the source of the current*; the IC only acts as a switch, to direct the current. The source of the current is the decoupling capacitor and/or the parasitic trace and load capacitance.

Only the transient (switching) current is important from a noise or EMC perspective, and the transient current flow does not depend on the load at the end of the line (see discussion associated with Fig. 5-19). Because the propagation time of the line is finite, the transient current does not know what the load impedance is until after the signal has already transversed the line.

The return current path is a function of the topology of the transmission line, stripline or microstrip, and on whether the logic transition is a high-to-low or a low-to-high transition. Also, in the case of a microstrip line, the return current path is a function of whether the trace is adjacent to a ground or power plane, and in the case of a stripline whether the trace is located between two ground planes, two power planes, or one ground and one power plane.

Figure 10-31 shows the circuit of a CMOS logic gate with its output signal trace located between a power and ground plane, which is a stripline configuration. Also shown in the figure are the load IC, the source's decoupling capacitor, the parasitic capacitance of the signal trace, and the load capacitance. Figures 10-32 to 10-37 show the logic current paths for the various possible configurations.

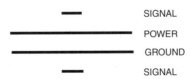

FIGURE 10-30. A common four-layer printed circuit board stackup.

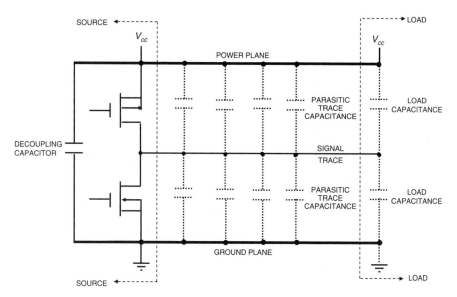

**FIGURE 10-31.** Circuit of a CMOS logic gate driving a stripline located between a power and ground plane.

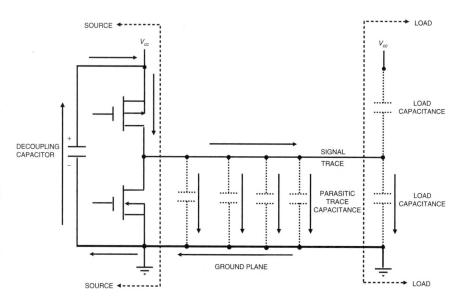

**FIGURE 10-32.** Current flow on a microstrip line adjacent to a ground plane, low-to-high transition.

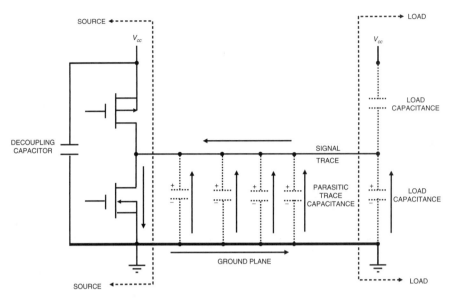

**FIGURE 10-33.** Current flow on a microstrip line adjacent to a ground plane, high-to-low transition.

### 10.7.1   Microstrip Line

Figure 10-32 shows the current path for a low-to-high transition on a micro-strip line adjacent to a ground plane. As can be observed, the current source is the decoupling capacitor. The current flows through the upper transistor of the CMOS logic gate, down the signal trace toward the load, through the trace to ground parasitic capacitance,* and returns to the decoupling capacitor on the ground plane.

Figure 10-33 shows the current path for a high-to-low transition on a microstrip line adjacent to a ground plane. As can be observed, the current source is the trace-to-ground parasitic capacitance. The current flows down the signal trace toward the driver IC, through the lower transistor of the CMOS driver, and returns on the ground plane. In this case, it is the lower transistor of the CMOS driver shorting out the trace-to-ground capacitance that produces the current flow. Note that in this case, the decoupling capacitor is not involved in the current path.

Figure 10-34 shows the current path for a low-to-high transition on a microstrip line adjacent to a power plane. As can be observed, the current source is the trace-to-power plane parasitic capacitance. The current flows down the power plane toward the source, through the upper transistor of the

---

* In this section, whenever I refer to the parasitic trace-to-ground capacitance, I mean the parasitic trace capacitance plus the load capacitance.

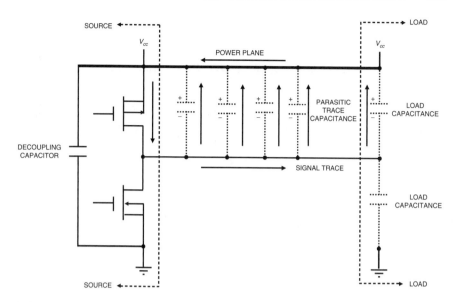

**FIGURE 10-34.** Current flow on a microstrip line adjacent to a power plane, low-to-high transition.

CMOS driver, and returns on the signal trace. In this case, it is the upper transistor of the CMOS driver shorting out the trace-to-power plane capacitance that produces the current flow. As in the previous case, the decoupling capacitor is not involved in the current path.

Figure 10-35 shows the current path for a high-to-low transition on a microstrip line adjacent to a power plane. As can be observed, the current source is the decoupling capacitor. The current flows through the power plane, through the trace-to-power plane capacitance, returns to the driver IC on the signal trace, and flows through the lower transistor of the driver back to the decoupling capacitor.

## 10.7.2   Stripline

Figure 10-36 shows the current path for a low-to-high transition on a stripline located adjacent to both a power and a ground plane. As can be observed, the current source is the decoupling capacitor *plus* the trace-to-power plane parasitic capacitance. The decoupling capacitor current (solid arrow) flows through the top transistor of the CMOS driver, down the signal trace toward the load, through the parasitic trace-to-ground plane capacitance, and returns to the decoupling capacitor on the ground plane. The trace-to-power plane capacitance current (dashed arrow) flows on the power plane back toward the driver IC, through the top driver transistor, and returns on the signal trace. Notice that for this configuration, current flows on both the

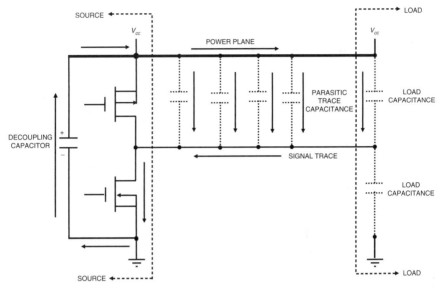

**FIGURE 10-35.** Current flow on a microstrip line adjacent to a power plane, high-to-low transition.

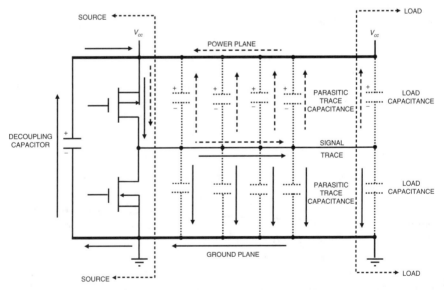

**FIGURE 10-36.** Current flow on a stripline adjacent to both a ground and a power plane, low-to-high transition.

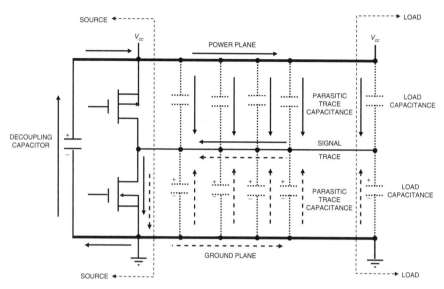

**FIGURE 10-37.** Current flow on a stripline adjacent to both a ground and a power plane, high-to-low transition.

ground and the power planes, and that the current in both planes flows in the same direction, in this case, from the load toward the source.

Figure 10-37 shows the current path for a high-to-low transition on a stripline located adjacent to both a power and ground plane. As can be observed, the current source is the decoupling capacitor *plus* the trace-to-ground plane capacitance. The decoupling capacitor current (solid arrow) flows down the power plane, through the trace-to-power plane capacitance, back to the driver IC on the signal trace, and through the bottom transistor of the CMOS driver back to the decoupling capacitor. The trace-to-ground plane capacitance current (dashed arrow) flows on the signal trace back toward the driver IC, through the bottom driver transistor, and returns on the ground plane. Notice that for this configuration, current also flows in the same direction on both the ground and the power planes; but in this case, the current direction is from the source toward the load.

For the case of a stripline referenced to two ground planes, the current source and current path are the same as the case of a microstrip line referenced to a ground plane (Figs. 10-32 and 10-33), except that in the stripline case, each plane carries only one half the total current. For the case of a stripline referenced to two power planes, the current flow is the same as the case of a microstrip line referenced to a power plane (Figs. 10-34 and 10-35), except in the stripline case, each plane carries only one half the total current.

In all the stripline cases, the current flows in two different signal loops, and the current in the two loops flows in the opposite direction, counter clockwise (CCW) in one loop and clockwise (CW) in the other loop, and in addition each loop contains only one-half the total current. Therefore, radiation from the two loops is reduced and tends to cancel each other. Hence, the stripline configuration produces considerably less radiation than a microstrip line. In addition, the two planes provide shielding to what radiation does occur, which also reduces the emission even more.

### 10.7.3  Digital Circuit Current Flow Summary

From the above examples, it can be concluded that it makes no difference to the current whether the reference plane, or planes, are ground or power. In all the above cases, the current has no trouble returning directly to the source through a small loop area. In none of the cases does the current have to go out of its way or flow through a larger loop area to return to the source. *Therefore, it does not matter what the reference plane is, power or ground.*

Table 10-4 summarizes the results, listing the current source and the return current path for each of the 10 configurations discussed above.

Note that in all cases that involve a low-to-high transition, the current enters the driver IC through the power pin and exits the driver IC via the signal pin,

**TABLE 10-4.  Digital Logic Current Flow Summary.**

| Configuration | Reference Plane | Transition | Current Source | Return Current Path |
|---|---|---|---|---|
| Microstrip | Ground | Low to high | Decoupling capacitor | Ground plane |
| Microstrip | Ground | High to low | Parasitic trace capacitance | Ground plane |
| Microstrip | Power | Low to high | Parasitic trace capacitance | Power plane |
| Microstrip | Power | High to low | Decoupling capacitor | Power plane |
| Stripline | Power and ground | Low to high | Decoupling and parasitic trace capacitance | Power and ground plane |
| Stripline | Power and ground | High to low | Decoupling and parasitic trace capacitance | Power and ground plane |
| Stripline | Ground and ground | Low to high | Decoupling capacitor | Ground planes |
| Stripline | Ground and ground | High to low | Parasitic trace capacitance | Ground planes |
| Stripline | Power and power | Low to high | Parasitic trace capacitance | Power planes |
| Stripline | Power and power | High to low | Decoupling capacitor | Power planes |

and in all cases that involve a high-to-low transition, the current enters the driver IC through the signal pin and exits the driver IC via the ground pin. This occurs regardless of the configuration or what reference plane or planes are used.

## SUMMARY

- Because of its high switching speed, digital logic is a prime source of radiated emission.
- Digital systems require different techniques for grounding than analog systems.
- The noise voltage needed to cause an EMC problem is about three orders of magnitude less than that required for a signal integrity problem.
- With regard to noise control, the single most important consideration in the layout of a digital logic system is the minimization of the ground inductance.
- Ground inductance in digital systems can be minimized by using a ground grid or plane.
- A ground plane will have an inductance that is typically two orders of magnitude less than that of a trace.
- Although ground plane inductances are small, they are still not negligible.
- The reason that the ground plane inductance is so small is because the current can spread out in the plane.
- As the distance between the trace and the return plane is increased, the return plane inductance will also increase.
- The lowest impedance high-frequency signal return path is in a plane directly under the signal trace. Therefore, digital logic currents will return in the plane adjacent to the trace and will spread out a distance to either side of the trace.
- Following is a list of transmission line topologies listed in order of the least amount of return current spread:
  - Stripline
  - Asymmetric stripline
  - Microstrip
- For stripline, 99% of the return current will be contained in a distance of $\pm$ three times the height of the trace above the plane.
- For a microstrip line, 97% of the return current will be contained in a distance of $\pm$ 20 times the height of the trace above the plane.
- Reducing the height of the trace above its reference plane will:
  - Decrease the inductance of the plane
  - Decrease the voltage drop in the plane

- Decrease the radiated emission
- Decrease the crosstalk between adjacent traces
- For large trace heights, typically greater than 0.010 in, the inductance of the return plane will represent the predominate impedance of the plane.
- Below some critical height, however, the ground plane resistance will be the predominate impedance.
- This critical height is a function of frequency, and decreases as frequency increases.
- Current is fed into and out of a plane at a via. The constriction of the ground plane current near the via increases the plane inductance.
- Many engineers are confused about how and where digital logic currents flow, and what is the source of the digital logic current.
- Only the transient (switching) current is important from a noise or EMC perspective.
- The source of the digital logic current is not the logic gate.
- The source of the current is either (1) the decoupling capacitor and/or (2) the parasitic trace and load capacitance.
- It makes no difference to the current flow, and hence the inductances, if the reference plane or planes are power or ground.
- For similar dimensioned structures, stripline radiation is significantly less than microstrip radiation.

## PROBLEMS

10.1 What is the basic ground topology that produces a low-inductance structure?

10.2 a. The width of a ground trace affects what parameter, and is important at what frequencies?
b. The griding of a ground affects what parameter, and is important at what frequencies?

10.3 What percentage of the signal return current is contained in the portion of the reference plane located between plus and minus twice the trace height from the centerline of the trace:
a. For the case of a microstrip line?
b. For the case of a stripline?

10.4 For the case of microstrip lines, how is the crosstalk between two adjacent traces related to:
a. The height of the trace above the reference plane?
b. The separation of the two traces?

10.5 An asymmetrical stripline is spaced 0.005 in from one reference plane and 0.015 in from the other reference plane:

a. What percentage of the signal current will return on the close plane?

b. What percentage of the signal current will return on the far plane?

10.6 Why does the ground plane voltage measurement shown in Fig. 10-18 have to be made on the top of the ground plane? It would be much easier to make the measurement on the bottom of the ground plane.

10.7 Consider the case of a 0.010-in wide microstrip line.

a. What will be the minimum ground plane impedance at 600 MHz?

b. At what trace height will it occur?

10.8 What are the two possible sources for the digital logic current?

10.9 Name two reasons why stripline produces less radiation than a microstrip line?

10.10 a. Draw a diagram of the digital logic current flow on a stripline located between two power planes for the case of a low-to-high transition.

b. For a low-to-high transition, what is the source of the current?

c. For a high-to-low transition, what is the source of the current?

# REFERENCES

Barna, A. *High Speed Pulse and Digital Techniques*. New York Wiley, 1980.

Blakeslee, T. R. *Digital Design with Standard MSI and LSI*, 2nd ed. New York, Wiley, 1979.

German, R. F. Use of a Ground Grid to Reduce Printed Circuit Board Radiation. *6th Symposium and Technical Exhibition on Electromagnetic Compatibility*, Zurich, Switzerland, March 5-7, 1985.

Holloway, C. L. Expression for Conductor Loss of Strip-Line and Coplanar-Strip (CPS) Structures. *Microwave and Optical Technology Letters*, May 5, 2000.

Holloway, C. L. and Hufford, G. A. Internal Inductance and Conductor Loss Associated with the Ground Plane of a Microstrip Line. *IEEE Transactions on Electromagnetic Compatibility*, May 1997.

Holloway, C. L. and Kuester, E. F. Closed-Form Expressions for the Current Density on the Ground Plane of a Microstrip Line, with Applications to Ground Plane Loss. *IEEE Transactions on Microwave Theory and Techniques*, vol. 43, no. 5, May 1995.

Holloway, C. L. and Kuester, E. F. Net Partial Inductance of a Microstrip Ground Plane. *IEEE Transactions on Electromagnetic Compatibility*, February 1998.

Holloway, C. L. and Kuester, E. F. Closed-Form Expressions for the Current Densities on the Ground Plane of Asymmetric Stripline Structures. *IEEE Transactions on Electromagnetic Compatibility*, February 2007.

Johnson, H. W. and Graham, M. *High-Speed Signal Propagation, Advanced Black Magic*, Upper Saddle River, NJ, Prentice Hall, 2003.

Leferink, F. B. J. and van Doorn, M. J. C. M. Inductance of Printed Circuit Board Ground Planes. *IEEE International Symposium on Electromagnetic Compatibility*, 1993.

Ott, H. W. Ground — A Path for Current Flow. *IEEE International Symposium on Electromagnetic Compatibility*, 1979.

Paul, C. R. *Introduction to Electromagnetic Compatibility*, 2nd ed. (Chapter 3, Signal Spectra—the Relationship between the Time Domain and the Frequency Domain). New York, Wiley, 2006.

Smith, T. S. and Paul, C. R. "Effect of Grid Spacing on the Inductance of Ground Grids," *1991 IEEE International Symposium on Electromagnetic Compatibility*, Cherry Hill, NJ, August 1991.

## FURTHER READING

Grover, F. W. *Inductance Calculations, Working Formulas and Tables*. Research Triangle Park, NC, Instrument Society of America, 1973.

Mohr, R. J. "Coupling Between Open and Shielded Wire Lines Over a Ground Plane," *IEEE Transactions on Electromagnetic Compatibility*, September 1967.

# PART II
# EMC Applications

# 11 Digital Circuit Power Distribution

As was discussed in Chapter 4, the characteristics of an *ideal* direct current (dc) power distribution system are as follows:

1. To supply a constant dc voltage to the load
2. Not to propagate any ac noise generated by the load
3. To have a 0 Ω ac impedance between power and ground

Ideally, the power distribution layout should be the same as, and parallel to, the ground system. From a practical point of view, however, this is not always possible or necessary. Because power-supply noise can often be controlled by proper power supply decoupling, a power grid or power plane distribution system, although desirable, is not as important as a proper ground system. If a compromise is necessary, it is often better to use the available board space to provide the best ground system possible and control power-supply noise by other means.

A number of authors have actually extolled the virtues of removing power planes from a printed circuit board (PCB) (Leferink and van Etten, 2004; Janssen, 1999).

## 11.1 POWER SUPPLY DECOUPLING

Power supply decoupling is a means to disassociate a circuit's function from the power bus serving the circuit. This provides two beneficial effects as follows:

1. It reduces the effect of one integrated circuit (IC) upon another (inter-IC coupling).
2. It provides a low impedance between power and ground, so that the IC operates as intended by its designers (intra-IC coupling).*

---

*The proper operation of electronic circuits and ICs, both analog and digital, depend on a low, ideally zero, ac impedance between power and ground. The assumption is that both power and ground are at the same ac potential.

---

*Electromagnetic Compatibility Engineering,* by Henry W. Ott
Copyright © 2009 John Wiley & Sons, Inc.

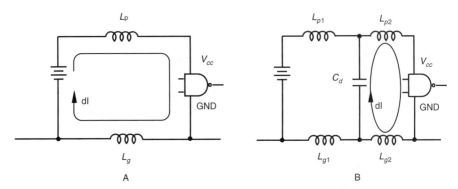

**FIGURE 11-1.** Transient power-supply current (*A*) without and (*B*) with a decoupling capacitor.

When a logic gate switches, a current transient *dI* occurs in the power distribution system as shown in Fig. 11-1A. This current transient flows through both the ground and power traces. The transient current flowing through the power and ground inductances produces a noise voltage that appears between the $V_{cc}$ and ground terminals of the logic gate. In addition, the transient current flowing around a large loop makes an efficient loop antenna.

The magnitude of the power supply voltage transient can be reduced by decreasing the inductances $L_p$ and $L_g$, and/or by decreasing the rate of change of the current (*dI/dt*) that flows through these inductances. The inductance can be reduced, but not eliminated, by using power and ground planes or grids, as discussed in Chapter 10. The loop area and inductance can both be minimized by supplying the transient current from another source, such as a capacitor or capacitors, located near the logic gate as shown in Fig. 11-1B. The noise voltage across the logic gate is then a function of the capacitor $C_d$ and the traces between it and the gate (represented by $L_{p2}$ and $L_{g2}$ in Fig. 11-1B). The number of capacitors, their type, value, and placement with respect to the IC are important in determining their effectiveness.

Decoupling capacitors, therefore, serve two purposes. First, they provide a source of charge close to the IC so that when the IC switches, the decoupling capacitors can supply the required transient currents through a low-impedance path. If the capacitor cannot supply the required current, then the magnitude of the voltage bus will dip, and the IC may not function properly. The second purpose of the decoupling capacitor is to provide a low ac impedance between the power and ground rails, which effectively shorts out (or at least minimizes) the noise injected back into the power/ground system by the IC.

## 11.2  TRANSIENT POWER SUPPLY CURRENTS

Two different power supply transient currents occur when a digital IC switches. These currents are shown in Fig. 11-2. First, when the logic gate switches from

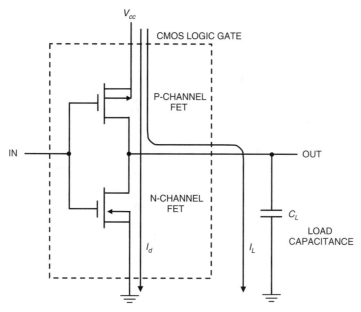

$V_{cc}$

CMOS LOGIC GATE

P-CHANNEL
FET

IN

OUT

N-CHANNEL
FET

$C_L$

LOAD
CAPACITANCE

$I_d$

$I_L$

**FIGURE 11-2.** Transient currents produced when a CMOS logic gate switches.

low to high, a transient current $I_L$ is needed to charge the load capacitance $C_L$. This current only occurs in logic gates connected to outputs. The second current results from the totem pole output structure of the IC, and it occurs on both the high-to-low and the low-to-high transition. Part way through the switching cycle, both transistors are partially on, which produces a low impedance across the power supply and draws a transient current $I_d$ as shown in Fig. 11-2. This dynamic internal current occurs even in gates that are not connected to external loads.

These transient power supply currents are drawn every time the logic circuit changes state. Which of these two currents predominate depends on the type of IC. For digital ICs with a lot of input/output (I/O) drivers, such as clock drivers, buffers, or bus controllers, the transient load current $I_L$ will predominate. For large application-specific integrated circuits (ASICs), microprocessors, or other devices with a large amount of internal processing, the dynamic internal current $I_d$ will predominate. The magnitude of these currents can be estimated by the following means.

### 11.2.1  Transient Load Current

The magnitude of the transient load current can be determined by considering the effect of the capacitive load $C_L$ on the IC, as shown in Fig 11-2. The transient current through the load capacitance, during the switching cycle, will be $I_L = (C_L V_{cc}) / t_r$. For more than one load connected to the IC, multiply $C_L$

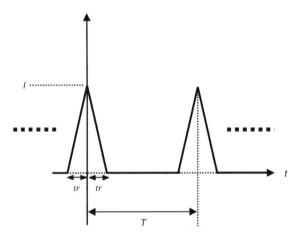

**FIGURE 11-3.** Isosceles triangular wave.

by the number of loads $n$, thus giving the magnitude of the transient load current as follows:

$$I_L = \frac{nC_L V_{cc}}{t_r},\qquad(11\text{-}1)$$

where $C_L$ is the capacitance of one load, $n$ is the number of loads on the IC, $V_{cc}$ is the IC supply voltage, and $t_r$ is the rise time of the output waveform. A typical complementary metal–oxide semiconductor (CMOS) gate, which acts as a load, has an input capacitance of 7 to 12 pF.

Experience has shown that the waveshape of the transient current pulse can be approximated by a triangular wave (see Fig. 11-3) with a peak amplitude determined by Eq. 11-1, and a pulse width equal to twice the rise time $t_r$ (Archambeault, 2002 or Radu et al., 1998). This transient load current will normally only occur on the low-to-high transition because, as was indicated in Table 10-4, the power supply or decoupling capacitor is the source of the transient load current only on the low-to-high transition.*

For example, a device operating from a 5-V power supply with a 1-ns rise time output pulse, driving 10 CMOS loads with 10 pF of input capacitance each will have a transient I/O current of 500 mA.

## 11.2.2    Dynamic Internal Current

It is often harder to obtain estimates of the shoot-through current than the transient load current. Many IC data sheets provide no information on

---

*The one exception to this is a stripline located adjacent to both a power and ground plane; in which case, the decoupling capacitor will be the source of the current on both the high-to-low and the low-to-high transitions.

the transient power supply currents drawn by the IC. Some CMOS device data sheets, however, provide information on the internal power dissipation, by providing the value of an equivalent power dissipation capacitance $C_{pd}$. The power dissipation capacitance can be thought of as an equivalent internal capacitance that can be used to estimate the magnitude of the shoot-through current; this approach is similar to the way the load capacitance $C_L$ was used to determine the magnitude of the transient load current. This capacitance is often specified as so much capacitance per gate. The total power dissipation capacitance will then be the capacitance per gate $C_{pd}$ multiplied by the number of gates switching simultaneously (that may be difficult to determine). Other IC data sheets actually list a value for the dynamic power supply current $I_{ccd}$, which is specified in units of A/MHz.

The magnitude of the dynamic internal current $I_d$ can then be estimated by using one of the following two equations:

$$I_d = I_{ccd} f_0 \tag{11-2a}$$

or

$$I_d = \frac{n C_{pd} V_{cc}}{t_r}, \tag{11-2b}$$

where $I_{ccd}$ is the dynamic power supply current, $f_o$ is the clock frequency, $C_{pd}$ is the power dissipation capacitance per gate, $n$ is the number of gates switching, $V_{cc}$ is the supply voltage, and $t_r$ is the switching time of the gate.

The waveshape of this dynamic current pulse can also be approximated by a triangular wave associated with each edge of the clock, and having a peak amplitude determined by Eq. 11-2a or 11-2b. The pulse width is equal to twice the rise time $t_r$ (Archambeault, 2002, or Radu et al., 1998). Because the dynamic internal current occurs on every transition of the clock (low to high and high to low), the dynamic internal current will contain harmonics that are multiples of twice the clock frequency.

### 11.2.3    Fourier Spectrum of the Transient Current

For an isosceles-triangular wave as shown in Fig. 11-3, the amplitude of the current in the $n$th harmonic is given by (Jordan, 1985, p. 7–10)

$$I_n = \frac{2 I t_r}{T} \left( \frac{\sin\left(\frac{n \pi t_r}{T}\right)}{\frac{n \pi t_r}{T}} \right)^2, \tag{11-3}$$

where $I$ is the amplitude of the triangular wave, $t_r$ is the rise (fall) time, $T$ is the period, and $n$ is the harmonic number. Figure 11-4 shows a log-log plot of

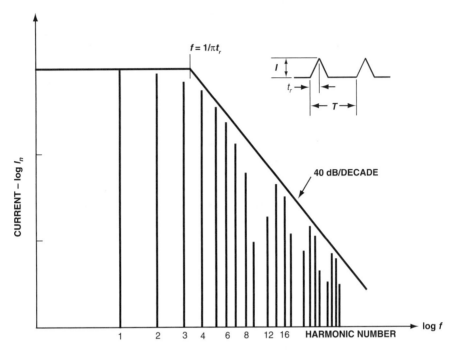

**FIGURE 11-4.** Envelope of Fourier spectrum of an isosceles triangular wave.

the envelope of the harmonics for a wave with a $t_r/T$ ratio of 0.1. As can be observed, above a frequency of $1/\pi\ t_r$ the harmonics fall off with frequency at a 40 dB per decade rate, and both odd and even harmonics are present.

Table 11-1 lists the percentage of current present in the first six harmonics versus the $t_r/T$ ratio. Note that as the rise time becomes faster (smaller $t_r/T$), the percent current in the fundamental decreases, and the current has a slower falloff with frequency (harmonic number). For a $t_r/T$ ratio of 0.1 or less, the fundamental frequency will contain less than 20% of the current.

**TABLE 11-1.  Percent Current in the First Five Harmonics versus the $t_r/T$ Ratio for an Isosceles-Triangle Waveform.**

| $t_r/T$ | Harmonic | | | | | |
|---|---|---|---|---|---|---|
| | 1 | 2 | 3 | 4 | 5 | 6 |
| 0.05 | 10% | 10% | 9% | 9% | 8% | 7% |
| 0.1 | 19% | 18% | 15% | 12% | 8% | 5% |
| 0.2 | 35% | 23% | 10% | 2% | 0% | <1% |
| 0.3 | 44% | 15% | 2% | 2% | 3% | <1% |

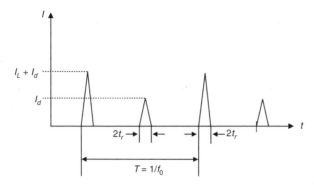

**FIGURE 11-5.** The waveshape of the total transient current drawn by an IC.

The $1/\pi t_r$ frequency will occur at a frequency equal to $k$ times the fundamental frequency, where $k$ (which may be a noninteger) is equal to

$$k = \left(\frac{1}{\pi}\right)\left(\frac{T}{t_r}\right). \tag{11-4}$$

The heavy line in Table 11-1 indicates between which harmonics the $1/\pi\ t_r$ frequency falls.

### 11.2.4  Total Transient Current

The total transient current drawn by the IC is the sum of the transient load current $I_L$ and the dynamic internal current $I_d$. Figure 11-5 shows the waveshape of the total transient current. The transient load current spikes occur at the clock frequency, and the dynamic internal currents spikes occur at twice the clock frequency.

Archambeault (2002, pp. 127–129) has demonstrated a good correlation between measured and predicted $V_{cc}$-to-ground noise voltage using the current waveshape shown in Fig. 11-5.

## 11.3  DECOUPLING CAPACITORS

Effective power supply decoupling has become increasingly more difficult to achieve as the result of increasing clock frequencies and faster rise times. Ineffective decoupling can lead to excessive power bus noise as well as to excessive radiated emission.

Many designers are decoupling digital logic ICs by placing a single 0.1- or 0.01-μF capacitor adjacent to the IC. This method is the same one that has been used on digital logic ICs for the last 50 years, so it still must be the correct approach—right? After all, how much has IC technology changed in the last 50 years? I think it is interesting that this method has worked for as many years as it has. No one should be surprised that, possibly, we are now to the point that a new approach to digital IC decoupling is required.

It is important to understand that decoupling is *not* the process of placing a capacitor adjacent to an IC to supply the transient switching current as shown in Fig. 11-6A; rather it is the process of placing an *L–C* network adjacent to an IC to supply the transient switching current as shown in Fig. 11-6B. All decoupling capacitors have inductance in series with them. Therefore, the decoupling network is a series resonant circuit. As can be observed from Fig. 11-7, the inductance comes from three sources, as follows:

1. The capacitor itself
2. The interconnecting PCB traces and vias
3. The lead frame inside the IC

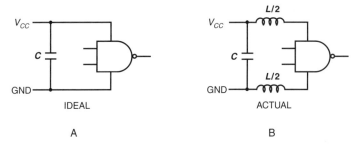

IDEAL                          ACTUAL

A                                B

**FIGURE 11-6.** (*A*) Ideal decoupling network and (*B*) actual decoupling network.

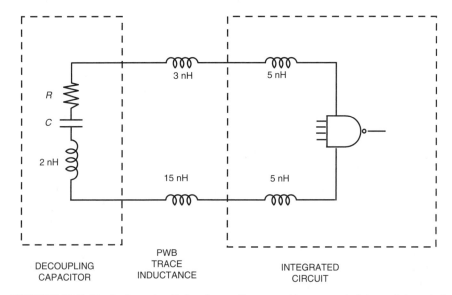

DECOUPLING          PWB          INTEGRATED
CAPACITOR           TRACE         CIRCUIT
                   INDUCTANCE

**FIGURE 11-7.** Equivalent circuit for decoupling capacitor connected to an integrated circuit, for a DIP package with power and ground on diagonally opposite pins of the IC.

Although Fig. 11-7 is for a dual in-line package (DIP), the total inductance value is still representative of that obtained with other IC packages. Figure 11-7 assumes that the capacitor is placed at one end of the DIP, close to one of the pins (power or ground), and about an inch away from the other pin.

The internal inductance of a surface mount technology (SMT) capacitor itself is typically 1–2 nH, the interconnecting PCB traces and vias add 5 to 20 nH or more according to the layout, and the internal lead frame of the IC may have 3 to 15 nH of inductance according to the type of IC package. The inductances of the interconnecting PCB traces, however, are usually the only parameters under the system designer's control.

It is extremely important therefore to minimize the inductance of the traces between the IC and the decoupling capacitor. The printed circuit traces should be as short as possible and should be placed as close together as possible to minimize their loop area. Notice also from Fig. 11-7 that the decoupling capacitor contributes the least amount of inductance. Therefore, it is not the major problem.

From the above, we observe that the total inductance can vary from a low of about 10 nH to a high of 40 nH. Typically it is in the range of 15 to 30 nH. It is this inductance that limits the effectiveness of the decoupling network. It is very important to remember this fact—*we are placing an L–C network between the power and ground, not a capacitor!*

Because of this combination of capacitance and inductance, the decoupling network will, at some frequency, become resonant. At the resonant frequency, the magnitude of the inductive reactance is equal to the magnitude of the capacitive reactance and the network is a very low-impedance and an effective bypass. Above the resonant frequency, the circuit becomes inductive and its impedance increases with frequency.

The resonant frequency of a series $L–C$ circuit is

$$f_r = \frac{1}{2\pi\sqrt{LC}}.$$

(11-5)

Figure 11-8 shows the impedance versus frequency of various decoupling capacitors when in series with 30 nH of inductance. The top horizontal scale also shows the equivalent digital logic rise time that corresponds to the frequency shown on the bottom horizontal scale. The relationship between rise time and frequency is $t_r = 1/\pi f$.

The point where the capacitive reactance is equal to the inductive reactance is the resonant frequency of that combination of capacitor and inductor. The impedance at resonance decreases to a low value (equal to just the series resistance of the network), because the capacitive reactance cancels the inductive reactance and leaves only the resistance.

As can be observed in Fig. 11-8, the commonly used 0.1-μF capacitor reso-nates with 30 nH of inductance at 3 MHz, and a 0.01-μF capacitor resonates

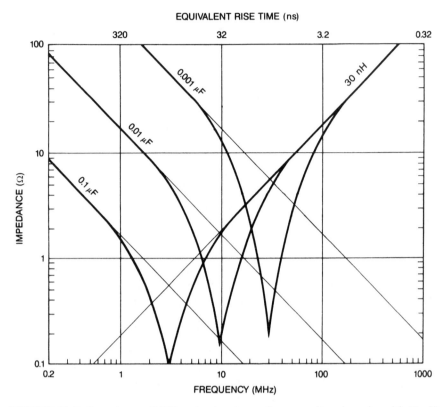

**FIGURE 11-8.** Impedance of different value decoupling capacitors in series with 30 nH of inductance.

at 9 MHz,. which is not impressive in this day of 100+ MHz clocks. Above about 50 MHz, the impedance of the decoupling network is dominated by the 30 nH of inductance regardless of what value capacitor is used. If a good layout was achieved and the inductance was half of the 30 nH shown in Fig. 11-8, the resonant frequency would only increase by the $\sqrt{2}$, or a factor of 1.41. Therefore, a 0.1-µF capacitor will resonate at about 4 MHz, and a 0.01-µF capacitor will resonate at about 13 MHz when in series with 15 nH of inductance.

Figure 11-8 clearly demonstrates that *placing a single capacitor adjacent to an IC is not an effective way to decouple digital logic at frequencies above 50 MHz, regardless of the capacitor value or placement.* Figure 11-8 also shows that for frequencies less than about 50 MHz, it is possible to tune the decoupling network, by choosing different value capacitors to place the resonant dip at a problem frequency. But this same approach will not work above 50 MHz.

Figure 11-9 shows measured values of the $V_{cc}$-to-ground noise voltage versus frequency on a board using 0.01-µF decoupling capacitors. The measurements were made between the power and ground pins of the IC, with and without decoupling capacitors on the board. In the frequency range of 20 to 70 MHz,

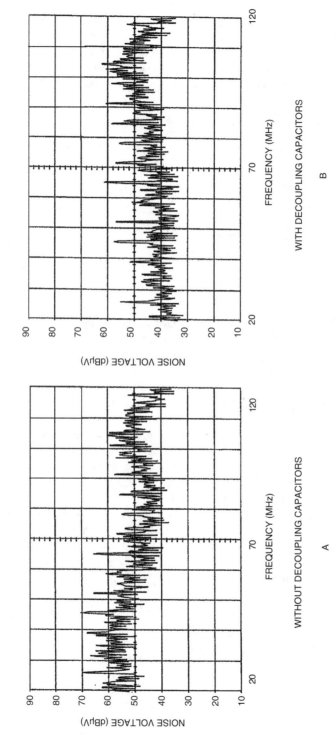

**FIGURE 11-9.** Measured $V_{cc}$-to-ground noise voltage versus frequency for a printed circuit board with and without decoupling capacitors.

435

the presence of the decoupling capacitors significantly decreased the magnitude of the $V_{cc}$-to-ground noise voltage. However, in the frequency range of 70 to 120 MHz, the noise voltage was the same, with and without the decoupling capacitors, this result indicates that the capacitors were ineffective in this frequency range, which confirms the results predicted by Fig. 11-8.

## 11.4 EFFECTIVE DECOUPLING STRATEGIES

Contamination of the $V_{cc}$ bus by clock harmonics that result from ineffective decoupling can cause signal integrity problems and can be the cause of excessive radiated emission. Possible solutions to the high-speed decoupling problem are as follows:

1. Slow down the rise time
2. Decrease the current transient
3. Decrease the inductance in series with the capacitor
4. Use multiple capacitors

The first two approaches listed above go against the advancement in technology and will not provide a long-term solution to the problem. Decreasing the inductance in series with the decoupling capacitor is desirable and should always be done, to whatever extent possible; however, this by itself will not solve the high-speed decoupling problem. Even if the total inductance in series with the capacitor was somehow reduced to only 1 nH, which is an unlikely possibility, a decoupling network with a 0.01-µF capacitor would have a resonant frequency of only 50 MHz. Therefore, it is not possible to move the resonant frequency of a single capacitor decoupling network up to a frequency of a few hundred megahertz using any realistic value of capacitance.

Table 11-2 lists the resonant frequency of various values of capacitors when in series with 5, 10, 15, 20, and 30 nH of inductance.

The conclusion that we can draw from Table 11-2, is that in most cases, it is impossible to raise the resonant frequency of the decoupling $L-C$ network to be above the clock frequencies commonly used in most digital electronics today, no less above the frequency of the harmonics. With only 10 nH of inductance, a 1000-pF capacitor will only have a resonant frequency of 50 MHz.

**TABLE 11-2.  Resonant Frequency (in MHz) of Various Value Capacitors When in Series With the Listed Value of Inductance.**

| Capacitor (µF) | 5 nH | 10 nH | 15 nH | 20 nH | 30 nH |
|---|---|---|---|---|---|
| 1 | 2.3 | 1.6 | 1.3 | 1 | 0.9 |
| 0.1 | 7.1 | 5 | 4.1 | 3.6 | 3 |
| 0.01 | 22.5 | 16 | 13 | 11 | 9 |
| 0.001 | 71.2 | 50 | 41 | 36 | 30 |

At frequencies below the decoupling network resonance, the two most important considerations are (1) to have sufficient capacitance (see Eq. 11-12) to provide the required transient current, and (2) to provide an impedance that is low enough to short out the noise current generated by the IC (see Section 11.4.5). However, above resonance the most important consideration is to have a low enough inductance, such that the decoupling network is still a low impedance and shorts out the noise current. Therefore, a decoupling network can still be effective above resonance, if a way can be found to lower its inductance sufficiently.

### 11.4.1  Multiple Decoupling Capacitors

*No single decoupling capacitor network will provide a low enough inductance.* Therefore, the real solution to the high-frequency decoupling problem lies in the use of multiple decoupling capacitors. Three approaches have been proposed. They are as follows:

1. The use of multiple capacitors all of the same value.
2. The use of multiple capacitors of two different values.
3. The use of multiple capacitors of many different values, usually spaced a decade apart. For example, a 1 μF, 0.1 μF, 0.01 μF, 0.001 μF, 100 pF, etc.

### 11.4.2  Multiple Capacitors of the Same Value

When a number of identical $L$–$C$ networks are connected in parallel as shown in Fig. 11-10, the total capacitance $C_t$ becomes

$$C_t = nC,  \tag{11-6a}$$

where $C$ is the capacitance of one of the networks and $n$ is the number of networks in parallel.

The total inductance $L_t$ of $n$ identical $L$–$C$ networks in parallel is

$$L_t = \frac{L}{n},  \tag{11-6b}$$

where $L$ is the inductance of one of the networks. By referring to Eq. 10-8, we can conclude that Eq. 11-6b will only be correct if the mutual inductance between the inductances of the individual networks are negligible compared with their self-inductance. Therefore, the $L$–$C$ networks must be physically separated from each other.

Notice from Eqs. 11-6a and 11-6b that when identical $L$–$C$ networks are paralleled, the capacitance (the good parameter) is multiplied up by the number of networks, and the inductance (the bad parameter) is divided down by the

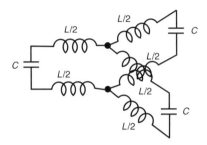

**FIGURE 11-10.** Three identical *L–C* networks in parallel.

number of networks. Therefore, the inductance of the network can be reduced to any desirable value just by using a sufficient number of *L–C* networks in parallel.

The requirements for effectively paralleling multiple *L–C* networks are as follows:

1. Make all the capacitors the same value; this way they share the current equally.
2. Each capacitor must feed the IC through a different inductance than the others. Therefore, they cannot be located together, because of the mutual inductance that would occur. Spread them out.

Rather than just considering the resonant frequency, a better approach to analyzing the effectiveness of a decoupling network is to consider the IC as a noise current generator as shown in Fig. 11-11. The decoupling network can then be designed to be a low impedance across the frequency band of interest, to short out the noise current and to prevent it from contaminating the power bus. A maximum allowable value for the decoupling network impedance, which is sometimes referred to as a *target impedance*, can be specified, for example, 100 to 200 mΩ, and the impedance of the chosen decoupling network can be calculated to determine over what frequency range the impedance is less than the target impedance. See Section 11.4.5 for more information on determining target impedance.

Figure 11-12 is a plot of the impedance versus frequency for various numbers (1, 8, 64, 512) of identical *L–C* networks in parallel. In all cases, the *total* capacitance equals 0.1 µF and the inductance in series with each capacitor is 15 nH.

Although the high-frequency impedance is reduced dramatically by the use of multiple capacitor networks, the low-frequency impedance is not—only the frequency of the resonant dips shift. Actually, at low frequencies, where the single capacitor network was resonant, the impedance is actually higher when multiple capacitors are used than when only one capacitor is used. This is because the total capacitance, in this case 0.1 µF, is not large enough to be a low impedance at these frequencies.

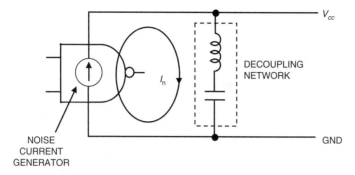

**FIGURE 11-11.** The IC as a noise current generator.

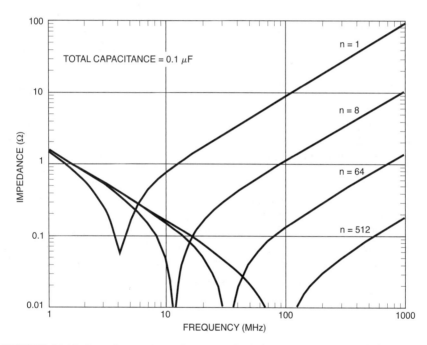

**FIGURE 11-12.** Impedance versus frequency for a decoupling network consisting of from 1 to 512 equal-value capacitors when in series with 15 nH of inductance. In all cases, the total capacitance equals 0.1 μF.

Figure 11-13 is similar to Fig. 11-12, except in this case the multiple capacitors add up to a *total* capacitance of 1 μF (instead of 0.1 μF). As a result, the low-frequency impedance is significantly reduced. For the case of 512 capacitors, Fig. 11-13 shows that the impedance of the *L–C* decoupling network is less than 0.2 Ω from 1 MHz to 1000 MHz. With 64 capacitors, the impedance is less than 0.5 Ω from less than 1 MHz to 350 MHz. Notice

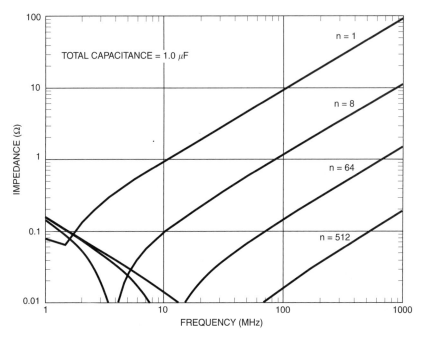

**FIGURE 11-13.** Impedance versus frequency for a decoupling network consisting of from 1 to 512 equal-value capacitors when in series with 15 nH of inductance. In all cases, the total capacitance equals 1.0 μF.

also that the resonant frequency of the decoupling networks in Fig. 11-13 are lower than the resonant frequency of the networks in Fig. 11-12, but the decoupling effectiveness is improved (lower impedance over a wider frequency range). Therefore, *the use of a large number of equal value capacitors is an effective way to provide a low-impedance decoupling network that is effective over a large frequency range.* This strategy can be very effectively used when decoupling a large IC.

### 11.4.3 Multiple Capacitors of Two Different Values

Two different decoupling capacitor values are sometimes recommended, based on the theory that the large value capacitor will provide effective low-frequency decoupling, and the small value capacitor will provide effective high-frequency decoupling. If two different capacitor values are used, there will be two distinct resonant dips as shown in Fig. 11-14, which is good.

Although the above is true, when capacitors of different values are placed in parallel, a potential problem exists with respect to the parallel resonance (sometimes referred to as an "antiresonance") that occurs between the two networks. Figure 11-14 shows a plot of the impedance of a 0.1-μF capacitor in

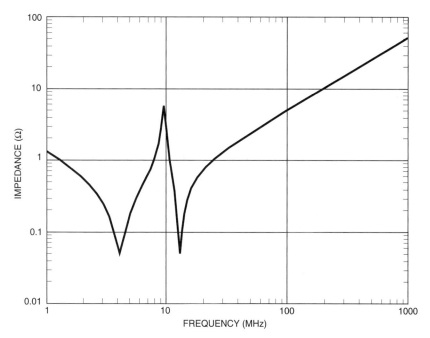

**FIGURE 11-14.** Impedance versus frequency for a decoupling network consisting of a 0.1 and a 0.01-µF capacitor, both in series with 15 nH of inductance.

parallel with a 0.01-µF capacitor, which both have 15 nH of inductance in series with them. Figure 11-14 clearly shows the two resonant dips produced by the two different value capacitors in series with 15 nH of inductance, one at 4.1 MHz and the other at about 13 MHz. Notice, however, that an impedance spike occurs at about 9 MHz, which is bad. This effect, as described by Paul (1992), is caused by the parallel, or antiresonance, between the two networks.

Why this occurs can best be understood by referring to Fig 11-15A, which shows two $L$–$C$ decoupling networks with different value capacitors connected between $V_{cc}$ and ground. Assume that $C_1 >> C_2$ and $L_1 = L_2$. Let $f_{r1}$ be the frequency where the capacitor $C_1$ is resonant with inductor $L_1$, and let $f_{r2}$ be the frequency where capacitor $C_2$ is resonant with inductor $L_2$.

Below the frequency at which either of the networks is resonant ($f < f_{r1}$), both networks look capacitive, and the total capacitance is equal to the sum of the two capacitors, which for all practical purposes is just equal to the large capacitor. Therefore, the small capacitor has no, or very little, effect on the performance of the decoupling network.

Above the frequency where both networks are resonant ($f > f_{r2}$), both networks look inductive and the total inductance is equal to the two inductors in parallel, or one half the inductance. This improves the decoupling at frequencies above $f_{r2}$.

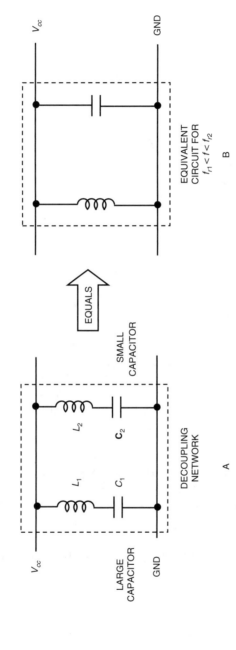

**FIGURES 11-15.** (*A*) Decoupling network with two different value capacitors, and (*B*) the equivalent circuit of *A* for a frequency *f* where $f_{r1} < f < f_{r2}$

At a frequency between the resonances of the two networks ($f_{r1} < f < f_{r2}$), however, the network with the larger capacitor has become inductive, and the network with the smaller capacitor is still capacitive. The equivalent circuit of the two networks is therefore a capacitor in parallel with an inductor as shown in Fig. 11-15B, or a parallel resonant circuit, the impedance of which is large at resonance, which produces the resonant spike shown in Fig. 11-14.

The exact shape, amplitude, and location of the parallel resonant peak will vary as a function of the ratio of the two capacitor values, the equivalent series resistance (ESR) of the capacitors, and the PCB layout. If the two capacitor values are within a two-to-one ratio, the amplitude of the resonant peak will be reduced to an acceptable value, because it will occur in the resonant dip. For example, even the same nominal value capacitors will have different values, because of their tolerance. The resonant peak produced between two capacitors that have a 20%, or even 50%, tolerance will not cause a problem. The major problem with the antiresonance occurs when the capacitors have values that are an order of magnitude or more apart.

Therefore, we can conclude that when two different value capacitors are used in each of the decoupling networks that:

- The small capacitor network will have *no effect* on the decoupling performance at frequencies below which the large capacitor network is resonant.
- The decoupling will *be improved* at frequencies above which both capacitor networks are resonant, because the inductance will decrease.
- The decoupling will actually *be worse* at some frequencies between these two resonance frequencies, because of the impedance spike caused by the parallel resonant network, which is bad.

For example, Bruce Archambeault (2001) provides information on the decoupling effectiveness for various decoupling strategies by using a network analyzer to measure the power to ground plane impedance of a test PCB. For the case of two different value decoupling capacitors, he concludes:

> There is no noticeable improvement in the high frequency decoupling performance when a second capacitor value is added.   In fact the decoupling performance is worse in the frequency range where much of the typical noise energy exists (50-200 MHz).

Archambeault's data show that at some frequencies, between the resonant frequencies of the two decoupling networks, the noise increased by as much as 25 dB, in the frequency range of 50 to 200 MHz when two different capacitor values were used, as compared with the results when all the capacitors were of the same value.

### 11.4.4  Multiple Capacitors of Many Different Values

Capacitors of many different values (typically spaced a decade apart) are sometimes also recommended, based on the theory that the multiple impedance dips produced by the resonances of the different value capacitors is advantageous, because it will provide a low impedance at many frequencies.

However, when decoupling capacitors of many different values are used, additional impedance spikes are also produced. Figure 11-16 shows the impedance plot for the case when four different value decoupling capacitors are used, all in series with 15 nH of inductance. The solid line plot is for the case of a 0.1-μF, a 0.01-μF, a 0.001-μF, and a 100-pF capacitor all in parallel. Each capacitor is in series with 15 nH of inductance. As can be observed four resonant dips exist, one for each value of capacitor. However, the antiresonance also produces three resonant spikes in the impedance plot. If some clock harmonics should fall on or near the frequency of these spikes, the power to ground noise will actually increase. Some people consider this approach as the equivalent of playing "Russian Roulette," because one must hope that none of the clock harmonics fall on or near any of the resonant spikes. Notice also, as shown in Fig. 11-16, that the amplitude of the resonant peaks increases with frequency, whereas the impedance of the resonant dips remains the same.

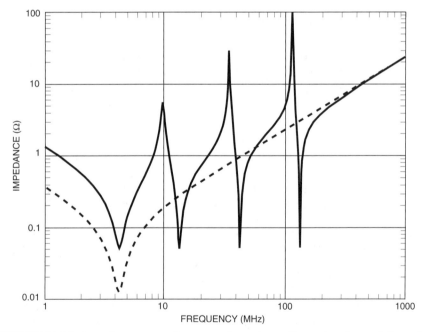

**FIGURE 11-16.** Impedance versus frequency for a decoupling network consisting of a 0.1, a 0.01, and a 0.001 μF plus a 100-pF capacitor (solid line), and a network that consists of four 0.1-μF capacitors (dashed line). For both networks, the capacitors are in series with 15 nH of inductance.

If, however, instead of four different value capacitors, four 0.1-μF capacitors had been used, the impedance plot would be as shown by the dashed line in Fig. 11-16. Both plots are for the same number of capacitors. The lower impedance at low frequency is the result of the greater total capacitance provided by the four parallel 0.1-μF capacitors. The absence of the resonant spikes is the result of all of the capacitors being of the same value. At frequencies above 200 MHz, the results are the same for the case of the four 0.1-μF capacitors and the four different value capacitors. This results from the fact that at these frequencies only the inductance in series with the capacitors matters, and in both cases four inductances are in parallel. However, below 200 MHz the results are better for the case of the four 0.1-μF capacitors, except at the few frequencies that correspond to the resonant dips in the multiple-value capacitor case. Which impedance plot would you rather have for your product?

My preference for effective high-frequency decoupling, therefore, is to use multiple capacitors, all of the same value. This approach works well and has fewer pitfalls than either of the schemes using multiple capacitors of different values.

### 11.4.5    Target Impedance

To be an effective decoupling network, the impedance of the network must be kept below some target value over the frequency range of interest. If this can be accomplished, then the location of the resonant frequency does not matter. If the target impedance were 200 mΩ, then for the case of the 64-capacitor configuration, which was depicted in Fig. 11-12, the range over which the impedance would have been below 200 mΩ is from about 8 MHz to 130 MHz.

However, using a target impedance that is constant across the frequency range is overly restrictive and not necessary. From Eq. 11-3 and Fig. 11-4, we know that the amplitude of the harmonics of a triangular wave fall off at a rate of 40 dB per decade above a frequency of $1/(\pi \, t_r)$. Therefore, the target impedance can rise above this frequency without increasing the noise voltage. If the target impedance is allowed to rise at a rate of 20 dB per decade above this corner frequency (as shown by the solid line in Fig. 11-17), then the noise will still decrease at a 20 dB per decade rate beyond this frequency. This approach greatly simplifies the decoupling network design and minimizes the required number of capacitors.

Using this approach, one can easily estimate the number of decoupling capacitors needed to provide effective high-frequency decoupling. The minimum number of capacitors $n$ is equal to

$$n = \frac{2L}{Z_t t_r},$$
(11-7)

where $L$ is the inductance in series with each capacitor, $Z_t$ is the low-frequency target impedance, and $t_r$ is the switching (rise/fall) time of the logic. Using at

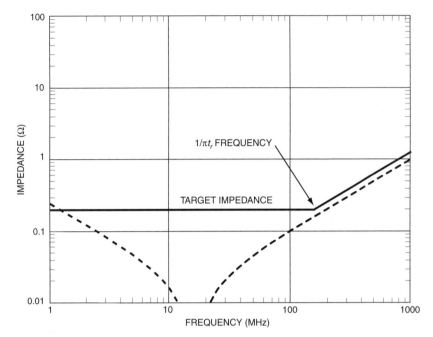

**FIGURE 11-17.** Target impedance (solid line). Impedance of a decoupling network consisting of 64, 0.01-μF capacitors, each in series with 10 nH of inductance (dashed line).

least this number of capacitors will keep the high-frequency impedance of the decoupling network at or below the target impedance.

The key to an optimum decoupling design is to know what inductance to use in Eq. 11-7. As far as the power-to-ground noise on the IC chip itself is concerned, the total inductance (decoupling capacitor, PCB traces, and IC lead frame) must be considered. However, nothing can be done at the PCB level about the inductance of the IC lead frame. Besides, when we measure the power-to-ground noise on an IC we measure it at the pins of the IC, not at the chip. In addition, as far as contaminating the PCB power bus with noise, it is the voltage at the IC-to-PCB interface (i.e., the IC pins) that matters, not the noise voltage at the chip itself.

Therefore, the objective of decoupling should be to minimize the $V_{cc}$-to-ground noise voltage at the pins of the IC. To accomplish this, one can neglect the internal inductance of the IC. Therefore, we only have to consider the decoupling capacitor inductance and the PCB trace (including via) inductance. The internal inductance of a good SMT capacitor will be 1.5 nH or less. The PCB trace inductance will be about 10 nH/in, and the inductance of a through-via on a 0.062-in-thick PCB will be about 0.8 nH.

The low-frequency target impedance is often determined by considering the magnitude of the total transient current and the allowable variation of the supply voltage. Therefore, we can write

$$Z_t = \frac{k dV}{dI},$$   (11-8)

where $dV$ is the allowable power supply transient voltage variation, $dI$ is the amplitude of the transient power supply current drawn by the IC, and $k$ is a correction factor discussed in the next paragraph.

Let us assume that our concern is the $V_{cc}$-to-ground noise spikes and their effect on the operation of the circuit. We can determine from Table 11-1 that no more than about 50% of the current is contained in the frequencies below the $1/\pi\, t_r$ frequency. Therefore, the low-frequency target impedance could be increased by a factor of two, because the $dI$ used in Eq. 11-8 was the total transient current. Therefore, in this case we can use $k = 2$ in Eq. 11-8. This approach will limit the total noise voltage spike to less than the target impedance times the total transient current.

For example, consider the case of decoupling a large IC having a rise/fall time of 2 ns operating from a 5-V supply, where it is desirable to keep the maximum power supply deviation to no more than 5% of the supply voltage. Assume also that the total transient current drawn by the IC is 2.5 A. From Eq. 11-8 with $k = 2$, we get a low-frequency target impedance of 200 m$\Omega$. Assume that each decoupling capacitor has 10 nH of inductance in series with it. The target impedance will then be 200 m$\Omega$ up to the $1/(\pi\, t_r)$ frequency, which for a 2-ns rise time is equal to 159 MHz, above which the target impedance can be allowed to increase at a 20 dB per decade rate. This target impedance is shown as the solid line in Fig. 11-17.

From Eq. 11-7 we determine that 50 capacitors would be necessary to meet the 200-m$\Omega$ target impedance at 159 MHz. If the lowest frequency of interest were 2 MHz, the total decoupling capacitance necessary to meet the 200-m$\Omega$ target impedance, at 2 MHz, would be 400 nF. Dividing 400 nF by 50, gives 8000 pF for the minimum value of each capacitor. The total capacitance must also meet the transient current criteria of Eq. 11-12, which in this case it does. If the designers decide to use 64, 0.01-$\mu$F capacitors, the impedance versus frequency plot will be as shown by the dashed line in Figure 11-17. The decoupling network will then have an impedance below the target impedance at all frequencies above 1.3 MHz.

### 11.4.6   Embedded PCB Capacitance

Taking the concept of using a large number of equal value capacitors to its limit, one can conclude that the ideal decoupling configuration is an infinite number of infinitesimal capacitors (i.e., instead of using discrete capacitors, one should use distributed capacitance).

Could this result possibly be achieved by taking advantage of the interplane capacitance between the power and ground planes of a PCB? To be effective above about 50 MHz, a power-ground plane capacitance of about 1000 pF/in$^2$ would be required. Standard layer spacings of 0.005 to 0.010 in, however, provide capacitances that are 1/5 to 1/10 of this value.

Therefore, if advantage is to be taken of the concept of distributed capacitance, then a new PCB structure, that has additional capacitance needs to be developed. This increase in distributed capacitance could possibly be achieved in one of the following two ways:

1. Reduce the layer spacing
2. Increase the dielectric constant of the PCB material

***11.4.6.1 Effective Decoupling Capacitance Area.*** The effective decoupling capacitance, however, is not the capacitance of the total power-ground-plane pair. Because of the finite velocity of propagation, of electromagnetic energy the effective interplane capacitance is only the capacitance of an area that is located within a finite radius $r$ of the IC being decoupled, as shown in Fig. 11-18. Charge that is stored in the interplane capacitance farther away than this cannot be moved fast enough to arrive at the IC during the switching transient.

The velocity of propagation of electromagnetic energy in a dielectric was given in Eq. 5-14 and is repeated here:

$$v = \frac{c}{\sqrt{\varepsilon_r}}, \qquad (11\text{-}9)$$

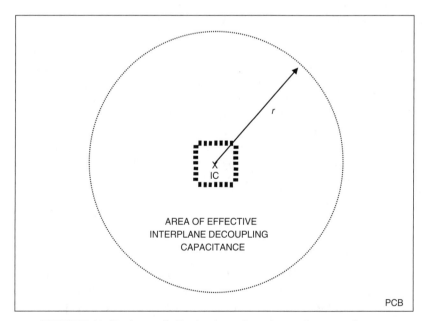

**FIGURE 11-18.** Area of effective interplane decoupling capacitance.

where $c$ is the speed of light (12 in/ns) and $\varepsilon_r$ is the relative dielectric constant of the material.

The radius of the effective capacitance area is equal to

$$r = vt = \frac{12t}{\sqrt{\varepsilon_r}},$$
(11-10)

where $r$ is the radius in inches, $t$ is the time required to move the charge in nanoseconds, and $\varepsilon_r$ is the relative dielectric constant of the material. If we want to move charge on an FR-4 epoxy-glass PCB ($\varepsilon_r = 4.5$) in 0.5 ns, then $r$ would be equal to approximately 3 in.

The effective capacitance of a power-ground-plane pair is

$$C = \frac{\varepsilon A_e}{s},$$
(11-11)

where $\varepsilon$ is the dielectric constant, $s$ is the spacing of the planes, and $A_e$ the effective area of the planes.

The capacitance is proportional to the effective area (which is determined by the velocity of propagation and the rise/fall time of the IC), the spacing between the planes, and the dielectric constant. From Eq. 11-11 we can observe that *decreasing the layer spacing is an effective way of increasing the board's effective decoupling capacitance.*

However, if the dielectric constant of the PCB material is increased, to increase the capacitance, Eq. 11-9 tells us that the velocity of propagation will decrease. Thus, if the dielectric constant is increased, then the radius of the effective area decreases by the square root of $\varepsilon$, and the area of the effective capacitance circle decreases by the square of this, or proportional to $\varepsilon$. Therefore, the area $A_e$ decreases proportional to the increase in $\varepsilon$. Hence from Eq. 11-11 we conclude that *changing the dielectric constant of the PCB has no effect on the effective decoupling capacitance,* because $\varepsilon$ increases as much as the effective area $A_e$ decreases.

Therefore, the effective interplane decoupling capacitance remains the same regardless of the dielectric constant of the PCB material, and the only effective way to increase the effective capacitance is to place the planes closer together. Hence, no advantage is gained on a large, high-speed digital logic PCB, by using exotic (read expensive) high dielectric materials to increase the embedded decoupling capacitance, and the only effective way to increase the effective decoupling capacitance is to place the power and ground planes closer together, or to use multiple power and ground planes.*

---

\* In the case of a PCB with an area less than the area of the effective interplane decoupling capacitance circle, the whole board area will contribute to the effective decoupling capacitance, and in that case increasing the dielectric constant of the material will increase the decoupling capacitance. The $\varepsilon_r$ of the high dielectric constant material must be used to calculate the propagation delay used to determine the radius of the effective decoupling capacitance circle.

***11.4.6.2 Practical Implementation of Embedded Capacitance.*** In 1989–1990 Zycon, (Zycon since became Hadco and now is Sanmina), developed a special PCB laminate with a 2-mil spacing between layers using standard FR-4 epoxy glass as the dielectric. This laminate known as ZBC-2000® provides 500 pF/in$^2$ of interplane capacitance.* By using two sets of power and ground planes made from this laminate in a PCB, one can obtain the desired 1000 pF/in$^2$ of capacitance.

Although Sanmina (Howard and Lucas, 1992) and Unisys (Sisler, 1991) have patents on the 2-mil thick embedded capacitance printed circuit boards, the technology is readily available and multisourced. Sanmina's trade name for this technology is Buried Capacitance.® Although this technology has been available for over 15 years, it is just now becoming popular. Conversion to an embedded capacitance PCB is simple, because no new artwork is required. Only the layer stackup is changed. It is therefore very easy to try out this technology by having two sets of prototype boards made, one the standard way and the second using the Buried Capacitance® layers, and then the performance of the two can be directly compared. The most common layer stackup is shown in Fig. 11-19.

Other layer stackups are also possible using the embedded capacitance approach, as discussed in Section 16.4.2. Because of the use of two power-ground-plane pairs, a four-layer board becomes a six-layer board, and a six-layer board becomes an eight-layer board when embedded capacitance is used. Although the ZBC-2000® laminate is slightly more expensive that standard

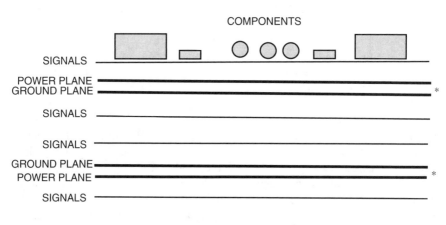

**FIGURE 11-19.** Stackup of a typical embedded capacitance printed circuit board.

---

* ZBC-2000 is a registered trademark of Sanmina-SCI, San Jose, CA.

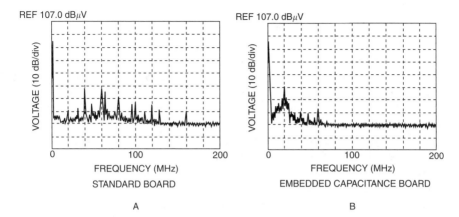

**FIGURE 11-20.** $V_{cc}$-to-ground noise voltage from 0 to 200 MHz on (A) a standard printed circuit board, and (B) on an embedded capacitance printed circuit board. (© 1991 UP Media Group)

PCB laminate, the primary cost increase comes from having to print the two extra PCB layers.

Figure 11-20 shows the $V_{cc}$-to-ground noise voltage from 1 to 200 MHz on an embedded capacitance PCB using the stackup of Fig. 11-19 and on an otherwise identical standard (nonembedded capacitance) board that has only one power and one ground plane (Sisler, 1991). Comparing the two plots in Fig. 11-20 shows that the embedded capacitance board has much reduced noise at all frequencies above 30 MHz and virtually no measurable noise above 60 MHz. For the data shown in Fig. 11-20, the standard board had 135 decoupling capacitors on it, whereas the embedded capacitance board had no decoupling capacitors on it. Both boards contained the same bulk decoupling capacitors. (see Section 11.8)

The decoupling, however, was worse on the embedded capacitance board at frequencies below 30 MHz, in particular at 20 MHz. This is the result of the embedded capacitance board not having sufficient total capacitance to be effective at these low frequencies. This result is similar to that shown in Fig. 11-12, where increasing the number of capacitors reduced the high-frequency impedance of the power distribution network but did not reduce the low-frequency impedance. To reduce the low-frequency impedance, additional capacitance is required as demonstrated by comparing Figs. 11-12 and 11-13.

For the case shown in Fig. 11-20B, replacing four of the original 135 decoupling capacitors (thereby increasing the total capacitance) reduced the low-frequency noise voltage on the embedded capacitance board to what it was on the standard board.

Subsequent measurements, by others, on embedded capacitance printed circuit boards have shown them to provide effective decoupling at frequencies as high as 5 GHz, with few or no discrete decoupling capacitors on the board.

I am not aware of anyone who has made measurements above 5 GHz, but there is no reason to suspect that embedded capacitance approach would not be effective at frequencies even greater than this.

Embedded capacitance boards have other advantages also. The power and ground plane inductances are significantly reduced because of the close spacing of the planes, as well as because of the use of multiple power and ground planes. Resonance problems are also minimized because the embedded capacitance power/ground plane sandwich tends to be a low-Q structure, as a result of its low $L/C$ ratio (see Eq. 10-3). In addition, by removing 90% or more of the decoupling capacitors and their associated vias, board routing is greatly simplified, and in many cases the board can be reduced in size. Also, in some applications, the removal of most of the discrete decoupling capacitors can eliminate the need for surface mounting components on both sides of the board.

Embedded capacitance boards tend to have similar power-to-ground noise voltages at all locations on the board, whereas standard boards, which use discrete decoupling capacitors, have power-to-ground noise voltages that are dependent on the location where the measurement is taken.

I am convinced that in the future, some form of embedded capacitance printed circuit boards will become the standard in the industry, because they provide a simple, effective way to provide high-frequency power supply decoupling using a minimum number of discrete capacitors. This transition to the common use of embedded capacitance may be accelerated when the Sanmina and Unisys patents expire in a few years. Theoretically, distributed capacitance decoupling becomes more effective as the frequency is increased, whereas discrete capacitor decoupling networks become less effective as frequency is increased.

### 11.4.7   Power Supply Isolation

The only true solution to the high-frequency decoupling problem is the use of a multiplicity of capacitors either discrete or embedded into the PCB.

Another approach, although it does not solve the fundamental decoupling problem, can and has been used to minimize the detrimental side effects of poor decoupling. The objective is to prevent the noisy power plane, which is the result of ineffective decoupling, from contaminating the rest of the PCB.

This objective can be accomplished by isolating, or segmenting, the noisy portion of the power plane from the rest of the PCB's power plane, and by feeding power to the isolated plane through a π-filter, as shown in Fig. 11-21. Note that only the power plane is split, not the ground plane. This approach does not reduce the noise or improve the decoupling effectiveness on the isolated power plane, but it prevents this noise from contaminating the main power plane.

This approach is most effective when only a small number of circuits operate at high frequencies. The high-frequency circuit, or circuits, are then isolated

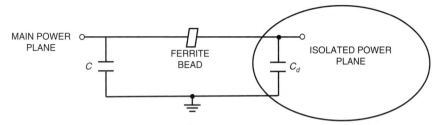

**FIGURE 11-21.** Filter used to feed power to an isolated power plane.

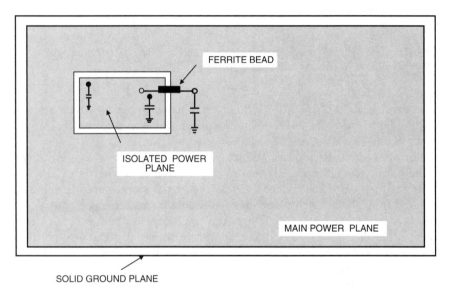

**FIGURE 11-22.** Example of an isolated power plane.

from the remainder of the circuits and powered from an isolated power plane as shown in Fig. 11-22. For example, if a microprocessor is the only chip operating at a high clock frequency, then the microprocessor and its clock oscillator might be powered from the isolated power plane. Another example would be to segregate the clock oscillator and clock drivers from the remainder of the circuitry and power them from an isolated power plane. Multiple isolated power planes could also be used for different circuits.

Figure 11-23 shows the $V_{cc}$-to-ground noise voltage, from 20 to 120 MHz, on a PCB with an isolated plane powering the microprocessor and its oscillator circuit. Figure 11-23A shows the $V_{cc}$-to-ground noise voltage on the isolated plane that powers the microprocessor, and Fig. 11-23B shows the $V_{cc}$-to-ground noise voltage on the main power plane. In Fig. 11-23A, we observe a large amount of clock noise resulting from ineffective decoupling of the microprocessor. Comparing Figs. 11-23A and 11-23B, we observe that the clock

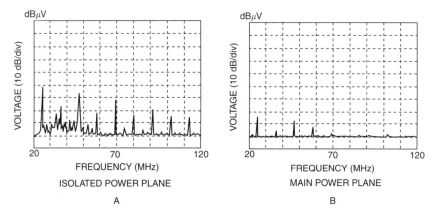

**FIGURE 11-23.** $V_{cc}$-to-ground noise voltage from 20 to 120 MHz on (*A*) an isolated power plane, and (*B*) on the main power plane.

noise on the main $V_{cc}$ plane is virtually eliminated above 60 MHz, and it is much-reduced below 60 MHz when compared with the noise on the isolated power plane.

Note that when this approach is used, no signal traces can cross over the split in the power plane, on an adjacent layer, as this would interrupt the return current path as discussed in Section 16.3. All signal traces connecting to the ICs powered from the isolated plane must be routed on layers adjacent to a solid plane (power or ground). Therefore, this approach has routing restrictions, whereas the distributed capacitance approach discussed previously performs better and actually simplifies routing as the result of the eliminated capacitors and their associated vias.

The isolated power plane approach is sometimes implemented by printing the isolated plane on one of the signal layers, instead of splitting the actual power plane. In this case, the isolated power plane is often referred to as a power island or "power puddle." The results are the same with either the split power plane or the power island approach.

## 11.5   THE EFFECT OF DECOUPLING ON RADIATED EMISSIONS

The transient power supply currents drawn when digital logic ICs switch produce radiated emissions by the following three different mechanisms:

1. The transient power supply currents flowing around the loop between the IC and the decoupling capacitor will cause the loop to radiate.
2. The transient power supply currents flowing through the impedance of the ground will produce a ground noise voltage that will excite any cables connected to the system, which causes them to radiate.

3. The $V_{cc}$-to-ground noise voltage couples, via the power bus, to other ICs
(e.g., I/O drivers) and eventually ends up on power and signal cables,
which causes them to radiate.

Effective decoupling can minimize all three of the above radiation mechanisms.

Figure 11-24A shows an IC and its decoupling capacitor $C_1$. The radiation
from the decoupling loop is proportional to the loop area times the magnitude
of the current in the loop (see Eq. 12-2). The closer the capacitor is to the IC
and the smaller the loop area, the less the radiation.

In addition, the transient power supply current also produces a voltage drop
in the ground, which is shown as $V_g$ in Fig. 11-24A. This ground noise voltage
will excite any cables connected to the system, which causes common-mode
radiation from the cables as shown in Fig. 12-2 and discussed in Section 12.2.

Figure 11-24B shows the same IC as in Fig. 11-24A, but this time it has two
identical decoupling capacitors $C_1$ and $C_2$; one capacitor is located on each side
of the IC. Because the two capacitors are equal in value and are located
symmetrically with respect to the IC, they will each provide half of the current
required by the IC when it switches. Therefore, the radiation from the loop that
contains capacitor $C_1$ will be cut in half, or reduced by 6 dB. The current in the
loop that contains $C_2$ will contain the other half of the transient current, and it
will also produce an equal amount of radiation.

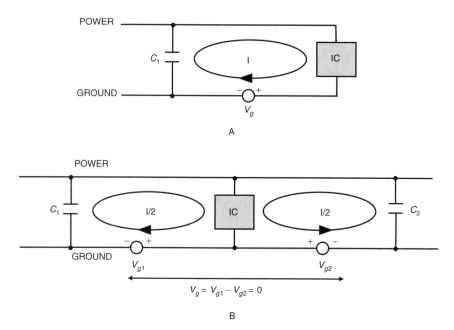

**FIGURE 11-24.** Current in the IC to decoupling capacitor loop, for the case of (A) one
decoupling capacitor, and (B) two identical decoupling capacitors.

Notice, however, that the current in the loop associated with $C_1$ is clockwise, whereas the current in the loop associated with $C_2$ is counterclockwise. Therefore, the radiation from these two loops do not add; rather, the radiation subtracts, and the two loops tend to cancel each other. Even if the cancellation is not perfect, the emission is still significantly reduced. Let us assume that the cancellation is only 12 dB, then the total reduction of the loop radiation will be 18 dB (6 dB from the reduced current in the loops and 12 dB from the canceling loops).

Notice also in Fig. 11-24B that two ground noise voltages are produced by the transient power supply currents, but they are of opposite polarity, and the net voltage across the section of ground plane that contains the IC is zero. Therefore, the common-mode radiation from any cables connected to the board will also be significantly reduced.

Therefore, using two decoupling capacitors on an IC significantly reduces the radiation, both common-mode and differential-mode, which results from the transient power supply currents. To decrease the radiation an equal amount again would require the addition of two more capacitors, which would bring the total to four. To then again reduce the radiation an equal amount would require the addition of four more capacitors, which would bring the total to eight capacitors, and so on. Each time, the number of capacitors must be doubled.

Where then is the largest return on investment? Clearly, the return comes by adding the second capacitor. This result leads to the question, why ever use less than two capacitors to decouple an IC, because it provides so much improvement for so little additional cost?

The above discussion leads to the following recommendations with respect to decoupling of digital ICs:

- Use a minimum of two capacitors on a DIP. Locate the capacitors on opposite ends of the package.
- Use a minimum of four capacitors on a small square IC package (e.g., a quad flat pack), one located on each of the four sides.
- Many large ICs that draw large transient power supply currents (e.g., microprocessors) will usually require many more decoupling capacitors (in some cases the number of capacitors is in the hundreds). Analyze the decoupling requirements of these ICs individually.

In addition, effective decoupling as discussed above and in Section 11.4 will minimize the magnitude of the $V_{cc}$-to-ground noise voltage, which reduces the contamination of the $V_{cc}$ bus, and significantly reduces the radiation caused by coupling this noise to other ICs on the circuit board.

## 11.6   DECOUPLING CAPACITOR TYPE AND VALUE

Decoupling capacitors must supply high-frequency currents; therefore, they should be low-inductance, high-frequency capacitors. For this reason, multi-layer ceramic capacitors are preferred.

For effective decoupling at low frequencies (below the decoupling network resonance), the total decoupling capacitance must satisfy the following two requirements: (1) the total capacitance must be large enough to have an impedance below the target impedance at the lowest frequency of interest (see Section 11.4.5) and (2) the capacitance must be large enough to supply the transient current required by the IC when it switches, while keeping the supply voltage within the required tolerance. The minimum value of capacitance to satisfy this latter requirement is

$$C \geq \frac{dIdt}{dV},$$

(11-12)

where $dV$ is the allowable transient voltage drop of the supply voltage caused by the current transient of $dI$ occurring in time $dt$. For example, if an IC requires a transient current of 500 mA for 2 ns, and one wishes to limit the power supply voltage transient to less than 0.1 V, then the capacitor must have a value of at least 0.01 μF.

As previously stated, only the low-frequency effectiveness of the decoupling network is affected by the value of capacitance, and the high-frequency performance is determined solely by how low the inductance can be driven. The inductance is determined by how many capacitors are used and the inductance in series with each capacitor. The internal inductance of a multilayer SMT capacitor is primarily determined by its package size. A 1206 SMT package capacitor will have about twice the inductance of a 0602 capacitor. The 1206 capacitor will have an inductance of about 1.2 nH, whereas the 0603 capacitor will have an inductance of about 0.6 nH.

Therefore, a good approach to choosing a decoupling capacitor is to use the smallest package size practical for the application, and then to use the largest value capacitor readily available in that package size. Choosing a smaller value capacitor provides no improvement in the high-frequency performance, but it will degrade the low-frequency effectiveness.

## 11.7 DECOUPLING CAPACITOR PLACEMENT AND MOUNTING

The decoupling capacitor must be placed as close to the IC as possible to keep the loop area, and inductance, as small as possible.

The question is often asked, should the decoupling capacitor be connected directly to the IC power and ground pins with traces or should the decoupling capacitor be directly connected to the power and ground plane? The answer is, it depends. Both configurations can be effective if the traces are kept short thus minimizing the inductance. In addition to the trace length, the length of any vias must also be considered in determining the inductance. My preference, on a purely digital board, is usually to connect both the capacitors and the ICs directly to the power and ground planes.

However, connecting the capacitor directly to the power and ground pins of the IC can save precious space on a crowded PCB because of the reduced number of vias required. As long as the loop area is kept small, connecting the capacitor directly to the power and ground pins of the IC is a perfectly acceptable approach. Therefore, there is no simple answer to the question: Should the decoupling capacitor be directly connected to the IC or to the power and ground plane? Each design has to be assessed individually to determine which will be the best approach.

Minimizing the inductance means minimizing the loop area between the capacitor and the IC, which is crucially important. The capacitor layout providing the minimum loop area will perform the best.

Figure 11-25 shows the approximate inductance between a 0805 SMT decoupling capacitor's mounting pads and a power-ground plane pair, which is located in the center of the board stackup. This does not include the inductance of the capacitor itself. It is just the inductance of the mounting

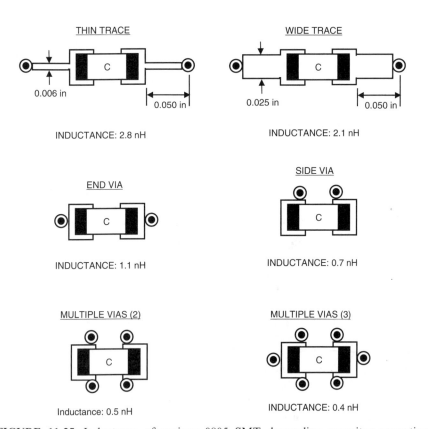

**FIGURE 11-25.** Inductance of various 0805 SMT decoupling capacitor mounting configurations.

pads, the traces, and the vias. Notice that the inductance can vary from almost 3 nH to less than 0.5 nH, based solely on the method of mounting the capacitor.

Multiple vias will reduce the inductance from the mounting pad to the power-ground-plane pair (see Section 10.6.4). However, the vias take up a lot of board space. Placing vias that carry opposite direction currents close together will also reduce the inductance as a result of mutual coupling. This is why the side via configuration's shown in Fig. 11-25 have less inductance that the end via configurations.

When one or two decoupling capacitors are used on an IC, their placement is critical. However, when a large number of capacitors are used, their exact placement becomes less important than when only one or two capacitors are used. Just spread them out around the IC, and try to place them symmetrically (or evenly) with respect to the IC.

Most designers, including the author, usually decouple on an individual IC basis. That is, the designer will determine the appropriate number and value of capacitors required for each individual IC. Another approach, however, is to decouple the PCB *globally*, by evenly distributing a large number of equal value capacitors across the entire board and connecting them directly to the power and ground planes, irrespective of the location of the individual ICs. This global approach works best when used on a circuit board that contains ICs all having similar transient current requirements, whereas the individual IC decoupling approach often works best when ICs having widely varying transient current requirements are on the board.

Other decoupling capacitor placement considerations are as follows:

- Is the PCB single or double sided surface mount?
- Is the IC package a peripheral pin package of a grid array?
- How many power and ground planes does the PCB have, what is the stackup, and what is the spacing between the planes?
- How many power and ground pins does the IC have?

## 11.8   BULK DECOUPLING CAPACITORS

When the logic switches, the IC decoupling capacitors will give up some of their charge. Before the logic switches a second time, the IC decoupling capacitors must be recharged. The recharging current occurs at a considerably lower frequency than in the case of the individual IC decoupling capacitors, and it is supplied by a bulk decoupling capacitor, or capacitors, located on the printed circuit board. The individual IC decoupling capacitors must work at the switching speed of the logic—basically that associated with the rise and fall times. The bulk decoupling capacitors usually have at least a half clock cycle to recharge the individual decoupling capacitors; therefore, they operate at twice the clock frequency or less.

The value of the bulk capacitor is not critical, but it should be greater than the sum of all the values of the IC decoupling capacitors that it serves. One bulk decoupling capacitor should be located where power comes onto the board. Other bulk capacitors should be located strategically around the board. It is better to err on the side of too much bulk decoupling capacitance than on the side of too little.

The bulk decoupling capacitors will typically have values in the 5- to 100-µF range (10 µF being a typical value) and should have a small equivalent series inductance. In the past, tantalum electrolytic capacitors were common. However, as capacitor technology improved, multilayer ceramic capacitors are becoming more common in this application. Aluminum electrolytic capacitors have inductances an order of magnitude or more higher than tantalum and should not be used.

Because the bulk decoupling capacitors have a capacitance value different from the individual IC decoupling capacitors, an antiresonance impedance spike will be caused by the two different capacitor values. However, because the bulk capacitors have a large value of capacitance, the impedance spike occurs at a much lower frequency and lower amplitude, and usually it is not a problem. A small amount of equivalent series resistance in the bulk capacitor can actually be helpful, because it provides a degree of damping thus reducing the amplitude of any resonant peaks that may be present.

## 11.9   POWER ENTRY FILTERS

External noise may be conducted on to the circuit board, and internal noise may be conducted off of the board on the dc power leads. High-frequency power-supply transient currents should be confined to the digital logic board, and these currents not allowed to flow out onto the dc power-supply wiring. Therefore, a power entry filter should be standard design practice. Figure 11-26 shows the circuit of a typical power entry filter. It should contain both a

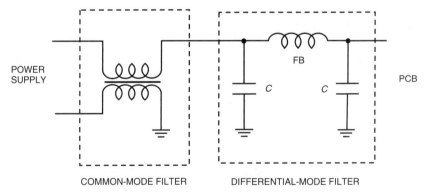

FIGURE 11-26. A typical power entry filter.

differential-mode and a common-mode section. The differential-mode filter is usually a π-filter with a ferrite bead or inductor as the series element.

Typical values for the filter elements are 0.1 to 0.01 μF for the capacitors and a ferrite bead whose impedance is between 50 and 100 Ω across the frequency range of interest. In this application, it is important to avoid saturation of the ferrite bead by the dc current. If an inductor is used, then typical values would be from 0.5 to 5 μH.

The common-mode filter element is usually a common-mode choke located on the printed circuit board or a ferrite core on the dc power cable.

## SUMMARY

- Decoupling capacitors are needed to supply, through a low-inductance path, some or all of the transient power supply current required when an IC logic gate switches.
- Decoupling capacitors are needed to short out, or at least reduce, the noise injected back into the power ground system.
- Decoupling is not the process of placing a capacitor adjacent to an IC to supply the transient switching current; rather it is the process of placing an L–C network adjacent to the IC to supply the transient switching current.
- The value of the decoupling capacitor(s) is important for the low-frequency decoupling effectiveness.
- The value of the decoupling capacitor(s) is not important at high frequencies.
- At high frequencies, the most important criteria is to reduce the inductance in series with the decoupling capacitors.
- Effective high-frequency decoupling requires the use of a large number of capacitors.
- In most cases, the use of single value decoupling capacitors perform better than when multiple value capacitors are used.
- For *optimum* high-frequency decoupling, discrete capacitors should not be used at all; rather, a distributed capacitor PCB structure should be used.
- The number one rule of decoupling is to have the current flow through the smallest loop possible.

## PROBLEMS

11.1   What will be the magnitude of the transient load current for a CMOS device with a 500-ps rise time, operating from a 3.3-V supply when feeding two CMOS gates? Assume that each CMOS gate has an input capacitance of 10 pF.

11.2    Name three mechanisms by which ineffective decoupling can lead to radiated emissions.

11.3    What are the three sources of inductance in series with a decoupling capacitor?

11.4    What two factors determine the upper frequency at which the decoupling will be effective?

11.5    What determines the lowest frequency at which the decoupling will be effective?

11.6    What two criteria must be satisfied when determining the minimum value for the total decoupling capacitance for an IC?

11.7    What should be the minimum number of decoupling capacitors used on an IC? Why?

11.8    What are the two requirements for effectively paralleling multiple capacitors?

11.9    Name three advantages of using multiple equal value capacitors for decoupling.

11.10   Which of the three answers to Problem 11.9 do NOT apply if unequal value decoupling capacitors are used?

11.11   What problem is caused when different value decoupling capacitors are used?

11.12   It is desired to provide effective decoupling for a digital IC with a 3-ns switching speed at all frequencies above 1 MHz. Using a low-frequency target impedance of 100 m$\Omega$ and assuming that each capacitor has 5 nH of inductance in series with it.
a. How many equal-value discrete capacitors must be used?
b. What is the minimum value for each of the decoupling capacitors?

11.13   A large microprocessor draws a total transient current of 10 A from a 3.3-V supply. The logic has a rise/fall time of 1 ns. It is desirable to limit the $V_{cc}$-to-ground noise voltage peaks to 250 mV, and each decoupling capacitor has 5 nH of inductance in series with it. The decoupling will be done with a multiplicity of equal value capacitors, and should be effective at all frequencies above 20 MHz.
a. Draw a plot of the target impedance versus frequency.
b. What is the minimum number of decoupling capacitors required?
c. What is the minimum value for each of the individual decoupling capacitors?
d. Could larger value capacitors be used just as effectively?

11.14   An IC draws a 1-A transient power supply current in 1 ns. What minimum value decoupling capacitance would be required to prevent the supply voltage from decreasing more than 0.1 V?

11.15   A $10 \times 12$-in. embedded capacitance FR-4 epoxy glass PCB has a power-to-ground plane capacitance of 500 pF per square inch. The logic circuits have rise times of 300 ps. What will be the effective interplane decoupling capacitance? Assume that the charge from the decoupling capacitance must arrive at the IC during the period of the rise time.

# REFERENCES

Archambeault, B. "Eliminating the MYTHS About Printed Circuit Board Power/ Ground Plane Decoupling." *ITEM*, 2001.

Archambeault, B. *PCB Design for Real-World EMI Control*. Boston, MA, Kluwer Academic Publishers, 2002.

Howard, J. R. and Lucas, G. L., Inventors; Zycon Corp., assignee. Capacitor laminate for use in capacitive printed circuit boards and methods of manufacture. U.S. patent 5,079,069, January 7, 1992.

Janssen, L. P. "Reducing The Emission of Multi-Layer PCBs by Removing the Supply Plane." *1999 Zurich EMC Symposium,* Zurich, Switzerland, 1999.

Jordan, E. C. *Reference Data For Engineers*. Indianapolis, IN, Howard W. Sams, 1985.

Leferink, F. B. J. and van Etten, W. C. "Reduction of Radiated Electromagnetic Fields by Removing Power Planes." *2004 IEEE International Symposium on Electromagnetic Compatibility*, Santa Clara, CA, August 9–13, 2004.

Paul, Clayton R. "Effectiveness of Multiple Decoupling Capacitors." *IEEE Transactions on Electromagnetic Compatibility*, May 1992.

Radu, S. , DuBroff, R. E. , Hubing, T. H., and Van Doren, T. P. "Designing Power Bus Decoupling for CMOS Devices." *1998 IEEE EMC Symposium Record*, Denver, CO, August 1998.

Sisler, J., inventor; Unisys Corp., assignee, Method of making multilayer printed circuit board, U.S. patent 5,010,641, April 30, 1991.

Sisler, J. "Eliminating Capacitors From Multilayer PCBs." *Printed Circuit Design*, July 1991.

# FURTHER READING

Hubing, T. H., et al. "Power-Bus Decoupling With Embedded Capacitance in Printed Circuit Board Design." *IEEE Transactions on Electromagnetic Compatibility*, February, 2003.

Wang, T. "Characteristics of Buried Capacitance™." *EMC Test and Design*, February 1993.

# 12 Digital Circuit Radiation

In today's regulatory environment, electromagnetic compatibility (EMC) engineering plays an important roll in bringing electronic products to market. Often the functional performance of a product is not the primary problem in meeting product introduction schedules; passing the required EMC tests are.

Controlling the emission from digital systems cost effectively can be as complicated and as difficult as designing complex digital logic. Emission control should be treated as a design problem from the start, and it should receive the necessary engineering resources throughout the design process.

This chapter models the radiated emission and outlines the parameters on which radiation depends. It also provides a method for predicting the radiated emission as a function of the electrical characteristics of the signals and the physical properties of the system. Knowing the parameters that affect radiation helps to develop techniques to minimize it.

Radiation from digital electronics can occur as either *differential mode* or *common mode*. Differential-mode radiation is the result of the *normal operation* of the circuit and results from current flowing around loops formed by the conductors of the circuit, as shown in Fig. 12-1. These loops act as small loop antennas that radiate predominately magnetic fields. Although these signal loops are necessary for circuit operation, their size and area must be controlled during the design process to minimize the radiation.

Common-mode radiation, however, is the result of *parasitics in the circuit* and results from undesired voltage drops in the conductors. The differential-mode current that flows through the ground impedance produces a voltage drop in the digital logic ground system. When cables are then connected to the system, they are driven by this common-mode ground potential, forming antennas, which radiate predominately electric fields as shown in Fig. 12-2. Because these parasitic impedances are not intentionally designed into the system or shown in the documentation, common-mode radiation is often harder to understand and control. During the design, steps must be taken to provide methods for handling the common-mode emission problem.

*Electromagnetic Compatibility Engineering*, by Henry W. Ott
Copyright © 2009 John Wiley & Sons, Inc.

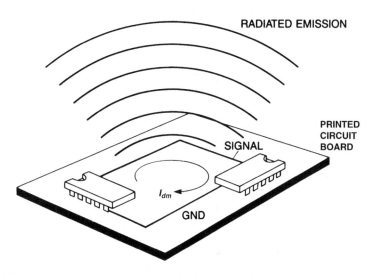

**FIGURE 12-1.** Differential-mode radiation from printed circuit board (PCB).

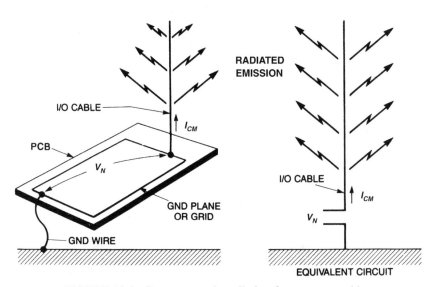

**FIGURE 12-2.** Common-mode radiation from system cables.

## 12.1 DIFFERENTIAL-MODE RADIATION

Differential-mode radiation can be modeled as a *small loop* antenna.* For a small loop of area $A$, carrying a current $I_{dm}$, the magnitude of the electric field $E$

---

* A small loop is one whose circumference is less than one quarter of a wavelength.

measured in *free space* at a distance *r*, in the far field, is equal to (Kraus and Marhefka, 2002, p. 199, Eq. 8)

$$E = 131.6 \times 10^{-16} (f^2 A I_{dm}) \left(\frac{1}{r}\right) \sin\theta, \qquad (12\text{-}1)$$

where $E$ is in volts/meter, $f$ is in hertz, $A$ is in square meters, $I_{dm}$ is in amperes, $r$ is in meters, and $\theta$ is the angle between the observation point and a perpendicular to the plane of the loop.

A small loop, where the perimeter is less than one quarter wavelength, is one in which the current is everywhere in phase. For larger loops, the current is not all in phase, and therefore, some current may subtract from, rather than add to, the overall emission.

As shown in Fig. 12-3, the free-space antenna pattern for a small loop is a torus (doughnut shape). The maximum radiation is from the sides of the loop and occurs in the plane of the loop. Radiation nulls occur in the direction normal to the plane of the loop. Because the electric field is polarized in the plane of the loop, the maximum electric field will be detected by a receive antenna, which is polarized in the same direction, as shown in Fig. 12-3.

As the loop perimeter increases beyond one quarter wavelength, the radiation pattern of Fig. 12-3 no longer applies. For a loop whose perimeter equals a wavelength, the radiation pattern rotates by 90° so that the maximum radiation

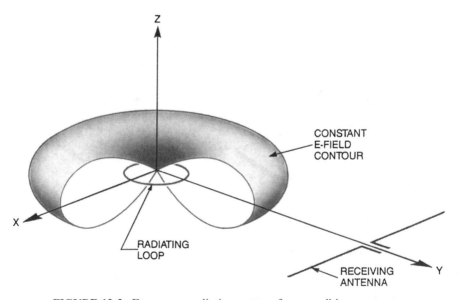

**FIGURE 12-3.** Free-space radiation pattern for a small loop antenna.

occurs normal to the plane of the loop. Therefore, the radiation nulls of the small loop become the maxima for a large loop.

Although Eq. 12-1 was derived for a circular loop, it can be applied to any planar loop because, for small loops, the magnitude and pattern of the radiation is insensitive to the shape of the loop and depends only on its area (Kraus and Marhefka, 2002, p. 212). *All small loops having equal area radiate the same regardless of their shape.*

The first term in Eq. 12-1 is a constant that accounts for the properties of the media—in this case free space. The second term defines the characteristics of the radiation source, which is the loop. The third term represents the decay of the field as it propagates away from the source. The last term accounts for the angular orientation of the measuring point with respect to a perpendicular to the plane of the loop, the angle off of the z-axis in Fig. 12-3.

Equation 12-1 is for a small loop located in free space, with no reflecting surfaces nearby. Most EMC measurements of radiation from electronic products, however, are made in an open area, over a ground plane, not in free space. The ground provides a reflective surface that must be accounted for. The reflective ground plane can increase the measured emission as much as 6 dB (or a factor of two). To account for this reflection, Eq. 12-1 must be multiplied by a factor of two. Correcting for the ground reflection and assuming that the observation is made at a distance $r$ in the plane of the loop ($\theta = 90°$), one can rewrite Eq. 12-1 for measurement in an *open area* as

$$E = 263 \times 10^{-16} \left( f^2 A I_{dm} \right) \left( \frac{1}{r} \right). \tag{12-2}$$

Equation 12-2 shows that the radiation is proportional to the current $I$, the loop area $A$, and the *square* of the frequency $f$. The frequency-squared term provides job insurance for future EMC engineers.

For a *3-m measuring distance*, Eq. 12-2 can be rewritten as

$$E = 87.6 \times 10^{-16} \left( f^2 A I_{dm} \right). \tag{12-3}$$

Therefore, differential-mode (loop) radiation can be controlled by

1. Reducing the magnitude of the current
2. Reducing the frequency or harmonic content of the current
3. Reducing the loop area

For a current waveform other than a sine wave, the Fourier series of the current waveshape must be determined before substitution into Eq. 12-3.

### 12.1.1  Loop Area

In the design of digital systems, the primary way to control differential-mode radiation is to minimize the area enclosed by current flow. This means placing signal leads and their associated ground-return leads close together, which is especially important for clocks, backplane wiring, and interconnecting cables.

For example, if 25 mA of current at a frequency of 30 MHz is flowing around a 10-cm$^2$ loop, the electric field strength measured at a distance of 3 m will be 197 μV/m. This field strength is almost twice the allowable limit for a commercial Class B product.

The maximum loop area that will not exceed a specified emission level can be determined by solving Eq. 12-2 for the loop area $A$. Thus,

$$A = \frac{380Er}{f^2 I_{dm}}, \tag{12-4}$$

where $E$ is the radiation limit in microvolts per meter, $r$ is the distance between the loop and measuring antenna in meters, $f$ is the frequency in MHz, $I$ is the current in milliamps, and $A$ is the loop area in square centimeters.

For example, for 25 mA of current at 30 MHz, the maximum loop area that will limit the radiation to 100 μV/m at a distance of 3 m [the Federal Communications Commission (FCC)/Interational Special Committee On Radio Interference (CISPR) Class B limit] is 5 cm$^2$.

If we are to design systems that meet legal requirements for radiated emission, loop areas must be kept very small. Under such conditions Eq. 12-2 is applicable for predicting the radiated emission.

### 12.1.2  Loop Current

If the current in the loop is known, it is easy to use Eq. 12-2 to predict the radiation. However the current is seldom known accurately; therefore, it must be modeled, measured, or estimated. The current depends on the source impedance of the circuit that drives the loop, as well as the load impedance of the circuit that terminates the loop.

Loop current can be measured with a wide-band current probe. This measurement may require cutting the signal trace and adding a wire in series with the trace, just long enough to clip the current probe to.

### 12.1.3  Fourier Series

Because digital circuits use square waves, the harmonic content of the current must be known before calculating the emission. For a symmetrical square wave (actually a trapezoidal wave, because the rise and fall times are finite) as shown in Fig. 12-4, the current in the $n$th harmonic is given by Jordan (1985, p. 7-11) as

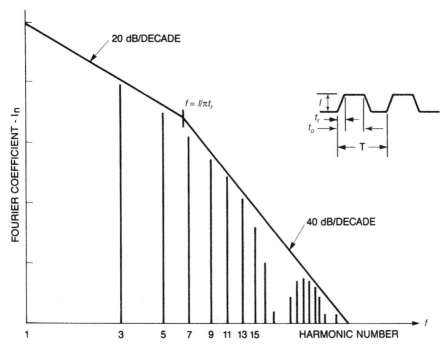

**FIGURE 12-4.** Envelope of Fourier spectrum of a 50% duty cycle trapezoidal wave.

$$I_n = 2Id\left[\frac{\sin(n\pi d)}{n\pi d}\right]\left[\frac{\sin\left(\frac{n\pi t_r}{T}\right)}{\frac{n\pi t_r}{T}}\right], \tag{12-5}$$

where $I$ is the amplitude of the square wave as shown in Fig. 12-4, $t_r$ is the rise time, $d$ is the duty cycle $[(t_r + t_o)/T]$, $T$ is the period, and $n$ is an integer from one to infinity. The units of $I_n$ will be the same as the units used for $I$, because the rest of the equation is dimensionless. Equation 12-5 assumes that the rise time equals the fall time. If it does not, the smaller of the two should be used for a worst-case result.

For a 50% duty cycle ($d = 0.5$), the first harmonic (fundamental) has an amplitude $I_1 = 0.64\ I$, and only odd harmonics are present. For the normal case where the rise time $t_r$ is much less than the period $T$, Fig. 12-4 shows the envelope of the harmonics for a symmetrical square wave. The harmonics fall off with frequency at a 20 dB per decade rate up to the frequency of $1/\pi\ t_r$, beyond which they fall off at a rate of 40 dB per decade. This shows that as the rise time increases (slows down), the current in the higher harmonics decreases.

The differential-mode radiated emission can be calculated by first determining the current content of each harmonic from Eq. 12-5, and then substituting this current and the respective frequency into Eq. 12-2. This calculation is then repeated for each harmonic frequency.

The frequency-squared term in Eq. 12-2 represents an increase in emission with frequency of 40 dB per decade. The result of combining Eq. 12-2 and Eq. 12-5 is that the radiated emission increases 20 dB per decade for frequencies less than $1/\pi\, t_r$, and it remains constant above this frequency as shown in Fig. 12-5.*

Figure 12-5 clearly illustrates the important effect that rise time has on radiated emission. It is the rise time that determines the breakpoint above which, the differential-mode radiation stops increasing with frequency. To minimize the emission, it is desirable to slow down the rise time of the signal as much as functionally possible.

As an example, the radiated emission measured at 3 m from a 6-MHz clock with a 4-ns rise time flowing around a loop with an area of 10 cm² (1.5 in²) is shown in Fig. 12-6. The loop is being driven by a square wave of current with a peak amplitude of 35 mA.

### 12.1.4 Radiated Emission Envelope

The radiated emission envelope shown in Fig. 12-5 can easily be estimated if the frequency, peak current, and rise time of a signal as well as the loop area are known. Because the shape of the radiated emission envelope is known, only the amplitude of the radiation at the fundamental frequency needs to be calculated. The Fourier coefficient for the fundamental frequency (assuming a 50% duty cycle square wave) is $I_1 = 0.64\, I$, where $I$ is the amplitude of the square wave current.

The amplitude of the radiation at the fundamental frequency can be calculated by substituting 0.64 times the amplitude of the square wave current into Eq. 12-2.

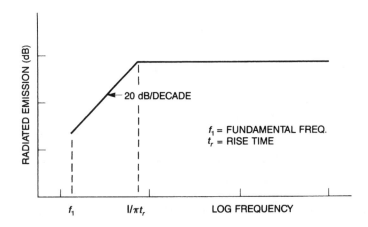

**FIGURE 12-5.** Differential-mode radiated emission envelope versus frequency.

---

*Actually, the high-frequency radiation will fall off above the frequency where the circumference of the loop becomes a quarter of a wavelength, because the assumption in Eq. 12-2 of a small loop is then no longer valid. This will occur at a frequency $f_{MHz} = 75/C_{meters}$, where $C$ is the circumference of the loop.

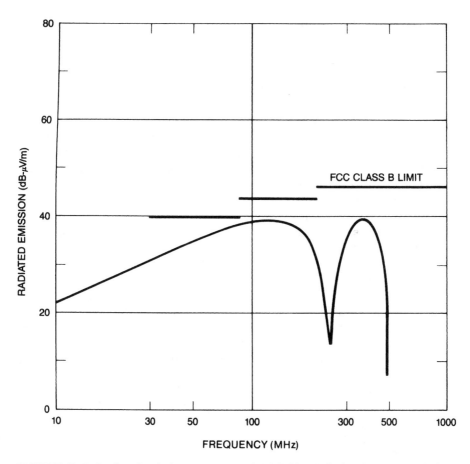

**FIGURE 12-6.** Radiated emission spectrum and FCC Class B limit. The spectrum is for a 6-MHz, 4-ns rise time, 35-mA clock signal in a 10-cm$^2$ loop.

The radiated emission at this frequency, in dB $\mu$V/m can be plotted on semi-log paper. A line increasing by 20 dB per decade is then drawn from this point, up to the frequency of $1/\pi t_r$, beyond which a horizontal line is drawn. This then represents the envelope of the differential-mode radiated emission.

## 12.2   CONTROLLING DIFFERENTIAL-MODE RADIATION

### 12.2.1   Board Layout

The place to start controlling differential-mode radiated emission is with the layout of the printed circuit board. To be cost effective, consideration must be given to emission control when the board is initially being designed.

When laying out a printed circuit board (PCB) to control emission, one should minimize the loop areas formed by the signal traces. Trying to control the area of the loops formed by *all* the signal and transient power supply currents can be a formidable job. Fortunately, it is not necessary to handle each loop individually. The most critical loops should be individually analyzed; however, most other noncritical loops (which are most of the loops) can be controlled by just using good PCB layout practices.

The most critical loops are those that operate at the *highest frequency* and where the *signal is periodic*, that is, it has the same waveshape all the time.

Why periodicity is so important can be understood by considering the following. A loop of a certain size (area), which carries a certain amount of current, can only radiate so much energy. If all that energy is at one frequency, or a very limited number of frequencies, then the amplitude at each frequency will be high. If, however, the energy is spread out over the frequency spectrum, then the energy at any one frequency will be much lower. Because the clock is usually the highest frequency signal in a system, and is periodic, most of its energy is concentrated in only a few narrow frequency bands that consist of the fundamental plus the lower order odd harmonics. Therefore, the amplitude at each of these frequencies will be large.

Figure 12-7A shows the radiated emission spectrum from a typical digital circuit. Figure 12-7B shows the emission from the same circuit with only the clock signal operating. The maximum emission is about the same in both cases, yet 95% or more of the circuitry was turned off for the case of Fig. 12-7B. In almost all cases, the emission from the clock harmonics exceeds the emission from all the other circuits.

Clock signals should be routed first on a PCB, and every effort should be made to route them in a manner that produces the absolute smallest loop area possible. The length of the clock trace should be minimized, as well as the number of vias used. On a multilayer PCB, clocks should be routed on a layer adjacent to a solid (not split) ground or power plane. The spacing between the clock trace layer and the return plane should be as small as possible.

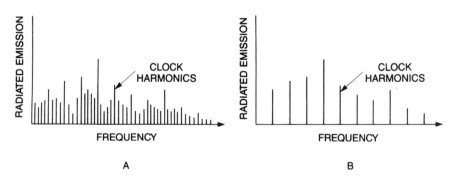

**FIGURE 12-7.** Typical radiated emission spectrum from a digital circuit: (*A*) with all circuits operational; (*B*) with only clock circuit operational.

On double-sided boards, clock traces should have adjacent ground return traces. *Get paranoid about clock routing!*

Clocks, of course, are not the only periodic signals in a system. Many other strobe or control signals are also periodic. In a microprocessor-based system, some critical, periodic signals are clocks (CLK), address latch enable (ALE), row address strobe (RAS), and column address strobe (CAS). Throughout this book, when the term clock is used, it refers not only to a clock signal but any high-frequency periodic signal.

To prevent the clock from coupling to cables that leave the PCB, the clock circuitry should be located away from the input/output (I/O) cables or circuitry.

To minimize crosstalk, clock traces should not be run parallel to data bus or signal leads for long distances. Crosstalk on digital logic boards is covered in more detail by Johnson and Graham (1993), Paul (1985), and Catt (1967).

Address buses and data buses are the second concern after clocks, because they are often terminated and carry large currents, and the radiated emission is proportional to current. Although they are usually not as important as clocks, their loop areas should be examined and kept to a minimum, by routing them adjacent to a plane on multi-layer boards.

On double-sided boards, at least one signal return (ground) trace should be provided adjacent to each group of eight data or address leads. This return trace is best placed adjacent to the least significant bit, because this usually carries the highest-frequency current. Most other miscellaneous signal loop areas can be controlled by using a ground grid or plane, which is also required to minimize the amount of internally generated noise (see Chapters 10 and 11).

Line and bus drivers can also be offenders because they often carry high currents. However, because of the random nature of their signals they generate broadband noise with less energy per unit bandwidth. Bus and line drivers should be located close to the lines they drive. Drivers for cables leaving the PCB should be located close to the connectors. Line driver integrated circuits (ICs) used to drive off-board loads should not also be used to drive other circuits on the board.

Another significant source of radiated emission can be the transient power supply currents required during digital logic switching, because these loops, although usually small, can carry very large currents. These loop areas can be controlled by proper power supply decoupling as discussed in Chapter 11.

The differential-mode emission is proportional to frequency squared and is controlled by minimizing the loop area, and loop area is primarily a function of PCB layout. Over the last 5 or 10 years, clock frequencies have increased dramatically, probably by a factor of 10 or more. Therefore, the differential-mode emission has increased 100-fold, or more. PCB technology, which determines the ability to print smaller loops, however has improved very little in the same time period—possibly by a factor of two.

Therefore, the emission problem has increased 100-fold, and our ability to deal with it by, printing smaller loops, had increased at most twofold. We are clearly losing the battle with respect to our ability to control differential-mode radiated emission. Hence, if we are to control this emission, short of having to shield the PCB, we must come up with some other, possibly unconventional, means to reduce the emission. Two common approaches involve canceling loops and spread-spectrum clocks.

### 12.2.2 Canceling Loops

If we cannot print loops small enough, can we find a simple way to print two loops that cancel each other? Consider the case of a clock trace and its ground return path as shown in Fig. 12-8A. The emission from this loop will be a function of the area of the loop and the current in the loop. If this represents the closest that one can print traces as a result of PCB technology, then we cannot reduce the emission beyond this point, short of shielding the PCB.

Consider, however, the layout shown in Fig. 12-8B. Here, we have a clock trace with two ground return traces, one on each side. Hence we have two loops, each of which has the same area as the loop shown in Fig. 12-8A. If the two return (ground) traces are symmetrically located with respect to the clock trace, then the return current will split between the two paths. Therefore, the lower loop in Fig. 12-8B will have only half as much current as the loop in Fig. 12-8A and will radiate only one half as much, or 6 dB less.

Of course, the other half of the radiation is in the upper loop of Fig. 12-8B, and it will radiate just as much as the lower loop. Notice, however, that the current in the upper loop is counterclockwise, whereas the current in the lower loop is clockwise. Therefore, the radiation from the top loop does not add to the radiation from the bottom loop; rather it cancels it. The cancellation will not be perfect; however it is very good. The layout of Fig. 12-8B will therefore radiate 20+ dB less than that of Fig. 12-8A.

The traces shown in Fig. 12-8 could be considered as being printed on the same layer of the PCB, or on different layers. The latter case would then represent a clock printed on a PCB layer located between two ground plane layers.

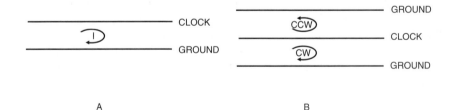

**FIGURE 12-8.** Clock trace (*A*) with a single ground return trace, and (*B*) with two ground return traces.

### 12.2.3 Dithered Clocks

Another approach to reducing the radiated emission, without reducing the loop area, is to spread the emission out in the frequency spectrum, in which case the amplitude of the emission at any one frequency will be reduced. This spreading of the emission can be accomplished by using a dithered, or spread-spectrum clock. Basically what we are doing is frequency modulating the clock.

Clock dithering intentionally varies the clock frequency by a small amount, at a low-frequency rate. This spreads the clock energy out in the frequency domain and therefore lowers the peak amplitude of the emission at any individual frequency (because each small frequency band contains less energy). Using an optimum design, the emission can be reduced about 15 dB by this method. Typical results are in the 10 to 14 dB range. It should be noted that if the fundamental frequency is shifted by 100 kHz, the frequency of the second harmonic shifts by 200 kHz and the third harmonic by 300 kHz, and so on.

The degree of reduction is a function of the dithering waveshape as well as the frequency deviation. The optimum modulating waveform is none of the standard waveforms, rather it is a special waveform as shown in Fig. 12-9 and often is referred to as the "Hershey Kiss." This modulating waveform, however, is patented by Lexmark (Hardin, 1997). For more information on the characteristics of various dithering waveforms, see Hoekstra (1997).

The "Hershey Kiss" is not much different than a triangular wave. The triangular modulating waveform produces only a few dB less reduction than the "Hershey Kiss." Therefore, most clock dithering circuits use a triangular waveform.

The specifications of a typical spread-spectrum clock might read as follows:

- Modulation waveform: Triangular
- Modulation frequency: 35 kHz

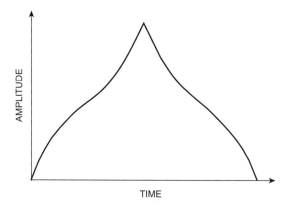

TIME

**FIGURE 12-9.** Optimum modulating waveform for a spread-spectrum clock.

- Frequency deviation: 0.6 %
- Modulation direction: Downward
- Emission reduction: 12 dB

It should be noted that some systems cannot tolerate a spread-spectrum clock, but most can. Many personal computers and printers use spread-spectrum clocking. If absolute real-time accuracy is required, a spread-spectrum clock may be a problem. Most phase locked loops, however, work just fine with a spread-spectrum clock.

Two basic approaches to clock dithering have been used, one is *center spread* and the other is *downward spread*. In the center spread approach, the clock is dithered both above and below its unmodulated frequency. In the downward spread approach, the clock is only spread downward from its normal frequency. The reduction in emission is the same in either case. The advantage of the downward spread is that it is less likely to cause timing margin problems, because the clock frequency is not increased.

Figure 12-10 shows the radiated emission level from the third harmonic of a 60-MHz clock, both with and without clock dithering. In this case, the dithered clock is downward spread and shows a reduction in emission of 13 dB.

Reducing the loop area, or providing canceling loops, only controls the differential-mode emission and has no effect on the common-mode emission that may also be caused by the same clock. Dithering the clock frequency, however, reduces both radiation modes, because it changes the characteristics of the source of the radiation (i.e., the clock). Therefore, a dithered clock approach provides significant leverage by reducing *all* the radiation modes associated with the dithered clock signal.

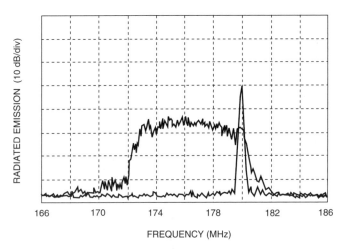

**FIGURE 12-10.** Frequency spectrum of the third harmonic of a 60-MHz clock with and without dithering.

## 12.3   COMMON-MODE RADIATION

Differential-mode radiation can be controlled in the design and layout of the PCB. However, common-mode radiation is harder to control and often determines the overall emission performance of a product.

The most common form of common-mode radiation emanates from the cables of the system. The radiated frequencies are determined by the common-mode potential (usually the ground voltage), as was shown in Fig. 12-2. For the case of common-mode radiation, it matters not what the purpose of the cable is, only that it is connected to the system and somehow referenced to the system ground. The frequencies radiated are not related to the intentional signals in the cable.

Common-mode emission can be modeled as a dipole, or monopole, antenna (the cable) driven by a noise voltage (the ground voltage). For a *short dipole* antenna of length $l$, the magnitude of the electric field strength measured, in the far field, at a distance $r$ from the source is (Balanis, 1982, p. 111, Eq. 4-36a)

$$E = \frac{4\pi \times 10^{-7}(fl I_{cm})\sin\theta}{r}, \qquad (12\text{-}6)$$

where $E$ is in volts/meter, $f$ is in hertz, $I$ is the common-mode current on the cable (antenna) in amperes, $l$ and $r$ are in meters, and $\theta$ is the angle from the axis of the antenna that the observation is made. The maximum field strength will occur perpendicular to the axis of the antenna where $\theta = 90°$.

The free-space antenna pattern for a small dipole antenna is the same as that for a small loop antenna (Fig. 12-3), with the dipole located along the 'Z'-axis. For a monopole over an infinite reference plane, the pattern and amplitude will be the same as the dipole but will only exist in the upper half-plane.

Equation 12-6 is valid for an ideal dipole antenna with a uniform current distribution. For a real dipole antenna, the current goes to zero at the open ends of the wire. For a small antenna, the current distribution is linear across the length of the antenna. Therefore, the average current on the antenna is only one half the maximum current.

In practice, a more uniform current distribution can be achieved if a metal cap is placed at the open end of a dipole or monopole as shown in Fig. 12-11. This increases the capacitance at the ends, which draws more current up to the end and produces a nearly uniform current distribution along the antenna. The resulting antenna is referred to as a capacitor loaded or top-hat antenna (Stutzman and Thiel, 1981, p.81). This configuration is approximated when the antenna (cable) connects to another piece of equipment. The top-hat antenna then closely approximates the ideal uniform current antenna model, and Eq. 12-6 is applicable.

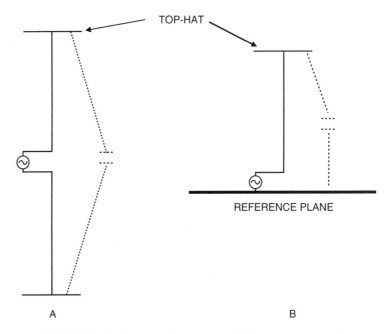

**FIGURE 12-11.** Capacitor loaded dipole (A) and monopole (B).

For the case where the observation is made at a distance $r$ perpendicular to the axis of the antenna ($\theta = 90°$), using MKS units, Eqs.12-6 can be rewritten as

$$E = \frac{12.6 \times 10^{-7}(flI_{cm})}{r}. \tag{12-7}$$

Equation 12-7 shows that the radiation is proportional to the frequency, the length of the antenna, and the magnitude of the common-mode current on the antenna. The primary method of minimizing this radiation is to limit the common-mode current, none of which is required for the normal operation of the circuit.

Therefore common-mode (dipole) radiation can be controlled by the following methods:

1. Reducing the magnitude of the common-mode current
2. Reducing the frequency or harmonic content of the current
3. Reducing the antenna (cable) length

For a current waveform other than a sine wave, the Fourier series of the current must be determined before substitution into Eq. 12-7 (See Section 12.1.3).

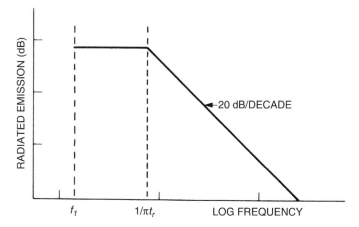

**FIGURE 12-12.** Common-mode radiated emission envelope versus frequency.

The frequency term in Eq. 12-7 represents an increase with frequency of 20 dB/decade. Therefore, the net result of combining Eq. 12-7 with Eq. 12-5 for the Fourier coefficients is that the envelope of the common-mode emission spectrum is flat up to a frequency of $1/\pi t_r$ and it decreases at a rate of 20 dB/decade above that frequency.

Figure 12-12 shows the envelope of the common-mode emission versus frequency. Comparing Figs. 12-5 and 12-12 we can conclude that common-mode emission is more likely to be a problem at low frequencies, and differential-mode emission is more likely to be a problem at high frequencies. For rise times in the 1-to 10-ns range, most of the common-mode radiated emissions will occur in the 30-to 300-MHz frequency band.

Solving Eq. 12-7 for $I$ gives

$$I_{cm} = \frac{0.8Er}{fl}, \qquad (12\text{-}8)$$

where $E$ is the electric field strength in $\mu V/m$, $I_{cm}$ is in microamps, $f$ is in MHz, and $r$ and $l$ are in meters.

Table 12-1 lists the approximate maximum allowed common-mode current in a 1-m-long cable at 50 MHz to not exceed the specified radiated emission regulatory limit.

**TABLE 12-1. Maximum Allowable Common-Mode Current in a 1-m-Long Cable at 50 MHz.**

| Regulation | Limit | Distance | Maximum Common-Mode Current |
|---|---|---|---|
| FCC Class A | 90 $\mu V/m$ | 10 m | 15 $\mu A$ |
| FCC Class B | 100 $\mu V/m$ | 3 m | 5 $\mu A$ |
| MIL-STD 461 | 16 $\mu V/m$ | 1 m | 0.25 $\mu A$ |

The ratio of the differential-mode current to the common-mode current needed to produce the same radiated emission can be determined by setting Eq. 12-2 equal to Eq. 12-7 and solving for the ratio of the currents. Therefore

$$\frac{I_{dm}}{I_{cm}} = \frac{48 \times 10^6 l}{fA}, \tag{12-9a}$$

where $I_{dm}$ and $I_{cm}$ are, respectively, the differential-mode and the common-mode currents needed to produce the same radiated emission. If the cable length $l$ is equal to 1 m, the loop area is equal to 10 cm$^2$ (0.001 m$^2$) and the frequency is 48 MHz, then Eq. 12-9a reduces to

$$\frac{I_{dm}}{I_{cm}} = 1000 \tag{12-9b}$$

From Eq.12-9b we observe that it takes about three orders of magnitude more differential-mode current than common-mode current to produce the same radiated field. In other words, the common-mode radiation mechanism is much more efficient than the differential-mode radiation mechanism, and a common-mode current of a few microamps can cause the same amount of radiation as a few milliamps of differential-mode current.

For a long cable ($l > \lambda/4$), Eqs. 12-6 and 12-7 that were derived for short cables overestimate the radiation. This can be corrected for by using the $\lambda/4$ prediction at all frequencies above where the cable is a quarter-wavelength long. To understand the reason for this, see the discussion of electrically long cables in Section. 2.17. For *long cables* ($l > \lambda/4$), Eq. 12-7 can then be rewritten as

$$E = \frac{94.5 I_{cm}}{r}, \tag{12-10}$$

where $I_{cm}$ is the common-mode current and $r$ is the measuring distance in meters. Notice, that for long cables the envelope of the common-mode radiated emission is not a function of the cable length or frequency, but only of the common-mode current in the cable.

## 12.4   CONTROLLING COMMON-MODE RADIATION

As is the case for differential-mode radiation, it is desirable to limit both the rise time and frequency of the signal to decrease the common-mode emission.

In practice, cable length is dictated by the distance between the components or equipment being interconnected, and is not under the EMC designer's control. In addition, when the cable length reaches one quarter wavelength, the emission no longer continues to increase with cable length (see Fig. 2-50 and Eq. 12-10), because of the presence of out-of-phase currents.

Therefore, the only parameter in Eq. 12-7 that is completely under the designer's control is the common-mode current. The common-mode current can be thought of as a "control knob" on the radiated emission. *No common-mode current is required for normal system operation.*

The net common-mode current on a cable can be controlled by the following techniques:

1. Minimizing the common-mode source voltage, normally the ground potential
2. Providing a large common-mode impedance (choke) in series with the cable
3. Shunting the current off the cable
4. Shielding the cable
5. Isolating the cable from the PCB ground, for example, with a transformer or optical coupler

The common-mode suppression technique used must be such that it affects the common-mode current (often the clock harmonics) and not the functionally required differential-mode signals on the cable. In the past, the frequency of most I/O signals were considerably lower than the clock frequency, which made the above easy to achieve. Today, however, many types of I/O signals are at frequencies as high as or in some cases actually higher that the clock frequency [e.g., universal serial bus (USB) or Ethernet], thus considerably complicating the requirement that the common-mode suppression technique not interfere with the desired signal.

### 12.4.1  Common-Mode Voltage

The first step in controlling the common-mode radiation is to minimize the common-mode voltage that drives the antenna (cable). This usually involves minimizing the ground voltage, which means minimizing the ground impedance, as was discussed in Chapter 10. The use of ground planes or ground grids is a very effective way to do this. Also, the importance of avoiding slots in ground planes cannot be overemphasized (see Section. 16.3.1) as these can significantly increase the ground impedance.

The proper choice of where and how the circuit ground and the enclosure are connected together is also important in determining the amount of the common-mode voltage available to drive common-mode current out on the cables (see Section 3.2.5). The farther away that the circuit-to-chassis ground connection is from the point where the cables terminate on the board, the more likely it is that there is a large noise voltage between the points. As will be discussed in the next section, the reference or return plane for common-mode currents on the external cables is the enclosure. This is shown in Fig. 12-13A. Therefore, the circuit ground in the I/O area of the PCB should be at the same potential as the enclosure. To accomplish this, the two grounds must be connected together in this area. To be effective, the impedance (inductance) of this connection must be extremely low across the complete frequency range of interest, which usually requires multiple connections.

Even when the ground voltage is minimized, it is usually not sufficient to control the common-mode radiated emission. It only takes a few millivolts, or less, of ground voltage to drive 5 µA of common-mode current out on a cable. Therefore, additional common-mode emission control techniques are usually also required.

### 12.4.2   Cable Filtering and Shielding

If the common-mode current cannot be reduced sufficiently by controlling the ground voltage, then the common-mode noise must be kept off the cable by some form of filtering, or the radiation from the cable must be eliminated by shielding of the cable.

Cable shielding and termination was discussed in Chapter 2. An example of how the cable shield termination affects the shielding effectiveness of a cable is shown in Fig. 12-13.

Figure 12-13A shows a product in a shielded enclosure, with an unshielded cable leaving the enclosure. The common-mode noise voltage $V_{cm}$ (usually ground noise) drives a common-mode current $I_{cm}$ out on the unshielded cable. This current is not returned on the cable; rather, it returns to the source through the parasitic capacitance $C$ between the cable and the enclosure. This uncontained current flowing through parasitic capacitance represents radiation, as discussed in Appendix D. If one measures the common-mode current on the cable it will be $I_{cm}$, and the radiated emission will be determined by Eq. 12-7.

Figure 12-13B is similar to Fig. 12-13A except, in this case, a properly terminated shielded cable is used, with the shield making a 360° connection to the enclosure. The common-mode noise voltage $V_{cm}$ still drives a current $I_{cm}$ out on the center conductor of the cable, the same as in Fig. 12-13A. However, in this case, the shield blocks the parasitic capacitance between the center conductor and the enclosure. The current therefore flows through the parasitic capacitance between the center conductor and the shield, returning on the inside surface of the shield. Because the shield has a good 360° electrical connection to the enclosure, the current returns to the source through the enclosure. The net common-mode current on the cable in this case is zero, and there is no cable radiation.

Note, that in Fig. 12-13B, the shield does not prevent the radiation just by its presence; rather it works by returning the current that would otherwise radiate through a very small loop directly to the source. If the shield carries no current, it will not be effective in preventing the radiation. Because the shield must carry the return current to be effective, the method by which the shield is terminated is critical to the performance of the shield.

Figure 12-13C is similar to Fig. 12-13B, except in this case the shield is not terminated to the enclosure at all. The common-mode voltage $V_{cm}$ still drives a

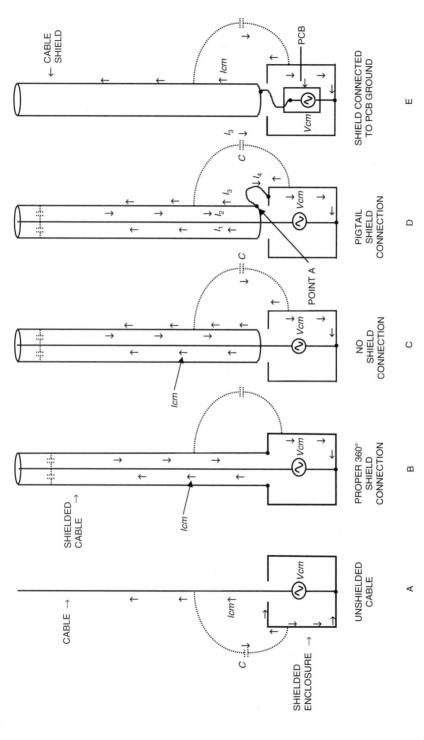

**FIGURE 12-13.** The effect of cable shield termination on the common-mode cable current, and hence on the radiated emission from the cable.

483

current of $I_{cm}$ out on the center conductor of the cable, and the presence of the shield blocks the parasitic capacitance between the center conductor and the enclosure. The current therefore flows through the parasitic capacitance between the center conductor and the shield, and down the inside surface of the shield to the bottom on the cable. Up to this point, the configurations of Figs. 12-13B and 12-13C behave exactly the same.

However because in Fig. 12-13C the shield is not terminated, the current cannot flow back on the enclosure to the source. The current therefore turns around and flows up the outside of the shield, through the parasitic capacitance $C$ between the outside surface of the shield and the enclosure, through the enclosure back to the source.* If one were to measure the common-mode current on the cable, it would be $I_{cm}$, and the radiated emission will be determined by Eq. 12-7. Therefore, the configuration of Fig. 12-13C has the same radiated emission as the configuration of Fig. 12-13A, despite the presence of the shield.

In Fig. 12-13D, the shield is terminated to the enclosure with a pigtail. The analysis of the common-mode current flow is the same as that of Fig. 12-13C, up to the point where the current arrives at the bottom of the inside surface of the shield (point $A$ in Fig. 12-13D). At point $A$, there will be a current division, with part of the current $I_2$ going up the outside of the shield as $I_3$ and through the parasitic capacitance between the outside of the shield and the enclosure, and the remainder of the current $I_4$ flowing on the pigtail back to the enclosure.

The current $I_1$ on the center conductor and the current $I_2$ on the inside surface of the shield will be equal and opposite and will cancel, which leaves the current $I_3$ on the outside surface of the shield as the net common-mode current on the cable. The radiated emission can then be determined by substituting $I_3$ into Eq. 12-7. The longer the pigtail, the higher its impedance will be and the larger will be the current $I_3$ and hence the radiation. The length of the pigtail is effectively a rheostat that can be used to adjust the radiated emission from this configuration. Because for any cable shield terminated with a pigtail, a current division will occur at point $A$, we can conclude that any shield terminated with a pigtail must radiate, and the only question is how much?

Notice from Fig. 12-13D, that the current path involving the outside of the shield is capacitive, whereas the current path involving the pigtail is inductive. Therefore, if the common-mode current on the cable is a square wave rather than a sine wave, the higher frequency harmonics will choose the path that involves the outside surface of the shield, and they will radiate, whereas the lower frequency harmonics will choose the path that involves the pigtail and will not radiate.

Figure 12-13E shows a case where the shield is terminated to the PCB circuit ground, instead of to the enclosure. It should be kept in mind that *the shield on*

---

* As a result of the skin effect, at high frequency, all currents on conductors are surface currents. Therefore, the outside surface of the shield, for all practical purposes, is a separate conductor independent from that of the inside surface of the shield.

*a shielded cable is unshielded.* Therefore, the common-mode noise voltage $V_{cm}$ excites the shield and drives a common-mode current $I_{cm}$ out onto the shield. The shield then radiates just like the case of the unshielded cable shown in Fig. 12-13A. In the case shown in Fig. 12-13E, no center conductor is even needed because the shield has become the problem. If one were to measure the common-mode current on the cable, it would be $I_{cm}$ with or without the center conductor present.

Although cable shields should be terminated to the enclosure (Fig. 12-13B) not to the PCB ground (Fig. 12-13 E), there are economic advantages to mounting the I/O connectors on the PCB, not on the enclosure. To be effective however the connector backshell must still make a 360° contact to the enclosure. One-way to accomplish this is shown in Fig 12-14 where the connector is mounted on the PCB, but when the PCB is placed in the enclosure the connector backshell is screwed to the enclosure using an electromagnetic interference (EMI) gasket, or metal spring fingers, thus making a 360° contact.

Filtering of the I/O cables can be accomplished by adding a high impedance in series with the common-mode noise (e.g., a common-mode choke or ferrite core), or by providing a low-impedance shunt (a capacitor) to divert the common-mode noise to "ground." But what ground? These shunt capacitors must be connected to a "clean" ground usually the enclosure, not to the "dirty" logic ground. As was the case of the I/O connectors discussed above, there are economic advantages to mounting these I/O cable filter capacitors on the PCB. This is discussed in the next section. The various configurations of common-mode filters were discussed at length in Section 4.2.

### 12.4.3  Separate I/O Grounds

If I/O connectors and/or cable filter capacitors are to be mounted on the PCB, then access is needed to the enclosure ground on the PCB. Unless consideration

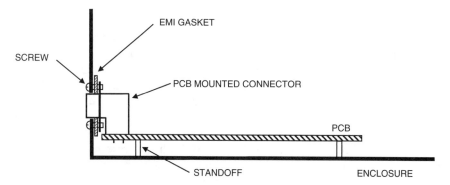

**FIGURE 12-14.** PCB mounted connector backshells *must* make 360° contact with the enclosure.

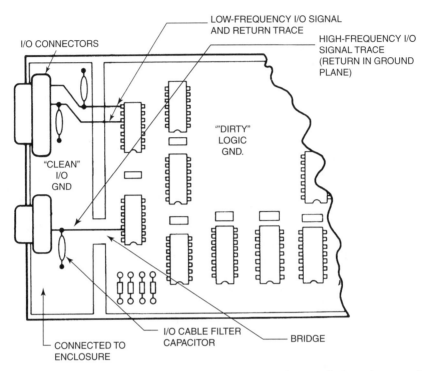

**FIGURE 12-15.** Digital PCB with a separate "clean" I/O ground plane that contains only I/O cable filter capacitors and connectors.

is given to this early in the design, such a ground will not be available when and where it is needed.

This access can be accomplished by placing all of the I/O connectors in one area of the board and providing in this area a separate "I/O ground plane" that has a low-impedance connection to the enclosure and connects to the digital logic ground at only one point. In this way, none of the noisy digital logic ground current will flow through the "clean" I/O ground and contaminate it.

Figure. 12-15 shows one possible implementation of this concept. To avoid contaminating the I/O ground plane, the only components connected to it should be I/O cable filter capacitors and I/O connector backshells. This ground must have a low-impedance connection to the enclosure. The I/O ground plane should have multiple connections to the enclosure to minimize its inductance and provide a low impedance connection. If this connection between the I/O ground and the enclosure ground is not made, or if it does not have a low enough impedance (at the frequencies involved), the I/O ground plane will not be effective and, as a matter of fact, the emissions from the cables may actually increase with this design approach. When implemented correctly, however, this approach works well and has been used successfully in many commercial products.

If properly designed, the traces that cross the I/O ground plane slot will not be a problem. Low-frequency I/O signals (less than 5 to 10 MHz) should be connected between the driver IC and the connector using two traces (a signal trace and a companion return trace) as shown in Fig. 12-15. Therefore, the signal return current is in the companion trace and not in the ground plane. The companion return trace only connects to the I/O connector pins and not to the I/O ground plane.

High-frequency I/O signals (greater than 5 to 10 MHz) can be routed as a single signal trace adjacent to the ground plane, as long as the trace is routed across a bridge as shown in Fig. 12-15. This approach provides an uninterrupted path, under the trace, for the return current, which is similar to the approach shown in Fig. 17-1B with respect to mixed-signal PCB layout. The bridge should be wide enough to accommodate the required number of traces plus 20 times the height of the trace above the ground plane (0.1 in for a trace height of 0.005 in) on each side. This will make the bridge wide enough to accommodate 97% of the return current, see Table 17-1.

The reason for using a bridge instead of a companion return trace for high-frequency signals is to reduce the impedance (inductance) of the return path. As was indicated in Section 10.6.2, a ground plane has an inductance that is two orders of magnitude less than that of a trace (approximately 0.15 nH/in for a plane versus 15 nH/in for a trace).

The I/O ground should be thought of as an extension of the enclosure. The enclosure is the reference or return plane for the common-mode currents on the external cables as was shown in Fig. 12-13A, and it is further discussed in Appendix D. The enclosure can be thought of as the high frequency reference for the product. I like to think of this approach as wrapping a section of the metal enclosure around and onto the PCB.

Another way to visualize this approach is as if the PCB ended where the logic ground ends and the connectors hang off the board on short wires. Capacitors are then connected between each signal wire and the enclosure to shunt high-frequency noise from the cables. Except that instead of hanging on a wire attached to the board, the capacitors and connectors are, for convenience, actually mounted on the PCB and connected to the I/O ground plane, which is really just an extension of the enclosure.

The key to making this all work is to have a *low inductance* connection between the I/O ground and the enclosure. This fact cannot be overemphasized.

The PCB's power plane should not be allowed to extend into the I/O ground area. The power plane will usually contain high-frequency logic noise, and if extended into the I/O area it can couple this noise to the I/O signal and ground conductors. I usually print two I/O ground planes in this area, one on the PCB ground plane layer and the other on the PCB power plane layer of the board.

I previously emphasized the fact that you should not have slots (or splits) in the ground plane, because any single-ended signal traces crossing the slot will have an interrupted return current path. Why, then, am I now recommending putting a slot in the ground plane? Well, I am really not! The digital logic ground plane is a solid uninterrupted plane, and the I/O plane is really an extension of the enclosure that just happens to be printed on the PCB. And as previously suggested, the two are tied together, in this case at one point, at the "bridge" shown in Fig. 12-15. This bridge should only be as wide as necessary to route the required number of high-frequency signals across.

This concept can be applied to any system configuration, even in large multiboard systems. The important point is that somewhere in the system there should be a clean I/O ground that is connected to the chassis. All unshielded cables should be decoupled to this ground before leaving the system. In large systems, the I/O ground might even be a separate PCB located at the cable entrance and containing only connectors and I/O cable filter capacitors. Once the noise is removed from the cables, their routing must be controlled carefully to avoid coupling noise back into the cables. Therefore, the clean I/O ground should be located at the point where the cables leave/enter the system.

The effectiveness of the I/O cable filter capacitors depends on the common-mode source impedance of the driving circuits. Sometimes, better results can be obtained by using a series resistor, ferrite, or inductor in addition to, or instead of, the cable filter capacitor. Filter pin connectors with built-in shunt capacitors and/ or series inductive elements (usually ferrites) behave similarly and can also be used.

### 12.4.4   Dealing With Common-Mode Radiation Issues

Because so little common-mode current is required to produce a radiated emission problem, almost all cables will have such problems unless something is done early in the design process to avoid it. A few microamps of common-mode current on a cable is enough to cause an emission problem, and it only takes a few millivolts, or less, of ground noise to produce.

It is good practice, therefore, at the beginning of a new design to make a list of *all* the cables, including power cables, and document what technique will be used to reduce or eliminate the common-mode current on that cable. Then during the prototype stage, the common-mode currents on all cables can easily be measured in the development laboratory, using the technique discussed in Section 18.3, and the results compared with the appropriate limit determined by Eq. 12-8. This allows the designer to evaluate the effectiveness of the common-mode suppression designed into the product prior to the final radiated emission testing.

### SUMMARY

- Emission control should be considered during the initial design and layout of a product.
- The differential-mode radiated emission is proportional to frequency squared, loop area, and differential-mode current in the loop.
- The primary way to control the differential-mode radiation is to reduce the loop area.

- PCB technology (the ability to print smaller loops) is not keeping up with the increased radiation that results from the frequency-squared term in the differential-mode radiation equation.
- When loops cannot be made small enough, other unconventional techniques such as dithered clocks and canceling loops may be necessary.
- The most critical signals are those:
  - That have the highest frequency
  - That are periodic
- The common-mode radiated emission is proportional to frequency, cable length, and common-mode current in the cable.
- The harmonic content of a square wave is determined by its rise time, not its fundamental frequency.
- The primary way to control common-mode radiation is by reducing, or eliminating, the common-mode current on the cables.
- Only a few microamps, or less, of common-mode current is required on a cable to fail radiated emission requirements.
- Common-mode radiation can be reduced by
  - Reducing the ground noise voltage
  - Filtering the I/O cables
  - Shielding the I/O cables
- Both common- and differential-mode radiation can be decreased by reducing the frequency and/or slowing down the rise time of the signals.
- I/O connector backshells and cable filter capacitors must be connected to the enclosure ground, not the circuit ground.
- Cable shields should make 360° connections to the enclosure.
- Most common-mode radiated emission problems occur below 300 MHz, and most differential-mode radiated emission problems occur above 300 MHz.

## PROBLEMS

12.1 a. Differential-mode radiation can be modeled using what type of antenna?

b. Common-mode radiation can be modeled using what type of antenna?

12.2 What is the shape of the emission envelope for differential-mode and for common- mode radiation if the source of the emission is an isosceles-triangular-shaped current pulse as shown in Fig. 11-3?

12.3 A 100 MHz, 50% duty cycle square wave current has an amplitude of 0.5 A and a rise/fall time of 0.5 ns. What is the amplitude of the 5th harmonic?

12.4 Name three techniques for reducing the differential-mode radiation from a product other than reducing the loop area, frequency, or current?

12.5 Which is the more efficient radiating structure, a small loop or a small dipole?

12.6 A PCB trace 4 in long and spaced 0.062 in from the return conductor is carrying a 10-MHz, 3.18-ns rise time clock signal. Assume the current in the loop is 50 mA.

    a. Determine the radiated field strength in dB $\mu$V/m at the fundamental frequency, when measured at a distance of 3 m from the board.

    b. Plot the envelope of the radiated emission from 10 to 350 MHz.

    c. What is the worst-case margin against the FCC Class B limit, and at what frequency does it occur?

12.7 a. Above what frequency will the radiated emission of Problem 12.6 begin to fall off as a result of the loop no longer being small?

    b. What will be the slope of the fall off?

12.8 If a small circular and a small rectangular loop both have the same area and carry the same current at the same frequency, which will produce the greater radiated emission?

12.9 An increase in clock frequency will have a larger effect on which radiation mode, differential or common mode?

12.10 The 75 MHz component of the common-mode current on a 0.5-m-long cable is measured as 50.8 $\mu$A. What will be the electric field strength at a distance of 3 m from the cable?

12.11 What is the maximum length cable that can be connected to a system with 25 mV of ground noise at 94 MHz, without exceeding the FCC radiated emission limit for a Class A product? Assume the cable has a common-mode impedance of 200 $\Omega$ at 94 MHz.

12.12 The FCC Class B radiated emission limit at 75 MHz is 100 $\mu$V/m, measured at a distance of 3 m. What would be the maximum allowable common-mode current in a 0.5-m-long cable not to exceed the limit?

## REFERENCES

Balanis, C. A. *Antenna Theory, Analysis and Design*, New York, Harper & Row 1982.

Catt, I. "Crosstalk (Noise) in Digital Systems." *IEEE Transactions on Electronic Computers*, December 1967.

Hardin, K. B. inventor; Lexmark International Inc., assignee. Spread Spectrum Clock Generator. U.S. patent 5,631,920. May 20, 1997.

Hoekstra, C. D. "Frequency Modulation of System Clocks for EMI Reduction," *Hewlett-Packard Journal*, August 1997.

Johnson, H. W. and Graham, M. High-Speed Digital Design, Englewood Cliffs, NJ, Prentice Hall, NJ, 1993.

Jordan, E. C. *Reference Data For Engineers*. Indianapolis, IN, Howard W. Sams, 1985.

Kraus, J. D. and Marhefka, R. J. *Antennas*, 3rd ed. New York, McGraw Hill, 2002.

Paul, C. R. "Printed Circuit Board EMC." *6th Symposium on EMC*, Zurich, Switzerland, March 5-7, 1985.

Stutzman, W. L. and Thiel, G. A. *Antenna Theory and Design*, New York, Wiley, 1981.

## FURTHER READING

Gardiner, S. et al. "An Introduction to Spread Spectrum Clock Generation for EMI Reduction." *Printed Circuit Design*, January 1999.

Hardin, K. B., Fessler, J. T., and Bush, D. R. "Spread Spectrum Clock Generators for Reduction of Radiated Emission." *1994 IEEE International Symposium on Electromagnetic Compatibility,* Chicago, IL, August 1994.

Mardiguian, M. *Controlling Radiated Emission by Design*, 2nd ed. Boston, MA, Kluwer Academic Publishers, 2001.

Nakauchi, E. and Brasher, L. "Techniques for Controlling Radiated Emission Due to Common-Mode Noise in Electronic Data Processing Systems." *IEEE International Symposium on Electromagnetic Compatibility*, September 1982.

Ott, H. W. "Controlling EMI by Proper Printed Wiring Board Layout." *6th Symposium on EMC*, Zurich, Switzerland, March 5-7, 1985.

# 13 Conducted Emissions

Conducted emission regulations are intended to control the radiation from the public alternating current (ac) power distribution system,* which results from noise currents conducted back onto the power line. Normally, these currents are too small to cause interference directly with other products connected to the same power line; however, they are large enough to cause the power line to radiate and possibly become a source of interference, for example to AM radio. The conducted emission limits exist below 30 MHz, where most products themselves are not large enough to be very efficient radiators, but where the ac power distribution system can be an efficient antenna. *The conducted emission requirements, therefore, are really radiated emission requirements in disguise.* The FCC's conducted emission limits are listed in Tables 1-5 and 1-6 and plotted in Fig. 18-12.

In addition, some products are nonlinear loads on the ac power distribution system. They have nonsinusoidal input current waveforms, rich in harmonics and can have a detrimental effect on the performance of the power distribution system. For this reason, the European Union has regulations limiting the harmonic emission from electronic products.[†] These limits are listed in Table 18-3.

Because it is the product's power supply that directly interfaces to the ac power line, the design of the power supply and the power-line filter has a large influence on the conducted and harmonic emissions. This is especially true in the case of switched-mode power supplies and variable speed motor drives.

## 13.1 POWER LINE IMPEDANCE

Conducted emission regulations limit the common-mode noise voltage that a product can conduct back onto the ac [and in some cases direct current (dc)]

---

*Automotive, military and some other regulations also have conducted emission limits applicable to dc power lines.

[†]In addition, IEEE Std 519, *Recommended Practices and Requirements for Harmonic Control in Electrical Power Systems*, gives limits for harmonic injection into the power distribution system for individual customers (facilities), not for individual equipment. It is widely used in the United States. to limit harmonics fed back into the power distribution system by individual customers.

---

*Electromagnetic Compatibility Engineering,* by Henry W. Ott
Copyright © 2009 John Wiley & Sons, Inc.

power line. This voltage is measured from the hot conductor to ground, and from neutral conductor to ground. In the case of a dc power line, the measurement would be made from the positive conductor to ground and the negative conductor to ground. It is the power line impedance that converts the product's conducted noise current into a noise voltage.

Figure 13-1 is a plot of the maximum and minimum impedance of the ac power line in the frequency range of 100 kHz to 30 MHz (Nicholson and Malack, 1973). These data are the result of measurements made on 36 unfiltered commercial ac power lines at different locations across the United States. As can be observed, the power line impedance values range from about 2 to 450 Ω.

Because of this wide variability of power line impedances, it would be difficult to obtain repeatable conducted emission test results. To produce repeatable results, the impedance seen by a product, looking back into the ac power line, would have to be stabilized or fixed. This is accomplished by using a line impedance stabilization network.

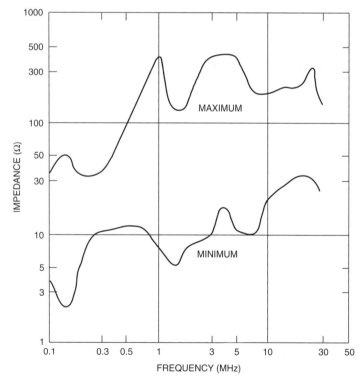

**FIGURE 13-1.** Measured impedance of the 115-V ac power line. © *IEEE 1973, reproduced with permission.*

### 13.1.1    Line Impedance Stabilization Network

During the conducted emission test, a line impedance stabilization network (LISN)* is placed between the product and the actual power line in order to present a known impedance to the product's power line terminals over the frequency range of 150 kHz to 30 MHz. One LISN is inserted in the hot side of the power line, and one is inserted in the neutral side of the power line. On a three-phase power line, three LISNs would have to be used.

The circuit of a 50-μH LISN used for most conducted emission testing is shown in Fig. 13-2. The 1-μF capacitor $C_2$ on the power line side of the LISN shorts out the variable impedance of the actual power line so that it does not influence the test results. The 50-μH inductor $L_1$ provides a rising impedance with frequency similar to that shown in the power line impedance measurements of Fig. 13-1.

Capacitor $C_1$ is used to couple the conducted emission measuring instrument to the power line. The 1000-Ω resistor $R_1$ discharges the LISN capacitors when the LISN is removed from the power line. This is necessary to prevent the charged capacitors in the LISN from becoming a shock hazard. The measuring instrument [spectrum analyzer or radio frequency (rf) receiver] places a 50-Ω shunt across resistor $R_1$ and limits the LISN's rising impedance versus frequency. It is important to realize that the 50-Ω input impedance of the measuring instrument is really part of the LISN impedance.

The impedance looking into the equipment under test (EUT) port of the LISN is shown in Fig. 13-3. As can be observed, the impedance has a value

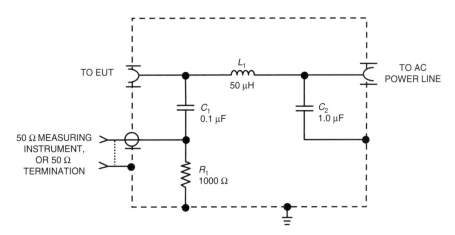

**FIGURE 13-2.** Circuit of a 50-μH LISN used for conducted emission testing.

---

*The LISN is sometimes referred to as an artificial mains network (AMN).

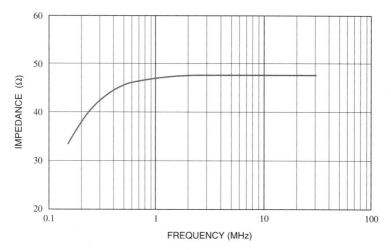

**FIGURE 13-3.** Impedance looking into the EUT terminals of a 50-µH LISN.

close to 48 Ω* over most of the frequency range from 0.15 to 30 MHz. Only below 500 kHz does the impedance drop significantly below this value.

The increase of the LISN impedance with frequency is caused by the LISN's 50-µH inductor. The LISN impedance approximates the average value of the actual measured power line impedance shown in Fig. 13-1. As a result, it is a good choice for a standardized impedance to use when performing conducted emission measurements. Because the LISN impedance is close to 50-Ω over most of the frequency range from 500 kHz to 30 MHz, the LISN is often modeled as just a 50-Ω resistor.

## 13.2  SWITCHED-MODE POWER SUPPLIES

Today's highly sophisticated switched-mode power supplies (SMPS) operate at efficiencies of 85% or greater and are only a fraction of the size and weight of comparable linear supplies. However, all these advantages come with many disadvantages. For one, switched-mode power supplies are a major source of both conducted and radiated emissions. They conduct large noise currents (both common-and differential-mode) back onto the power line at harmonics of the switching frequency.

In addition, the full wave rectification of the power line voltage feeding a capacitive input filter results in current spikes on the power line at the peaks of the voltage cycle as the filter capacitor recharges. Because current is not drawn over the entire cycle, the current waveform has a large amount of harmonic

---

*The parallel combination of the 50-Ω termination on the measuring port and the 1000-Ω resistor $R_1$.

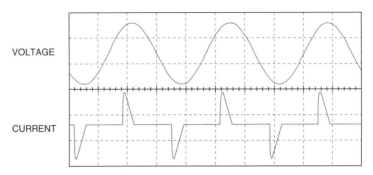

**FIGURE 13-4.** Switched-mode power supply input waveforms. Top trace, voltage; bottom trace, current.

distortion. The resultant current waveshape is rich in odd harmonics (3rd, 5th, 7th, 9th, etc.) and can cause overheating of the power company's transformers. In three-phase power distribution systems it produces excessive neutral conductor currents. Figure 13-4 shows typical voltage and current waveforms present on the input of a switched-mode power supply. The pulsating ac current drawn by the power supply not only contains many higher frequency harmonics but also has a much larger peak amplitude than a sine wave would have for the same power rating. According to the power line impedance and the peak current drawn from the line, the voltage waveshape may also be distorted and have a flat top at the peaks (not shown in Fig. 13-4).

Many different SMPS topologies exist,* but in this chapter we will use the flyback converter as an example of the electromagnetic capability (EMC) problems associated with SMPS. Figure 13-5 is a simplified schematic of the flyback converter SMPS.

In the flyback converter, a full-wave bridge rectifies the ac voltage. The full-wave rectified voltage is filtered by the capacitor $C_F$ producing a dc voltage that approaches the peak value of the ac waveform. A pulse width modulation (PWM) controller applies a variable duty-cycle square wave to the input of the switching transistor. The variable duty-cycle of this voltage (which is determined by the output load) provides regulation for the output voltage.[†] The switching transistor produces a square wave of voltage across the transformer, which is then stepped down in the secondary winding

---

*Some of the more common topologies are the buck converter, the boost converter, the buck-boost converter, the flyback converter, the forward converter, the half bridge converter, the full bridge converter, and the resonant converter (Hnatek, 1989, Chapter 2).

[†]The output voltage is sampled and fed back (often through an opto-isolator) to the PWM controller. This feedback circuit is not shown in Fig. 13-5. Regulation is achieved by varying the duty-cycle of the square wave drive to the switching transistor in response to changes in the output voltage of the supply.

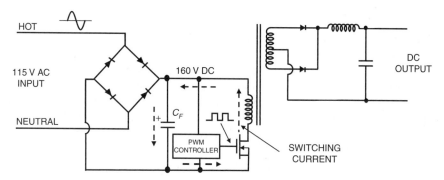

**FIGURE 13-5.** Simplified circuit of a flyback converter switched-mode power supply.

and then, rectified and filtered to produce the dc output voltage. The transformer can have multiple secondary windings, thereby producing a number of different dc voltages.

Because the transformer operates at the switching frequency of the power supply (typically 50 kHz to 1 MHz), it can be made much smaller and lighter than a 50- or 60-Hz transformer. The power-switching device is either a bipolar transistor or a metal oxide semiconductor field effect transistor (MOSFET). Because the transistor is switching a square wave (with typically a 25- to 100-ns rise time), it spends little time in the linear region, which minimizes its power dissipation and thereby accounts for the high efficiency of the design.

This SMPS configuration has multiple sources of noise, however. Some of these are the result of the normal operation of the circuit (differential-mode noise), whereas others are the result of circuit's parasitic capacitances (common-mode noise). The power supply generates both common-mode and differential-mode noise currents at harmonics of the switching frequency.

To obtain an idea of the magnitude of the conducted emission problem encountered with a SMPS, we can compare the operating levels inside the power supply to the regulatory conducted emission limits. For the case of a 115-V input, the SMPS generates a 160-V square wave connected directly to the ac power line, because no primary side (ac power line side) transformer is used to isolate the switcher from the line. The FCC's Class B conducted emission limit (between 500 kHz and 5 MHz) is 631 μV (56 dB above a microvolt). The ratio of 160 V to 631 μV is 253,566 or 108 dB. Hence, the allowed emission levels are about a quarter millionth of the operating levels within the power supply. Therefore, the operating signal within the power supply must be suppressed by more than 110 dB to comply with the regulatory limit. Although the inherent design of the power supply provides some degree of noise suppression, in almost all cases an additional power-line filter will also be required for regulatory compliance.

### 13.2.1 Common-Mode Emissions

The major contributor to common-mode emission is the primary side parasitic capacitance to ground. The three contributors to this capacitance are switching transistor to heat sink capacitance, transformer interwinding capacitance, and stray primary side wiring capacitance as shown in Fig. 13-6.

The largest single contributor is usually the switching transistor to heat sink capacitance. This capacitance can be reduced by (1) using an insulating thermal washer containing a Faraday shield between the transistor and heat sink (2) using a thicker cermanic washer (such as beryllium oxide) or (3) not grounding the heat sink. A Faraday shield thermal insulating washer consists of a copper shield between two thin layers of insulating materials. To be effective, the copper shield must be connected to the source terminal of the switching transistor. In the case of a bipolar switching transistor, it would be connected to the emitter terminal. Several suppliers manufacture thermal washers containing Faraday shields.

If the heat sink is electrically floated from ground, for safety reasons the heat sink must be protected from the possibility of anyone touching it. In the event that the insulating thermal washer between the switching transistor and the heat sink were to break down, the heat sink would be at the ac line potential and present an electrical shock hazard.

The second contributor to this parasitic capacitance is the transformer's interwinding capacitance. Because designers want physically small transformers, the primary and secondary windings are placed close together, which maximizes the interwinding capacitance. Using a transformer that separates the windings further or one that contains a Faraday shield can reduce this

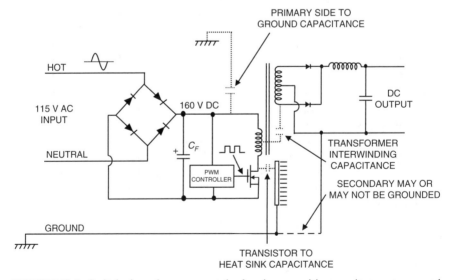

**FIGURE 13-6.** Switched-mode power supply showing parasitic capacitances to ground.

capacitance. The disadvantage of a Faraday shielded transformer is increased cost and possibly increased size.

Careful component placement, with careful wiring and/or printed circuit board (PCB) layout can minimize the third contributor, the primary side wiring capacitance.

Adding the LISN* and redrawing the circuit of Fig. 13-6 showing just the common-mode conducted emission path gives the circuit shown in Fig. 13-7. Note that to the common-mode current, the LISN impedance looks like 25 $\Omega$, the two 50-$\Omega$ resistors in parallel. The circuit of Fig. 13-7 can be simplified even more by representing the switching transistor as a square wave voltage generator having a peak amplitude equal to the dc voltage across the filter capacitor $C_F$. This simplified common-mode equivalent circuit of the SMPS is shown in Fig. 13-8.

From Fig. 13-8, we can determine that the power supply has a high source impedance, which is equal to the magnitude of the capacitive reactance of $C_P$.

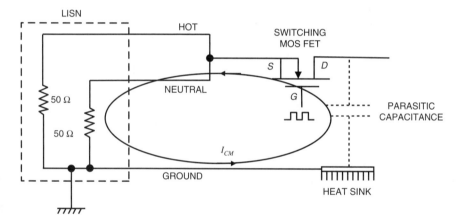

**FIGURE 13-7.** Common-mode equivalent circuit of a switched-mode power supply.

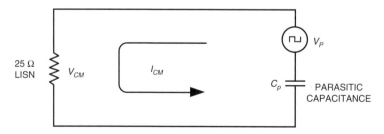

**FIGURE 13-8.** Simplified common-mode equivalent circuit of switched-mode power supply.

---

*In this chapter, we will represent the LISN impedance as a 50-$\Omega$ resistor.

The common-mode current, and therefore the LISN voltage, are determined primarily by the magnitude of this parasitic capacitance. Typical values for $C_P$ range from about 50 pF to as much as 500 pF.

From the circuit of Fig. 13-8, we can calculate the magnitude of the common-mode voltage $V_{CM}$ across the LISN resistance as

$$V_{CM} = 50\pi f C_P V(f), \tag{13-1}$$

where $V(f)$ is the magnitude of the voltage source $V_P$ at the frequency $f$ (see problem 13-1).

Because the voltage source is a square wave, the Fourier spectrum (Eq. 15-2) can be used to determine the harmonic content of the voltage $V(f)$. From Fig. 12-4, we know that for a square wave, the envelope of the Fourier spectrum decreases at a rate of 20 dB per decade up to a frequency of $1/\pi t_r$, where $t_r$ is the rise/fall time of the switching transistor, and beyond that it falls off at a rate of 40 dB per decade (see Fig. 12-4).

The frequency term in Eq. 13-1 represents a rising amplitude of 20 dB per decade. Therefore, the result of combining Eq. 13-1 and the Fourier spectrum $V(f)$ is that the common-mode conducted emission voltage $V_{CM}$ is flat up to a frequency of $1/\pi t_r$, and it decreases at a rate of 20 dB/decade above that frequency as shown in Fig. 13-9. The curve of Fig. 13-9 bounds the conducted emission; however, the actual emissions only exist at harmonics of the fundamental frequency $F_0$.

Because the emission is flat up to a frequency of $1/\pi t_r$ and then falls off at a rate of 20 dB/decade, we only need to calculate the emission at a single point, for example the fundamental frequency, to plot the complete envelope. From the information in Section 12.1.3 on the Fourier series, we know that the

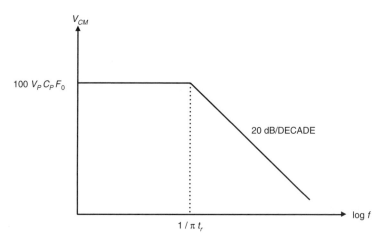

**FIGURE 13-9.** Envelope of the common-mode conducted emission versus frequency.

amplitude of the fundamental frequency is 0.64 $V_P$. Substituting the fundamental frequency $F_0$ for $f$ and *0.64 $V_P$* for *V(f)* in Eq. 13-1 gives the following expression for the amplitude of the common-mode conducted emission at the fundamental frequency:

$$V_{CM} = 100 V_P F_0 C_P \qquad (13\text{-}2)$$

For a rise time of 100 ns, the breakpoint of the plot in Fig. 13-9 is at 3.18 MHz. Realizing that $V_P$ is fixed by the power line voltage, we can conclude from Eq. 13-2 that once a fundamental frequency for the switching power supply is selected, the only remaining parameter under the designer's control, to reduce the common-mode conducted emission, is the parasitic capacitance $C_P$.

From Fig. 13-9, we also observe that slowing down the rise time of the switching transistor, which has the undesirable effect of increasing its power dissipation, does not reduce the maximum amplitude of the common-mode conducted emission. The only effect that slowing down the rise time will have is to move the breakpoint to a lower frequency. This will decrease the high-frequency emission but not the maximum emission at the lower frequencies.

**Example 13-1.** For the case where $V_P = 160$ V, $C_P = 200$ pf, $t_r = 50$ ns, and $F_0 = 50$ kHz, the envelope of the common-mode conducted emission will have a magnitude of 160 mV from 150 kHz to 6.37 MHz and then fall off at the rate of 20 dB/decade. This result is approximately 48 dB above the allowable emission for a Class B product. Therefore, a power line filter with 50 dB or more of common-mode attenuation will be required to bring the product into compliance.

## 13.2.2  Differential-Mode Emissions

The normal operation of the power supply consists of the switching transistor driving a current at the switching frequency around the loop that consists of the switching transistor, the transformer, and the filter capacitor $C_F$ as shown in Fig. 13-5. As long as the switching current flows through this loop, internal to the power supply, it will produce no differential-mode emission.

The primary purpose of the capacitor $C_F$, however, is to filter the full-wave rectified ac line voltage. The filter capacitor is therefore a large-value, high-voltage capacitor (typically 250 to 1000 μF with a voltage rating of 250 V or more), and it is far from an ideal capacitor. It typically has a significant equivalent series inductance (ESL) $L_F$ and equivalent series resistance (ESR) $R_F$. As a result of these parasitic impedances, not all the switching current will flow through the capacitor $C_F$. Rather, a current division will occur at the capacitor terminals, with some of the switching current flowing through the capacitor and the remainder flowing through the full-wave bridge rectifier out onto the power line as shown in Fig. 13-10. The switching current flowing out onto the power line is a differential-mode noise current that flows

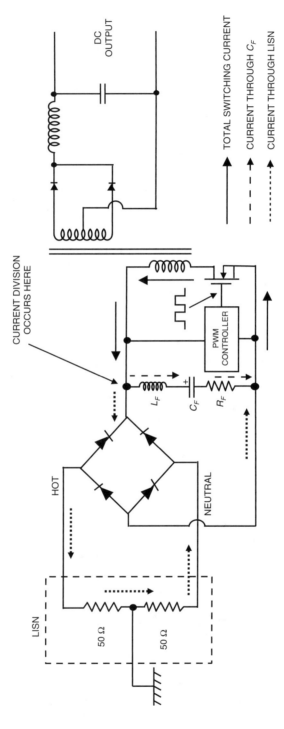

**FIGURE 13-10.** Switched-mode power supply, showing the differential-mode current path. Note that, a current division occurs at the terminals of capacitor $C_F$.

502

through the LISN. Note that to the differential-mode current, the LISN looks like 100 Ω (the two 50-Ω resistors in series). The circuit of Fig. 13-10 can be simplified even more by replacing the switching transistor with a current generator $I_P$ and eliminating the bridge rectifier. This simplified differential-mode equivalent circuit, which shows only the differential-mode conducted emission current path, is shown in Fig. 13-11.

From the circuit of Fig. 13-11, we observe that the power supply has a low differential-mode source impedance, which results from the large value of the input ripple filter capacitance $C_F$. The differential-mode current, and therefore the LISN voltage, are both determined primarily by the parasitics ($L_F$ and $R_F$) and the mounting of the filter capacitance $C_F$. Improper mounting will add additional inductance in series with the capacitor.

From the circuit of Fig. 13-11, we can calculate the differential-mode voltage $V_{DM}$ across the LISN resistance. Capacitor $C_F$ was chosen to be a low impedance at twice the power line frequency (100 or 120 Hz); it can therefore be assumed that the capacitive reactance at the conducted emission frequencies, which are three or more orders of magnitude higher, will be close to zero. For example, the capacitive reactance of an ideal 250-µF capacitor at 50 kHz is 0.01 Ω. Therefore, the parasitics $L_F$ and $R_F$ will be the dominant impedances at the switching frequency.

**13.2.2.1 Effect of the Filter Capacitor's ESL.** Neglecting $R_F$ for the moment and assuming the capacitive reactance of $C_F$ equals zero at the frequencies of interest, we can calculate the differential-mode noise current $I_{DM}$ as

$$I_{DM} = \frac{j2\pi f L_F I(f)}{100 + j2\pi f L_F}, \qquad (13\text{-}3)$$

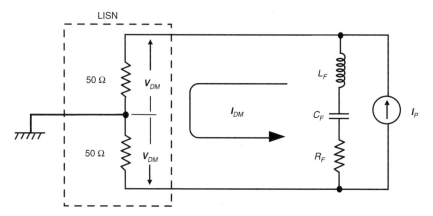

**FIGURE 13-11.** Simplified differential-mode equivalent circuit of switched-mode power supply.

where the current $I(f)$ is the magnitude of the current source $I_P$ at the frequency $f$.

For $2 \pi f L_F \ll 100$ (a reasonable assumption),* Eq. 13-3 reduces to

$$I_{DM} = \frac{j2\pi f L_F I(f)}{100} \tag{13-4}$$

The differential-mode LISN voltage $V_{DM}$ is equal to 50 Ω times $I_{DM}$; therefore, the magnitude of the LISN voltage can be written as

$$V_{DM} = \pi f L_F I(f) \tag{13-5}$$

We can determine the harmonic content of $I(f)$ by approximating the primary switching current by a square wave and then using the Fourier spectrum given by Eq. 12-5. From Fig. 12-4, we know that the envelope of the Fourier spectrum of a square wave decreases at a rate of 20 dB per decade up to a frequency of $1/\pi t_r$, where $t_r$ is the rise/fall time of the switching transistor, and beyond that frequency falls off at a rate of 40 dB per decade.

Because of the frequency term, Eq. 13-5 has a rising amplitude of 20 dB per decade with respect to frequency. Therefore, the result of combining Eq. 13-5 and the Fourier spectrum of the current $I(f)$ is that the envelope of the differential-mode conducted emission is flat up to a frequency of $1/\pi t_r$, and it decreases at a rate of 20 dB/decade above that frequency as shown in Fig. 13-12. Figure 13-12 is a plot of the differential-mode radiated emission envelope versus frequency, neglecting the effect of the capacitor's ESR.

Because the emission is flat up to a frequency of $1/\pi t_r$ and then falls off at a rate of 20 dB/decade, we only need to calculate the emission at a single point, for example the fundamental frequency, to plot the complete envelope. From Section 12.1.3 on the Fourier series, we know that the amplitude of the fundamental frequency of a 50% duty cycle square wave is $0.64 V_P$. Substituting the fundamental frequency $F_0$ for $f$ and $0.64 I_P$ for $I(f)$ in Eq. 13-5 gives

$$V_{DM} = 2F_0 L_F I_P \tag{13-6}$$

for the amplitude of the differential-mode conducted emission at the fundamental frequency, where $I_P$ is the peak value of the noise current generator.

Realizing that $I_P$ is determined by the power rating of the supply, we can conclude from Eq. 13-6 that once a fundamental frequency for the power supply is chosen, the only parameter under the designer's control, to reduce the

---

*For example, for $f$ = 500 kHz and $L_F$ = 50 nH, $2 \pi f L_F$ will equal 0.16 Ω.

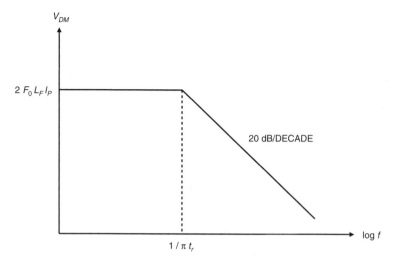

**FIGURE 13-12.** Envelope of the differential-mode conducted emission versus frequency, neglecting the equivalent series resistance $R_F$ of the input ripple capacitor.

maximum value of the differential-mode conducted emission, is the parasitic inductance $L_F$ of the input ripple filter capacitor.

**Example 13-2.** Let $I_P = 4$ A, $L_F = 30$ nH, $t_r = 50$ ns, and $F_0 = 50$ kHz. Neglecting the ESR of the input ripple filter capacitor, the envelope of the differential mode conducted emission will have a magnitude of 12 mV from 150 kHz to 3.18 MHz and then fall off at the rate of 20 dB/decade. This value is approximately 26 dB above the allowable emission for a Class B product. Therefore, a power line filter with 30 dB or more of differential-mode attenuation will be required to bring the product into compliance.

Comparing the results of Example 13-2 to those of Example 13-1, we observe that in this case the common-mode noise voltage is more than 20 dB greater than the differential-mode noise voltage, and it is the predominate emission. Equation 13-2 shows that the common-mode conducted emission is directly proportional to the magnitude of the switching voltage, whereas Eq. 13-6 shows that the differential-mode conducted emission is directly proportional to the magnitude of the switching current. Therefore, in high-voltage low-current supplies, the common mode emission will predominate, but in low-voltage high-current supplies the differential mode emission will predominate.

From Equations 13-2 and 13-6, we can determine that the criteria for the common-mode emission to be predominate is

$$V_P > \frac{L_F I_P}{50 C_P}. \tag{13-7}$$

Equation 13-7 was obtained by substituting Eqs. 13-2 and 13-6 into the equation $V_{CM} > V_{DM}$ and solving for $V_P$. If the inequality of Eq. 13-7 is not satisfied, then the differential-mode emission will predominate.

***13.2.2.2 Effect of the Filter Capacitor's ESR.*** The differential-mode emission envelope shown in Fig. 13-12 neglected the ESR of the input ripple filter capacitor. At power line frequency (50 or 60 Hz), the capacitance reactance is the dominant impedance of this capacitor. Above about 1 MHz, the inductive reactance becomes the dominant impedance of the capacitor. At frequencies somewhere in between, the resistance will be the dominant impedance. For the resistance to be dominant, its value must be greater than the inductive reactance of $L_F$. The criteria for this to be true is that

$$f < \frac{R_F}{2\pi L_F}. \tag{13-8}$$

The effect of the ESR on the differential-mode conducted emission is to add another lower frequency break point, at a frequency of $R_F/2\pi L_F$, to the differential-mode conducted emission envelope as shown in Fig. 13-13. Below this break point, the emission rises as frequency decreases at the rate of 20 dB/decade. For example, a filter capacitor with an ESL of 30 nH and an ESR of 0.1 Ω the low-frequency break point will occur at a frequency of 531 kHz.

The commercial Class B (see Fig. 1-4) and military conducted emission limits both have a 20 dB/decade rising slope with decreasing frequency below 500 kHz. The commercial Class A limits have a step increase of 6 dB in allowable

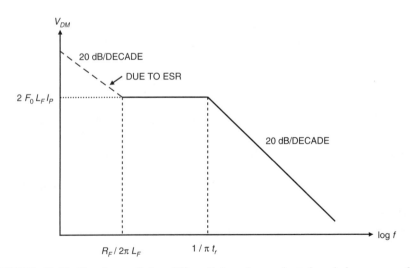

**FIGURE 13-13.** Envelope of the differential-mode conducted emission versus frequency, including the effect of both the input filter capacitor's ESR and ESL.

emission below 500 kHz. This tends to minimize the harmful effect of the rising differential-mode emission with decreasing frequency caused by the ESR of the filter capacitor. As long as the low-frequency breakpoint is at a frequency of 500 kHz or less, the ESR of the input filter capacitor should not be a problem. The criteria for this break point to be equal to or less than 500 kHz is that

$$R_F \leq \pi 10^6 L_F. \tag{13-9}$$

From the above discussion, it is clear that the input ripple filter capacitor should have both a low series inductance and low series resistance.

A similar procedure to that described above can be used for determining the envelope of the differential-mode emissions for other power converter topologies, where the switching current cannot be represented by a square wave. That is, determine the Fourier spectrum of the current waveshape and combine that with Eq. 13-5 to determine the envelope of the differential-mode emission. For example, in some SMPS topologies, the switching current would be represented more accurately as a triangular wave. Clayton Paul in Chapter 3, Signal Spectra—the Relationship between the Time Domain and the Frequency Domain, of his book *Introduction to Electromagnetic Compatibility* has a good discussion on the method of determining the Fourier spectrum of any arbitrary waveform (Paul, 2006).

Some power converter designs use two filter capacitors in series for the input ripple filter capacitor. This is done to increase the voltage rating of the capacitor or to create a voltage doubler configuration so that the supply can work off of either a 115 V or 230 V ac power line. This approach, however, has the disadvantage of halving the value of the total capacitance (the good parameter), while doubling the value of both the ESR and ESL (the bad parameters), which increases the differential-mode conducted emissions.

We can obtain the overall noise equivalent circuit of a SMPS by combining the common-mode equivalent circuit of Fig. 13-8 with the differential-mode equivalent circuit of Fig. 13-11. The result is shown in Fig. 13-14. The differential-mode source impedance shown in Fig. 13-14 is the impedance of the power supply's ripple filter capacitor, which includes its ESL and ESR. The common-mode impedance is the impedance of the power supply's primary side (ac line side) capacitance to chassis or ground. If a two-wire line cord were used to power the supply, the ground conductor shown between the power supply and LISN in Fig. 13-14 would not exist.

The figure also shows both the common-mode and differential-mode current paths. As can be observed, the differential-mode noise currents only flow through the power leads (phase and neutral), whereas the common-mode noise currents also flows through the external ground reference plane. Notice also that, in this case, the common-mode and differential-mode currents add together when flowing through the LISN impedance that is connected to the phase (hot) conductor and subtract from each other when flowing through the LISN

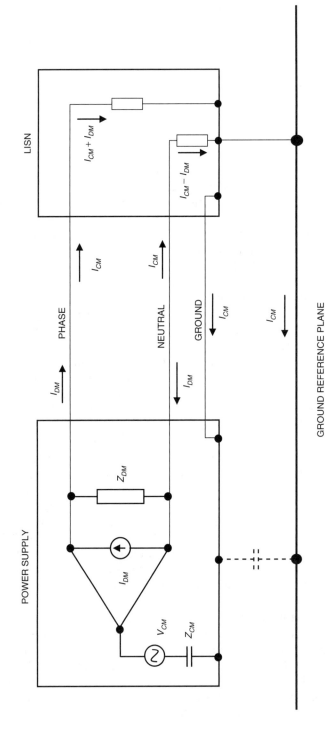

**FIGURE 13-14.** Noise equivalent circuit of switched-mode power supply and LISN showing both common-mode and differential-mode noise sources and currents.

impedance connected to the neutral conductor. As discussed in Section 18.6.1 knowledge of this provides a means to determine, by measurements, whether the predominant noise is caused by common- or differential-mode emissions.

The power supply's enclosure may be grounded or not. If it is not grounded, then the common-mode ground current will return through the capacitance of the power supply to ground as shown in Fig. 13-14. An open frame power supply (no metallic enclosure) is a good example of an ungrounded power supply where the capacitance to ground would constitute the common-mode return current path.

### 13.2.3  DC-to-DC Converters

The above discussion of SMPS is also applicable to dc-to-dc converters. A simplified circuit for a generic dc-to-dc converter would be the same as shown in Fig. 13-5 without the full-wave bridge rectifier. Most, if not all, of the theory and conclusions are applicable. The following example applies the previously developed SMPS theory to the case of a dc-to-dc converter.

**Example 13-3.** Consider a 28-V, 20-A dc input converter with 100 pF of primary side parasitic capacitance, which operates at 100 kHz and has a switching transistor rise time of 100 ns. The input capacitor has an ESR of 0.05 $\Omega$ and an ESL of 20 nH. From Eq. 13-2, the common-mode conducted emission will be 28 mV from 398 kHz to 3.18 MHz, and falling at a rate of 20 dB per decade above that frequency. From Eq. 13-6, the differential mode emissions will be 80 mV from 398 kHz to 3.2 MHz. Below 398 kHz, the emission will rise at 20 dB per decade, and above 3.18 MHz the emission will fall at the rate of 20 dB per decade.

Note, that in this case the differential-mode emission is greater than the common-mode emission because the supply is a low-voltage high-current power supply. This conclusion could also have been arrived at by evaluating Eq. 13-7.

### 13.2.4  Rectifier Diode Noise

Other noise sources are present in SMPS that should also be addressed. One of these noise sources are the diodes used in the rectification process. When a diode is forward biased, charge is stored in its junction capacitance. When the diode is turned off (reverse biased), this charge must be removed (Hnatek, 1989, p. 160). This is referred to as diode reverse recovery; it produces a sharp negative spike on the voltage waveform when the diode turns off, which can produce substantial ringing and be a source of high frequency, differential-mode noise.

Some diodes are fast-recovery diodes and turn off sharply. Other diodes are soft-recovery diodes and turn off slowly. Fast-recover diodes are usually preferred by power supply designers because they dissipate less power and are therefore more efficient. Fast-recovery diodes, however, produce higher frequency noise spectra than soft-recovery diodes.

The primary offenders in this respect are the secondary side rectifiers, because these diodes operate at a much higher current level than the primary rectifiers. These noise pulses can be conducted out of the power supply secondary and/or can be coupled back through the switching transformer to the primary side of the supply. In both cases, the diode noise manifests itself as a differential-mode conducted emission.

One solution to this high-frequency noise produced by the diodes switching, is to place a spike-reducing snubber network across each of the rectifier diodes, as shown in Fig. 13-15 (Hnatek, 1989, p. 190).* The snubber network consists of a series *R–C* circuit. Typical values might be 470 pF and 10 Ω. The snubber provides a current path to discharge the stored charge in the diode's junction capacitance when the diode is switched off. Because high-frequency currents will circulate in the snubber-to-diode loop, the area of this loop should be kept as small as possible.

Considering the fact that one or the other of the two rectifier diodes in Fig. 13-15 is always on, the two snubber-networks are effectively connected across the transformer's secondary winding. In some cases, especially low-power converters, a single snubber network placed across the transformer secondary winding can be used in place of a snubber network across each individual rectifier diode.[†]

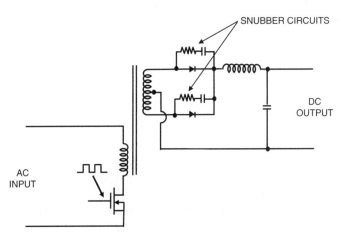

**FIGURE 13-15.** Snubber circuits placed across power supply's secondary side rectifier diodes.

---

*An R–C snubber network may also be used across the switching transistor to reduce ringing.
[†]Although snubber networks can be very effective in reducing ringing and noise there are also undesirable side effects associated with them, such as increased power dissipation and/or increased voltage and current across or through the devices they are used on.

Another approach is to add a small ferrite bead in series with each rectifier diode. This increases the high-frequency impedance in series with the rectifier and reduces the magnitude of the high-frequency ringing current. The most effective approach often is to combine the ferrite bead and the snubber circuit.

The concept of using a small ferrite bead to reduce ringing in a SMPS can be extended to other switching devices in the power supply as well. A small ferrite bead placed in series with any diode, rectifier, or switching transistor is often an effective way to dampen the ringing produced when these devices switch.

## 13.3  POWER-LINE FILTERS

As the previous discussion and examples have demonstrated, a power-line filter is almost always necessary for a SMPS to comply with the regulatory conducted emission requirements. The filter must provide attenuation for both the common-mode and the differential-mode noise currents.

The power-line filter is a low-pass $L$–$C$ topology. The source (the power supply) and the load (the LISN) impedances determine the exact configuration of the filter. Because filter attenuation is a function of impedance mismatch, the role of a power-line filter is to maximize the mismatch between the source and load impedances (Nave, 1991, p. 43).

For common-mode noise, the power supply is a high-impedance source (a small parasitic capacitance), and the LISN is a low-impedance load (a 25-$\Omega$ resistance). For maximum attenuation, the high-impedance filter element (the inductor) should face the low-impedance load (the LISN), and the low-impedance filter element (the capacitor) should face the high-impedance source (the power supply). Figure 13-16 shows the general topology of a power-line filter. The two line-to-ground capacitors ($C_1$ and $C_2$), and the common-mode choke $L_1$ form the common-mode section of a low-pass $L$–$C$ filter.

The maximum value of the line-to-ground capacitors is limited because of leakage requirements imposed by various safety agencies. Excessive leakage

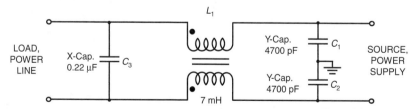

**FIGURE 13-16.** Generic power-line filter topology, including typical component values.

current to ground is considered a shock hazard and therefore is regulated. Worldwide leakage requirements vary from 0.5 mA to 5 mA according to the application and the safety agency.* For example, Underwriters Laboratory (UL) has a 0.5-mA leakage requirement for most consumer products. In a 115-V system, this limits the maximum value of the filter's line-to-ground capacitors to 0.01 μF.

These capacitors $C_1$ and $C_2$ in Fig. 13-16 are referred to as Y-capacitors and must be of a type approved and listed by a safety agency such as UL for use in a line-to-ground application. Similarly, the line-to-line capacitor $C_3$ (referred to as an X-capacitor) must be of a type approved for use in a line-to-line application.

To obtain the large inductance required to suppress the lower order harmonics of the switching frequency, $L_1$ is wound on a high permeability core. To prevent core saturation resulting from the large ac power line currents, the two windings of the inductor are wound on the same core, which forms a common-mode choke. Because the power line currents are in opposite directions in each winding, the magnetic flux produced in the core by these currents cancels.

### 13.3.1   Common-Mode Filtering

In actual practice, the line-to-ground capacitors usually have a value of one half the maximum allowable by the leakage requirements.[†] The common-mode noise voltage sees the two line-to-ground capacitors in parallel; therefore, the effective common-mode capacitance is equal to the sum of the two capacitor values. Starting with this capacitance, the value of the common-mode choke is then chosen to provide the required common-mode attenuation. Typical values for the choke are from 2 to 10 mH. If more than 10 mH is required to achieve the necessary attenuation, then multiple chokes should be used in series to limit the parasitic capacitance across the choke.

### 13.3.2   Differential-Mode Filtering

To differential-mode noise, the two Y-capacitors are connected in series. Therefore, the effective differential-mode capacitance is equal to only one half the value of one of the capacitors. This provides little differential-mode filtering, especially at low frequencies where it is needed the most, and the capacitor value cannot be increased because of the leakage requirements. These capacitors only contribute to the differential-mode attenuation above about

---

*Some types of medical equipment have leakage requirements as low as 10 μA. In these applications, no capacitors to ground can be used in the power-line filter.

[†]This is because the power-line filter cannot use up the product's entire allowable leakage requirement; some must be left for the power supply and the product itself. Therefore, these capacitors have a value that will contribute only one half the leakage requirement.

10 MHz where it is usually not required. Therefore, they are usually ignored with respect to differential-mode filtering.

To provide a significant amount of differential-mode capacitance, a line-to-line capacitor $C_3$ (X-capacitor) is added to the power line filter. Because this capacitor is not connected to ground, its value is not limited by leakage requirements. Typical values for this capacitor range from 0.1 to 2 μF. For safety reasons, a resistor, typically 1 MΩ, is sometimes added in parallel with this capacitor. This resistor is used to discharge the capacitors when power is removed.

When the power supply has a poor quality ripple filter capacitor, or two capacitors in series, a second X-capacitor located across the power line and located on the power supply side of the common-mode choke can be helpful.

### 13.3.3  Leakage Inductance

Leakage inductance of the common-mode choke is important in power line filters because it determines the degree of differential-mode inductance present. An ideal common-mode choke provides no differential-mode inductance. The direction of the differential-mode noise current is opposite in each winding, and all the magnetic flux in the core cancels.

The leakage inductance of a choke or transformer is the result of imperfect coupling between the two windings. All the flux produced by one winding does not couple to the other winding; hence, when differential-mode currents flow in the windings there is some leakage flux that is not canceled. This leakage causes the winding to have a small differential-mode inductance.

In a power-line filter, leakage inductance can be both good and bad. As a result of leakage inductance, each winding of the choke will have in series with it a small differential-mode inductance. This differential-mode inductance along with the X-capacitor forms an $L$-$C$ filter, which provides differential-mode filtering. Too much leakage inductance, however, can cause the common-mode choke to saturate at a low value of ac power current, and it is an undesirable characteristic. As is the case for many other things in life, a little bit is good, and too much is bad.

Common-mode chokes are generally designed and built to have a specific value of leakage inductance, such that they provide a useful degree of differential-mode filtering, and yet do not saturate when carrying the rated power line current. Typical power line chokes will have leakage inductances somewhere between 0.5 and 5% of their common-mode inductance.

The leakage inductance of a common-mode choke can easily be measured by shorting one of the windings and measuring the inductance across the other winding. If there were no leakage flux, the short across the one winding would be induced across the other winding, by transformer action, and the inductance would measure zero. Therefore, whatever inductance is measured by this test setup must be the result of leakage inductance.

The differential-mode filter in the circuit of Fig. 13-16 consists of the X-capacitor $C_3$ and the leakage inductance of choke $L_1$. As was the case for the common-mode filter, this differential-mode filter also has a low pass $L$-$C$ topology, where the source and load impedances determine the exact configuration. For differential-mode noise the power supply is a low impedance source (the large filter capacitor $C_F$), and the LISN is a high-impedance load (the 100 $\Omega$ of the LISN resistors). For maximum attenuation, the low-impedance filter element (the capacitor $C_3$) should face the high-impedance load (the LISN), and the high-impedance filter element (the leakage inductance of $L_1$) should face the low-impedance source (the power supply). This is exactly how the elements are arranged in Fig. 13-16. This power line filter topology along with a table of common-mode and differential-mode source and load impedances is shown in Fig. 13-17.

The common-mode filter is usually designed first and then the differential mode filter is designed by starting with the leakage inductance of the common-mode choke, and choosing a value for the line-to-line capacitor $C_3$, to provide the required attenuation.

If additional differential-mode attenuation is required, two additional discrete differential-mode inductors can be added to the filter as shown in Fig. 13-18. The differential-mode inductors are wound on low permeability

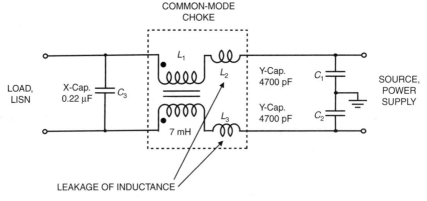

| LISN IMPEDANCE | MODE | POWER SUPPLY IMPEDANCE |
|---|---|---|
| LOW | COMMON-MODE | HIGH |
| HIGH | DIFFERENTIAL-MODE | LOW |

**FIGURE 13-17.** Power-line filter showing the leakage inductance of the common-mode choke, plus a table of the load and source impedances.

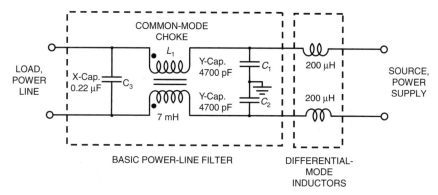

**FIGURE 13-18.** Power-line filter with the addition of two differential-mode inductors.

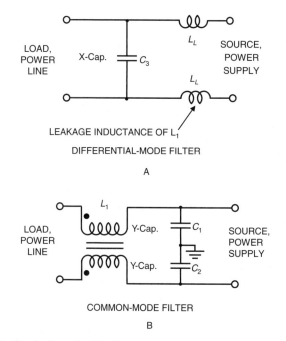

**FIGURE 13-19.** Equivalent circuits for a power-line filter: (*A*) differential-mode filter, and (*B*) common-mode filter. Note, $L_L$ is the leakage inductance of the common-mode choke $L_1$.

cores, so as not to saturate as a result of the large power frequency current flowing through them. Values for these differential-mode inductors are typically a few hundred microhenries.

The power-line filter shown in Fig. 13-17 can be separated into its common-mode and differential-mode components as shown Fig. 13-19. Figure 13-19A shows the circuit of the differential-mode filter, and Fig. 13-19B shows the circuit of the common-mode filter. Inductance $L_L$ in Fig. 13-19A represents the leakage inductance of the common-mode choke $L_1$.

Figure 13-20 is an overall noise equivalent circuit of a switched-mode power supply, power line filter, and LISN. The leakage inductance of the common-mode choke is represented by $L_L$. Capacitors $C_Y$ and choke $L_{CM}$ filter the common-mode noise and capacitor $C_X$ combined with the leakage inductance $L_L$ form the differential-mode filter.

From Fig. 13-20, we can observe that the connection between the power-line filter's shield (or filter's ground if the filter is not in a metallic enclosure) is in series with the Y-capacitors. Hence, any inductance in this connection will degrade the effectiveness of the Y-capacitors as a common-mode filter element.

### 13.3.4   Filter Mounting

Figure 13-21 shows a commercial power-line filter enclosed in a metal enclosure. The performance of this filter is as much, if not more, a function of how and where it is mounted, and how the leads are routed, as it is of the electrical design of the filter. Figure 13-22 shows three common problems associated with the mounting of a power-line filter that significantly decreases its effectiveness.

First, the filter is not mounted close to the point where the power line enters the enclosure; therefore, the exposed power line can pick up noise from electric and magnetic fields inside the enclosure. The filter cannot remove any noise picked up by the ac power line after the filter.

Second, the wire grounding the filter to the enclosure has a large inductance, which decreases the effectiveness of the Y-capacitors in the filter. The filter manufacturers mount the internal Y-capacitors to produce an absolute minimum inductance connection to the filter's enclosure, by using the technique shown in Fig. 14-11C.

Third, capacitive coupling occurs between the noisy power-supply-to-filter wiring and the ac power line. Do not route the filter-input leads close to output dc power leads, as this will maximize the parasitic capacitance coupling.

Figure 13-23 shows a properly mounted power-line filter that overcomes all three of the above problems. The filter is mounted where the ac power line enters the enclosure to prevent field coupling to the filtered power line. The metal enclosure now also blocks any capacitive coupling from the filter input cable and the filtered power line.

The filter is mounted such that the filter's metal case makes direct contact with the enclosure, which eliminates any additional inductance in series with the internal Y-capacitors. Any wire between the filter's case and the enclosure will decrease the effectiveness of the filter as a result of its inductance. Even short wires are too inductive and should be avoided.

The cable between the filter and the power supply should be routed close to the enclosure to minimize any pickup. The filter's input leads should also be kept away from any signal cables (especially digital cables) and should not

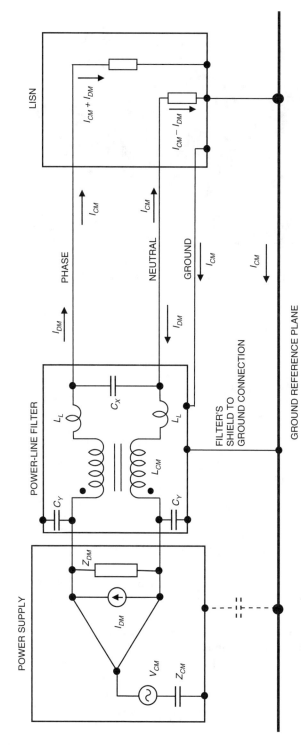

**FIGURE 13-20.** Noise equivalent circuit of switched-mode power supply, power-line filter, and line impedance stabilization network.

517

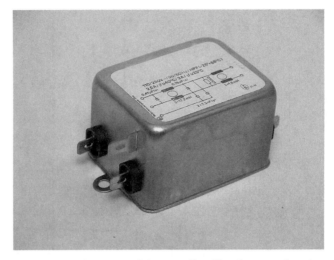

**FIGURE 13-21.** A commercial power-line filter in a metal enclosure.

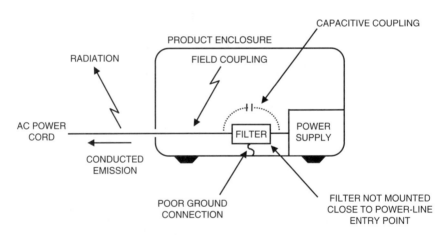

**FIGURE 13-22.** Improper power-line filter mounting and grounding.

be routed over, or near, a digital logic PCB. An additional improvement over the arrangement shown in Fig. 13-23 is to mount the power supply directly adjacent to the power line filter.

The above discussion points out the advantages of a power-line filter having an integral ac power cord connector as shown in Fig 13-24. This configuration forces the filter to be mounted where the power cord enters the enclosure, and when the filter's metal flange is screwed or riveted to the enclosure (on an unpainted, conductive surface) the Y-capacitors will be properly grounded. *The importance of proper power-line filter mounting and wiring cannot be overemphasized.*

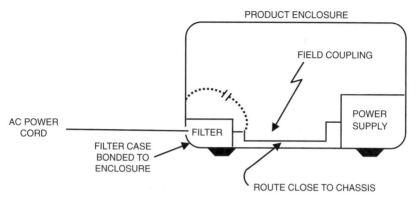

**FIGURE 13-23.** A properly mounted, and grounded power-line filter.

**FIGURE 13-24.** Commercial power-line filter with integral power cord connector.

### 13.3.5 Power Supplies with Integral Power-Line Filters

Some switched-mode power supplies have the power-line filter built into the supply on the same PCB as the power converter. This is usually done to reduce size and costs. An on-board filter can often be built for half the cost of a commercial filter in a separate enclosure (Nave, 1991, p. 29). However, this arrangement often violates some, if not all, rules for proper filter mounting and wiring discussed above. The three most common problems with this arrangement are as follows:

1. Long traces (too much inductance) connecting the Y-capacitors to the enclosure. In many cases, the Y-capacitor ground traces are a few inches long on the PCB, plus possibly a 1-in or even longer, metal standoff to the enclosure. That amounts to a significant length of conductor grounding the Y-capacitors and makes them ineffective as high-frequency filter element.

2. Magnetic coupling to the unshielded common-mode choke. Because PCB-mounted filters are seldom shielded, significant magnetic field coupling may occur between the switching transformer, as well as PCB current loops containing large $di/dt$ signals, and the common-mode choke. This couples the power supply noise directly to the choke, bypasses the Y-capacitors, and allows the noise to be conducted out onto the power line with much reduced attenuation. This problem can be overcome by proper layout and orientation of the common-mode choke on the board, or by placing a shield over the choke or power line filter portion of the board. The shield must be made of steel or other magnetic materials, not aluminum. For additional information on reducing magnetic field coupling, see Section 13.7.

3. Input and output traces to the filter, routed in such a way as to maximize the parasitic capacitance between the two, thus coupling noise around the filter to the power line.

Many times when the filter is integral with the power supply, it is necessary to add a second commercial filter to the product for it to pass the conducted emission tests. *Magnetic field coupling and high ground inductance are both significant contributors to the reduced effectiveness of on-board filters.* Filters integral with the power supply can be effective, but only if all the issues previously discussed relating to proper filter mounting and layout are considered during the design process.

### 13.3.6   High-Frequency Noise

The power-line filter is optimized to control the harmonics of the switching power supply, and it is not as effective at controlling high-frequency noise ($>$ 10 MHz). The high- frequency attenuation of the power supply noise is limited primarily by the interwinding capacitance of the common-mode choke and the inductance in series with the Y-capacitors. High-frequency noise fed back into the filter (e.g., from harmonics of the digital logic) can appear on the ac power line as conducted emissions, if below 30 MHz, or as radiated emissions if above 30 MHz. The best way to deal with this problem is at the source, the digital logic PCB. Include both a common-mode and differential-mode filter on the PCB, where the dc voltage comes onto the board, and design the board with minimum ground noise and optimum decoupling as covered in Sections 11.4 and 11.9.

Because most of this high-frequency noise is common mode, a small ferrite bead located at the dc output of the power supply (on the dc cable) can be effective in controlling noise above 30 MHz. As a last resort, a power supply transformer with a Faraday shield can be used to block the common-mode noise from feeding back from the output of the power supply to the power supply input.

A ferrite bead located on the ac power line between the filter and the power supply, located close to the filter end of the cable, can also be used effectively to keep high-frequency noise out of the filter.

Another problem that can occur in noise-sensitive applications is high-frequency noise on the dc output of the power supply interfering with the product that the supply is powering. For example, the case of a switching power supply used to power a sensitive radio receiver or low-level amplifier.

In addition to ripple at the switching frequency (or twice the switching frequency if a full wave rectifier is used in the dc output circuit), noise on the output leads of switch-mode power supplies consist of large amplitude narrow voltage spikes (the spacing of which is related to the switching frequency) often followed by substantial ringing. The amplitude of the voltage spikes typically can range from 50 mV to 1V peak to peak. The ringing typically has a frequency anywhere from 5 to 50 MHz. This high-frequency noise is usually both differential mode and common mode, and it may require additional high-frequency filtering on the dc output side of the supply.

High-frequency filtering can be obtained on the dc output, by adding ferrite beads in series with both conductors and by the addition of a high-frequency differential-mode capacitor, with very short leads, between the two dc output leads as shown in Fig. 13-25. The capacitor value should be chosen to have an impedance less than a few ohms at the lowest frequency of interest. A 1000-pF capacitor is usually satisfactory, if filtering is only required above 30 MHz. Below 30MHz, a 0.01-µF capacitor would be preferred. The high-frequency filtering is completed by adding capacitors to bypass the output leads to ground (chassis) to add high-frequency common-mode attenuation.

The ferrite bead material should be chosen to provide about 50 Ω of impedance at the lowest frequency of interest and the ferrite must be capable of carrying the output current without saturation. The use of a ferrite bead in each of the output leads provides both common-mode and differential-mode filtering. Multiple beads should be used in each lead to increase the series impedance if additional filtering is required.

At high frequency (greater than 30 MHz), radiation from the dc output cable may cause radiated emission problems. This can be reduced by placing a ferrite bead (common-mode choke) around both conductors of the cable.

Another approach when powering sensitive loads with a SMPS is to use a different switching topology, for example the quasi-resonant, zero-current, zero-voltage switching topology. With this topology, the switching current waveform is a half-sine wave that switches at a zero-current and zero-voltage point; it generates much less noise with a much-reduced spectral content.

522

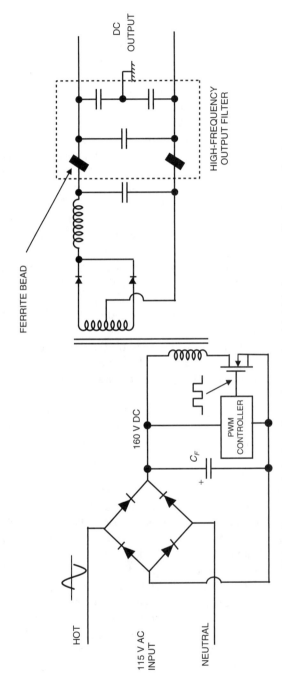

**FIGURE 13-25.** Switched-mode power supply with high-frequency filtering on the dc output.

## 13.4   PRIMARY-TO-SECONDARY COMMON-MODE COUPLING

Another mechanism by which common-mode currents are generated in switched-mode power supplies is shown in Fig. 13-26. This is the result of the input and output circuits both being grounded external to the power supply. In this case, the switching transistor acts as a noise voltage generator to produce a large $dV/dt$ on the primary winding of the power transformer; the magnitude of this voltage varies between the peak switching voltage and zero. This $dV/dt$ drives a common-mode current $I_{CM}$ through the transformer's interwinding capacitance $C_T$ to the power supply's output side ground, and back to the input side ground, as shown in Fig. 13-26. Even if the output circuit is not grounded, this loop may still exist as the result of parasitic capacitance between the output circuit and ground.

This external common-mode current can be eliminated or at least minimized by providing a smaller loop, therefore less inductance, internal to the power supply for the current to flow through. This result can be accomplished by adding a "bridge capacitor" between the primary common and the secondary ground inside the power supply, as shown in Fig. 13-27 (Grasso and Downing, 2006). This capacitor must be safety agency approved because it connects between one of the ac input power lines and the secondary ground. Usually, a Y-capacitor is used (the same type of capacitor used in the common-mode filter), with a value of 1000 to 4700 pF. To be effective, the bridge capacitor must be placed on the PCB in a location that minimizes the trace inductance (use short, wide traces) in series with it, and the traces must maintain a small loop for the common-mode current.

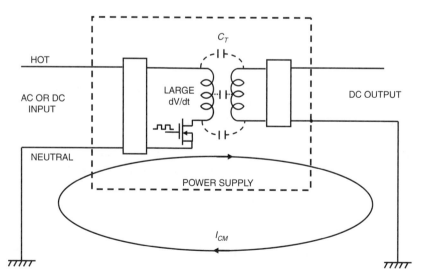

**FIGURE 13-26.** Example of $dV/dt$ coupling between power supply's primary and secondary sides.

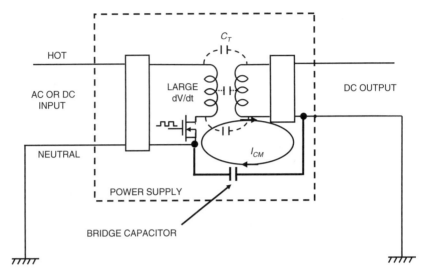

**FIGURE 13-27.** Use of a bridge capacitor to bypass the external common-mode ground loop.

Other ways to eliminate or minimize this problem are by using a transformer that contains a Faraday shield, which effectively reduces the interwinding capacitance, or by adding a common-mode choke in the dc output-leads to reduce the common-mode current.

### 13.5  FREQUENCY DITHERING

Yet another approach to reducing both common-mode and differential-mode emissions from a SMPS is to vary the switching frequency over a narrow range to spread the energy out over a larger frequency band, which decreases the peak amplitude at any one frequency. This technique is similar to the digital circuit clock dithering discussed in Section 12.2.3. Several integrated circuit (IC) pulse-width-modulation power supply controllers have this feature built in. The modulation waveform is usually a triangular wave with a frequency of a few hundred hertz and a maximum deviation of a few kilohertz.

In products with a low leakage requirement (e.g., 10 μA for some medical equipment) where Y-capacitors cannot be used in the power-line filter (or if used, the value of the capacitors severely restricted), the use of frequency dithering and/or a bridge capacitor are extremely useful techniques to reduce the common-mode emissions.

### 13.6  POWER SUPPLY INSTABILITY

Interactions between a power supply and the power-line filter can, under some conditions, cause the power supply to become unstable. A switched-mode power supply has a negative input impedance,* and if not properly terminated by the power-line filter it can actually become unstable.

It is easy to demonstrate that a regulated power supply has a negative input impedance, by analyzing how such a supply behaves when the input voltage varies. Consider the case of a power supply feeding a constant load impedance. If the output voltage is constant, then the output current will be constant and therefore the output power will be constant. The input power must then also be constant. If the input power is constant and the input voltage decreases, then the input current must increase to keep the power constant. This behavior, of the input current increasing when the voltage decreases, is that of a negative resistance.

When a power supply is operated directly off the low-impedance ac power line, it will normally be stable. However, when the impedance of a power-line filter is inserted between the power supply and the power line it has the potential of, causing the supply to become unstable.

Consider what could happen if a power supply having a negative input impedance were connected to a power-line filter with a positive output impedance. If the source voltage decreases slightly, then the power supply will have to draw more current to keep the power constant. The increased current will cause an increased voltage drop across the filter's output impedance, which lowers the input voltage to the power supply. This in turn will cause the power supply to draw even more current and will lower the input voltage even more and so on and so on. Hence instability!

It can be shown that for stability, *the magnitude of the filter's positive output impedance must be less than the magnitude of the power supply's negative input impedance* (Nave, 1991, p.121). For a more detailed discussion of power supply stability, I would direct the reader to Chapter 6 of *Power Line Filter Design for Switched-Mode Power Supplies*, by Mark Nave.

## 13.7   MAGNETIC FIELD EMISSIONS

Switched-mode power supplies produce strong magnetic fields that have the potential for causing many problems. Other than some military and automotive products, magnetic field emission has no regulatory limits. However, magnetic fields can have a detrimental effect on the power supply operation and on the operation of other circuitry in their vicinity.

Within a power supply, the major sources of magnetic fields are (1) current loops with large rates of change of current $di/dt$ and (2) the switching transformer. Because effective low-frequency magnetic field shielding is difficult to achieve (see Chapter 6), it is better to control the magnetic fields at their source—avoid or minimize generating them.

Critical loops, those with large $di/dt$, should be carefully laid out to minimize their area. Figure 13-28 shows the two most critical loops in a switched-mode

---

*This is true at frequencies within the power supply's feedback loop bandwidth.

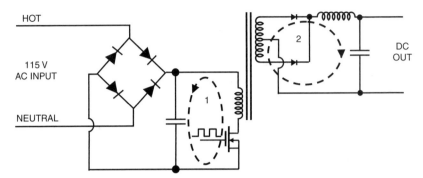

**FIGURE 13-28.** Critical loop areas in a switched-mode power supply. (1) The switching transistor loop (primary loop) and (2) the rectifier loop (secondary loop).

power supply. These are the switching transistor loop (primary loop) and the rectifier loop (secondary loop). Of the two, the rectifier loop usually has the higher *di/dt* and is therefore often the most important with respect to magnetic field generation. Magnetic field radiation from a loop is proportional to (1) the loop area and (2) the *di/dt* in the loop. The easiest parameter for the power supply designer to control is the loop area. This can be achieved by careful PCB layout and routing. The currents in the primary and secondary loops should be separated from each other and each should be made as small as possible.

When the power-line filter is located on the same board as the power supply, magnetic field coupling often occurs between the power transformer and the power-line filter. Many commercial power supplies use E-core transformers because of their low-cost and ease of manufacturing. These transformers, however, can have significant amounts of magnetic field leakage. One method of reducing the transformer's magnetic field leakage is by the use of a toroidal core, instead of an E-core. Toroidal core transformers have much less leakage flux but are more difficult to manufacture and are therefore more expensive.

The easiest way of reducing the leakage flux from an E-core transformer is by the use of a shorted turn (sometimes called a "belly band"). A shorted turn is a wide band of copper that wraps around the transformer's windings (see Nave, 1991, p. 180). The shorted turn is coupled magnetically only to the leakage flux, and it acts as a low-impedance high-current secondary winding. The current induced into the shorted turn produces a magnetic field, which is opposite in polarity to the leakage flux and thereby cancels most of the original leakage.

Whenever an on-board power-line filter is used, care must be taken to separate the filter and the power transformer as much as possible to minimize magnetic field coupling. Separation is a effective method of reducing near field coupling, because in the near field, the field strength falls off as the cube of the distance from the source.

Magnetic fields can also couple to other circuits on the PCB and cause noise in these signals. The best protection to these signals is to reduce their loop areas. Routing a signal trace and its return trace close together to minimize the loop area is one simple technique. Twisting wires or transposing traces on a PCB is another effective method of providing protection from magnetic field coupling. One sensitive trace that should be carefully protected is the voltage feedback signal that controls the regulation of the power supply's output voltage. If this signal runs close to the power transformer, then the signal trace and its ground return trace should be transposed periodically on the board, to simulate a twisted pair, which reduces its magnetic field susceptibility.

Figure 13-29 shows an interesting case of how magnetic field coupling inside a power supply can produce a common-mode noise current on the dc output conductors. Magnetic field coupling to the crosshatched area of the figure can induce a common-mode voltage $V_{CM}$ into the dc output wire bundle. This voltage will produce a common-mode noise current on the dc output cable.

A novel solution to this problem is to connect a wire (sometimes referred to as a "chassis wire") from the chassis near where the dc output cable bundle connects to the PCB, to the chassis near the output connector and route the wire adjacent to the output wire bundle as shown in Fig. 13-30. This method significantly reduces the loop area into which the stray magnetic field can induce a common-mode voltage, which reduces the voltage and consequently reduces the common-mode output noise current.

The chassis-wire technique is just a variation of the technique of using a shield grounded at both ends of a cable to prevent magnetic field coupling, as was discussed in Section 2.5.2. The chassis wire is a "poor man's" attempt to simulate a shield on the output wire bundle. In fact, this technique could be improved on by actually shielding the dc output wire bundle and grounding the shield to the chassis at both ends.

Capacitors from the individual dc output leads and connected to the enclosure could also be used to minimize this problem. However, to be effective

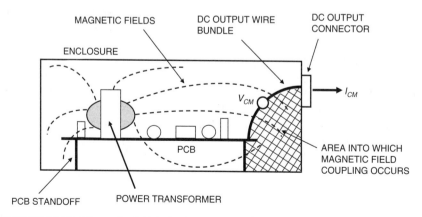

**FIGURE 13-29.** Magnetic field inducing a common-mode voltage into the dc output conductors of a power supply.

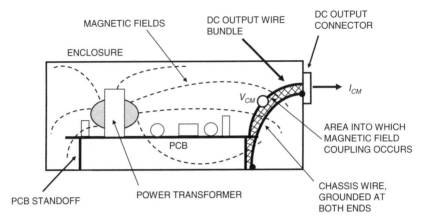

**FIGURE 13-30.** Using a "chassis wire" to reduce common-mode currents on the dc output of a power supply.

they must be located at the output connector and are considerably more expensive than just a wire. A capacitor filter pin connector works well in this application.

As a last resort, if shielding of the power supply is required to reduce the magnetic field emissions, use steel or other magnetic materials, not aluminum.

## 13.8   VARIABLE SPEED MOTOR DRIVES

High-power, solid-state, variable speed motor drives are becoming popular in industry because of their ability to easily control the speed of ac induction motors, and their high efficiency. They are also commonly used in hybrid and electric vehicles. Prior to the introduction of variable speed motor drives, the only practical way to vary the speed of an ac induction motor was mechanical with gears, belts, or pulleys. In a variable speed motor drive, speed control is accomplished by varying both the frequency and amplitude of the current driving the motor. As a result, variable speed motor drives are sometimes also referred to as variable frequency drives.

Variable speed drives are based on the principle that the synchronous speed of an ac induction motor is determined by the ac supply frequency and the number of poles in the motor's stator. The relationship between these parameters is as follows:

$$RPM = \frac{120 \times f}{p}, \qquad (13\text{-}10)$$

where *RPM* is the motor speed in revolutions per minute, $f$ is the supply frequency, and $p$ is the number of poles in the motor. An induction motor will operate at a speed about 4% less than the synchronous speed as a result of slip.* As can be observed from the above equation, the only convenient way to vary the speed of an ac motor electrically is by changing the applied frequency.

Motor torque, however, is a function of the ratio of the applied voltage to the applied frequency. Therefore, to maintain a uniform torque the voltage must also be proportionally adjusted whenever the frequency is varied. The objective is to maintain a constant voltage to frequency ratio. For example, if a motor designed to operate at 230 V, 60 Hz, is operated from a 30-Hz supply, the voltage must be reduced to 115 V.

Most variable speed drives are designed for a 230- or 460-V, 3-phase motors that range from 1/4 hp to 1000 hp. Using a variable speed motor drive, a motor can be started at a frequency of a few hertz and at a low voltage. This has the advantage of avoiding the high-inrush current normally associated with motor starting, when full utility voltage and frequency are applied. With this approach, the motor can provide 100% of its rated torque at virtually zero speed. The drive then can increase the frequency and voltage proportionately, to accelerate the motor at a controlled rate without drawing excessive current from the power line. Variable speed motor drives typically can vary their output frequency from about 2 Hz to 400 Hz.

A simplified block diagram of a basic 3-phase variable speed motor drive is shown in Fig. 13-31. The rectifier converts the three-phase input frequency to a dc voltage, which is stored on a capacitor in the dc link. The inverter then converts the dc voltage to a series of high-frequency, constant voltage pulses. The output pulse width is modulated in such a way that when filtered by the inductance of the motor windings it produces a quasisinusoidal current waveform, the amplitude of which varies in proportion to the frequency of the sinusoidal current. The wider the voltage pulse the larger the motor current will be, and the narrower the voltage pulse the smaller the motor current will be. Figure 13-32 shows voltage and current output waveforms for a typical pulse-width modulated inverter. An integral microprocessor in the drive controls the overall operation of the inverter and adjusts the output voltage pulse width appropriately.

The rate at which the pulses switch ON and OFF is the switching or carrier frequency. The higher the switching frequency the smoother the current waveform and the greater the resolution will be. This is similar to a digital-to-analog converter, the more bits the better the resolution and the smoother the created analog waveform. The current waveform will also have some high-frequency

---

*Slip is the difference between the motor (rotor) speed and the rotating magnetic field in the stator. Slip is what actually causes a motor to rotate as the rotor tries to catch up with the stator's rotating magnetic field. A 4-pole 60-Hz motor will have a synchronous speed of 1800 rpm (Eq. 13-10) but will typically operate at a speed of 1750 rpm as the result of slip.

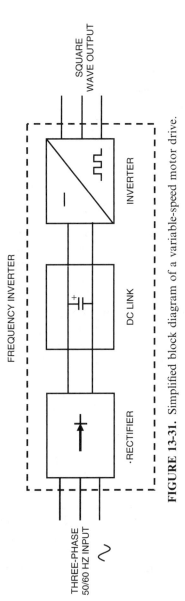

**FIGURE 13-31.** Simplified block diagram of a variable-speed motor drive.

noise riding on it (not shown in Fig. 13-32), which consists of harmonics of the switching frequency.

Several different switching devices are available to accomplish this, but insulated gate bipolar transistors (IGBTs) are most commonly used today. Figure 13-33 shows a circuit diagram for the power output stage of a pulse-width-modulated, variable speed motor drive operating from a three-phase ac power source and driving a three-phase motor. These drives have many characteristics in common with switched-mode power supplies, except that there is no transformer to step down the output voltage, and no secondary side rectifier to convert the secondary voltage to dc. Variable-speed motor drives are

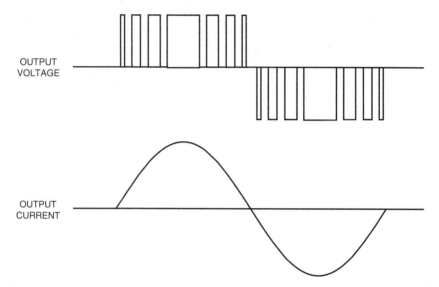

**FIGURE 13-32.** Variable-speed motor drive output waveforms. Top trace, pulse-width-modulated voltage. Bottom trace, filtered motor current.

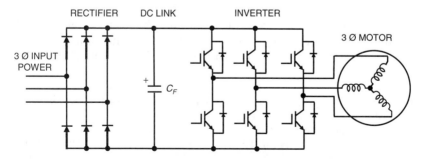

**FIGURE 13-33.** Circuit of a pulse-width-modulated variable-speed motor drive.

a source of broadband conducted and radiated emissions, and a major producer of harmonic distortion on the input ac power line.

An additional advantage of these motor drives is that a three-phase ac motor can be powered off of a single-phase ac source, or even a dc source as in the case of hybrid and electric vehicles. In all cases, the input is or is first converted to a dc voltage, then it is chopped into a pulse-width-modulated square wave that drives the three-phase motor with a variable frequency and amplitude quasi-sinusoidal current.

As shown in Fig. 13-33, the inverter portion of the drive consists of three pairs of IGBT transistors; each pair drives one phase of the three-phase motor. One transistor of each pair pulls the output high to the positive dc rail, and the other one pulls the output low to the negative dc rail, this is similar to the "totem pole" output of a digital logic gate. To understand the operation, however, we need only consider the operation of one pair of transistors.

Figure 13-34 is a simplified circuit of the basic variable speed motor drive, which shows only one pair of switching transistors. The power source provides a dc voltage to capacitor $C_F$, either directly from a dc source or by rectifying a single- or three-phase ac source. When the upper transistor turns ON (the lower one is OFF), the drive outputs a positive voltage pulse. When the lower transistor turns on (the upper one is OFF), the drive outputs a negative voltage pulse. This switching ON and OFF of the square wave output voltage typically occurs at a rate somewhere between 2 and 15 kHz. Using an analogy to the switched-mode power supply, the motor winding is similar to the transformer in the power supply, and the IGBTs are the switching transistors.

The major noise problems associated with variable speed motor drives are common-mode conducted emissions and radiated emissions from the motor cable. The common-mode conducted emission is created in a manner similar to that previously discussed in Section 13.2.1 for the case of the SMPS.

The primary side (input) conducted emission can usually be controlled by using a two-stage power line filter, often which contains two common-mode choke sections as shown in Fig. 13-35.*

The secondary (output) side common-mode currents are often more of a problem in variable speed motor drives and often more difficult to control. A large $dV/dt$ appears at the junction of the two IGBTs as shown in Fig. 13-34 and is a noise source that will drive a common-mode current $I_{CM}$ out on the motor cable. Because of the parasitic capacitance between the motor windings and motor housing (typically 100 to 500 pF) the common-mode current will return on the external ground back to the IGBT drive circuit.

---

*Two stages are used to provide increased attenuation. Two common-mode chokes are preferred in this case, as opposed to just increasing the inductance of the single choke for two reasons. One, an increased value single choke is more likely to saturate as the result of the power line current and second, a larger value single choke would have increased inter-winding capacitance and therefore worse high-frequency filter effectiveness than the case with two chokes as was discussed in the Section 13.3.1 on Common-Mode Filtering.

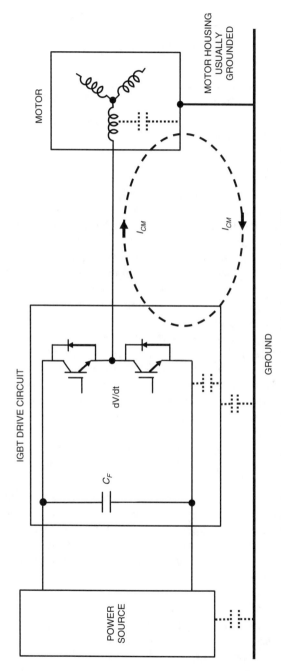

**FIGURE 13-34.** Simplified diagram of a variable-speed motor drive output circuit.

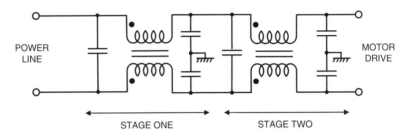

**FIGURE 13-35.** A two-stage power-line filter for use with a variable-speed motor drive that has a single-phase ac input.

In practice, any of the three components (power source, IGBT drive circuit, or motor) may or may not be intentionally grounded. Even if they are not grounded, parasitic capacitance will complete the common-mode current loop. This situation is similar to the switched-mode power supply shown in Fig. 13-26, less the transformer. In both cases, the switching device is the source of the $dV/dt$ that drives the common-mode current through the parasitic capacitance to ground. In the power supply case, the common-mode current flows through the transformer's interwinding capacitance. In the motor drive case, the current flows through the motor winding-to-ground capacitance.

Conceptually, it is always better to try and reduce the source of the noise, which in this case is the $dV/dt$ of the switches. Reducing $dV/dt$ by slowing the rise time of the switch is not, however, desirable from a functional point of view, because it reduces the efficiency of the drive as the result of the switching transistors remaining in the linear region for a longer period of time and therefore dissipating more power. Slowing down the rise time will only decrease the common-mode conducted emission at frequencies above the $1/\pi t_r$ frequency. Keep in mind, however, that slowing down the switches, even slightly, can have a significant effect on the higher frequency harmonics. Also, it is fairly easy to implement by inserting an $R$–$C$ filter, or in some cases just a ferrite bead in the gate drive to the IGBTs.

Another approach is to provide a return path other than the external ground for the common-mode current. Adding a ground return wire to the motor cable that connects the motor housing to the switch-common (negative rail) would be one solution, but often it is impractical to implement. Either the motor housing must be floated off of ground, or the switch common must be connected to ground; both occurrences are unlikely.

A more practical approach would be to run a ground wire between the motor housing and the variable speed drive enclosure and then to connect the drive enclosure to the switch-common through a capacitor as shown in Fig. 13-36. This technique is similar in concept to the use of a bridge capacitor in a switched-mode power supply shown in Fig. 13-27.

The ground wire can be implemented as a shield on the motor cable. Not only is this a practical approach, but also it has the added advantage that the

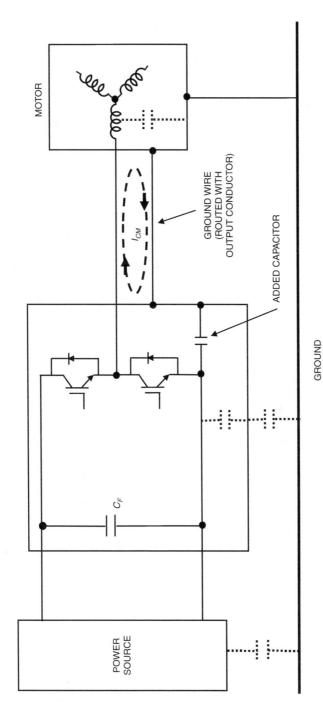

**FIGURE 13-36.** Ground wire added between the motor housing and the drive enclosure, plus a capacitor from drive enclosure to the negative rail of the switch to provide a return path for the common-mode current.

535

shield will also reduce any radiated emission from the motor drive cable, hence doing double duty. In addition, adding a ferrite core to the cable (close to the drive) will increase the common-mode impedance of the cable at high frequencies without affecting the lower frequency differential-mode motor drive current. The ferrite will decrease the common-mode current as well as reduce the radiation from the cable. The ferrite core approach can be used on a shielded or unshielded motor drive cable.

Another approach is to insert inductors (often called line reactors or $dV/dt$ chokes) in series with the output of each motor driver transistor pair. This approach slows down the motor current rise time without increasing power dissipation in the switches, as would occur if the switches themselves were slowed down. The slowed down rise time reduces the spectral content of the common-mode current in the cable. The disadvantage is that these inductors must be physically large to handle the motor current without saturating.

Figure 13-37 is a block diagram of a variable speed motor drive with both common-mode current and harmonic (see next section) suppression. The cable shield and the ferrite core on the shielded cable are controlling output common-mode current and cable radiation. The power-line filter controls the conducted emission on the power line side of the variable speed motor drive.

## 13.9   HARMONIC SUPPRESSION

The European Union has requirements that limit the harmonic content of the current drawn by products that are directly connected to the public ac power distribution system. These limits are listed in Table 18-3.

The generation of harmonics is the result of the nonlinear behavior of loads connected to the ac power line. The major contributor to this problem is a full wave rectifier followed by a capacitor input filter. This combination is common in both switched-mode power supplies and variable speed motor drives. In this circuit, current is only drawn from the ac power line when the magnitude of the input voltage exceeds the voltage on the input filter capacitor. As a result, current is only drawn on the peaks of the voltage waveform as was shown Fig. 13-4. The resultant current waveform is rich in odd harmonics. Total harmonic distortion (THD) values of 70% to 150% are not uncommon under these circumstances.

To overcome this problem, some form of power factor correction circuitry is required. Power factor correction circuitry can be classified into two general categories, passive or active. Passive power factor correction uses only passive elements, usually an inductor. Whereas passive power factor correction is simple, it is difficult to achieve low levels of harmonic distortion, and the components may be physically large because they must operate at the ac line frequency.

Active power factor correction, however, uses a combination of passive and active elements. The active elements are usually a switching transistor

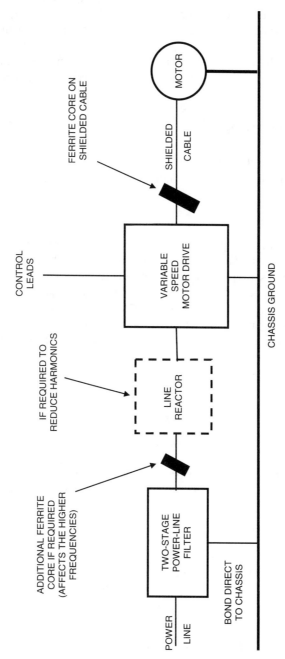

**FIGURE 13-37.** Block diagram of variable-speed motor drive with both conducted emission and harmonic suppression.

537

and an IC controller. The passive elements are usually a diode and inductor. The circuit operates at a high frequency, and therefore the inductor can be smaller than one that would operate at the ac power-line frequency. Active power factor correction can achieve low levels of THD, in some cases 5% or less.

To reduce the harmonic content, the current pulse must be spread out over a larger portion of the voltage waveform's cycle. Three possible ways to do this are as follows:

- Use an inductive input filter instead of a capacitive input filter after the rectifier
- Use active power factor corrector circuitry
- Add power factor correction inductors on the ac side of the rectifier

### 13.9.1   Inductive Input Filters

The use of an inductive input filter will spread out the current waveform, and in many cases reduce the harmonics sufficiently to make the product compliant. Figure 13-38 shows a typical input waveforms for a SMPS having an inductive input filter. The inductor limits the $di/dt$, and therefore it slows down the rise of the current waveform. In addition, it spreads out the current pulse and by decreases the peak amplitude, which reduces the total harmonic content of the waveform. Compare the current waveform of Fig. 13-38 to that of Fig. 13-4.

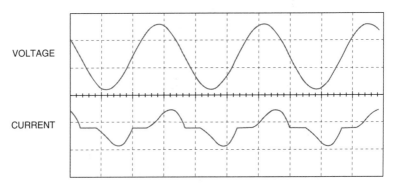

**FIGURE 13-38.** Input waveforms for a switched-mode power supply with an inductive input filter. Top trace, voltage; bottom trace, current.

### 13.9.2   Active Power Factor Correction

In the case of a switched-mode power supply, a much better solution to the harmonics problem is to use an active power factor corrector (PFC) circuit.

This circuit can be physically small and can reduce the current distortion to a low level. Several PFC circuits are available, which will spread the current pulse out across the complete cycle. Figure 13-39 shows one common approach to active power factor correction. This circuit uses a boost converter topology to produce an active harmonic filter. This circuit is a discontinuous current mode PFC circuit and is often used on power supplies under a few hundred watts. Some IC manufacturers combine the functions of a PFC controller and the PWM power supply controller into a single IC.

The PFC controller monitors both the full wave rectified input voltage and current. The controller turns the PFC switch ON and OFF continuously at a high-frequency rate (tens of kilohertz), this draws a triangular waveform input current through the PFC inductor; the peak amplitude of each triangular pulse is controlled to be proportional to the input voltage. The envelope of the triangular pulses therefore is sinusoidal, and the peak amplitude of the sinusoid varies with the power demands of the supply. The voltage and current at the input to the PFC circuitry, point *A* in Fig. 13-39, during one full cycle of the ac power line's input voltage is shown in Fig. 13-40.

Because the triangular current pulses flow continuously through the PFC inductor, the average current is a sinusoid with a peak value of one half of the peak of the triangular pulses. The power-line filter's differential-mode inductance is then used to average these triangular pulses, which produces a sinusoidal current at the input of the power-line filter; the amplitude of which is proportional to the power demand of the SMPS.

When the PFC switch is ON, the triangular current stores energy in the PFC inductor. The PFC inductor then charges the power supply's filter capacitor $C_F$ through the diode $D_1$ during the time that the PFC switch is OFF (see Fig. 13-39). The diode is needed to prevent the filter capacitor $C_F$ from discharging back into the PFC circuit when the PFC switch is turned ON.

### 13.9.3  AC Line Reactors

A power factor inductor on the ac side of the rectifier produces results similar to that of an inductive input filter. However, when the inductor is on the ac side of the power supply, two inductors must be used for a single-phase line and three inductors for a three-phase line, in order not to unbalance the power line and convert common-mode noise into differential-mode noise. Seldom is this approach practical in the case of switched-mode power supplies.

The use of ac line reactors (inductors) on the ac power line, however, is a common and effective approach to solving the harmonic problem for variable-speed motor drives. They can be used on single-phase or three-phase ac power lines. A block diagram of a variable speed motor drive with an ac line reactor to control harmonics was shown in Fig. 13-37. A line reactor is also effective in protecting the drive from transient overvoltages on the ac power line.

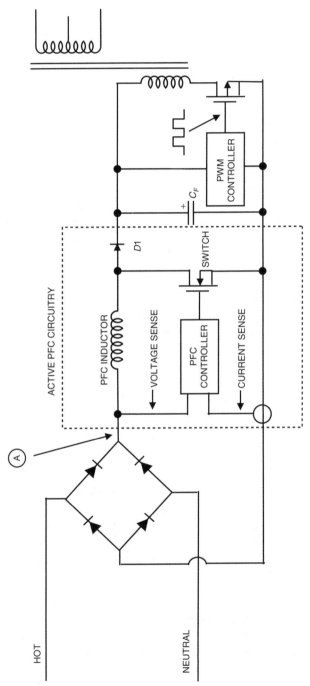

**FIGURE 13-39.** A switched-mode power supply with active power factor correction.

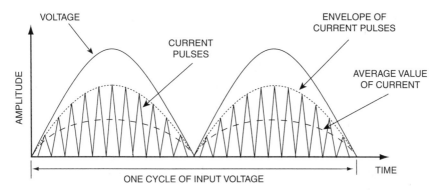

**FIGURE 13-40.** Active power factor corrector circuit input voltage and currentwaveforms. Waveforms shown are at point A in Fig. 13-39.

## SUMMARY

- The ac power line impedance in the frequency range of 100 kHz to 30 MHz can vary from about 2 to 450 Ω.
- Common-mode emissions from SMPS are proportional to the input voltage to the power supply.
- Common-mode emissions from SMPS are proportional to the value of the primary-side parasitic capacitance to chassis or ground.
- Differential-mode emissions from SMPSs are proportional to the magnitude of the input current, and therefore they are a function of the power rating of the supply.
- The power supply element that has the largest influence on the differential-mode emission is the input ripple filter capacitor.
- Differential-mode emissions above approximately 500 kHz are proportional to the ESL, and those below about 500 kHz to the ESR of the input ripple filter capacitor.
- Both common and differential-mode emissions are directly proportional to the switching frequency of the power supply.
- The power-line filter elements that affect the common-mode attenuation are the Y-capacitors and the common-mode choke.
- The power-line filter elements that affect the differential-mode attenuation are the X-capacitor and the leakage inductance of the common-mode choke.
- The effectiveness of a power-line filter is as much a function of how and where the filter is mounted, and how the leads are run to it, as it is of the electrical design of the filter.
- When the power-line filter is on the same PCB as the power supply, magnetic field coupling, and high ground inductance in series with the Y-capacitors can significantly reduce the effectiveness of the filter.

- Interactions between a power supply and the power-line filter can, in some cases, cause instability of the power supply.
- To reduce magnetic fields in power supplies:
  - Minimize the area of all loops containing a large $di/dt$.
  - Use a toroidal core transformer instead of an E-core transformer.
  - Add a shorted turn to an E-core transformer.
  - Shield the unit with steel or other magnetic material, not aluminum.

- A bridge capacitor can be effective in reducing certain types of common-mode conducted emission in switched-mode power supplies.
- Intentionally dithering (varying) the frequency of the power supply is another method of reducing the conducted emissions, both common- and differential-mode.
- The primary generator of harmonics on the ac power line is a full wave rectifier followed by a capacitor input filter. This configuration is common in the following:
  - Switched-mode power supplies
  - Variable speed motor drives.

- In many respects, variable speed motor drives behave very similar to, and have the same EMC problems as, switched-mode power supplies.
- Conducted emissions on the ac input to variable speed motor drives is usually controlled with a power line filter, often having two common-mode sections.
- Common-mode emissions on the output of variable speed motor drives is often best controlled by adding a ground wire and capacitor between the motor housing to the drive common, which provides an alternative return path for the common-mode noise current.
- The following methods can be used to reduce harmonic currents on the ac power line in switched-mode power supplies and variable speed motor drives:
  - Use an inductive input filter instead of a capacitive input filter,
  - Use active power factor correction circuitry, or
  - Add a power factor correction inductor (line reactor) on the ac power line.

## PROBLEMS

13.1  a. Derive Eq. 13-1.
   b. What assumption must be made in the derivation of Eq. 13-1?
   c. Demonstrate that the assumption(s) of Problem 13-1b is reasonable.

13.2    To comply with a 0.5-mA leakage requirement, what is the maximum value of the power-line filter's line-to-ground capacitance (Y-capacitor) for a 230 V, 50-Hz system?

13.3    List three significant differences between the current waveform of Fig. 13-4 and Fig. 13-38?

13.4    A switched-mode power supply has the following characteristics:
Switching frequency: 100 kHz
Maximum switching current: 4 A
Input voltage: 115 V ac (dc voltage on input ripple filter capacitor is 160 V)
    Input ripple filter capacitor characteristics:
        Capacitance: 470 μF,
        Voltage rating: 250 V,
        ESL: 30 nH
        ESR: 0.1 Ω
    Switching transistor rise time: 75 ns
    Primary (input) circuit parasitic capacitance to ground: 150 pF
Draw a log-log plot of the envelope of the common-mode conducted emission from 150 kHz to 30 MHz.

13.5    For the power supply of Problem 13-4, make a log-log plot of the envelope of the differential-mode conducted emission from 150 kHz to 30 MHz?

13.6    In a 12-V, 10-A input dc-to-dc converter, would you expect common-mode or differential-mode conducted emission be predominate? The input ripple filter capacitor has the characteristics listed in Problem 13-4, and the primary circuit to ground parasitic capacitance is 100 pF.

13.7    What is the significance of leakage inductance in the common-mode choke of an ac power-line filter?

13.8    In a properly mounted power-line filter, what parameter limits the high-frequency common-mode effectiveness of the filter?

13.9    What would be the maximum value of the power-line filter's Y-capacitors be, for a product that must meet a 10-μA leakage requirement, when connected to a 115-V, 60-Hz power line?

13.10   What two parameters of a switched-mode power supply, other than the switching frequency and peak current rating, have the most effect on the differential-mode conducted emission?

13.11   Which has the largest effect on the maximum value of the common-mode conducted emission from a switched-mode power supply, the switching frequency or the rise time of the switching transistor?

## REFERENCES

Grasso, C. and Downing, B. "Low-Cost Conducted Emissions Filtering in Switched-Mode Power Supplies." *Compliance Engineering*, 2006 Annual Reference Guide.

Hnatek, E. R. *Design of Solid State Power Supplies,* 3rd ed. New York, Van Nostrand Reinhold, 1989.

Nave, M. J. *Power Line Filter Design for Switched-Mode Power Supplies.* New York, Van Nostrand Reinhold, 1991.

Nicholson, J. R. and Malack, J. A. "RF Impedance of Power Lines and Line Impedance Stabilization Networks in Conducted Interference Measurements." *IEEE Transactions on Electromagnetic Compatibility,* May 1973.

Paul, C. R. *Introduction to Electromagnetic Compatibility*, 2nd ed. New York, Wiley, 2006.

## FURTHER READING

EN 61000-3-2. *Electromagnetic Compatibility (EMC)—Part 3-2: Limits—Limits for harmonic current emissions (equipment input current ≤ 16 A per phase).* CENELEC, 2006.

Erickson, R. W. and Maksimovic, D. *Fundamentals of Power Electronics.* 2nd ed. New York, Springer, 2001.

Fulton, D. "Reducing Motor Drive Noise." Part 1, *Conformity*, July 2004.

IEEE Std. 519. *Recommended Practices and Requirements for Harmonic Control in Electrical Power Systems.* IEEE, 1992.

Schneider, L. M. "Noise Source Equivalent Circuit Model for Off-Line Converters and its Use in Input Filter Design." *IEEE 1993 International Symposium on Electromagnetic Compatibility*, Washington, DC, August 1983.

Severns, R. *Snubber Circuits For Power Electronics*, ebook, www.snubberdesign.com/snubber-book.html, Rudolf Severns, 2008.

# 14 RF and Transient Immunity

Since 1996, interest in the subject of electromagnetic immunity has increased as a result of the European Union's EMC regulations for commercial products that covers both emissions and immunity. Whereas, digital circuits are the primary contributors to produce radiation (see Chapter 12), low-level analog circuits are the primary concerns with respect to radio frequency (rf) susceptibility. Digital circuits, however, are much more likely to be susceptible to high-voltage transients such as electrostatic discharge (ESD) than to rf fields. Much of the material already covered in this book (e.g., the chapters on cabling and shielding) is equally applicable to emission or immunity.

Immunity is defined as the ability of a product to operate without degradation in the presence of an electromagnetic disturbance. The inverse of immunity is susceptibility, which is the tendency of a device to malfunction or exhibit degraded performance in the presence of an electromagnetic disturbance.

This chapter covers the design of electronic systems for immunity. Radio frequency immunity, transient immunity, and immunity to power-line disturbances are all addressed. It should be obvious that not all equipment needs to be designed for the same degree of immunity. In selecting an appropriate immunity level, one should consider the use of the product, the potential consequences of a malfunction, the expectation of the user, the electromagnetic environment in which the product will be used, as well as any applicable regulatory requirements.

Even if a product is not required to meet immunity requirements (e.g., a commercial product that is only being marketed in the United States), it would still be wise to design and test it for immunity to avoid field failures and to keep the user satisfied. All products, therefore, should be designed and tested for at least a minimum degree of conducted, radiated, transient, and power-line immunity.

## 14.1 PERFORMANCE CRITERIA

One issue that arises with respect to immunity requirements and testing (that is not an issue with respect to emissions) is what constitutes a failure. We should all be able to agree that if during immunity testing a product is damaged or

*Electromagnetic Compatibility Engineering,* by Henry W. Ott
Copyright © 2009 John Wiley & Sons, Inc.

becomes unsafe, that would constitute a failure. However, short of actual damage or the product becoming unsafe, there is a lot of room for different interpretations as to what constitutes a failure. For example, if during immunity testing of a television set, the display breaks vertical hold and the screen display rolls vertically once or twice, is that a failure? Different people might have different answers to that question.

To their credit, the European Union's immunity standards define three failure criteria. Each immunity test then specifies which of the three criteria is applicable to that test. The three criteria are as follows (EN 61000-6-1, 2007):

*Criteria A*: The equipment shall continue to operate as intended *during and after the test*. No degradation of performance or loss of function is allowed.

*Criteria B*: The equipment shall continue to operate as intended *after the test*. After the test no degradation of performance or loss of function is allowed. During the test, degradation of performance is allowed. However, no change of the operating state or stored data is allowed.

*Criteria C*: *Temporary loss of function is allowed*, provided the function is self-recoverable or can be restored by the operation of controls.

Criteria A is applicable to rf immunity, Criteria B is applicable to transient immunity as well as in the case of some power line disturbances, and Criteria C is applicable in the case of severe power-line dips and interruptions.


## 14.2   RF IMMUNITY

Radio frequency interference (RFI) can be a serious problem for all electronic systems, including home entertainment equipment, computers, automobiles, military equipment, medical devices, and large industrial process control equipment.

Radio frequency immunity standards exist to control or limit a product's susceptibility to electromagnetic fields. At high frequencies, typically 50 MHz or greater, electromagnetic energy easily couples directly into equipment and/or its cables. At lower frequencies, typically 50 MHz or less, most products are not large enough to be efficient receptors of electromagnetic energy. As a result, the electromagnetic coupling almost always occurs to the cables at these frequencies. Cables will be most efficient receiving antennas when they are a quarter or a half wavelength long. At 50 MHz, a half wavelength is equal to 3 m.

Having to expose 3 m of cable to a uniform electromagnetic field is a difficult test to perform; it requires a large test chamber and expensive equipment. Therefore, analogous to the conducted emission case discussed in Chapter 13, the test is performed by injecting a voltage into the cable's conductors to simulate the electromagnetic field pickup. These are referred to as "conducted immunity test," and are really just radiated immunity tests in disguise.

Conducted rf immunity standards for commercial products typically require that the product operate properly without degradation (performance criteria A) when an rf voltage of 3 V (for residential/commercial products) or 10 V (for industrial equipment), 80% amplitude modulation (AM) from 150 kHz to 80 MHz is coupled common-mode into the alternating current (ac) power cables (EN 61000-6-1, 2007). The test also must be applied to signal cables, to direct current (dc) power cables and ground conductors if they are over 3 m in length. The voltage is applied as a common-mode voltage to the cable conductors. For unshielded cables the voltage is capacitively coupled into each conductor through a common-mode impedance of 150 $\Omega$ (50-$\Omega$ source impedance of the generator plus a 100-times-$n$ $\Omega$ resistor to each conductor, where $n$ is the number of conductors in the cable). For shielded cables, the voltage is coupled directly to the shield through a 150-$\Omega$ resistance (50-$\Omega$ source impedance of the generator plus a 100-$\Omega$ resistor).

Radiated rf immunity standards for commercial products typically require that the product operate properly without degradation (performance criteria A) when exposed to an electric field of either 3 V/m (for residential/commercial products) or 10 V/m (for industrial equipment), 80% AM modulated from 80 to 1000 MHz. Higher field strengths, up to 200 V/m, are applicable for automotive and military products.

## 14.2.1  The RF Environment

The electric field strength at a distance $d$ from a transmitter can easily be calculated. Assuming a small isotropic radiator (one that radiates equally in all directions), the power density $P$ at any distance $d$ from the source is equal to the effective radiated power (ERP) divided by the surface area of a sphere whose radius is equal to the distance $d$, or

$$P = \frac{ERP}{A_{sphere}} = \frac{ERP}{4 \cdot \pi \cdot d^2}. \qquad (14\text{-}1)$$

The power density $P$ (in watts per square meter) is equal to the product of the electric field $E$ times the magnetic field $H$. In the case of the far field, $E/H$ is equal to 120 $\pi$ (377) $\Omega$. Substituting this into Eq. 14-1 and solving for $E$ gives

$$E = \frac{\sqrt{30 \cdot ERP}}{d}, \qquad (14\text{-}2)$$

where the ERP is the transmitter power times the antenna gain, which is expressed as a numeric. For small handheld transmitter the antenna gain can usually be assumed to be unity. For a dipole antenna, the gain is equal to 2.14 dB or a factor of 1.28 over that of an isotropic radiator.

Equation 14-2 is applicable to FM transmission. For AM transmitters, multiply Eq. 14-2 by 1.6 to account for the modulation peaks (assuming 80% modulation).

For example, a 50,000-W FM station will produce an electric field strength of 0.77 V/m at a distance of 1.6 km (approximately 1 mi). However, a 600-mW cell phone produces an electric field strength of 4.24 V/m at a distance of 1 m. As can be concluded from this example, the nearby low-power transmitters often pose more of a threat to electronic equipment than do the high-power distant transmitters.

Industry Canada, after surveying the radio environment in Canada, concluded that the maximum electric field strength expected in urban and suburban areas ranged between 1 and 20 V/m in the frequency range from 10 kHz to 10,000 MHz (Industry Canada, 1990).

Digital circuits are usually not susceptible to radiated rf energy unless the field strength is above 10 V/m. Low-level analog circuits including voltage regulators, however, are often very susceptible to radiated rf fields in the 1 to 10 V/m level.

### 14.2.2  Audio Rectification

Radio frequency susceptibility usually involves *audio rectification*. Audio rectification is the unintentional detection (rectification) of high-frequency rf energy by a non-linear element, in a low-frequency circuit. When a modulated rf signal encounters a nonlinear device (like the base-to-emitter junction of a bipolar transistor), the signal is rectified and the modulation appears in the circuit. In the case of an unmodulated rf signal, a dc-offset voltage will be produced. In the case of a modulated rf signal, an ac voltage equivalent to the modulation frequency will appear in the circuit. The dc offset, or modulation frequency, is usually within the pass-band of the low-frequency analog circuit and hence may cause interference. The classic example of this is 27 MHz Citizen Band radio interfering with home high-fidelity or stereo systems.

Audio rectification can occur in audio/video circuits, such as stereo systems, telephones, microphones, amplifiers, televisions, and so on, as well as in low-frequency feedback control systems such as voltage regulators, power supplies, industrial process control systems, temperature and pressure sensors—and in some rare cases even in digital circuits. In the former cases, the demodulated rf signal is usually heard or seen, in the latter cases the demodulated rf signal produces a dc or low-frequency offset voltage in the control system that upsets the control function.

For audio rectification to be a problem the following two things must happen:

- First, the rf energy must be picked up.
- Second, it must be rectified.

Eliminate either of the above, and audio rectification does not occur.

The rf energy is usually picked up by the cables—and in some very high-frequency cases, by the circuit itself. In most cases, the detection occurs in the first p-n junction that the rf energy encounters. In rare cases, the detection can be caused by the rectifying properties of a bad solder joint or a poor ground connection.

The most critical circuits are usually low-level analog circuits such as amplifiers and linear voltage regulator circuits.


### 14.2.3   RFI Mitigation Techniques

Both radiated rf and conducted rf immunity are dealt with using the same techniques, because they are both forms of radiated electromagnetic coupling. Figure 14-1 shows an example of a typical circuit that should be protected from RFI. It consists of a sensor, an unshielded cable, and a printed circuit board (PCB). The cable normally picks up the rf energy, both common-mode and differential-mode. The sensor circuitry and/or the PCB circuitry then rectifies the signal.

The cable could be protected from picking up the rf energy by using twisted pairs (differential-mode), as well as common-mode chokes (common-mode), or by shielding (both modes). For many products the most susceptible frequencies are those at which the cables are resonant. Proper filtering at the sensor and/or PCB ends of the cable, and/or at the device doing the rectification can bypass the rf energy and thus eliminate the problem.

RFI mitigation techniques can and should be applied:

- At the device level
- To the cables
- To the enclosure


*14.2.3.1 Device Level Protection.*   RFI suppression should start at the device level and can then be supplemented with enclosure and cable-level protection.

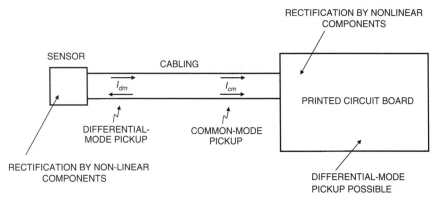

**FIGURE 14-1.** Radio frequency immunity example.

The most critical circuits are the ones that operate at the lowest signal levels, and those located closest to the input/output (IO) cables.

Keep all critical signal loop areas as small as possible, especially the input circuit and the feedback circuit of low-level amplifiers. Sensitive ICs should be protected with rf filters directly at their inputs. A low-pass *R–C* filter consisting of a series impedance (ferrite bead,* resistor, or inductor) and a shunt capacitor should be used at the input to the sensitive device, as shown in Fig. 14-2, to divert the rf currents away from the device and prevent audio rectification.

An effective RFI filter can be made using a series element with an impedance of 50 to 100 Ω and a shunt element (usually a capacitor) with an impedance of a few ohms or less, both determined at the frequency of interest.

Resistors can be used for the series element where the dc voltage drop is acceptable. At frequencies above 30 MHz, ferrite beads work well and do not have any applicable dc voltage drop.

Below about 10 MHz, inductors may have to be used because the series element should have an impedance of 50 to 100 Ω. Let us consider the case where the series impedance is 62.8 Ω. This value is chosen because it results in an easy method of determining the appropriate inductor value. The magnitude of the inductive reactance can then be written as

$$X_L = 2\pi f L = 62.8 \tag{14-3}$$

or

$$f L = 10. \tag{14-4}$$

Therefore, choose an inductor such that the product of the frequency and the inductance equals 10, for example, 10 μH at 1 MHz or 1 μH at 10 MHz.

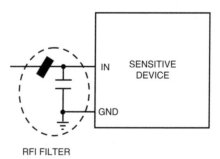

RFI FILTER

**FIGURE 14-2.** RFI filter on the input to a sensitive device.

---

* Ferrite beads act as small ac resistors with virtually 0 Ω impedance from dc to about 1 MHz. They are most effective above 30 MHz.

**TABLE 14-1.   Impedance of Ideal Capacitors (Ω).**

| Freq. (MHz) | 0.047 μF | 0.01 μF | 4700 pF | 1000 pF | 470 pF | 100 pF |
|---|---|---|---|---|---|---|
| 0.3 | 11.3 | 53 | 112.9 | 530 | 1129 | 5300 |
| 1.0 | 3.3 | 15.9 | 33.9 | 159 | 339 | 1590 |
| 3.0 | 1.1 | 5.3 | 11.3 | 53 | 113 | 530 |
| 10 | 0.3 | 1.59 | 3.4 | 15.9 | 33.9 | 159 |
| 30 | 0.11 | 0.53 | 1.1 | 5.3 | 11.3 | 53 |
| 100 | 0.03 | 0.16 | 0.34 | 1.6 | 3.4 | 15.9 |
| 300 | 0.01 | 0.05 | 0.11 | 0.53 | 1.13 | 5.3 |
| 1000 | 0.003 | 0.02 | 0.03 | 0.16 | 0.34 | 1.6 |

In some cases, just a shunt capacitor is effective, with the series element being the inductance of the cable or PCB trace.

Table 14-1 lists impedance values of ideal capacitors (no series inductance or resistance) at various frequencies. Over the frequency range of 80 to 1000 MHz (the frequency range that the European Union requires radiated immunity tests be performed), 1000 pF is an effective value for an rf filter capacitor, with impedance ranging from 0.16 to 1.99 Ω. For the lower frequency conducted immunity problems, larger value capacitors may be required.

Because the shunt element of the filter should have an impedance of a few ohms or less, let us consider the case where the impedance is 1.6 Ω. The magnitude of the capacitive reactance will then be

$$X_C = 1/2\pi fC = 1.6 \qquad (14\text{-}5)$$

or

$$fC = 0.1. \qquad (14\text{-}6)$$

Therefore, choose a capacitor such that the product of the frequency and the capacitance equals 0.1, for example, 1000 pF at 100 MHz.

For example, consider the case of an L-filter inserted in a signal line that has a low source impedance and a large load impedance. If the filter has a 62.8-Ω series element and a 1.6-Ω shunt element, it will provide an attenuation of 32 dB.

To be effective, the shunt capacitor must be mounted with *short leads, virtually zero.* This is extremely important and cannot be overemphasized. The shortcoming of many filters is the inductance in series with the shunt capacitor. This inductance includes the PCB traces used to connect the capacitor between the sensitive device input and ground, as well as the inductance of the capacitor itself.

For the case of a bipolar transistor amplifier, rectification of the rf energy usually occurs in the base-to-emitter junction. Figure 14-3 shows an example of

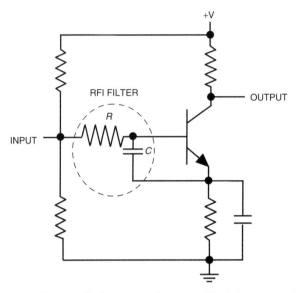

**FIGURE 14-3.** RFI filter applied to a transistor amplifier's base-to-emitter junction.

an *R–C* filter connected between the base and emitter of the transistor. In Figures 14-3 through 14-5, the series filter element is shown as a resistor; however, the series element could also be a ferrite bead or an inductor as previously discussed.

In the case of most integrated circuit amplifiers, the input transistor's base-to-emitter junction is not accessible at the pins of the integrated circuit (IC). In these cases an rf filter must be applied on the input to the device to keep the rf energy from entering the package. Figure 14-4 shows such an RFI filter applied to the input of an operational amplifier.

Some IC amplifiers (e.g., the AD620 instrumentation amplifier) do provide direct access to the input transistors base and emitter, so that the user can apply rf filtering directly across the base to emitter junctions as shown in Fig. 14-5.

A remote sensor, as was shown in Fig. 14-1, that contains nonlinear devices (transistors, diodes, solid state amplifiers, etc.) also may require protection to prevent audio rectification. The sensor's wires act as the receiving antenna, and the sensor's p-n junctions act as the rectifiers. In such cases both the sensor and PCB end of the cable will require filters to prevent audio rectification. Figure 14-6 shows RFI filtering applied to an optical encoder consisting of a light-emitting dicode (LED) and phototransistor. The filter uses ferrite beads for the series elements and capacitors for the shunt elements.

A multilayer PCB with a power and ground plane will provide significantly greater rf immunity than a single- or double-sided board. This is the result of the lower ground impedance provided by the planes, as well as the smaller loop areas present. Effective high-frequency power supply decoupling as well as

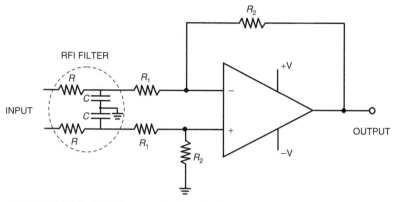

**FIGURE 14-4.** RFI filter applied to the input of an operational amplifier.

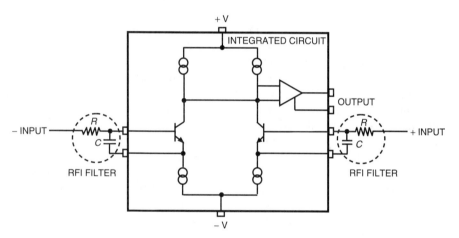

**FIGURE 14-5.** IC instrumentation amplifier with RFI filters applied directly across the input transistors base to emitter junctions.

sufficient bulk capacitance between power and ground, even on low-frequency analog circuits, is important for good rf immunity.

For direct field-induced noise, an external image plane* can be added to a one- or two-layer PCB with similar results to that of an in-board ground plane. The plane can be made of foil or thin metal and placed as close to the PCB as possible. At rf frequencies, currents induced into the image plane produce canceling fields to the direct field induction. The field cancellation effect will

---

*A conducting plane placed beneath and parallel to a PCB. Because the result achieved with the conducting plane is caused entirely by the equivalent current images induced into the plane, it is called an image plane (German, Ott, and Paul, 1990).

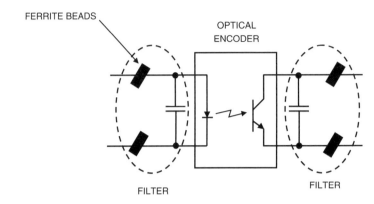

**FIGURE 14-6.** Filtering of an optical encoder to eliminate audio rectification.

occur even if the image plane is not connected to the PCB. Mounting of a PCB close to a metallic chassis can have a similar effect.

Circuit ground should have a low impedance connection to the chassis in the I/O area (see Sections 3.2.5 and 12.4.3) to divert any rf energy present to the chassis. High-frequency boards may require additional circuit ground to chassis connections. However, these ground connections are *in addition to the ones in the I/O area*, not in place of them.

Voltage regulators, which include three terminal regulators, have also demonstrated susceptibility to rf fields. This can be dealt with by adding rf filter capacitors directly on the input and output of the regulator. A value of 1000 pF is usually adequate. A small ferrite bead added to the regulator's input and output leads will increase the effectiveness of the filter. These capacitors are in addition to the larger value capacitors required for proper operation and/or stability of the regulator and should be connected directly to the common pin of the regulator as shown in Fig. 14-7.

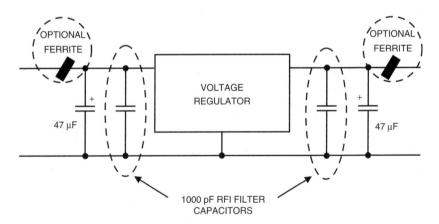

**FIGURE 14-7.** Protecting a voltage regulator from rf interference.

***14.2.3.2 Cable Suppression Techniques.*** In most cases, rf energy is picked up by the cables. Minimizing this pickup is therefore an important aspect of good rf immunity design. In the case of shielded cables, use a good quality high-coverage braid, multiple braids, or braid over foil with proper termination, no pigtails or drain wires (see Section 2.1.5). Do not use a spiral-shielded cable for rf immunity. Although cable connectors with metal backshells that make 360° contact are the ideal shield termination methods, other less-expensive approaches can also be effective. Figure 14-8 shows one possible way to terminate a shielded cable effectively by using a metal cable clamp. The cable clamp should be located as close as to where the cable enters the enclosure as possible.

Think of a cable shield as an extension of the shielded enclosure. Therefore, how effective the shield is has a lot to do with how well the cable shield is terminated to the enclosure. Use a 360° termination to the enclosure, not to circuit ground.

If you do not shield I/O cables, then use twisted pairs, with filters at the point where the cables enter (or leave) the enclosure. This is in addition to the sensitive device filters discussed previously. Place the I/O filters as close to the I/O connectors as possible, and connect the shunt capacitors of the I/O filters to the enclosure not to circuit ground. Capacitor filter pin connectors, although more expensive, can also be effective in this respect. Common-mode chokes (ferrite cores) can be used on external cables to help reduce rf susceptibility.

For optimum rf immunity, ribbon cables and flex circuits should contain multiple ground conductors spread across the width of the cable. Ribbon cables and flex circuits containing a ground plane the full width of the cable are even better. Terminate the cable ground plane across as much of its width as

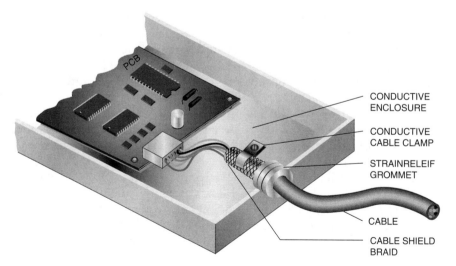

CONDUCTIVE ENCLOSURE

CONDUCTIVE CABLE CLAMP

STRAINRELEIF GROMMET

CABLE

CABLE SHIELD BRAID

**FIGURE 14-8.** A simple method of effectively terminating a shielded cable without a pigtail or coaxial connector.

possible. The effect of a cable ground plane can be simulated by adding copper tape, properly grounded, on one side of a flat cable.

Multiple ground conductors in cables and connectors reduce signal loop area and minimize ground differential between the ends of the cable. Ideally, the signal-to-ground conductor ratio should be 1 to 1, although a ratio of 2 to 1 also performs well, because it still provides a ground conductor adjacent to each signal conductor. With a 3-to-1 ratio, every signal conductor will be within two conductors of a ground conductor. A 3-to-1 ratio is a good design compromise, because it provides a reasonable number of ground conductors while limiting the total number of conductors. The higher the frequency of the signals in the cable, the more grounds should be used. Under no circumstance should the signal-to-ground conductor ratio exceed 5 to 1.

Internal cables can also contribute to susceptibility problems. The use of ferrite common-mode chokes and the routing of the cables close to the chassis, but away from seams and apertures, will improve the immunity. If cable shielding is used on internal cables, then the best approach is to terminate the shield to the chassis on both ends. For short internal cables runs, however, a shield grounded at only one end may be adequate, but proper termination should be used—no pigtails.

For products in nonshielded enclosures, cable filters can be connected to an image plane (German, Ott, and Paul, 1990). Image planes will also reduce direct rf pickup by the PCB. If no other option is available, connect the filters to the circuit ground plane, although this approach is not as effective as a connection to the enclosure or a separate image plane.

***14.2.3.3 Enclosure Suppression Techniques.***    High-frequency radiated rf fields can directly couple to a product's circuitry and internal cables. Thin shields made of aluminum, copper, or steel are effective, provided the apertures are properly controlled (see Section 6.10). The only exception to this is the case of low-frequency magnetic fields. For low-frequency magnetic fields ($<500$ kHz), thicker steel shields are required. For example, the shield for a switching power supply operating at 50 kHz should be made of steel not aluminum.

Apertures (seams, cooling holes, etc.) in shielded enclosures should be limited to a maximum linear dimension of 1/20 wavelength. This will provide about 20 dB of attenuation through the aperture. Under some conditions, the apertures may have to be limited to an even smaller dimension. Table 14-2 lists the approximate 1/20 wavelength dimension for various frequencies from 10 MHz to 5 GHz.

Because most rf immunity regulations require testing up to 1000 MHz, a 1.5-cm (0.6-in) maximum linear dimension is usually a good design criterion.

Note it is the maximum linear dimension, not the area of the aperture, that determines the rf leakage (see Section 6.10). A long thin slot (e.g., seam) will have much more leakage than a cooling hole with a much larger area but smaller maximum linear dimension.

To provide good electrical contact across enclosure seams, the mechanical design must provide for a conductive finish and adequate pressure between the

**TABLE 14-2. Approximate Dimensions Corresponding to 1/20 Wavelength as a Function of Frequency.**

| Frequency (MHz) | 1/20 Wavelength Dimension |
|---|---|
| 10 | 1.5 m (5 ft) |
| 30 | 0.5 m (1.6 ft) |
| 50 | 0.3 m (12 in.) |
| 100 | 0.15 m (6 in.) |
| 300 | 5 cm (2 in.) |
| 500 | 3 cm (1.2 in.) |
| 1000 | 1.5 cm (0.6 in.) |
| 3000 | 0.5 cm (0.2 in.) |
| 5000 | 0.3 cm (0.1 in.) |

mating surfaces (see Section 6.10.2). Most surface finishes require a pressure of $> 100$ psi to provide a low impedance contact.

Any unfiltered cables passing through a shield can carry rf energy from the outside of the shield to the inside or vice versa and reduce the effectiveness of the shield. All cables penetrating a shield must therefore be either filtered or shielded to avoid the transferring of energy from one environment to the other.

Digital circuits are also susceptible to rf energy but usually not unless the field strength is greater than 10 V/m. If digital susceptibility is observed, then the solution is similar to that described above for analog susceptibility. Decouple the rf energy from the cable and use a filter at the input to the sensitive digital circuit. The usual susceptibility problem with digital circuits, however, is the result of fast rise-time, high-voltage transient pulses.

## 14.3 TRANSIENT IMMUNITY

The European Union requires that products also be tested for high voltage transient immunity. There are basically three types of high voltage transients that electronic equipment designers need be concerned about. They are:

- ESD
- Electrical fast transient (EFT)
- Lightning surge

For high-voltage transients, the most susceptible circuits are digital control circuits, such as resets, interrupts, and control lines. If these circuits are triggered by a transient voltage, they can cause the entire system to change state.

The European Union's transient immunity standards for commercial products require that the product continue to operate as intended, without degradation or loss of function after the transient immunity test. During the test, degradation of performance is allowed. No change of operating state or

**TABLE 14-3. Characteristics of High-Voltage Transients.**

| Transient | Voltage | Current | Rise Time | Pulse Width | Pulse Energy |
|---|---|---|---|---|---|
| ESD | 4–8 kV | 1–10s A | 1 ns | 60 ns | 1–10s mJ |
| EFT (Single Pulse) | 0.5–2 kV | 10s A | 5 ns | 50 ns | 4 mJ |
| EFT (Burst) | 0.5–2 kV | 10s A | n/a | 15 ms | 100s mJ |
| Surge | 0.5–2 kV | 100s A | 1.25 μs | 50 μs | 10–80 J |

stored data is allowed, however (performance criteria B). In some critical applications (such as some types of medical devices), upset may not even be allowed during the ESD event (EN 60601-1-2, 2007).

Table 14-3 summarizes the characteristics of the three high-voltage transients. The two most important parameters are rise time and energy. Notice from Table 14-3 that ESD and EFT have similar rise times and energy levels. Surge, however, has a much slower rise time, microseconds instead of nanoseconds, and it contains three to four orders of magnitude more energy (Joules instead of milli-Joules). Therefore, ESD and EFT can be dealt with in a similar manner, but surge must often be dealt with differently.

### 14.3.1 Electrostatic Discharge

Two test methods are applicable to ESD, contact discharge and air discharge. In contact discharge, the *unenergized* discharge electrode is placed in contact with the equipment under test, and the discharge is initiated by a switch within the tester. In air discharge, however, the *charged electrode* of the tester is moved toward the equipment under test until a discharge (spark) occurs through the air. Contact discharge produces more repeatable results and is the preferred test method. Air discharge more closely simulates the actual ESD event but is not as repeatable, and it is only used where contact discharge cannot be applied, for example a product in a plastic enclosure.

The European Union's electrostatic discharge test requires a product to pass (performance criteria B) a $\pm 4$ kV contact discharge and a $\pm 8$ kV air discharge. The test generator has a source impedance of 330 Ω, which limits the 8-kV discharge current to 24.24 A. The discharges are applied to those points and surfaces of the equipment that are accessible during normal usage and customer's maintenance procedures.

ESD mitigation techniques are covered in Chapter 15.

### 14.3.2 Electrical Fast Transient

Deenergizing inductive loads such as relays or contactors will produce short bursts of high-frequency impulses on the electric power distribution system. Utility power factor corrector capacitor switching is another cause of oscillatory power-line transients.

Figure 14-9 shows the waveshape of the European Union's EFT/burst test impulse. It consists of a burst of 75 pulses repeated every 300 ms for a duration

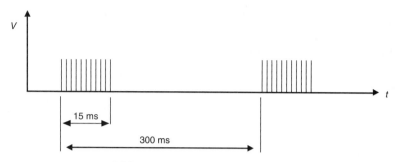

**FIGURE 14-9.** EFT test impulse.

of not less than 1 minute. Each individual pulse has a 5-ns rise time and a 50-ns pulse width with a repetition frequency of 5 kHz. For residential/commercial products, the amplitude of the individual pulses is $\pm 1$ kV on ac power lines and $\pm 0.5$ kV on dc power as well as signal and control lines and is applied as a common-mode voltage. For industrial products, the EFT pulses go up to $\pm 2$ kV on the ac and dc power lines and $\pm 1$ kV on signal lines. The test generator has a 50-$\Omega$ source impedance. The test is only required on signal, control lines, and ground conductors if they exceed 3 m in length.

### 14.3.3  Lightning Surge

The European Union's surge requirements are not intended to simulate a direct lightning strike to the ac power line. Rather, they are intended to simulate voltage surges on the power line caused by a nearby lightning strike or the downing of a utility pole resulting from an accident or storm. Voltage surges can also be caused by the inductance of the power line when high current loads are suddenly switched OFF.

The European Union's surge test generator is designed to produce a 1.25-μs rise time and a 50-μs pulse width (between 50% amplitude points) voltage surge into an open circuit, as well as a 8-μs rise, 50-μs pulse width (50% amplitude points) current surge into a short circuit. The test generator has an effective source impedance of 2 $\Omega$. The voltage surge only has to be applied to ac and dc power lines, both common- and differential-mode, not the signal lines. On the ac power line, the voltage level is $\pm 2$ kV line to ground and $\pm 1$ kV line to line. On the dc power line, the voltage level is $\pm 0.5$ kV line to ground and line to line. A $\pm 0.5$-kV pulse must also be applied to any and all ground conductors.

Most high-voltage transient disturbances, with the exception of ESD, are applied to the cables; ESD is applied to the enclosure. Table 14-4 summarizes the EFT and surge voltage level for residential/commercial products* and

---

* For industrial products the EFT pulses go up to $\pm 2$ kV on the ac and dc power lines and $\pm 1$ kV on signal lines EN 61000-6-2, 2005.

**TABLE 14-4.  Applicability Matrix for EFT and Surge Testing (Residential and Commercial Environments).**

| Cable | EFT | Surge | Surge |
|---|---|---|---|
| | Common-Mode | Common-Mode | Differential-Mode |
| AC Power Lines | ±1 kV | ±2 kV | ±1 kV |
| DC Power Lines[a] | ±0.5 kV | ±0.5 kV | ±0.5 kV |
| Signal Lines[a] | ±0.5 kV | n/a | n/a |
| Ground Conductor | ±0.5 kV | n/a | n/a |

[a]Applicable only to cables whose length exceeds 3 m.

specifies how the test voltage is applied, common-mode or differential-mode. (EN 61000-6-1, 2007)

### 14.3.4  Transient Suppression Networks

Some desirable characteristics of transient protection networks are listed below.

- Limit the voltage
- Limit the current
- Divert the current
- Operate fast
- Be capable of handling the energy
- Survive the transient
- Have a negligible affect on the system operation
- Fail safe
- Have minimal cost and size
- Require minimal or no maintenance

In most cases, not all the above objectives can be met simultaneously.

The general configuration of a transient voltage suppression network is shown in Fig. 14-10. The network consists of both a series and shunt element. Although the shunt element could be a linear device, such as a capacitor, in most cases it is a nonlinear breakdown device that has a large impedance during normal circuit operation and a much lower impedance when a high voltage transient is present, thus shunting the transient current to ground. This result can be achieved with a breakdown or clamping device such as a zener diode or gas tube, or a voltage variable resistor. In the former case, the voltage across the shunt device will be approximately constant after the transient exceeds the breakdown voltage of the device. In the latter case, the shunt resistance will be approximately constant after the transient exceeds the breakdown voltage of the device.

The series element is used to limit the transient current through the shunt device, and as a result of the voltage divider composed of $Z_1$ and $Z_2$, it will

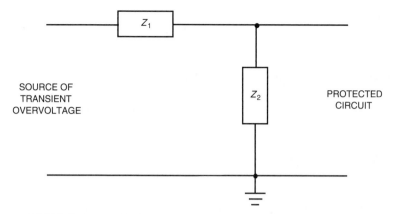

**FIGURE 14-10.** Single-stage transient voltage protection network.

reduce the voltage applied to the protected circuit. The series element could be a resistor, inductor, or ferrite; but in some cases, it consists of the parasitic inductance and resistance of the conductors as well as the source impedance of the transient generator. In power circuits, the impedance of the fuse or circuit breaker would also be part of the series impedance.

It is important to understand that a series element must exist somewhere in the circuit, otherwise the current through the shunt device after breakdown would be infinite. Therefore, it is important that whenever transient suppressors are used the designer considers where and what constitutes the series impedance. It may be a separate component, the source impedance of the transient generator, or the impedance of the wiring or PCB traces to the device.

Effective transient suppression requires the following three-pronged approach:

- First, divert the transient current
- Second, protect the sensitive devices from damage or upset
- Third, write transient hardened software (this subject is covered in Section 15.10)

Many techniques used for transient suppression are similar to those discussed previously for RFI immunity. Because most high-voltage transient disturbances are applied to the cables, cable protection becomes an important aspect of transient immunity design.

### 14.3.5   Signal Line Suppression

I/O signal cable protection can be achieved with the addition of transient voltage suppression (TVS) diodes where the cables enter the product. A TVS diode is similar to a zener diode but with a larger p-n junction area that is proportional to its transient power rating. This increased junction area also increases the capacitance of the diode, which adds capacitive loading of the signal line and could, in some cases, have a detrimental effect on the normal operation of the circuit.

Transient voltage suppression diodes (sometimes referred to as silicon avalanche diodes) are available in both unidirectional and bidirectional configurations. The three most important parameters of TVS diodes are as follows:

- Reverse standoff voltage
- Clamping voltage
- Peak pulse current

In selecting a TVS diode, the reverse standoff voltage must be greater than the maximum operating voltage of the circuit being protected to ensure that the diode does not clip the circuit's maximum operating or signal voltage. The maximum clamping voltage represents the peak voltage that will appear across the diode when subjected to the peak pulse current. This is the voltage that the protected circuit must be able to withstand, without damage, during the transient.

The peak pulse current is the maximum transient current that the diode can tolerate without damage. Adding series impedance to the circuit ($Z_1$ in Fig. 14-10) such as a resistor, prior to the TVS diode will help to limit the peak pulse current through the diode.

TVS diodes must have a low inductance connection to the chassis to effectively divert the transient energy away from the susceptible circuits. For example, the voltage across the series combination of a 9-V breakdown TVS diode with ½ in of PCB trace on each side when hit with a 10 A/ns transient pulse will be 159 V. This is based on a trace inductance of 15 nH/in. A typical transient pulse can produce a voltage drop of 6 V/mm (150 V/in) across a trace on a PCB. Proper layout is, therefore, critical to obtain optimum performance.

Figure 14-11 shows three possible layouts for a TVS diode on a PCB. Figure 14-11A shows a typical mounting arrangement with appreciable

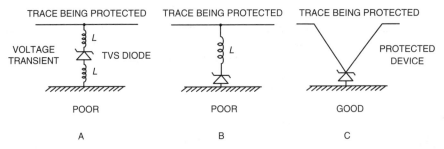

**FIGURE 14-11.** Mounting configurations for TVS diodes. (*A*) and (*B*) show improper mounting, and (*C*) shows proper mounting.

inductance in both of the leads to the diode, which makes the diode ineffective in limiting the transient voltage on the trace. In Fig. 14-11B, the diode is mounted adjacent to the ground reducing the diode-to-ground inductance to almost zero. A trace is then routed from the diode to the trace being protected, which again adds significant inductance in series with the TVS diode and makes it equally ineffective. Figure 14-11C shows the diode again mounted adjacent to the ground but instead of routing a trace from the diode to the trace being protected, the trace being protected is routed to the diode resulting in virtually zero inductance in series with the diode and effective transient voltage protection.

Figure 14-12 shows an RS-232 interface protected against transient over-voltages with four bipolar TVS diodes. Note that even the ground conductor is protected with a TVS diode. This is often required because the internal connection between circuit ground and chassis ground may not be located at the point of cable entry and may not have a low inductance.

Microprocessor resets, interrupts, and any other control inputs that can change the operating state of a digital device should be protected against false triggering by fast rise time transients by adding a small capacitor or $R$–$C$ network (50 to 100 $\Omega$, 500 to 1000 pF) similar to that shown in Fig. 14-2 to the IC input order to reduce its susceptibility to sharp narrow transient pulses (1 to 50 ns), such as those generated by ESD or EFTs.

A multilayer PCB with power and ground planes will provide greater transient immunity than a single-or double-sided board. This is the result of the lower ground impedance and the low inductance in series with the interlayer capacitance. Embedded PCB capacitance technology, as discussed in Section 11.4.6, can also be used to effectively increase the transient immunity. Sufficiently large bulk decoupling capacitance is also effective in improving transient immunity, because it will decrease the change in the supply voltage

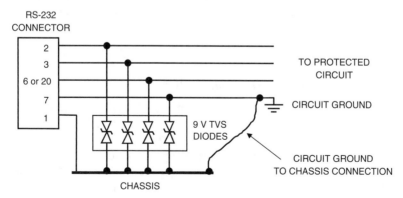

**FIGURE 14-12.** An RS-232 input protected against transient overvoltages with four TVS diodes.

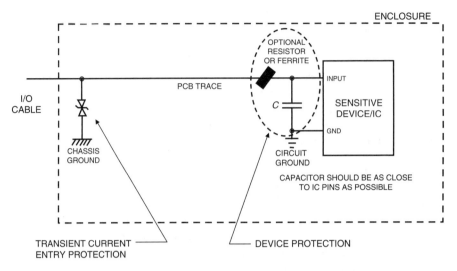

**FIGURE 14-13.** Transient protection device grounding.

produced by a transient charge.* The relationship between voltage, charge, and capacitance is shown in Eq. 15-1. As can be observed from that equation, for a fixed change in charge $Q$ produced by a transient current, the amount that the voltage will vary is determined by the capacitance. The larger the capacitance, the less the voltage will change when additional charge is injected into the system.

Figure 14-13 shows a combination of transient current entry protection and sensitive device hardening. Notice that the transient suppressor on the input cable is grounded to the chassis, because its purpose is to divert the transient current away from the PCB. The protective filter on the sensitive device's input, however, is grounded to the circuit ground, because its purpose is to minimize or eliminate any transient voltage from appearing between the device's input pin and ground pin.

### 14.3.6   Protection of High-Speed Signal Lines

High-speed I/O interfaces with data rates of 100 MB/s or more universal serial bus, such as (USB) 2.0, high-speed Ethernet, and IEEE Std 1394 (FireWire and iLink), present a special problem with respect to high-voltage transient protection. To avoid affecting the desired signal, capacitive loading on many of these interface lines must be kept to less than a few pF. Most TVS diodes and other transient suppression devices, with the exception of gas tubes, have too

---

* Charge $Q$ is equal to the integral of $I\, dt$.

much capacitance. Gas tubes, however, are too slow responding to be used for ESD or EFT protection. Most TVS diodes have capacitance in the tens to hundreds of pF range. Special low-capacitance TVS diodes are available with capacitances in the 1- to 10-pF range. These diodes are usually useable with data rates up to about 100 MB/s but not above.

Special polymer voltage variable resistors (VVR) have been developed specifically for these high-speed applications.* They are bipolar devices with typical capacitances in the range of 0.1 to 0.2 pF, off-state resistances of $10^{10}$ $\Omega$, and an on-state resistance of a few ohms. The triggering voltage, however, is considerably higher than most TVS diodes. Voltage variable resistors are crowbar devices, where the clamping voltage after turn on is lower that the triggering voltage. Typical triggering voltage is 150 V, with a clamping voltage of about 35 V, and a peak transient currents rating of 30 A. These polymer VVRs can be used on data lines with signal frequencies up to 2 GHz.

Figure 14-14 shows examples of the use of a combination of TVS diodes and VVRs to provide transient protection to three different high-speed interfaces. The TVS diodes are used on the dc power lines and the VVRs on the high-speed data lines. Notice, as was the case shown in Fig. 14-12, that the protection devices are connected to chassis ground.

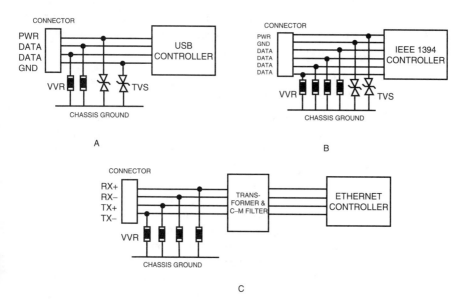

**FIGURE 14-14.** Transient voltage protection applied to high-speed; (*A*) USB, (*B*) IEEE Std 1394 (FireWire), and (*C*) Ethernet interfaces.

---

* Examples are Littlefuse's PulseGuard® and Cooper Electronics' SurgX.®

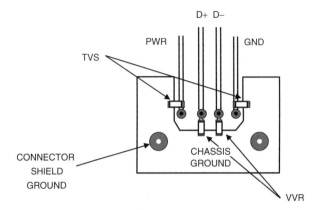

**FIGURE 14-15.** PCB layout for transient suppression devices protecting a high-speed USB port.

Figure 14-15 is an example of the optimum mounting of the transient suppression devices protecting a USB interface. The figure shows the bottom side of the PCB under the USB connector. The surface mount suppression devices are mounted to minimize any inductance in series with them, and they are connected to chassis ground, not circuit ground, to divert the transient current away from the circuitry. The PCB chassis ground plane should have direct and multiple connections to the actual chassis.

### 14.3.7 Power Line Transient Suppression

Transient voltage protection is often required at the ac or dc power entry point, which is usually the power-line filter. This is necessary to protect against EFT and surge. Automotive and some industrial process control equipment are in environments where large power-line spikes are common and therefore often require power-line transient protection.

Many power-line filters can handle the low energy transients such as ESD and EFT. Additional EFT or ESD suppression, if required, can often be obtained on the power line by the addition of a common-mode ferrite choke on the input cable to the power-line filter or power supply. Power-line filters are linear devices that proportionally attenuate high-frequency noise as well as power-line transients. Although optimized for controlling low-level high-frequency noise, they do also attenuate high voltage transients and will provide a degree of protection to both common-mode and differential-mode transients.

Surge, however, has a pulse energy level that is a thousand or more times greater than that of ESD or EFT and is another problem altogether. Transient protection for these high-energy pulses is often required prior to the power-line filter. Three types of nonlinear transient protectors are commonly used for high-power transients. They are as follows:

- TVS diodes
- Gas discharge tubes
- Metal oxide varistors (MOVs)

Transient voltage suppression diodes and MOVs are voltage-clamping devices. They operate by limiting the voltage to a fixed level. Once they turn on, they must dissipate the transient pulse energy internally. Gas discharge tubes, however, are *crowbar* devices. Once they turn on, the voltage across them drops to a very low value; therefore, their power dissipation is much reduced. They can handle extremely large currents.

TVS diodes are commonly used on signal lines (see Section 14.3.5) and on dc power lines. They do not have as much current-carrying or energy-dissipating capacity as MOVs; however, they are available with lower clamping voltages. Surge currents must usually be limited to less than 100 A to use TVS diodes. Their response times are in the picosecond range and can be used for ESD, EFT, and surge protection. They are the least robust of the three transient protection devices listed above.

Gas-discharge tubes are primarily used on telecom circuits. They are the slowest responding transient protection devices, with response times in the microsecond range. Therefore, they cannot be used for ESD or EFT protection. Because they are crowbar devices, they do not have to dissipate much energy internally and are the most robust of all three transient suppression devices. Often, they can withstand currents of tens of thousands of amperes.

Metal-oxide varistors are voltage variable (nonlinear) resistors fabricated from various mixtures of zinc oxides, in which the resistance decreases when the magnitude of the voltage across the device exceeds the threshold level (Standler, Chapter 8, 1989). A typical *V-I* curve for a MOV is shown in Fig. 14-16. As can be observed, the varistor is a symmetrical bipolar device that clamps both positive and negative voltages. The clamping voltage is usually defined as the voltage at which the current through the varistor is 1 mA. Because the device clamps the voltage, all the energy from the transient pulse must be dissipated in the device itself.

Traditional MOVs are most commonly used on ac power lines. Their response time is slower than TVS diodes and faster than gas tubes. They have response times in the hundreds of nanosecond range, which is fast enough for surge but not usually fast enough for EFT or ESD. They can tolerate surge currents in the hundreds, or thousands, of amperes range and can dissipate tens of Joules or more of energy.

MOVs degrade gradually when subjected to surge currents, which usually is not a problem because they are typically rated for a few million surges. All things considered, MOVs seem to be the best device for protecting electronic equipment from ac power line surges.

New multilayer MOVs are also available in surface-mount packages for PCB use. They have sub-nanosecond response times, with capacitances less

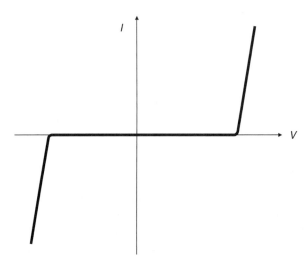

**FIGURE 14-16.** Current-voltage relationship of typical MOV.

than 100 pF. These devices are fast enough to be used for ESD and EFT protection, and they are available with clamping voltages in the 10 to 50 V range. Although their power dissipation (usually less than 1 J) and current ratings (usually less than 100 A) are not as high as traditional MOVs, there are many applications where their characteristics are appropriate.

The primary common-mode lightning (surge) protection for a facility, or structure, is provided by the power and telephone companies, where their wires enter the building. This is accomplished, in the case of the ac power line, by a combination of the power distribution transformer, plus the bonding together of the neutral and ground conductors and connecting them to earth ground at the service entrance panel as was shown in Fig. 3-1.

In the case of the telephone line, common-mode surge protection is provided at the protector block located close to the point the wires enter the facility. In most cases today, this consists of a pair of gas tubes connected from the two telephone line conductors to ground. This also provides differential-mode protection, but at twice the clamping voltage of the gas tubes.

Because the power and telephone lines already have common-mode surge protection, electronic products located within a facility or structure often only require additional differential-mode surge protection. This can easily be accomplished, in the case of the power (or telephone) line, with a single MOV connected between the hot and neutral lines, as shown in Fig. 14-17.

A commonly used power line surge protection circuit providing both differential-mode and common-mode protection is shown in Fig. 14-18. This circuit uses three MOVs, one from each line to ground and one between the hot and neutral lines. In many cases, this circuit is incorporated in ac power outlet strips that contain integral surge suppression. This approach however has a disadvantage when used downstream of the service entrance. The grounded

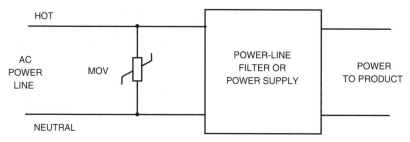

**FIGURE 14-17.** A single MOV used to provide differential-mode surge protection for an ac power line.

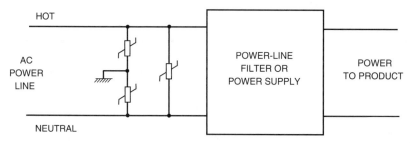

**FIGURE 14-18.** Three MOVs used to provide common-mode and differential-mode surge protection for a power line.

MOVs will draw large surge currents (hundreds of amperes or more)* into the facility and dump them onto the safety ground conductor.

The surge protection circuit of Fig. 14-18 is therefore not recommended for use in an electronic product or ac power outlet strip. As a result of the large surge currents induced into the ground conductor, a high-voltage differential will be created in the safety ground. For example, a 500-A surge current in a 100-ft long, 12-gauge ground conductor (1.59-mΩ/ft resistance) will produce a voltage drop of 80 V. This ground voltage will then appear across the signal conductors interconnecting various pieces of equipment, possibly causing damage.

The protection circuit of Fig. 14-17 does not have this problem, because no MOVs are connected to ground. This design provides only differential-mode surge protection, but it is adequate in many applications. In some cases, however, MOVs to ground will be required to meet the European Union's surge test requirement.

Isolation transformers (see Section 3.1.6) installed in the facility's ac power wiring are also effective in suppressing common-mode transients in the branch circuits into which they are installed. They do not, however, block differential-mode transients.

---

* For example, a 2-kV surge can produce currents as high as 1000 A.

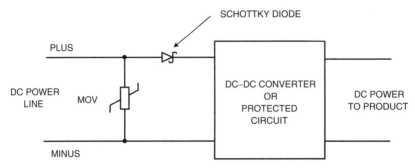

**FIGURE 14-19.** A MOV and a Schottky diode used to provide differential-mode surge protection for an dc power line.

Differential-mode surge protection for dc power lines can be provided with the circuit shown in Fig. 14-19. This is similar to the circuit of Fig. 14-17 with the addition of a Schottky diode to provide reverse polarity protection to the protected circuit. A Schottky diode is used because of its low forward-voltage drop. The diode must have a reverse sustain voltage in excess of the MOV's breakdown voltage.

Table 14-5 summarizes the characteristics of many common transient suppression devices.

### 14.3.8   Hybrid Protection Network

If protection is required for both fast rise time transients, such as EFT or ESD, and for high-energy, high-current transients such as surge, it is often difficult to design a single-stage network, as shown in Fig. 14-2, to satisfy all the necessary requirements. In this case, a two-stage hybrid transient suppression network should be considered (Standler, 1989, pp. 113, 236–242). A two-stage hybrid network consisting of a TVS diode plus a gas tube, or MOV, is shown in Fig. 14-20. The TVS diode turns on first, at a voltage below the maximum voltage rating of the circuit being protected. The TVS diodes absorb the initial transient energy allowing time for the gas tube or MOV to turn on. The current through the TVS diodes produce a voltage drop across the series resistors, which increases the potential across the gas tube, until its breakdown voltage is reached. The gas tube then turns on to absorb the bulk of the transient energy.

The resistor values are chosen such that, when combined with the transient pulse's source resistance, the TVS diode current is limited to a safe value. In addition, the resistors cause the voltage across the gas tube to increase above the TVS diode's breakdown voltage, which thereby allows the gas tube to turn on. In some cases, the required resistor value may be larger than what can be tolerated in the circuit during normal operation. In these cases, the series element does not have to be a resistor. For fast rise time transients, it could be an inductor (or possibly a ferrite) or a combination of resistor and inductor (or ferrite).

**TABLE 14-5. Typical Characteristics of Transient Suppression Devices.**

| Device | Type | Response Time | Capacitance | Advantages | Disadvantages | Types of Transients | Typical Applications |
|---|---|---|---|---|---|---|---|
| TVS Diodes | Clamping | <1 ns | >10 pF | • Low cost<br>• Low clamping voltage | • Limited power handling capability<br>• I <100 A | ESD, EFT, Surge[a] | • Signal lines<br>• DC power lines |
| Standard MOVs | Clamping | 100's ns | 10 to 10,000 pF | • Low cost<br>• Large power handling capability | • Higher clamping voltages<br>• Degrad with multiple surges | Surge | • AC power lines<br>• DC power lines |
| Multilayer MOVs (surfacemount) | Clamping | <1 ns | 10 to 2500 pF | • Lower voltage rating<br>• Fast switching<br>• Small size | • Lower power handling capability | ESD, EFT, Surge | • PCB applications |
| VVRs | Crowbar | <1 ns | <1 pF | • Very low capacitance<br>• Fast switching | • Higher clamping voltages | ESD, EFT | • High-speed signal lines |
| Gas Tubes | Crowbar | μs range | <1 pF | • Very high surge current capability<br>• Very reliable<br>• Most robust | • High cost<br>• Higher breakdown voltage<br>• Slow turn-on | Surge | • Telecom |
| L–C Filters | Linear | n/a | High | • May already exist in circuit<br>• Little power dissipated in filter | • Voltage not clamped<br>• May cause ringing | ESD, EFT, Surge | • Power supplies<br>• Signal lines |

[a] If current is limited.

571

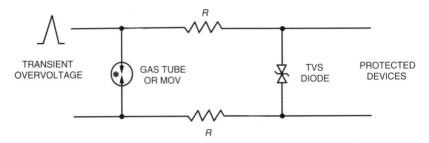

**FIGURE 14-20.** Two-stage hybrid transient protection network.

For example a 1-µH inductance will have a 100-V drop across it when exposed to a 1000-V/µs transient pulse, representative of a surge. A wire-wound resistor can sometimes be used as the series element to provide both resistance and inductance in one component.

## 14.4  POWER LINE DISTURBANCES

Other power-line disturbances of concern to equipment designers include voltage sags and interruptions. A voltage sag is a short-duration reduction in root mean square (rms) voltage, typically less than a few cycles, on the ac power line caused by faults in the power system and/or by turning on products with high inrush currents, such as motors and large heaters. Momentary interruptions are the complete loss of voltage, typically for a few seconds, which are usually caused as a result of action taken by the power utilities to clear transient faults on their system.

Table 14-6 lists the European Union's ac power line voltage dips and interruptions requirements, which are applicable to products that operate in residential, commercial, or light-industrial environments (EN 61000-6-1, 2007).

To ride through a sag or momentary interruption requires a source of stored energy. That can be accomplished by having sufficient capacitance in the output of the dc power supply. A product should be designed with sufficient capacitance to operate properly with a complete loss of ac voltage for at least 17 ms (approximately one cycle of the 60 Hz ac power-line frequency).

**TABLE 14-6.  Voltage Sags and Interruption Requirements for ac Power Lines.**

| Test | Voltage Reduction | Duration | Performance Criteria |
|------|-------------------|----------|----------------------|
| Voltage dip | 30% | 0.5 cycles | B |
| Voltage dip | 60% | 5 cycles | C |
| Voltage Interruption | >95% | 250 cycles | C |

The well-known current voltage relationship for a capacitor is

$$i = C\frac{dv}{dt},$$  (14-7)

which can be rewritten as

$$C = \frac{i}{dv/dt} = \frac{i \cdot dt}{dv}.$$  (14-8)

**Example 14-1**. A product draws 1 A of steady-state current from the power supply, and the voltage specification for proper operation is 12 V $\pm$ 3 V. What value capacitance is required in the output of the power supply for the product to be able to ride through a 10 ms complete loss of ac voltage on the input to the power supply? From Eq. 14-8 we get C = $10 \times 10^{-3}/3$ = 3,333 µF.

### 14.4.1  Power Line Immunity Curve

In the early 1980s, the Computer and Business Equipment Manufacturers Association (CBEMA), a predecessor to the Information Technology Industry Council (ITI) established a realistic power-line immunity profile curve for information technology equipment (ITE). This curve defines the ac power line voltage envelope that can be tolerated with no interruption or loss of function by most information technology equipment. In 2000, based on extensive additional research, the ITI updated the curve to that shown in Fig. 14-21. This curve has become a standard within the ITE industry as well as other industries, and usually, it is referred to as just the CBEMA curve (ITI, 2000).

Although not intended as a design specification for equipment, the curve is often used for that purpose. Manufacturers that want to produce a quality product often use the CBEMA curve as a guideline for power-line disturbance immunity. Power quality voltage testers are even available that continuously monitor the ac power-line voltage and record any events that exceed the CBEMA curve's "no interruption of function" limits.

The CBEMA curve is a voltage tolerance envelope applicable to single-phase 120-V, 60 Hz equipment that covers both transient and steady-state conditions. It is a semi-log plot of percent variation of ac voltage versus time duration. The plot contains three regions. The area to the left of the curve defines the voltage tolerance envelope in which a product should operate properly with no degradation in performance. In the voltage sag region to the lower right of the curve, the product may not operate properly but should not be damaged. The voltage surge region to the upper right of the curve is the prohibited region. If the equipment is subjected to these conditions, then the product may malfunction or possibly even be damaged.

The curve indicates that a product typically should operate indefinitely within a voltage variation of $\pm$10%. It should be able to withstand a voltage

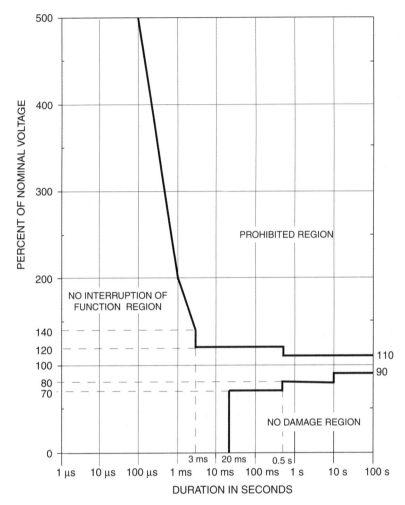

**FIGURE 14-21.** The CBEMA curve (revised in 2000) defines the power line voltage tolerance envelope applicable to single-phase, 120-V equipment. (© *Information Technology Industry Council, 2000)*

sag of 20% for 10 s, a drop of 30% for 0.5 s, and complete loss of voltage for as long as 20 ms and still operate properly.

The curve also indicates that a product typically will operate properly with a voltage surge of 20% for 0.5 s, a voltage increase of 40% for 3 ms, and an increase of 100% for 1 ms.

Voltage surges that occur in the region to the left of the curve, with time durations of less than 100 μs, typically are the result of nearby lightning strikes. In this portion of the curve, it is the energy of the transient that is most important, rather than the voltage amplitude. The intent is for the product to provide an 80-J minimum transient immunity in this region. This result is consistent with the European Union's surge test requirement, because the surge energy, as listed in Table 14-3, is less than 80 J.

Note that although the CBEMA curve is only applicable to 120-V, 60-Hz systems, the European Union's voltage sag and interruption requirements listed in Table 14-5 are consistent with the limits of the curve. The voltage dip of 30% for 0.5 cycles (10 ms for a 50 Hz power line) requirement falls within the no interruption of function region of the CBEMA curve, which is consistent with the European Union's (EU's) performance Criteria B. The voltage dip of 60% for 5 cycles (100 ms at 50 Hz) and more than 95% for 250 cycles (5 s at 50 Hz) fall within the no damage region of the curve and is consistent with the EU's performance Criteria C.

## SUMMARY

- Most rf susceptibility problems are the result of audio rectification.
- For audio rectification to be a problem, the following two things must occur:
     First, the rf energy must be picked up.
     Second, the rf energy must be rectified.
- RFI mitigation techniques should be applied at the device level, to the cables, and to the enclosure.
- Keep loop areas of sensitive signals as small as possible.
- Add RFI filters on inputs to sensitive devices.
- The filter's series element should have an impedance of 50 to 100 Ω at the frequency of interest.
- The filter's series element can be a resistor, inductor or ferrite.
- Choose an RFI filter's inductor such that the product of the frequency and the inductance equals ten.
- The RFI filter's shunt element should have an impedance of a few ohms or less at the frequency of interest.
- Choose the RFI filter's capacitor such that the product of the frequency and the capacitance equals 0.1.
- A 1000-pF capacitor is effective over the frequency range of 80 MHz to 1000 MHz.
- Use rf decoupling, even if the circuit operates at low frequency.
- Use multilayer PCBs with ground and power planes.
- Use balanced circuitry whenever possible, especially on the low-level inputs to amplifiers.
- Add high-frequency capacitive filters on the inputs and outputs of voltage regulators.

- Use high-quality shielded cables, high-coverage braid, or braid-over-foil shields.
- Properly terminate shield to enclosure, no pigtails or drain wires.
- Do not use spiral-shielded cables for rf immunity.
- If shielded cables are not used, filter the I/O leads where they enter/leave the enclosure.
- For ribbon cables, limit the signal to ground conductor ratio to 3 to 1.
- The maximum linear dimension of apertures in shielded enclosures should be less than 1/20 wavelength at the highest frequency of concern.
- Even thin shields made of aluminum or copper are effective at frequencies above 500 kHz.
- Steel shields should be used at frequencies below 500 kHz.
- The three most common high voltage transients effecting equipment are as follows:
    Electrostatic discharge
    Electrical fast transient
    Lightning surge
- Effective transient protection involves a three-prong approach, as follows:
    First, divert the transient current.
    Second, protect the sensitive devices from upset or damage.
    Third, write transient hardened software (see Section 15.10).
- Very low-capacitance polymer VVRs should be used for transient protection on high-speed interfaces with data rates greater than 100 MB/s.
- Similar techniques can usually be used to protect against ESD and EFT.
- Surge, however, has a pulse energy a thousand times greater than ESD or EFT and requires protection devices that can handle the additional energy.
- Surge protection is only required on ac and dc power lines.
- When protection for both sub-nanosecond transients and surge are both required, a two-stage hybrid protection network might be required.
- Table 14-5 summarizes the characteristics of various transient suppression devices.
- Power-line disturbances, such as voltage dips, interruptions, and surges, are a fact of life, and should be protected against.
- The CBEMA curve (Fig. 14-21) can be used as a power-line voltage tolerance immunity guideline for equipment.

## PROBLEMS

14.1   What is the electric field strength 1 km from an AM broadcast station with an effective radiated power of 10,000 W?

14.2   What is the electric field strength 1 m from a 1-W small hand–held FM transmitter?

14.3   An FM transmitter is driving 100 W into a dipole antenna. What is the electric field strength 10 m from the antenna?

14.4   Electromagnetic fields are usually measured or specified in terms of the electric field strength (V/m) or in terms of the power density (W/m$^2$ or mW/cm$^2$). Write an equation for the electric field strength $E$ (V/m) in terms of the power density $P$ (W/m$^2$), in the far field.

14.5   If the electromagnetic field of a transmitter has a power density of 10 mW/cm$^2$, measured in the far field, what is the electric field strength in V/m?

14.6   What is audio rectification?

14.7   a. What would be an appropriate value for the series inductor to use in an RFI filter to protect against susceptibility problems caused by a 5-MHz transmitter?

       b. What would be an appropriate value shunt capacitor to use with the above filter?

14.8   When calculating the attenuation of an $L$-filter, it is usually assumed that the circuit into which the filter is inserted has a low source impedance and a high load impedance.

       a. If the source impedance is not low, what affect will that have on the performance of the filter?

       b. If the load impedance is not high, what affect will that have on the performance of the filter?

14.9   What is the maximum linear dimension of a shielded enclosure's aperture that will provide a minimum of 20 dB of shielding at frequencies up to 650 MHz?

14.10  What will be the voltage drop across a 12-V TVS diode in series with 1-in of PCB trace (1/2 in on either side of the diode) when clamping a 2-A, 500 ps rise time ESD pulse?

14.11  a. What will be the impedance of a 470-pF filter capacitor in series with 3/4-in of PCB trace, at 300 MHz?

       b. What would be the impedance, at 300 MHz, of an ideal 470-pF capacitor (with no series inductance)?

14.12  a. The European Union's ESD test generator is set for 4-kV contact discharge. What will the peak current be when discharged into a short circuit?

       b. What will be the peak current from a 1-kV EFT pulse when applied to a short circuit?

       c. What would be the peak current from the EU's surge test generator when it is discharging a 2-kV surge into a short circuit?

14.13  The hybrid transient protection network shown in Fig. 14-20 uses a 150-V breakdown gas tube and a 12-V bidirectional TVS diode. The series resistors are 10 $\Omega$ each. A 1000-V, 1-$\mu$s rise time, 50-$\mu$s pulse width surge (with a 5 $\Omega$ source impedance) is applied to the network.

    a. What will be the maximum initial current through the TVS diode?

    b. What will be the maximum current through the TVS diode after the gas tube turns on?

    c. What will be the maximum current through the gas tube?

14.14 A product draws 0.5 A of steady-state current from the power supply, and the voltage specification for proper operation of the product states that the dc voltage should be $9V \pm 2V$. What value capacitance is required in the output of the power supply for the product to be able to ride through a 20-ms complete loss of ac voltage on the input to the power supply?

14.15 Your product draws 95 mA of current from the output of a SMPS. The product specification is that the dc voltage be $12 \pm 2V$. The power supply has 1200 $\mu$F of output capacitance. For how many cycles of the 60 Hz power line can the product still operate if ac power is lost to the input of the SMPS?

14.16 Would you expect a computer or printer that is operated off of a 120-V, 60-Hz power line to satisfy the EUs performance criteria A when subjected to:

    a. An 80 % voltage drop for 0.5 cycles?

    b. A 40% voltage dip for 5 cycles?

    c. A 150% voltage surge for 1 ms?

    d. Would you expect the product to be damaged under any of the above conditions?

14.17 Define a set of ac power line conditions (voltage amplitude and duration) under which you might suspect a piece of ITE equipment to not operate properly, or even possibly be damaged?

## REFERENCES

EN 60601-1-2. *Medical Electrical Equipment—Part 1-2: General Requirements for Safety—Collateral Standard: Electromagnetic Compatibility—Requirements and Tests*, CENELEC, 2007.

EN 61000-6-1. *Generic Immunity Standard for Residential, Commercial, and Light-Industrial Environments*, CENELEC, 2007.

EN 61000-6-2. *Generic Immunity Standard for Industrial Environments*, CENELEC, 2005.

German, R. F., Ott H. W., and Paul, C. R. "Effect of an Image Plane on Printed Circuit Board Radiation", *IEEE Electromagnetic Compatibility Symposium*, Washington, DC, August 21–23, 1990.

Industry Canada. *EMCAB 1, Issue 3.* "Immunity of Electrical/Electronic Equipment Intended to Operate in the Canadian Radio Environment (0.010–10,000 MHz)." Industry Canada, June 1990.

*ITI (CBEMA) Curve Application Note.* Information Technology Industry Council. 2000. Available at www.itic.org/archive/iticurv.pdf.

Standler, R. B. *Protection of Electronic Circuits from Overvoltage.* Wiley, New York, 1989. [Out of print]

Standler, R. B. *Protection of Electronic Circuits from Overvoltage.* Frederica, DE, Dover, 2002 [Paperback reprint of above book].

## FURTHER READING

AN-671. *Reducing RFI Rectification in Instrumentation Amplifiers.* Analog Devices, Norwood, MA, 2003.

Lepkowski, J. AND8229/D. *An Introduction to Transient Voltage Suppression Devices,* ON Semiconductor, 2005.

Lepkowski, J. and Johnson, J. *Zener Diode Based Integrated Filters, an Alternative to Traditional EMI Filter Devices,* ON Semiconductor.

General Electric. *Transient Voltage Suppression,* 5th ed. 1986.

# 15 Electrostatic Discharge

Static electricity is familiar to all of us as the static cling of clothing, as arcing occurring when touching a doorknob or other metal object, and as lightning. Static electricity was known to the ancient Greeks over 2000 years ago. In medieval times, magicians used electrostatic effects as part of their "bag of tricks." In our time, static electricity has been harnessed to perform many useful functions. Examples of products using this principle are electrostatic copiers, dust precipitators, air purifiers, and electrostatic spray painters.

However, uncontrolled electrostatic discharge (ESD) has become a hazard to the electronics industry. Since the early 1960s, it has been recognized that many integrated circuits (ICs), metal-oxide semiconductors (MOSs), discrete electrical parts such as film resistors and capacitors, and crystals are susceptible to damage from electrostatic discharge. As electronic devices become smaller, faster, and operate at lower voltages, their susceptibility to ESD will increase.

ESD control is a special case of the overall subject of EMC and transient immunity. As will be observed, many of the techniques used to decrease the susceptibility of a system to ESD are similar to those used to provide transient immunity and to control radiated emission.

## 15.1 STATIC GENERATION

Static electricity can be created in many different ways,* but the most common is by contact and subsequent separation of materials. The materials may be solids, liquids, or gases. When two nonconductors (insulators) are in contact, some charge (electrons) is transferred from one material to the other. Because charge is not very mobile in an insulator, when the two materials are separated, this charge may not return to the original material. If the two materials were originally neutral, they will now be charged, one positively and the other negatively.

---

* For example, triboelectric charging, induction charging, and piezoelectric effect.

---

*Electromagnetic Compatibility Engineering,* by Henry W. Ott
Copyright © 2009 John Wiley & Sons, Inc.

This method of generating static electricity is referred to as triboelectric charging. In ancient times, static electricity was generated by rubbing wool against amber. *Tribos* is the Greek word for rubbing, and *ēlektron* is the Greek word for amber, so triboelectric means "rubbing amber." Although we tend to think that rubbing is required to generate charge between two materials, that is not true, all that is actually required is that the materials come into contact and are then subsequently separated.

Some materials readily absorb electrons, whereas others tend to give them up easily. The triboelectric series is a listing of materials in order of their affinity for giving up electrons. Table 15-1 is a typical triboelectric series. The materials at the top of the table easily give up electrons and therefore acquire a positive charge. The materials at the bottom of the table easily absorb electrons and therefore acquire a negative charge. It should be kept in mind, however, that this series is only approximate.

When two materials are in contact, electrons will transfer from the material higher on the list to the material lower on the list. The degree of separation of the two materials in Table 15-1 does not necessarily indicate the magnitude of the charge created. The magnitude depends not only on the position of the materials in the triboelectric series but also on the surface cleanliness, pressure of the contact, amount of rubbing, surface area in contact, smoothness of surface, and the speed of separation. A charge can also be generated when

**TABLE 15-1.   Triboelectric Series**

| POSITIVE | |
|---|---|
| 1. Air | 18. Hard rubber |
| 2. Human skin | 19. Mylar®a |
| 3. Asbestos | 20. Epoxy glass |
| 4. Glass | 21. Nickel, copper |
| 5. Mica | 22. Brass, silver |
| 6. Human hair | 23. Gold, platinum |
| 7. Nylon | 24. Polystyrene foam |
| 8. Wool | 25. Acrylic |
| 9. Fur | 26. Polyester |
| 10. Lead | 27. Celluloid |
| 11. Silk | 28. Orlon |
| 12. Aluminum | 29. Polyurethane foam |
| 13. Paper | 30. Polyethylene |
| 14. Cotton | 31. Polypropylene |
| 15. Wood | 32. PVC (vinyl) |
| 16. Steel | 33. Silicon |
| 17. Amber | 34. Teflon®a |
| | NEGATIVE |

a Registered Trademark of DuPont

two pieces of the same material are in contact and subsequently separated, although in this case which of the two will acquire a positive charge and which a negative charge cannot be predicted. A good example of this is the opening of a plastic bag.

Charge is measured in coulombs, which is difficult to measure. Therefore we normally refer to the electrostatic potential (measured in volts) of an object, rather than its charge. The relationship between charge, voltage, and capacitance is

$$V = \frac{Q}{C}. \tag{15-1}$$

As two materials are separated, the charge imbalance $Q$ remains fixed; therefore, the product $VC$ is a constant. When the materials are close together, the capacitance is large; hence, the voltage is low. As the materials are separated, the capacitance decreases and the voltage increases. For example, if the capacitance is 75 pF and the charge is 3 μC, the voltage will be 40,000 V.

Triboelectric charging also occurs when an insulator is separated from a conductor, but not when two conductors are separated. In the latter case, as soon as separation starts, the charge returns to the original material, because the mobility of charge is large in a conductor.

Thus, both conductors and insulators can easily be charged by contact and separation with an insulator. Intimate contact is all that is required for electron transfer to occur. Rubbing tends to increase the pressure of the contact and bring more of the surface in contact and hence increases the charge transfer. Faster separation allows less time for charge reflow, which also increases the charge transfer and the subsequent voltage.

**TABLE 15-2.   Typical Electrostatic Voltages.**

| | Electrostatic Voltage | |
| --- | --- | --- |
| Means of Static Generation | 10% to 20% Relative Humidity | 65% to 90% Relative Humidity |
| Walking across carpet | 35,000 | 1500 |
| Walking on vinyl floor | 12,000 | 250 |
| Worker moving at bench | 6000 | 100 |
| Opening a vinyl envelope | 7000 | 600 |
| Picking up common polyethylene bag | 20,000 | 1200 |
| Sitting on chair padded with polyurethane foam | 18,000 | 1500 |

Table 15-2 from DOD-HDBK-263 lists typical electrostatic voltages that can be generated under various conditions. Notice the significant effect that humidity has on the voltages. The generation of 10 to 20 kV on common materials in the home and work environments is not unusual under low humidity (<20%) conditions. However, when the humidity is greater than 65% these voltages are limited to 1500 V or less.

*Static electricity is a surface phenomenon.* The static charge exists solely on the surface of the materials and not inside them. The charge on an insulator remains in the area where it is created and is not distributed within the material or over the entire surface of the material. Grounding an insulator will not eliminate the charge.

If charge is created on a conductor, the like polarity charges want to separate from each other and therefore will distribute themselves over the entire surface of the conductor, because the surface is the farthest that the charges can move away from each other. None of the charge will be inside the conductor, but it will exist only on the surface. Unlike an insulator, however, a charged conductor will lose its charge if grounded.

Electrostatic discharge is normally a three-step process as follows:

1. A charge is generated on an insulator.
2. This charge is transferred to a conductor by contact or induction.
3. The charged conductor comes near a metal object and a discharge occurs.

For example, when a person walks across a carpet, the soles of their shoes (insulators*) become charged as they make contact and separation from the carpet. This charge is then transferred to their body (a conductor) usually by induction. If the person then touches a metallic object (grounded or not), a discharge occurs. When the discharge occurs to an ungrounded object (e.g., a doorknob), the discharge current flows through the capacitance between the object and ground.

A charged insulator by itself is not directly an ESD threat. Since the charge on an insulator is not free to move, it cannot produce a static discharge. The danger from a charged insulator comes from its potential for producing a charge, usually by induction, onto a conductor, such as a person, which then is capable of a discharge.

### 15.1.1  Inductive Charging

An electrically charged object (insulator or conductor) is surrounded by an electrostatic field. If a neutral conductor is brought into the vicinity of a charged object, the electrostatic field will cause the balanced charges on the

---

*Some shoes are static dissipative (e.g., those with leather soles).

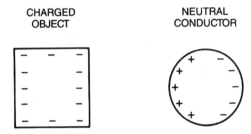

**FIGURE 15-1.** The charge on a neutral conductor separates in the presence of a charged object.

neutral conductor to separate as shown in Fig. 15-1. The polarity of charge opposite to that on the charged body will be on the surface of the neutral conductor nearest the charged body, and the same polarity charge will be on the surface farthest away. The conductor will remain neutral, however, with equal amounts of positive and negative charge. When the neutral object is moved away from the charged object, the positive and negative charges will recombine.

If, however, a connection is made between the neutral conductor and ground (e.g., the object is touched by a person or a grounded object) while in the

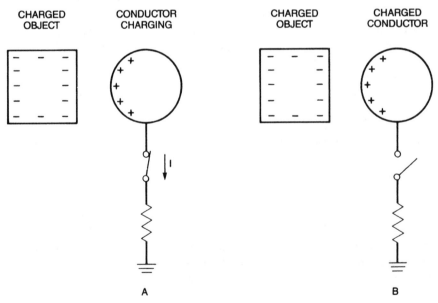

**FIGURE 15-2.** If the neutral conductor of Fig. 15-1 is momentarily grounded (A), the negative charge will bleed off and leave the conductor charged (B).

vicinity of the charged object, the charge on the side of the neutral conductor farthest away from the charged object will bleed off, as shown in Fig. 15-2A. Then, if the ground is removed, as in Fig. 15-2B, while the conductor is still in the vicinity of the charged object, the conductor will be charged without ever having come in contact with the charged object. The ground connection only has to be momentary, and it can have considerable impedance (100 kΩ or more).

### 15.1.2  Energy Storage

Although charge exists on the surface of an object, the energy (field) associated with the charge is stored in the object's capacitance. Normally, we think of capacitance as occurring between closely spaced parallel plates. However, all objects have a *free-space capacitance* of their own, the object itself being one of the plates and the second plate being located at infinity. This represents the minimum capacitance that an object can have. The free-space capacitance of even an irregularly shaped object is determined primarily by its surface area. Therefore, the free-space capacitance can be approximated by considering the simple geometry of two concentric spheres, one sphere having the same surface area as the object and the second sphere located at infinity.

The capacitance between two concentric spheres (Hayt, 1974, p. 159) is

$$C = \frac{4\pi\varepsilon}{\left(\dfrac{1}{r_1}\right) - \left(\dfrac{1}{r_2}\right)}, \qquad (15\text{-}2)$$

where $r_1$ and $r_2$ are the radii of the two spheres ($r_2 > r_1$), and $\varepsilon$ is the dielectric constant of the medium between the spheres.

For free space, $\varepsilon = 8.85 \times 10^{-12}$ F/m. If the radius of the outer sphere is allowed to go to infinity, then Eq. 15-2 reduces to

$$C = 111r, \qquad (15\text{-}3)$$

where $C$ is the capacitance in pF and $r$ is the radius of the sphere in meters.

Equation 15-3 represents the capacitance of an isolated body in space, and it can be used to estimate the minimum (free-space) capacitance of any object. The procedure is (1) to determine the surface area of the object for which you want to calculate the free-space capacitance, (2) to calculate the radius of a sphere having the same surface area, and (3) to calculate the capacitance from Eq. 15-3. Surface-area equations for various shaped objects are given in Fig. 15-3.

For example, a person has a surface area approximately equivalent to a 1-m diameter sphere; therefore, the free-space capacitance of the human body is about 50 pF. The earth has a free-space capacitance of slightly more than 700

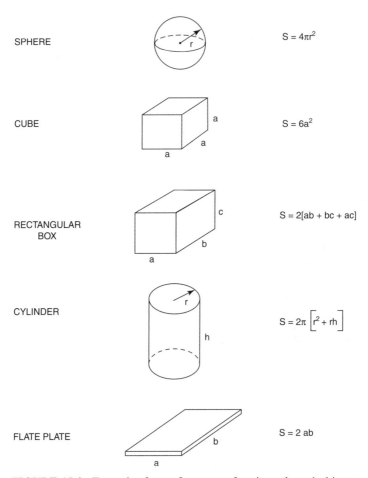

SPHERE     $S = 4\pi r^2$

CUBE     $S = 6a^2$

RECTANGULAR BOX     $S = 2[ab + bc + ac]$

CYLINDER     $S = 2\pi \left[ r^2 + rh \right]$

FLATE PLATE     $S = 2\,ab$

**FIGURE 15-3.** Formulas for surface area of various shaped objects.

µF, and an object the size of a marble has a free-space capacitance of slightly more than 1 pF.

Besides the free-space capacitance given by Eq. 15-3, additional parallel plate capacitance also exists because of the proximity of the object to other surrounding objects.

The capacitance between two parallel plates is equal to

$$C = \frac{\varepsilon A}{D}, \tag{15-4}$$

where $A$ is the area of the plates and $D$ is the distance between the plates.

The total capacitance of an object is then the combination of the free-space capacitance (Eq. 15-3) plus the parallel plate capacitance (Eq. 15-4) to adjacent objects.

## 15.2  HUMAN BODY MODEL

Humans are a prime source of electrostatic discharge. As previously discussed, it is easy for a person to build up a static charge. This charge can then be transferred from the person to a piece of sensitive electronic equipment in the form of an electrostatic discharge.

To model this human body discharge, we start with the capacitance of the human body. In addition to the 50 pF of free-space capacitance, the primary contributor to the capacitance of the body comes from the capacitance between the soles of the feet and ground. As shown in Fig. 15-4, this is about 100 pF (50 pF per foot). Additional capacitance of from 50 to 100 pF may exist because of the proximity of the person to other surrounding objects, such as structures, walls, and so on. Therefore, the capacitance of the human body varies between 50 and 250 pF.

The human body model (HBM) for ESD is shown in Fig. 15-5. The body capacitance $C_b$ is charged up to a voltage $V_b$ by triboelectric charging (or other means), and the discharge occurs through the body resistance $R_b$. The body resistance is important because it limits the discharge current. The body resistance can vary from about 500 to 10,000 $\Omega$, depending on which part of the body the discharge occurs from. If the discharge is from the tip of

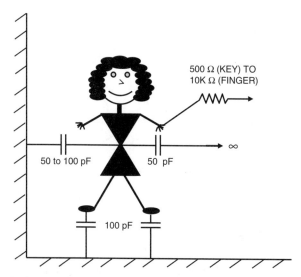

**FIGURE 15-4.** Human body capacitance and resistance.

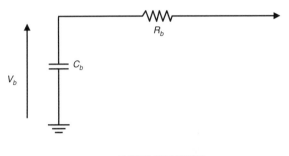

RANGE OF VALUES

| | |
|---|---|
| $C_b$ | 50 to 250 pF |
| $R_b$ | 500 to 10k $\Omega$ |
| $V_b$ | 0 to 20 kV |

**FIGURE 15-5.** Human body model.

the finger, then the resistance will be about 10,000 $\Omega$; if from the palm of the hand, about 1000 $\Omega$; if from a small metal object in the hand (e.g., a key or a coin), it will be about 500 $\Omega$. If, however, the discharge occurs from a large metal object in contact with the person, such as a chair or a shopping cart, the resistance can be as low as 50 $\Omega$.

The circuit of Fig. 15-5 is used in ESD testing to simulate the human body discharge. Various ESD test standards use different values for the components

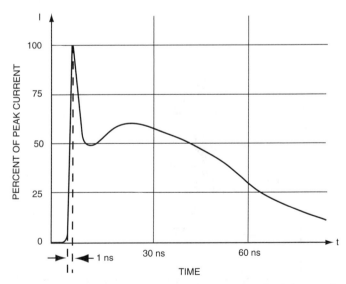

**FIGURE 15-6.** Typical waveform produced by a 150-pF, 330-$\Omega$ human body model discharge.

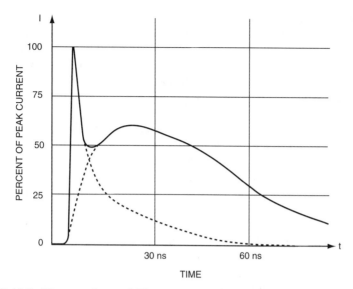

**FIGURE 15-7.** The waveform of Fig. 15-6 consists of two waveforms; the fast rise narrow pulse is the discharge of the free-space capacitance of the ESD tester's probe tip, and the slower wide pulse is the discharge from the 150-pF capacitor in series with the tester's ground strap inductance.

in the model. The most commonly used model consists of 150 pF and 330 $\Omega$ as specified in the European Union's basic ESD standard EN 61000-4-2.

Figure 15-6 shows the typical waveshape produced by a 150-pF, 330-$\Omega$ human body model discharge into a special 2-$\Omega$ test target specified in EN 61000-4-2. The rise time is 0.7 to 1 ns, and the peak current is 30 A for an 8-kV discharge, and 15 A for a 4-kV discharge.

This waveshape is actually the combination of two discharges as shown in Fig. 15-7. The fast rise time narrow pulse is the discharge of the free-space capacitance of the ESD tester's probe tip (the size and shape of which is defined in the standard), and the slower wide pulse is the discharge of the 150 pF capacitor in series with the tester's ground strap inductance (not shown in Fig. 15-5).

An actual discharge from a voltage of less than 3500 V will not be felt or sensed by the person involved. Because many electronic devices are sensitive to damage from discharges of only a few hundred volts, component damage can occur from a discharge that is not felt, heard, or seen. At the other extreme, discharges from potentials greater than 25 kV are painful to the person involved.

## 15.3  STATIC DISCHARGE

Charge accumulated on an object leaves the object by one of two ways, leakage or arcing. Because it is better to avoid arcing, leakage is the preferred way to discharge an object. Charge can leak off an object through the air, because of humidity. The higher humidity, the faster the charge will leak off the object. The charge on an object can also be counteracted by using an ionizer to fill the air with positive and negative charged ions. The opposite polarity ions will be attracted to the object and will neutralize the charge on it. The more ions present, the faster the charge will be neutralized.

Leakage from a *charged conductor* can be made to occur by intentionally grounding the object. This ground may be a hard ground (close to 0 Ω) or a soft ground (a large impedance, a few hundred thousand ohms to a few megohms) that will limit the current flow. Because the human body is conductive, grounding it with a conductive wrist strap, for example, will drain off the charge. However, grounding a person will not drain the static charge from his or her clothing (nonconductors), or from a plastic object held in the hand, such as a Styrofoam coffee cup. To remove the charge from these objects, ionization or high humidity (> 50%) must be used.

When grounding a person, a hard ground should be avoided because of the safety hazard that would exist if the person came in contact with the ac power line or another high voltage. The minimum impedance that should be used in grounding a person is 250 kΩ. Grounded wrist straps usually have a 1-MΩ resistance to ground. The higher the resistance, the longer it will take for the charge to bleed off the object.

### 15.3.1   Decay Time

Because the charge on an object may leak off over a period of time, an important parameter is the decay time—the time it takes for the charge to be reduced to 37% of its initial value. The decay time (sometimes called the relaxation time) is equal to (Moore 1973, p. 26, Eq. 15)

$$\tau = \frac{\varepsilon}{\sigma}, \tag{15-5}$$

where $\varepsilon$ is the dielectric constant for the material and $\sigma$ is the conductivity. The decay time can also be written in terms of the surface resistivity of the material and is

$$\tau = \varepsilon\rho. \tag{15-6}$$

From Eq. 15-6, we observe that the decay time can be used as an indirect method of measuring the resistivity of the material.

Because static electricity is a surface phenomenon, materials can be classified by their surface resistivity. Surface resistivity has the dimensions of ohms per square. It is equivalent to the resistance measured across a square section of the material. It does not matter what size the square is, the resistivity will be the

**TABLE 15-3. Surface Resistivity of Various Classes of Materials.**

| Material | Surface Resistivity ($\Omega$/Square) |
|---|---|
| Conductive | 0 to $10^5$ |
| Static dissipative | $10^5$ to $10^9$ |
| Antistatic[a] | $10^9$ to $10^{14}$ |
| Insulative[a] | $> 10^{14}$ |

[a] A surface resistivity of $10^{14}$ is high for the transition from antistatic to insulative. A more realistic value would be $10^{12}$ $\Omega$/square.

same. Surface resistivity is measured with a fixture having two electrodes that form the opposite sides of a square. As long as the spacing between the electrodes is the same as the length of the electrodes, the resistance will be the same regardless of the length of the electrodes. That is, if the two electrodes are 3 cm long, they must be placed 3 cm apart.

Based on surface resistivity, DOD-HDBK-263 classifies materials into four categories, as listed in Table 15-3.

Materials with surface resistivities of $10^9$ $\Omega$ per square or less can be discharged rapidly by grounding. If a charge already exists on an object, then it should be discharged slowly to limit the current and avoid damage.

Conductive materials are the fastest to dissipate charge and can be dangerous when used near already charged electronic devices. If a charged device should come in contact with a grounded conductive material, then it will be discharged rapidly with a large peak current, and damage may result.

Static-dissipative materials are preferred to conductive materials because charge dissipation occurs at a slower rate. Grounded static-dissipative materials can be used to prevent charge buildup and to discharge objects already charged safely.

Antistatic materials are the slowest to dissipate charge. Nevertheless, they are useful because they can dissipate charge faster than it is generated and therefore prevent an object from accumulating a charge. An example of this is a pink polyethylene bag. To prevent triboelectric charging, the surface resistivity of a material should not exceed $10^{12}$ $\Omega$ per square.

Neither static-dissipative nor antistatic materials will charge when separated from themselves or any other materials. They have similar applications and are sometimes grouped together. They are the preferred materials to use in an ESD-sensitive environment, such as a manufacturing line for electronic equipment.

Insulators do not dissipate charge but retain whatever charge they have. Examples are a polyethylene bag and Styrofoam packing material. These materials should not be allowed in an ESD-sensitive environment.

## 15.4   ESD PROTECTION IN EQUIPMENT DESIGN

*ESD protection should be part of the original system design and not added at the end, when testing indicates a problem exists.*
Effective ESD immunity design requires a *three-pronged approach.*

- First, prevent or minimize the entry of the transient currents by:
  - Effective design of the enclosure
  - Cable shielding
  - Providing transient protection on all conductors of unshielded external cables

- Second, harden sensitive circuits, such as:
  - Resets
  - Interrupts
  - Other critical control inputs

- Third, write transient hardened software capable of detecting, and if possible correcting, errors in the following:
  - Program flow
  - Input/output (I/O) data
  - Memory

For optimum ESD protection, all three of the above must be addressed.
Energy from a static discharge can be coupled to an electronic circuit in two ways:

1. By direct conduction
2. By field coupling, including
   a. Capacitive coupling
   b. Inductive coupling

Direct conduction occurs when the discharge current (typically tens of amperes) flows directly through the sensitive circuit. This may result in actual damage to the circuit.

You do not, however, need a direct discharge to cause an ESD problem. The fast rise time, large voltage, and high current associated with ESD produces intense electric and magnetic fields. These fields, although usually not causing damage, are strong enough to upset the operation of many electronic circuits, even at distances of a meter or more away from the actual discharge.

A circuit or system may be protected from a static discharge by any of the following:

1. Eliminating the static buildup on the source
2. Insulating the product to prevent a discharge
3. Providing an alternative path for the discharge current to bypass the sensitive circuits
4. Shielding the circuit against the electric fields produced by the discharge
5. Decreasing loop areas to protect the circuit from the magnetic fields produced by the discharge

The first three items in the above list deal with controlling the direct discharge, and the last two items deal with controlling the associated field coupling.

ESD-induced effects in electronic systems can be divided into the following three categories:

1. Hard errors
2. Soft errors
3. Transient upset

Hard errors cause actual damage to the system hardware (e.g., destruction of an IC). Soft errors affect system operation (e.g., a changed memory bit or program lockup) but do not cause physical damage. Transient upset does not cause an error, but the effect is perceptible (e.g., rolling of a CRT display, or momentary changing of a display reading).

The European Union's criteria for an ESD failure (Performance Criteria B) is as follows:

> The apparatus shall continue to operate as intended after the test. No degradation of performance or loss of function is allowed. During the test, degradation of performance is however allowed. No change of actual operating state or stored data is allowed.

In other words, transient upset is allowed, but no soft or hard errors are allowed.

The first step in designing equipment to be immune to ESD is to prevent the direct discharge from flowing through the susceptible circuitry. This can be accomplished either by insulating the circuit or by providing an alternative path for the discharge current.

If insulation is used, it must be complete, because a spark can enter through an extremely narrow air gap, such as a seam or the air gap surrounding the keys on a keyboard. For example, a discharge can occur through something as small as a pinhole.

In the case of a product in a metallic enclosure, the enclosure can be used as an alternative path for the ESD current. To divert the ESD current effectively from sensitive circuits, all metallic components of the enclosure must be bonded

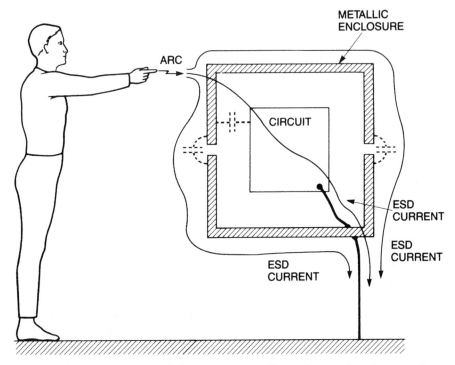

**FIGURE 15-8.** Electrostatic discharge to a metallic enclosure that does not have electrical contact across the seams.

together. If the enclosure is not electrically continuous, then a portion of the current may be forced to flow through the internal circuitry as shown in Fig. 15-8.

*The basic principle of ESD bonding and grounding is to use low-inductance multipoint bonding where ESD current is desired and single-point bonding where ESD current flow is not wanted.* Therefore, good high-frequency electrical continuity (multipoint) must be provided across all joints, seams, and hinges of the enclosure. For an improperly bonded enclosure, the ESD current paths are complex and unpredictable, flowing through the parasitic capacitance among the enclosure and internal circuitry.

## 15.5  PREVENTING ESD ENTRY

The three most common points of ESD entry are the enclosure, cables, and keyboards or control panels. The enclosure can be metallic or plastic, both have advantages and disadvantages. Different ESD mitigation approaches are, however, used in the two cases.

## 15.5.1  Metallic Enclosures

The major advantage of a metallic enclosure is that it can be used as an alternative path for the ESD current. The major disadvantage of a metallic enclosure is that it encourages a discharge to occur.

Consider the situation shown in Fig. 15-9, of a circuit insulated from and completely enclosed by a grounded metallic enclosure. The circuit has no connection to anything outside of the enclosure. Note that because of the large inductance of the ground conductor, most of the initial ESD current flows through the parasitic capacitance between the enclosure and ground. When a discharge occurs, the enclosure rises in potential as a result of charging this capacitance. In the case of a 10,000-V discharge, the enclosure may typically rise in potential to about 1000 to 2000 V.

Assuming a conductor inductance of 15 nH/in, a 6-ft ground conductor has about 1 μH of inductance. At 300 MHz, the spectral content of a 1-μs rise-time ESD pulse produces a ground conductor impedance of approximately 2000 Ω. The charging of the enclosure's capacitance to ground is what causes the potential of the enclosure to rise initially. The ground conductor then

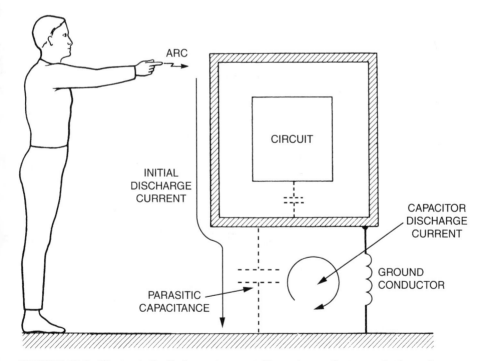

**FIGURE 15-9.** Electrostatic discharge to a metallic enclosure that completely encloses the circuit. The circuit has no external connections.

discharges the parasitic ground capacitance at a much slower rate, returning the enclosure to the ground potential.

When the enclosure rises in potential as a result of the discharge, the circuit inside the enclosure also rises to the same voltage. Therefore, there is no difference in potential between the enclosure and the circuit, or between different parts of the circuit, and the circuit is protected and perfectly safe.

Discontinuities in the enclosure (e.g., seams or holes) can cause differential voltages to appear on the enclosure as well as allowing ESD induced fields to couple to the inside of the enclosure, see Fig. 15.8. These enclosure voltages and fields can then couple to the circuit and affect its operation.

Two approaches can be used to solve this field-coupling problem. The first and best is to make the enclosure as complete as possible. The enclosure should be as continuous as possible with a minimum number of seams and apertures. To minimize ESD field coupling any apertures in the enclosure should have a maximum length of 25 mm (1 in).

The second approach is to minimize loop areas of the circuitry to minimize inductive coupling and/or to add internal shields to block the capacitive coupling between the enclosure and the circuit. This approach is discussed in more detail in the discussion associated with Fig. 15-11 and in Section 15.9.

The situation shown in Fig. 15.9 is not a practical configuration, because the circuit has no connection to anything outside of the enclosure. A more practical

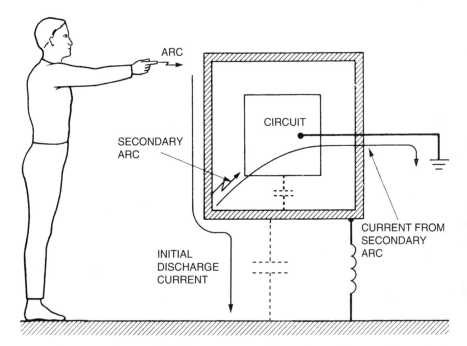

**FIGURE 15-10.** Electrostatic discharge to a metallic enclosure that contains a circuit with an external ground connection.

case is shown in Fig. 15-10, where the enclosed circuit has a connection to an outside ground. When a discharge occurs to the enclosure, the enclosure rises in potential as before. Now however, because of the external ground connection, the circuit remains at, or close to, the ground potential. Therefore, a large potential difference exists between the enclosure and the circuit, and a *secondary arc* as shown in Fig. 15-10 may occur between the enclosure and and the circuit. This secondary arc occurs without the current-limiting body resistance and can produce currents much larger (hundreds of amperes) than the primary arc, and therefore it is potentially much more destructive.

A similar effect occurs if the enclosure is ungrounded. In this case, instead of rising to about a few thousand volts, as was the case with a grounded enclosure, the enclosure may rise closer to the full potential of the discharge source. Therefore, it is desirable to ground all metallic enclosures for ESD protection.

The secondary arc can be prevented by (1) providing sufficient space between all metal parts and the circuit or (2) by connecting the circuit to the metallic enclosure, thus keeping it at the same potential as the enclosure. The spacing should be sufficient to withstand about 2000 V for a grounded enclosure and 15,000 V for an ungrounded enclosure.

The breakdown voltage for air is approximately 3000 V/mm (75,000 V/in) at standard temperature and pressure (STP). The breakdown voltage is approximately proportional to pressure and inversely proportional to absolute temperature. The safe clearance distance to prevent an arc is usually considered to be about one third of this or 1 mm/kV. Table 15-4 lists the safe clearance distances for various voltages.

Even without a secondary arc, the strong electric field produced between the metallic enclosure and the circuit can cause problems. Often, a second shield inside the enclosure, and around the sensitive circuits, will be needed to block the electric field coupling. This second, unexposed shield should be connected to the circuit common as shown in Fig. 15-11B.

If the circuit is connected to the enclosure, this connection should be a low-inductance connection made in the I/O area of the printed circuit board (PCB). This is similar to the circuit-to-chassis ground connection used to limit the

**TABLE 15-4.  Safe Clearance Distances Based on 1 mm/kV.**

| Voltage | Clearance (mm) | Distance (in) |
|---|---|---|
| 2000 | 2 | 1/16 |
| 4000 | 4 | 3/16 |
| 8000 | 8 | 5/16 |
| 10,000 | 10 | 3/8 |
| 15,000 | 15 | 9/16 |
| 20,000 | 20 | 13/16 |
| 25,000 | 25 | 1 |

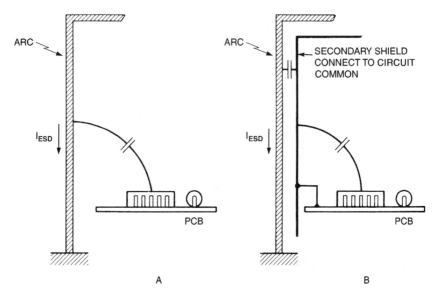

**FIGURE 15-11.** Capacitive coupling between a metallic enclosure and a circuit (A). A secondary shield (B) can be used to block the capacitive coupling between a circuit and a metallic enclosure.

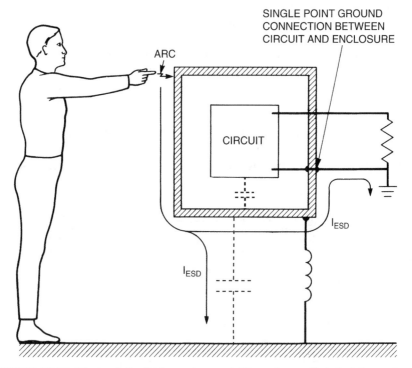

**FIGURE 15-12.** Electrostatic discharge to a metallic enclosure that contains a circuit having a single-point connection between the enclosure and the circuit.

common-mode radiation from the cables described in Section 12.4.3. *Therefore, the one technique can be used for two purposes.*

The configuration in Fig. 15-12 illustrates the result. When a discharge occurs to the enclosure, the enclosure rises in potential. However, because the circuit common is connected to the enclosure, the circuit potential rises with the enclosure, and no potential difference exists between points on the circuit or between the circuit and the enclosure.

What, however, has happened to the high voltage potential on the enclosure? It is transferred as a common-mode voltage to the interface cables and applied to whatever is at the other end of the cables. Therefore, the problem is transferred from the circuit in the enclosure to which the discharge is applied, to the circuits at the other end of the cables. If the cable is an alternating current (ac) power cable, momentarily applying a few thousand volts to it will not do any harm. But, if the cable is a signal cable connected to a low-level circuit, the circuit will most likely be damaged.

The situation, of course, can be reversed, with the discharge applied to the circuit at the far end of the cable and the damage being done to the circuit inside the first enclosure. I call this phenomena the "classic ESD problem," where a discharge is applied to box "A" and damage is done to the circuit in box "B," or vice versa.

Therefore, for a circuit enclosed in and connected to a complete conductive enclosure, the primary ESD problem involves the interface cables. These cables must be treated to prevent ESD damage.

### 15.5.2  Input/Output Cable Treatment

Cables become an ESD entry point as the result of (1) a direct discharge, (2) acting as an antenna, or (3) the classic two-box ESD problem discussed in the previous section. ESD entry on cables can be stopped or at least minimized by one or more of the following methods:

1. Use of cable shielding
2. Common-mode chokes
3. Transient voltage suppression diodes
4. Cable bypass filters

Use high-coverage braid, or braid-over-foil shielded cables for ESD protection. For optimum protection, the shield must be terminated with a 360° connection to the enclosure. The problem with foil shielded cables for ESD protection is that the shields are terminated using a drain wire (pigtail) and do not make 360° connection to the enclosure.

To demonstrate the importance of proper shield termination in ESD protection, consider the classic ESD problem of two boxes connected together with a shielded cable as shown in Fig. 15-13. One way to look at this approach is that it is an attempt to convert the two enclosures into one enclosure, by

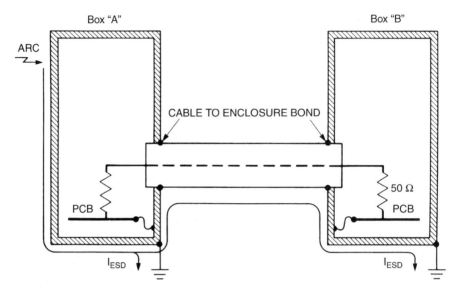

**FIGURE 15-13.** Two enclosures connected with a shielded cable in an attempt to turn the two into one continuous enclosure.

using the cable shield to tie the two enclosures together. The bonding of the shield to the enclosures is therefore the most important parameter that determines the ESD performance of this configuration during a discharge. Table 15-5 lists the voltage measured across the 50-$\Omega$ signal cable terminating resistor in box "B" when a 10,000-V discharge occurs to box "A", as a function of the shield termination method (Palmgren, 1981). In all cases, the shield had a 360° bond to one of the boxes, whereas the shield termination method to the second box was varied.

**TABLE 15-5.   The Effect of Shield Termination on ESD-Induced Voltage (from Palmgren, 1981).**

| Shield Termination Method | Induced Signal Voltage |
|---|---|
| No shield, or shield not connected to cabinet | > 500 |
| Drain wire ground connection | 16 |
| Shield soldered to connector; connector in contact with cabinet through jack screws only | 2 |
| Shield soldered to connector; a 360° contact between connector and cabinet | 1.25 |
| Shield clamped directly to cabinet with a 360° contact (no connector) | 0.6 |

When the shield was not used or the shield was not connected to the second box, the voltage across the 50-Ω resistor in box "B" was greater than 500 V, probably around 1000 V. Using a short drain wire (approximately 0.75 in), the voltage was reduced to 16 V. Had the drain wire been 3 to 4 in long, as is typically used, the measured voltage probably would have been closer to 75 or 100 V.

When the shield was soldered to the connector backshell 360°, but the backshell was only making contact to the enclosure through the two D-connector jackscrews, the voltage was reduced to 2 V. With the shield soldered to the connector backshell 360° and the two mating backshells making proper 360° contact, the measured voltage was 1.26 V, and finally with the connector removed completely and the shield clamped 360° directly to the cabinet the voltage was 0.6 V. Hence, the voltage induced into the 50-Ω load resistor in box "B," from a 10,000 V discharge to box "A" varied from > 500 V to 0.6 V (close to a thousand-to-one ratio), the only variable being the cable shield termination method used.

Ferrites can also provide very effective ESD protection, in addition to or in place of cable shielding. The spectral density of ESD is in the 100-to-500-MHz range, and this is just where most ferrites provide their maximum impedance. A ferrite or common-mode choke placed on an interface cable will cause most of the transient discharge voltage to be dropped across the

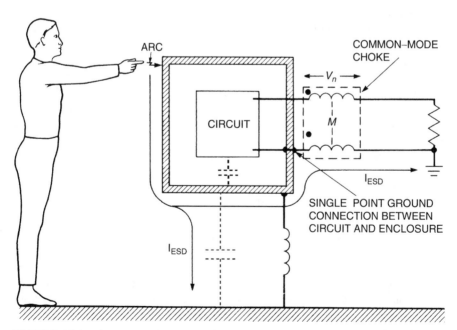

**FIGURE 15-14.** A common-mode choke can be used on the interface cable to drop the ESD-induced noise voltage (V$n$).

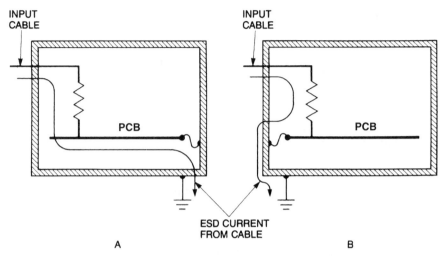

**FIGURE 15-15.** Improper connection between PCB and chassis (A) forces ESD currents to flow through the PCB. Proper connection (B) diverts ESD current away from PCB to the enclosure.

choke rather than across the circuit connected to the end of the cable. This is shown in Fig. 15-14. Because of the fast rise time of the ESD pulse, stray capacitance across the choke or ferrite must be minimized for it to be effective (see Section 3.6 on the high-frequency analysis of common-mode chokes). If the circuit is connected to the enclosure at only one point, that point should be where the cables enter/leave the enclosure as shown in Fig. 15-15B, not as shown in Fig. 15-15A.

If the input/output cables are not shielded, then *all* the conductors should have transient voltage protection devices, or filters, on them as was shown in Fig. 14-12. The protective devices must be fast enough ($\ll$ 1 ns) to turn on during the rising edge of the ESD pulse. The preferred devices are transient voltage suppression (TVS) diodes. These devices switch fast (typically $< 10^{-12}$s) and have large junction areas that can dissipate a significant amount of energy. Gas tubes or standard metal-oxide varistors (MOVs) are usually too slow for ESD protection, although the newer multilayer, surface-mount MOVs work faster and can be used for ESD protection as was discussed in Sections 14.3.5 and 14.3.6.

The TVS diodes will be even more effective against ESD if some series impedance can be added to the lines being protected, which forms a two-element L-network. The series elements should have an impedance of 50 to 100 $\Omega$ in the 100- to 500-MHz frequency range. Resistors and ferrites work well in this application. Use the ferrite when the circuit cannot tolerate the low-frequency or direct current (dc) voltage drop caused by the resistor.

You could also use an L-filter that consists of a series resistor (or ferrite) and a shunt capacitor, in place of the transient voltage protector. Whereas transient

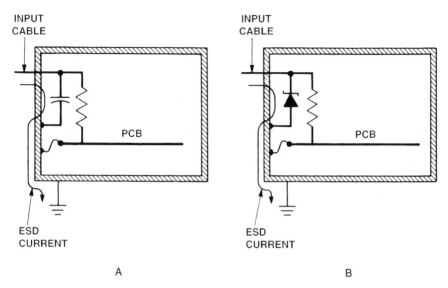

**FIGURE 15-16.** A properly grounded capacitor (A) or transient voltage suppression diode (B) used to divert the ESD current on the cable to the enclosure.

protectors are nonlinear devices that clamp the input voltage to a defined level, filters are linear and reduce the ESD transients by a proportional amount depending on their attenuation. Because most of the ESD energy is in the 100 to 500- MHz range, the filters should be designed to provide at least 40 dB of attenuation in this frequency range. The shunt element of the filter will usually be a capacitor (100 to 1000 pF), and the series element will be a resistor or a ferrite (50 to 100 $\Omega$). A ferrite core on the cable and a capacitive filter-pin connector directly mounted on the enclosure also makes a good L-filter for ESD protection.

These protection components should be placed so that the ground currents that they produce do not flow through the circuit ground; that is, they should connect to the enclosure or a separate I/O ground as shown in Fig. 15-16 (see Section 12.4.3).

The cable input protection methods just discussed will prevent component damage but may not prevent soft errors or transient upset, because noise voltages may still be present on the inputs. To prevent soft errors, these noise signals must be controlled by building additional noise immunity into the system. This can be done by additional filtering of the sensitive devices, use of balanced inputs, strobed input circuitry, or by transient hardened software design (see Sec. 15.10).

### 15.5.3  Insulated Enclosures

The major advantage of insulated enclosures is that they tend to prevent a discharge from occurring. However, unless *completely* insulated, there will be seams and apertures through which a discharge can occur.

In the case of a metallic enclosure, the chassis or enclosure can be used as a low-inductance alternative path to route the discharge current away from the internal circuitry. In the case of a nonmetallic enclosure, however, this low-inductance path does not exist, which makes, in many respects, ESD more difficult to control.

The primary disadvantages of an insulated enclosure are as follows:

- No convenient alternative path for direct discharge. *The ac power cord's green-wire ground is useless as an ESD ground*, see Section 15.7.
- No shielding for indirect discharge (field coupling).
- No convenient place to connect the following:
  - Cable shields
  - Connector backshells
  - Transient voltage suppressors
  - Input cable filters

So where should cable shields, transient voltage protectors, and I/O filters be connected when the product is in a plastic enclosure? There are three possibilities as follows:

1. To the circuit ground plane (poorest choice)
2. To a separate I/O ground plane as discussed in Section 12.4.3 (better choice)
3. To a separate large metal plate added to the bottom of the product (best choice)

When the circuit ground is used to divert the ESD current, a large ground voltage may be produced, and this voltage can cause damage or soft errors to occur, especially when solid ground planes are not used.

If all cables enter the system in the same area of the PCB, then a separate I/O ground plane (as discussed in Section 12.4.3) can be used to bypass the ESD cable currents. In this case, however, the I/O ground will not be connected to the enclosure, because there is no metallic enclosure exists. The ESD current will pass through the separate I/O ground plane and through the capacitance of the plane to actual ground, thus bypassing the circuitry. The effectiveness of this approach is a function of how large the I/O ground plane is and how much capacitance it has to ground.

The most desirable approach, however, is to have a separate ESD ground plate in the system to act as both a reference potential and a low-inductance

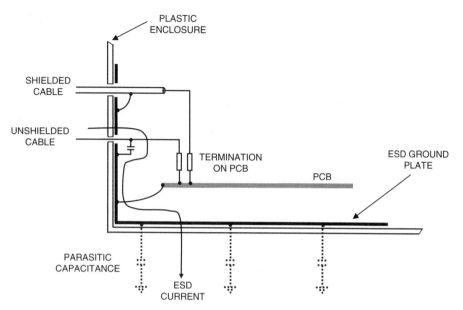

**FIGURE 15-17.** Use of a metal ESD ground plate to divert ESD current from the PCB.

path for ESD current flow. It is the free-space capacitance of this plate that provides the ground. Such an arrangement is shown in Fig. 15-17. When considering this approach, the question always asked is, how big should the ESD ground be? The answer is simple, the ESD ground should be as large as the enclosure to have as much capacitance to ground as possible. The ESD ground does not have to be thick or heavy, but it should be large—metal foil will work just fine, as would conductively coating the bottom inside surface of the plastic enclosure.

Another approach to diverting ESD currents away from sensitive circuits when a product is in a plastic enclosure is to use diode clamps to the power supply as shown in Fig. 15-18A. The diodes divert the surge current to the power bus and away from the input of the circuit to be protected. Dumping

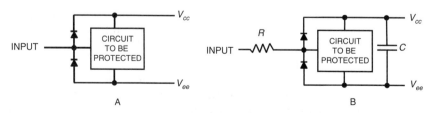

**FIGURE 15-18.** (A) Diodes clamps to the power supply rails. (B) The addition of (1) a series resistor and (2) additional bulk power supply capacitance.

the ESD current into the power bus can, however, cause the supply voltage to increase or decrease temporarlity. This change in supply voltage may be sufficient to upset, or in some cases even damage, the protected circuit or other circuits connected to the same power supply.

The diode approach, however, can be made to work well if two additional components are added. First, a series resistor (or ferrite) added in front of the diodes to limit the magnitude of the ESD current, and second, additional bulk capacitance (5 to 50 μF) added across the power supply rails. Both of these approaches are shown in Fig. 15-18B.

The voltage current relationship for a capacitor was expressed in Eq. 14-8 as

$$dV = \frac{idt}{C}.$$ 
(15-7)

Because *idt* represents charge $Q$, Eq. 15-7 can be rewritten as

$$dV = \frac{dQ}{C}.$$
(15-8)

Therefore, for a fixed amount of charge $dQ$ dumped into the power-ground system by an ESD event, the larger the capacitance $C$ the smaller the change in the voltage $dV$ will be.

When using a plastic enclosure, the electronics should be spaced away from the seams and apertures in the enclosure, as these are probable ESD entry points. Use a safe clearance distance of 1 mm/kV. If you cannot provide the required clearance distance then introduce an additional dielectric material between the aperture and the electronics.

For products in unshielded enclosures consideration should also be given to protecting internal cables. Although these internal cables are not likely to experience a direct ESD hit,they are vulnerable to ESD-related field coupling. Most field coupling to internal cables will be common mode, so ferrite cores over the overall cable are usually effective.

Ribbon cables are especially susceptible to ESD and should have ferrite chokes to minimize common-mode coupling. Ribbon cables are also susceptible to differential-mode coupling, which depends on the number of, and position of, the grounds (signal returns) in the cable. Ribbon cable should have a large number of grounds uniformly distributed across the cable to minimize the signal loop areas. Ideally, one ground for each signal conductor. However, this can usually be relaxed to one ground for every three signal conductors (a three-to-one signal-to-ground ratio), with a five-to-one signal-to-ground ratio being the maximum considered acceptable.

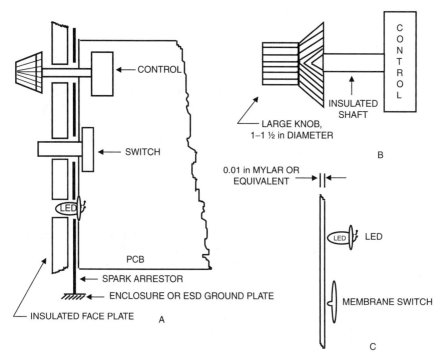

**FIGURE 15-19.** ESD suppression for keyboards and control panels, (A) a spark arrestor behind keyboard or control panel (B) a large knob and insulated shaft on a control and (C) complete insulation over keyboard/control panel.

### 15.5.4  Keyboards and Control Panels

Keyboards and control panels must be designed in such a way that a discharge will not occur, or if one does the current will flow through an alternative path and not go directly through the sensitive electronics. In many cases, a metal spark arrestor can be placed between the keys and the circuit, as shown in Fig. 15-19A, to provide an alternative path for any discharge current. This spark arrestor should be connected to the enclosure (if metallic) or a separate metal ESD ground plate.

Other protection methods shown in Fig. 15-19 are the use of an insulated shaft and/or large knob to prevent discharge to a control or potentiometer (Fig. 15-19B), and the use of an insulator over the entire keyboard, with no air gaps (Fig. 15-19C). The configuration shown in Fig. 15-19A provides an alternative path for the ESD current, whereas the configurations shown in Figs. 15-19B and 15-19C prevent a discharge from occurring.

## 15.6   HARDENING SENSITIVE CIRCUITS

After every effort has been made to prevent the entry of the ESD current, the second step is to harden the sensitive circuits. Because of the fast rise time of ESD, digital circuits are more prone to *upset* than analog circuits. However, both analog and digital circuits are equally vulnerable to ESD damage.

Resets, interrupts, and any other control inputs that can change the operating state of a device should be protected against false triggering by a fast rise time, narrow pulse width ESD transient. This can be accomplished by adding a small capacitor, or resistor/capacitor (or ferrite/capacitor) network (50 to 100 Ω, 100 to 1000 pF) to the IC input (see Fig. 14-2) to reduce its susceptibility to sharp narrow transient pulses such as those generated by ESD. This was discussed in Section 14.2.3.1, and it should be a standard design practice.

Multilayer PCBs provide an order of magnitude, or more, ESD immunity than double-sided boards. The improvement is the result of the lower power and ground impedance associated with the use of power and ground planes, and the smaller signal loop areas that result from the use of the planes.

Sufficiently large bulk decoupling capacitance is also effective in improving transient immunity, because it will decrease the change in the supply voltage produced by a transient charge, as was discussed in Section 15.5.3.

## 15.7   ESD GROUNDING

The first thing to remember about ESD grounding is that the ac power cord's green-wire ground is a high impedance at ESD frequencies. As previously stated, a typical conductor has an inductance of about 15 nH/in. Therefore, the ground conductor in a 6-ft-long ac power cord has an inductive reactance of 2000 Ω at 300 MHz. This does not include the inductance of all the ac power wiring behind the receptacle, in the walls of the building, before the green-wire conductor actually gets to "ground." That could easily be 50 ft or more in length, which adds another 17,000 Ω to the ground wire and gives a total power cord ground impedance of close to 20,000 Ω at 300 MHZ.

*The real ground or reference for ESD is the chassis (or metallic enclosure) or* ESD ground plate within the product, and its free-space capacitance as was shown Fig. 15-9. If the enclosure or ESD ground has only 26 pF of capacitance, then the impedance at 300 MHz will be 20 Ω. This represents three orders of magnitude less impedance than the impedance of the ac power cord.

Therefore, the way to divert the ESD current away from the circuit is to connect the input circuit transient voltage protectors, and/or filters, to the enclosure. If you do not have a metallic enclosure, then connect the transient protectors or filters to a separate ESD ground within your product. If you have neither a metal enclosure nor ESD ground, then about all you can do is try to limit the discharge current with resistors or ferrites, and connect the protectors or filters to the ground plane. If you have no ground plane, you are really in trouble.

Figure 14-13 showed an example of the combination of transient current entry protection and sensitive device hardening. Notice that the transient suppressor on the input cable is grounded to the chassis, because the purpose is to divert the transient ESD current away from the PCB. The protective filter on the input to the sensitive device, however, is connected to the circuit ground, because its purpose is to minimize or eliminate any transient voltage from appearing between the device's input and ground pins.

Diverting large amplitude ESD currents to ground, close to or internal to the product, produces strong magnetic fields that can affect the system adversely. Therefore, in some cases it is advantageous to add some resistance between the ESD entry point and ground, to reduce the magnitude of this current. This is often referred to as a *soft ground*. Usually a resistor of a hundred to a thousand ohms is adequate.

## 15.8   NONGROUNDED PRODUCTS

Can an electrostatic discharge occur to a nongrounded object? Yes, we probably all have observed a discharge to an ungrounded doorknob. In this case, what is the ESD current path? On a product with no external ground connection (e.g., a handheld calculator), the ESD current path will be from the entry point through the part of the product with the largest capacitance (i.e., lowest impedance) to ground. On many small handheld products, the part with the largest capacitance to ground is the printed circuit board. Having the ESD current flow through the PCB does not usually have a desirable outcome.

The solution is to provide an alternative path with lower impedance (larger capacitance) to ground for the ESD current to flow through. This is usually accomplished by adding an ESD ground plate to the product underneath the PCB. This plate blocks the capacitance of the actual PCB to ground while at the same time providing a large capacitance between the plate itself and ground for the discharge current to flow through, which is similar to what happens when a metallic enclosure is used. This approach is used to provide ESD protection in many small handheld devices, such as calculators.

Figure 15-20 shows the inside of the plastic clamshell enclosure of a handheld calculator. The left-hand image shows the printed circuit board, and the right-hand image shows the stainless-steel ESD ground plate in the bottom of the enclosure. In the left-hand image of Fig. 15-20, you can also see the bottom part of a second metal plate that is located between the PCB and the keyboard. This is a spark arrestor similar to that shown in Fig. 15-19A, which is used to provide an alternate path for the transient current that results from a discharge to the keyboard.

**FIGURE 15-20.** Clamshell enclosure for a handheld calculator showing the ESD ground plate (right-hand figure) and printed circuit board (left-hand figure).

## 15.9 FIELD-INDUCED UPSET

A direct discharge to the product is not necessary to produce an ESD problem. A discharge to a nearby object can produce strong electric and magnetic fields that can couple to the product and cause soft errors or transient upset. This effect can occur as the result of a discharge that occurrs as far as a few meters away.

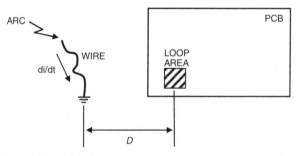

**FIGURE 15-21.** Inductive coupling resulting from a nearby discharge.

### 15.9.1  Inductive Coupling

The voltage induced into a loop by a transient current is equal to

$$dV = \left(\frac{2A}{D}\right)\left(\frac{di}{dt}\right),$$

(15-9)

where $A$ is the loop area in square centimeters, $D$ is the distance between the discharge and the loop in centimeters, and $di/dt$ is in amperes per nanoseconds. Figure 15-21 shows the typical configuration for inductive coupling. For example, consider the case where the susceptible loop area on the PCB is 10 cm$^2$, and the discharge occurs 5 cm (2-in) away. For an ESD transient current of 20 A/ns, the voltage induced into the 10 cm$^2$ loop will be 80 V. If the discharge occurred 1 m away, the induced voltage would be 8 V. The magnitude of these induced voltages can clearly cause circuit upset and in some cases actual damage.

It is interesting to note that none of the common ESD test standards require a test to simulate this type of inductive coupling.

### 15.9.2  Capacitive Coupling

Figure 15-22 shows a product in a plastic enclosure sitting on a table adjacent to a metal file cabinet. If a discharge occurs to the file cabinet, then a current will be injected into the product as a result of the parasitic capacitance between the product and the cabinet.

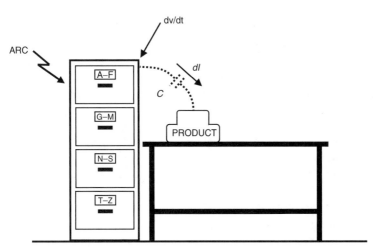

**FIGURE 15-22.** Capacitive coupling resulting from a discharge to a nearby metal object.

The transient current injected into the product will be equal to

$$dI = C\frac{dV}{dt},\tag{15-10}$$

where $C$ is in picofarads, and $dV/dt$ is in kilovolts per nanosecond, and $t$ is in nanoseconds.

For example, if 10 pF of capacitance (not much capacitance) exists between the product and file cabinet, a $dV/dt$ of 2000 V/ns on the file cabinet (resulting from a discharge to it), will inject a transient current of 20 A into the product, surely enough to cause a problem.

Most modern-day ESD standards do require a test that simulates this type of capacitive coupling. For example, the European Union's ESD standard EN 61000-4-2 requires not only direct discharges to the product but also discharges to both vertical and horizontal coupling planes located adjacent to the product. The product is placed on, but insulated from, a horizontal coupling plane. A 50 cm by 50 cm vertical coupling plane is placed 10 cm from each of the four faces of the product. Discharges are then made to both the vertical and horizontal coupling planes, as well as to the product.

## 15.10   TRANSIENT HARDENED SOFTWARE DESIGN

The third approach to ESD protection of a product is to write transient hardened software/firmware. In minimizing ESD problems, the role of properly designed software or firmware should not be overlooked. Software should be designed in such a way that if an ESD transient upsets the program, it does not lock up but recovers gracefully. Properly designed software can go a long way in eliminating or minimizing errors caused by ESD.

Two basic steps are involved in writing ESD-immune software:

- First, a fault must be detected.
- Second, the system must recover gracefully to a known stable state.

To do this, the software must be regularly checking for abnormal conditions. The object is to detect an error as quickly as possible before it has a chance to do any damage.

Software error-detecting techniques fall into three general categories:

1. Errors in program flow
2. Input-output errors
3. Memory errors

### 15.10.1    Detecting Errors in Program Flow

The most important aspect of writing noise-tolerant software is to guarantee the sanity of the program itself. Errors in program flow can be caused by changes in either the internal registers of the microprocessor (e.g., program counter) or a memory bit that is part of the program's instructions. Consequently, the program may lock up in an infinite loop from which it cannot escape. It may try to address instructions in nonexistent memory, or it may try to interpret data as an instruction. When writing fault-tolerant software, you should assume that an ESD event can set the microprocessor's program counter to an arbitrary value. This, by the way, is easier to do than you may think.

Detecting errors in program flow consists of regularly checking the program for either one of the following two conditions:

* Is the program taking too much time?
* Is the program operating in a valid range of memory?

Checking for these conditions is not difficult and may require only a few extra lines of code. Some techniques that can be used include watchdog (sanity) timers, software checkpoints, error traps, no-op/return codes, and by trapping of unused interrupt vector locations.

The most effective protection against the infinite loop is a sanity or watch-dog timer. Many microprocessors now contain integral sanity timers, or if not the timer can be implemented as an external circuit. The idea is to set the timer and have it count up to a specified number and then reset the microprocessor. The software is written to output a sanity pulse periodically that resets the timer before it times out. If everything is operating normally, the timer never times out and therefore never resets the microprocessor. If the processor gets locked up in an infinite loop, it does not output the sanity pulse, hence, the timer times out and resets the microprocessor and gets the system out of the infinite loop. The system may make an error during this process, but it is no longer locked up and has recovered gracefully. The code for the sanity pulse can be written as a subroutine that is used repeatedly by the main program. This involves only a few additional lines of code.

Software tokens are another approach. A token is added at the entry and exit points of a software module. The entry and exit tokens are set to the same value. If when exiting a module the exit token does not match the entry token, then you may have jumped into the routine from somewhere else. You can then exit to an error recovery program, which minimizes the possible damage and recovers gracefully.

If memory is partitioned is such a way that the program is limited to a certain range of memory, or the program memory is in read-only memory (ROM), traps can be written into the software to prevent the program from trying to access instructions outside the valid range of memory. Unused portions of program memory should be filled with "no op" (or similar) instructions with a jump to an error handling routine at the end. This way, if an inadvertent jump to unused or nonexistent memory occurs, the error-handling routine will be called.

Unused microprocessor hardware interrupt-vector locations are often the source of program-flow errors. If an ESD transient appears on the unused interrupt input, then it will cause a jump to the interrupt-vector location. If this location contains a program instruction or stored data, then unpredictable results will occur. The simple solution is to put a "return" or a jump to an error-handling routine in all unused interrupt-vector locations.

Once an error in program flow is detected, it becomes necessary to get the system back to a known and stable state, with as little damage as possible. This result can be accomplished by transferring control to an error-handling routine. The simplest error-handling routine is a system reset. In some instances, however, this brute force approach may be unacceptable. Error recovery consists of assessing the damage and then repairing the program as necessary. How this should be done depends on the specific system in question and is beyond the scope of this book.

### 15.10.2   Detecting Errors in Input/Output

Transient pulses on the input or output can cause incorrect information to be communicated in or out of the system. Output errors can be detected by echoing (sending back) the output and comparing the data with that which was sent.

Input errors can be controlled by software filtering the input data, and by checking the data for reasonableness. A simple software filtering technique is to read the input data $n$ times in succession, with a short delay between the readings and only accepting the data when the readings agree. This way, a valid input can be distinguished from a transient noise spike. For ESD protection, a delay of a few hundred nanoseconds between readings is sufficient. The degree of filtering is a function of the value chosen for $n$, and it can easily be adjusted. The larger $n$, the greater the input filtering will be. A value of 2 or 3 for $n$ usually provides adequate ESD protection.

Figure 15-23 shows the flow diagram of a subroutine that reads the input data until $n$ successive readings match before accepting the data. This same routine also periodically produces a sanity pulse. By ignoring short noise transients, the program acts as a low-pass filter on the input data.

Additionally, input data protection can be provided by checking the reasonableness of the data with a type and range check, before accepting the data. This way, extreme input errors can often be detected and flagged before they enter the system and propagate through it.

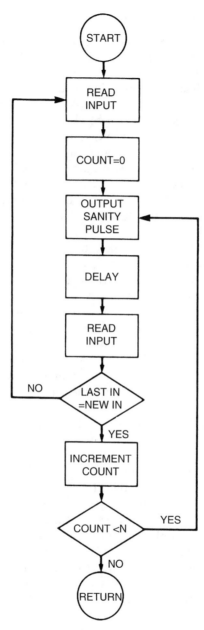

**FIGURE 15-23.** Software subroutine for filtering input data and outputting a sanity pulse.

### 15.10.3   Detecting Errors in Memory

Changes in memory that result from ESD-induced noise transients may not have an immediate effect. However, if left undetected, these errors can affect the system at a later time. To detect this type of error, all data taken from memory should be validated before being used. Many useful techniques exist for checking the validity of data. The simplest of these is the use of a single parity bit. Additional techniques involve the use of checksums, cyclical redundancy checks (CRC), and various error-correcting codes. All of these techniques can detect the presence of an error, and some can even correct it.

For example, by adding one parity bit per data word, all odd number of bit errors can be detected. The parity bit is set to 1 if the data word is odd or 0 if the data word is even. When reading data from memory, the system then can use this information to flag the data that does not pass the parity test and question its validity.

Error-correcting codes can both detect and in some cases correct certain types of errors. This is accomplished by adding extra data bits to each memory word. For example, by adding 6 extra data bits per 16-bit word, single- or double-bit errors can be detected, and single-bit errors can be corrected. The degree of data memory protection required is something that must be determined as part of the overall system specification.

Another simple approach to error detection is to store data in blocks and use checksums or cyclical redundancy checks. A checksum works by adding up the numbers in a block of data and storing the sum with the data. When the data are read back, the same operation is performed, and the result is compared with the stored checksum.

A more sophisticated approach to error detection in blocks of data is the use of a CRC. The idea behind the CRC is to treat the block of data as a single binary number (word) $w$ and divide it by another number (the key word) $k$. The quotient $q$ is ignored, and the remainder $r$ stored as the CRC. The novel aspect to this process is that it uses a simplified form of division. The method is not foolproof because there are many different numbers $w$ that will give the same remainder $r$ when divided by $k$. However, as the number of digits $n$ in the key word $k$ is increased, the probability of an error going undetected decreases. Assuming that the original binary word $w$ is random, the probability of not detecting a error is approximately equal to $1/n$. Therefore, if $n$ is made large enough, the chances of an error going undetected can be made very small.

It is interesting to note that a parity bit is a simple form of CRC that uses the decimal number 2 (binary 10) as the divisor $k$ (key word).

CRCs are popular because they are good at detecting bit errors caused by transients, easy to analyze mathematically, and simple to implement in digital systems. In both the UNIX and LINUX operating systems, there is a function "cksum" that automatically generates a 32-bit CRC for any given file. The CRC approach to error detection was originally described in a paper by Peterson and Brown (1961).

Yet another simple form of error detection is to store multiple copies of critical data, then compare the two copies when reading the data back from memory. Although simple, this method can be wasteful of memory.

## 15.11   TIME WINDOWS

In clocked digital systems, there are *time windows* during which ESD susceptibility varies. This is the result of the system performing different functions during different periods of time, and only some of these functions may be susceptible to ESD. For example, a computer system may be reading data from the hard drive during one time window but sending data to a peripheral device or performing a computation during another time window. Yet later, the computer may be refreshing the display or writing data to memory.

Therefore, when conducting ESD testing, the ESD discharges should be applied to the product over a sufficiently long period of time to cover all modes of operation of the equipment under test (EUT). This means a large number of discharges must be applied to each point on the EUT, possibly hundreds or even thousands. Also, the ESD events must not be synchronized to the EUT's operation in any way; they should be at random.

In addition, ESD susceptibility is inherently a statistical process. Therefore, to obtain repeatable results, the test criteria should have a sound statistical basis. This also will require many discharges. Also, the pass/fail criteria should be statistical—that is, the product exhibits no more than x% of nondestructive failures out of a specified large number of discharges. Present standards, however, use an absolute criterion of no failures allowed.

Neither of the above effects are taken into account in present-day commercial ESD test standards. For example, EN-61000-4-2 requires only 10 discharges to each test point on the EUT, and no failures are allowed. This is done to simplify and speed up the test process. However, this is not enough discharges to be statistically significant or to cover all possible time windows.

## SUMMARY

- ESD protection should be part of the original system design.
- A three-prong approach should be used for ESD protection, as follows:
  - First, prevent the entry of the direct discharge.
  - Second, harden the sensitive circuits.
  - Third, write transient-hardened software.

- ESD hardening of a system involves the electrical, mechanical, and software design.
- Digital circuits are more sensitive to ESD upset than analog circuits.

- Both digital and analog circuits are equally susceptible to ESD damage.
- The most critical inputs on digital ICs are as follows:
  - Resets
  - Interrupts
  - Control signals

- Sensitive devices should have transient protection filters on critical inputs.
- ESD protection filters can be:
  - A capacitor
  - A resistor capacitor
  - A ferrite capacitor

- Critical device transient protection filters should be connected to the device ground, not the enclosure.
- ESD has a spectral content in the 100- to 500-MHz frequency range.
- All exposed metal should be bonded together.
- For ESD protection, enclosure apertures should have a maximum dimension of 2.5 cm (1 in).
- Products in plastic enclosures should contain an "ESD ground" plate.
- Keyboards and control panels must be designed carefully to tolerate a static discharge.
- All cables must be treated for ESD protection, by either:
  - Shields
  - Transient voltage protectors
  - Filters

- If cable shields are used, 360° contact between the shield and the enclosure is essential.
- Transient voltage suppressors and cable input filters should be connected to the enclosure, not the circuit ground.
- Transient voltage suppressors must:
  - Switch fast ($< 1$ ns)
  - Be capable of handling the current
  - Have almost zero lead length to chassis ground

- For products in plastic enclosures, internal cables may also need ESD protection.
- Ribbon cables are especially susceptible to ESD.

- Ribbon cables should contain a large number of ground return conductors (a three-to-one signal-to-ground ratio or less is prefered) distributed uniformly across the cable.
- All loop areas on printed circuit boards should be kept as small as possible.
- Multilayer boards will be an order of magnitude less susceptible to ESD than double-sided boards.
- Ferrites can be effective in limiting ESD currents.
- Software if "glitched" should not lock up but should recover gracefully.
- Protection should be provided for the following three categories of software errors:
  - Program flow
  - Validity of input/output signals
  - Validity of data in memory

## PROBLEMS

15.1 Can an electrostatic discharge occur from a charged insulator?

15.2 A neutral conductor is brought into the vicinity of a charged object and then grounded.
   a. If the ground is removed and then the conductor is moved away from the charged object, will the conductor be charged?
   b. If the conductor is moved away from the charged object and then the ground is removed, will the conductor be charged?

15.3 If polyester and aluminum are rubbed together and then separated, what will be the polarity of the charge on each material?

15.4 a. Can you charge a conductor by rubbing it against another conductor?
   b. Can you charge a conductor by rubbing it against an insulator?

15.5 a. Can opposite polarity charges exist in different areas of an isolated insulator?
   b. Can opposite polarity charges exist in different areas of an isolated conductor?

15.6 Calculate the free-space capacitance for the following objects:
   a. A ½-in diameter ball bearing.
   b. A 5-ft diameter meterized Mylar® balloon.

15.7 What is the approximate free-space capacitance of a 0.2-m by 0.25-m flat rectangular plate?

**620**    ELECTROSTATIC DISCHARGE

15.8  What is the free-space capacitance of a rectangular metal enclosure whose dimensions are 0.2 m by 0.3 m by 0.4 m?

15.9  Explain why when measuring the surface resistivity of a material, it does not matter over what size square section of the material the resistivity is measured.

15.10  A product in a plastic enclosure uses diode clamps to the power supply to protect the input from ESD damage. The circuit has a $V_{cc}$-to-ground capacitance of 0.5 μF. An ESD event dumps a 20-A current for a period of 100 ns into the power-ground bus (assume that the current is a square wave).
a. What will be the change in the $V_{cc}$-to-ground voltage?
b. If an additional 10 μF of capacitance is added between $V_{cc}$ and ground, what will be the change in the $V_{cc}$-to-ground voltage?

15.11  What is the real ground or reference for ESD?

15.12  What is the advantage of a "soft ground" used for ESD protection?

15.13  A 5-cm$^2$ loop is located 10 cm from a 10-A/ns discharge. What will be the transient voltage induced into the loop?

15.14  A large metal object has 10 pF of capacitance to an adjacent product in a plastic enclosure. A discharge occurs to the metal object, and its voltage rises to 4000 V in 1 ns.What will be the transient current injected into the product?

15.15  Software ESD hardening should be provided to detect errors in what three general categories?

15.16  If a 16-bit word is used as the key in a CRC to detect a random error in a block of data in memory, then what is the probability that the error will be detected?

**REFERENCES**

DOD-HDBK-263. *Electrostatic Discharge Control Handbook*. Washington, DC, Department of Defense, May 2, 1980.

EN 61000-4-2. Electromagnetic Compatibility (EMC)—Part 4-2: Testing and Measurement Techniques—Electrostatic Discharge Immunity Test, 2001.

Hayt, W. H. *Engineering Electromagnetics*. New York, McGraw-Hill, 1974.

Moore, A. D. *Electrostatics and its Applications*. New York, Wiley, 1973.

Palmgren, C. M. "Shielded Flat Cables for EMI and ESD Reduction." *1981 IEEE International Symposium on Electromagnetic Compatibility*, Boulder, CO, August 18–20, 1981.

Peterson, W. W. and Brown, D. T. "Cyclic Codes for Error Detection." *Proceedings of the IRE*, January 1961.

# FURTHER READING

Anderson, D. C. "ESD Control to Prevent the Spark that Kills." *Evaluation Engineering*, July 1984.

Bhar, T. N. and McMahon, E. J. *Electrostatic Discharge Control*, New York, Hayden Book Co., 1983.

Boxleitner, W. *Electrostatic Discharge and Electronic Equipment*. New York, IEEE Press, 1989.

Calvin, H. , Hyatt, H. , and Mellberg, H. "A Closer Look at the Human ESD Event." *EOS/ESD Symposium*, 1981.

Gerke, D. and Kimmel, W. "Designing Noise Tolerance into Microprocessor Systems." *EMC Technology*, March-April, 1986.

Jowett, C. E. *Electrostatics in the Electronics Environment*. New York, Halsted Press, 1976.

Kimmel, W. D. and Gerke, D. D. "Three Keys to ESD Systems Design." *EMC Test & Design*, September 1993.

King, W. M. and Reynolds, D. "Personal Electrostatic Discharge: Impulse Waveforms Resulting from ESD of Humans Directly and through Small Hand-Held Metallic Objects Intervening in the Discharge Path." *1981 IEEE International Symposium on Electromagnetic Compatibility*, Boulder, CO, August 18-20, 1981.

Mardiguian, M. *Electrostatic Discharge; Understand, Simulate and Fix ESD Problems*. Interference Control Technologies, 1986.

Mardiguian, M. "ESD Hardening of Plastic Housed Equipment," *EMC Test & Design*, July/August 1994.

Sclater, N. *Electrostatic Discharge Protection for Electronics*. Blue Ridge Summit, PA, Tab Books, 1990.

Violette, J. L. N. "ESD Case History—Immunizing a Desktop Business Machine." *EMC Technology*, May/June 1986.

Wong, S. W. "ESD Design Maturity Test for a Desktop Digital System." *Evaluation Engineering*, October 1984.

# 16 PCB Layout and Stackup

In most products, the electronics are located on a printed circuit board (PCB), the design and layout of which is crucial to the functionality and electromagnetic compatibility (EMC) performance of the product. The PCB represents the physical implementation of the schematic.

The proper design and layout of a printed circuit board can mean the difference between the product passing or failing EMC requirements. Such things as component placement, keep out zones, trace routing, number of layers, layer stackup (order of layers and layer spacing), and return path discontinuities all are critical to the EMC performance of the board.

## 16.1  GENERAL PCB LAYOUT CONSIDERATIONS

### 16.1.1  Partitioning

Component placement is an important, but often overlooked, aspect of PCB layout that can have a significant impact on the board's EMC performance. Components should be grouped into logical functional blocks. Some of these blocks might be: (1) high-speed logic, clocks, and clock drivers; (2) memory; (3) medium- and low-speed logic; (4) video; (5) audio and other low-frequency analog circuits; (6) input/output (I/O) drivers; and (7) I/O connectors and common-mode filters, as shown in Fig. 16-1.

On a properly partitioned board, the high-speed logic as well as memory should not be located near the I/O area. The crystal or high-frequency oscillator should be located near the integrated circuits (ICs) that use them, and away from the I/O area of the board. The I/O drivers should be located close to the connectors, and the video and low-frequency analog circuits should have access to the I/O area without having to pass through the high-frequency digital sections of the board.

Proper partitioning will minimize trace lengths, improve signal quality, minimize parasitic coupling, and reduce both PCB emissions and susceptibility.

### 16.1.2  Keep Out Zones

Be particularly careful to keep the oscillators and/or crystals, as well as any other high-frequency circuitry, away from the I/O area. These circuits generate

---

*Electromagnetic Compatibility Engineering,* by Henry W. Ott
Copyright © 2009 John Wiley & Sons, Inc.

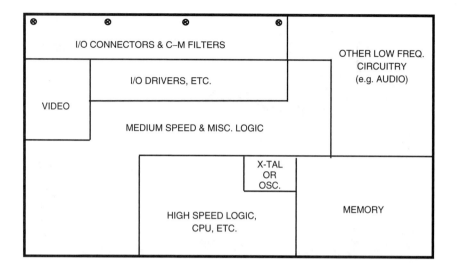

⊗   CIRCUIT GROUND TO CHASSIS CONNECTIONS

**FIGURE 16-1.** An example of proper printed circuit board partitioning.

high-frequency fields (both electric and magnetic) that can easily couple directly to the I/O cables, connectors, and circuitry, see Fig. 6-42. Experience has shown, that if board size permits, keeping these circuits at least 0.5 in. (13 mm) from the I/O area will minimize the parasitic coupling.

Route all critical signal traces (as defined in Section 16.1.3) away from the edges of the board to allow the return current to spread out under the trace as explained in Section 10.6.1. A good rule is to define a keep out zone, that is 20 times the signal-layer to return-plane spacing, around the periphery of the board. No critical signals should be routed in the keep out zone; see Fig. 16-2.

### 16.1.3   Critical Signals

Experience has shown that 90% of PCB problems are caused by 10% of the circuitry. This 10% of the circuitry should, therefore, be given the most consideration in the layout of the board. For *emissions*, the greatest problems are high-frequency (fast rise time) digital circuits with repetitive wave shapes, such as clocks, buses, and some control signals. These signals contain a multiplicity of large-amplitude, high-frequency harmonics. Clocks are usually the worst offenders, followed in order by buses and then repetitive control signals.

A metric that is useful in categorizing critical signals is the concept of "Signal Speed" (Paul, 2006, p. 805). Radiation of a signal is directly related to

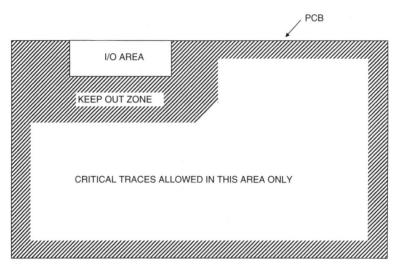

**FIGURE 16-2.** A PCB with a defined keep out zone for critical signals.

the high-frequency spectral content of its current. The high-frequency spectral content or signal speed is proportional to:

- The fundamental frequency $F_0$ of the signal
- The reciprocal of the rise/fall time $t_r$
- The magnitude of the transient drive current $I_0$ when the gate switches

Therefore, an effective metric for categorizing signal speed (in $A/s^2$) is

$$\text{Signal Speed} \approx (F_0 I_0)/t_r. \tag{16-1}$$

Repetitive, high-frequency signals with large currents and fast rise/fall times will have large spectra content. Hence, signal speed should be considered for all critical signals.

### 16.1.4   System Clocks

*Get paranoid about system clocks!* Keep the clock traces as short as possible and provide for optimum placement by routing them first. Locate crystals, oscillators, or resonators as close to the circuits that use them as possible. Add a ground plane on the component side of the board under the crystal, oscillator, and/or clock driver. Connect this plane to the main ground plane with multiple vias. This provides a termination for any stray capacitance (electric fields) from the crystal or oscillator, and it prevents the routing of other signals, on the top layer, under the crystal. If the crystal or oscillator has a metal case, ground it to this component-side ground plane, and provide a provision for a board level shield over this area in case it should be needed.

Small series damping resistors (or ferrite beads) should be added to all clock output traces with a frequency of 20 MHz or more. This will help reduce ringing and control reflections. This is recommended even on short clock traces, unless adding the resistor would increase the length of an already very short trace. A typical value resistor would be 33 Ω.*

Clock oscillators and drivers should also have ferrite beads in series with the $V_{cc}$ line to isolate the circuit from the main power distribution system.

## 16.2   PCB-TO-CHASSIS GROUND CONNECTION

A major source of radiation from electronic products is due to common-mode currents on the external cables. From an antenna theory perspective, a cable can be considered as a monopole antenna, with the enclosure being the associated reference plane (see Appendix E). The voltage driving the antenna is the common-mode voltage between the cable and the chassis. The reference for the cable radiation is therefore the chassis and not some external ground such as the earth.

Because the potential difference between the cable and chassis should be minimized, the connection between the PCB ground and the chassis becomes important. *The internal circuit ground should be connected to the chassis at a point as close to the location that the cables terminate on the PCB as possible.* This is necessary to minimize the voltage difference between the two. This connection must be a low-impedance connection at radio frequencies. Any impedance between the circuit ground and the chassis will produce a voltage drop, and will excite the cables with a common-mode voltage, which causes them to radiate.

The circuit-ground-to-chassis connection is often made with poorly placed metal stand offs, and it can have considerable high-frequency impedance. Seldom is this connection optimized for EMC purposes. *The design of this connection is critical to the EMC performance of the product.* The connection should be short, and there should be multiple connections to parallel the inductance of the connections, and, hence, decrease the radio frequency (rf) impedance. Figure 16-1 shows an example of multiple circuit-ground-to-chassis connections located in the I/O area of the PCB. This points out the advantage of having all the I/O located in one area of the board.

If metallic backshell connectors are used, the backshell should make a 360° direct electrical connection (via an EMC gasket or other means) to the enclosure. The connector backshell then can become part of the low-impedance

---

*If the trace is long (length in inches ≥ than three times the rise time in nanoseconds) use a series damping resistor value equal to the characteristic impedance of the transmission line minus the output resistance of the driver.

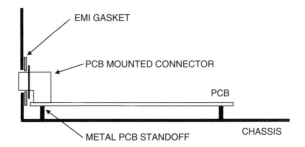

NOTE: THE CONNECTOR BACKSHELL <u>MUST</u> MAKE 360° CONTACT WITH THE CHASSIS.

**FIGURE 16-3.** I/O connector backshell making direct 360° electrical contact to the chassis.

connection between the PCB ground reference plane and the enclosure. This is shown in Fig. 16-3.

## 16.3   RETURN PATH DISCONTINUITIES

One of the keys to determining the optimum printed circuit board layout is to understand how and where the signal return currents actually flow. The schematic only shows the signal path, whereas the return path is implicit. Therefore, most PCB designers only think about where the signal current flows (obviously on the signal trace) and give little, or no, consideration to the path taken by the return current.

To address the above concern, one must remember how high-frequency return currents flow. The lowest impedance return path is in a plane directly underneath the signal trace (irrespective of whether this is a power or ground plane), because this provides the lowest inductance path (see Section 3-2). This also produces the smallest loop area.

Because of the "skin effect," high-frequency currents cannot penetrate a plane, and therefore, *all high-frequency currents on power and ground planes are surface currents* (see Section 10.6.1). This effect will occur at frequencies above 30 MHz for 1-oz. copper layers in a PCB.* Therefore, *a plane is really two conductors*. There can be a current on the top surface of the plane and there can be a different current, or no current at all, on the bottom surface of the plane.

Major EMC, and signal integrity (SI), problems occur when there are discontinuities in the return current path. These discontinuities cause the return current to flow in large loops, which increases the ground inductance and the radiation from the board—as well as increasing the crosstalk between adjacent

---

* This is based on the plane having a thickness of at least three skin depths. For 2-oz copper, this will occur at frequencies above 8 MHz and, for 1/2-oz copper, at frequencies above 120 MHz.

traces and causing waveform distortion. In addition, a return plane discontinuity on a constant impedance PCB will change the characteristic impedance of the trace and produce reflections. The three most common return path discontinuities that PCB designers must deal with are as follows:

- Slots or splits in the power and/or ground plane
- Signal traces changing layers, which causes the return currents to change reference planes
- Ground plane cutouts around connectors, or under ICs

### 16.3.1  Slots in Ground/Power Planes

When a trace crosses a slot in the adjacent power or ground plane, the return current must detour from underneath the trace to flow around the slot, as shown in Fig. 16-4A. This causes the current to flow through a much larger loop area. The longer the slot, the larger the loop area. Large current loops increase both the radiation and the inductance of the ground plane—both undesirable effects. *The most important thing that I can say about slots in ground planes, is do not have them!* If all PCB designers would follow this one simple rule, then a many EMC problems would be avoided. If you must have slots, make sure that no traces cross over them on adjacent layers. Ground plane slots and/or splits can increase the PCB radiation in excess of 20 dB.

Figure 16-4B shows a ground plane with multiple clearance holes for through-hole components and vias. If the holes overlap, they produce a slot and divert the return current the same as the slot shown in Fig. 16-4A.

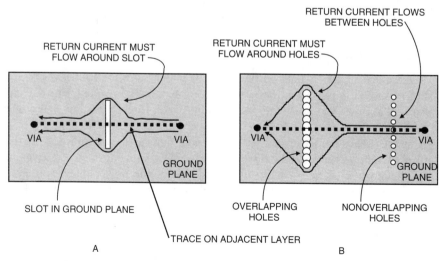

**FIGURE 16-4.** (*A*) Slotted ground plane. (*B*) Holes in ground plane.

**TABLE 16-1.    Increase in Ground Plane Voltage from a Slotted Ground Plane.**

| Length of Slot (in) | Ground plane voltage (mV) |
|---|---|
| 0 | 15 |
| 0.25 | 20 |
| 0.5 | 26 |
| 1 | 49 |
| 1.5 | 75 |
| Array of holes[a] | 15 |

[a] A linear array of 15 holes, each with a 0.052 in diameter, oriented perpendicular to the current flow covering a linear distance of 1 in. The holes were not overlapping, similar to the holes shown in Fig. 16-4B.

However, if the holes do not overlap, then the current flows between the holes and the holes do not significantly disrupt the return current path and, therefore, are not detrimental to the EMC performance of the board.

Table 16-1 lists measured values of ground plane voltage differential, with and without a slot in the plane.* The slot is oriented perpendicular to the direction of current flow, which is similar to that shown in Fig. 16-4A. The voltage measurements were made between two points on the ground plane 1 in. apart (1/2 in. on either side of the slot) and directly underneath the trace.

The measurements were made with a 10-MHz, 3-ns rise time signal flowing down the trace and returning in the ground plane. A larger voltage is indicative of increased ground plane impedance.

As can be observed clearly, the ground plane voltage increases with the slot length. For a slot length of 1.5 in, the ground plane voltage (and hence the impedance) increased by a factor of five (14 dB). The 1-in-long array of nonoverlapping holes, however, did not increase the ground plane voltage.

## 16.3.2    Split Ground/Power Planes

When a trace crosses a split[†] in the adjacent plane, as in the four-layer board example shown in Fig. 16-5, the return current path is interrupted. The current must find another way to get across the split, which forces it to flow in a much larger loop.

For the case of a trace crossing a split power plane as shown in Fig. 16-5, the return current will divert to the nearest decoupling capacitor to cross over to the solid ground plane; then on the other side of the power plane split, the current must find another decoupling capacitor to return to the power plane that is adjacent to the trace. The inter-plane capacitance, between the power

---

* For a discussion of the measurement technique used, see Appendix E, Section E.4.

[†] A split plane is one that is divided into completely separate regions or parts (see Fig. 16-6), whereas a slot is a finite narrow aperture in a plane (see Fig. 16-4A).

**FIGURE 16-5.** Signal trace crossing a split in the adjacent power plane. The solid arrow shows the signal current path, and the dashed arrow shows the return current path.

and ground plane, is too small to provide a sufficiently low impedance path, except in the case of frequencies considerably above 500 MHz. This much larger return current path significantly increases the inductance, and loop area, of the return path.

If in the above example, both the power and ground planes are split, then how will the return current get across the gap? In some instances, it may have to go all the way back to the power supply. The best solution to the split plane problem is to avoid crossing the split with any signal traces, especially critical signal traces. In the case of the above example, the signal should have been routed on the bottom signal layer adjacent to the solid ground plane. A continued discussion of split planes is in Section 17.1.

Many products today require multiple direct current (dc) voltages to operate. As a result, split power planes are becoming a common occurrence. One must realize, however, that split planes will require routing restrictions to avoid having traces that cross over the split.

Five approaches are available for dealing with the problem created by split power planes. They are as follows:

- Split the power plane and live with the routing restrictions.
- Use a separate solid power plane for each dc voltage.
- Use a "power island" for one or more of the voltages. A power island is a small isolated power plane on a signal layer (usually on the top or bottom layer of the board) under one or more ICs.
- Route some (or all) of the dc voltages as a trace on a signal layer.
- As a last resort, add stitching capacitors where the trace crosses the split plane.

Each approachs has its advantages and disadvantages. The power island approach is most useful when a dc voltage is used only by one or more ICs located adjacent to each other.

Although signal traces should not be run across a split in an adjacent plane, design constraints, and cost considerations, sometimes make it necessary to do so, especially in the case of power planes. If one absolutely must route a signal trace across a split power plane, place a few small stitching capacitors to bridge the split between the two sections of the power plane, one on either side of the

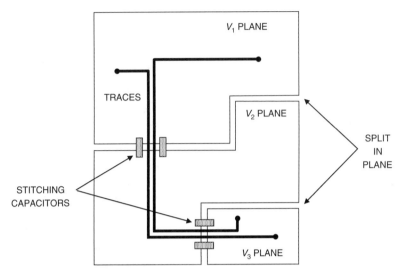

**FIGURE 16-6.** Stitching capacitors used across splits in the power plane to provide a return path for the signal currents of traces crossing over the split.

traces as shown in Fig. 16-6. This technique will provide high-frequency continuity across the split, while maintaining dc isolation between the two sections of the plane. The capacitors should be located within 0.1 in. of the trace and have a value of 0.001 to 0.01 uF according to the frequency of the signal.

This solution is far from a ideal; however, because the return current must now flow through a via, a trace, a mounting pad, a capacitor, a mounting pad, a trace, and finally a via to the other section of the split plane. This adds about 5 nH, or more, of additional inductance (impedance) in the ground return path, but it is better than the alternative of doing nothing.

Five nanohenries will have an impedance of 3 Ω at 100 MHz and 16 Ω at 500 MHz. These impedances are orders of magnitude greater than that of a solid (non-split) plane. Archambeault (2002, p. 76, Fig. 5-7) presents data showing a 28-dB reduction in radiated emissions when one stitching capacitor is used and a 32-dB reduction when two stitching capacitors are used (one on each side of the trace), at 300 MHz. This result is compared with a 37-dB reduction when the plane was solid, no split.

### 16.3.3  Changing Reference Planes

When a signal trace changes from one layer to another, the return current path is interrupted because the return current must also change reference planes, as shown in Fig. 16-7.

The question then becomes how does the return current flow from one plane to another? As was the case for the split planes mentioned previously, the

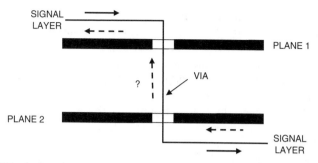

**FIGURE 16-7.** A signal trace routed on two layers adjacent to two different planes. How does the return current (dashed arrow) get from the bottom of plane 2 to the top of plane 1? The solid arrow shows the signal current path, and the dashed arrow shows the return current path.

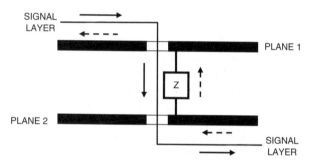

**FIGURE 16-8.** Return path impedance resulting from the signal trace changing layers. The solid arrow shows the signal current path, and the dashed arrow shows the return current path.

interplane capacitance is not large enough to provide a low impedance path, so the return current will have to flow through the nearest decoupling capacitor, or plane-to-plane via to change planes. Changing reference planes obviously increases the loop area and is undesirable for all the reasons previously stated for split planes. Changing reference planes effectively adds impedance (inductance) in the return path as shown in Fig. 16-8.

One solution to this problem is to avoid switching reference planes for critical signals (such as clocks), if at all possible. If you must switch references from a power plane to a ground plane, then you can place an additional decoupling capacitor* adjacent to the signal via to provide a high-frequency current return path between the two planes. This solution is not ideal, however, because this adds considerable additional inductance in the return path (typically about 5 nH).

---

* Use the same value as the other decoupling capacitors on that power supply.

Note that if the two reference planes are of the same type (either both power or both ground), then you can use a plane-to-plane via (ground-to-ground, or power-to-power) instead of a capacitor immediately adjacent to the signal via. This approach is much better, because the added inductance (hence impedance) of a via is much less than that of a capacitor and its mounting. *It is highly recommended that either a capacitor or via be added whenever critical signals change reference planes.*

Present PCB designers have mostly ignored this issue. Boards are usually laid out with no special concerns about the transitioning of signals to different layers, and the boards have worked and met EMC requirements—possibly because most boards already contain many decoupling capacitors. These existing decoupling capacitors minimize the problem without the designer taking any special precautions. One can only speculate, however, as to how much better existing boards could have been if this phenomenon was considered, and corrected, as part of the board design.

Figure 16-9 shows measured radiation from a four-layer test PCB with a single 30-cm-long signal trace (Smith, 2006). The stackup was similar to that shown in Fig. 16-7, with both planes being ground planes. Figure 16-9A shows the emission when the trace is confined to a single layer only, and Fig. 16-9B shows the emission when the trace makes a single transition from the top to bottom layer of the board halfway along its length. As can be observed, the emissions are significantly higher for the case where the signal layer makes a transition from the top to the bottom of the board. At 247 MHz (diamond marker in Fig. 16-9B), the emission is almost 30 dB greater for the case where the signal transitions from the top to bottom layer, versus the case where the signal is routed on a single layer. Above about 2 GHz, the interplane capacitance is sufficient to reduce the impedance of the return path, and hence, the radiation in both cases are about equal.

The data presented in Fig. 16-9 were taken by exciting the trace with the output of a spectrum analyzer's tracking generator and sweeping the frequency range up to 3 GHz. The two ground planes were connected together at four places, two at the load end and two at the source end. No plane to plane capacitors were on the

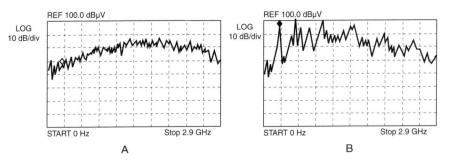

**FIGURE 16-9.** Radiated emission from a four-layer PCB with (A) the signal trace on a single layer only and (B) with the signal trace making a single transition from the top to bottom layer of the board. © 2006 Douglas C. Smith.

test board. Had additional plane-to-plane capacitors or plane-to-plane vias been used, the difference between the emissions, shown in Fig. 16-9A and 16-9B, would not have been so dramatic. The above example, however, clearly shows that transitioning a signal trace between layers of a PCB, such that it references two different planes, introduces a significant discontinuity into the signal return path and increases the radiated emission significantly.

### 16.3.4   Referencing the Top and Bottom of the Same Plane

Whenever a signal switches layers and references first the top and then the bottom of the same plane, how does the return current transition from the top to the bottom of that plane? Because of the skin effect, the current cannot flow through the plane; it can only flow on the surface of the plane.

To drop a signal via through a plane, a clearance hole (anti-pad) must be provided in the plane; otherwise the signal would be shorted to the reference plane. The inside surface of the clearance hole provides a surface connecting the top and bottom of the plane and provides the path for the return current to flow from the top to the bottom of the plane as depicted in Fig. 16-10. There-fore, when a signal passes through a via and continues on the opposite side of the same plane, a return current discontinuity does not exist. *This is, therefore, the preferred way to route a critical signal* if two routing layers must be used.

High-speed clocks and other critical signals should be routed (in order of preference) as follows:

- On only one layer adjacent to a plane.
- On two layers that are adjacent to the same plane.
- On two layers adjacent to two separate planes of the same type (ground or power) and connect the planes together with plane to plane vias wherever the signal trace changes layers.
- On two layers adjacent to two separate planes of different types (ground and power) and connect the planes together with capacitors whenever the signal trace changes layers, and hence reference planes.
- On more than two layers. Preferably this should not be used at all.

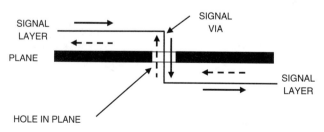

**FIGURE 16-10.** A signal trace routed on two layers adjacent to the same reference plane. The inner surface of the hole in the plane provides the path for the return current.

### 16.3.5  Connectors

Another location where return current discontinuities often occur is in the vicinity of connectors. If copper is removed from the ground plane under the connector as shown in Fig. 16-11A, then the return current has to go around the cutout area creating a large loop and therefore a noisy area on the PCB. The bigger (longer) the connector is, the worse this problem becomes. The solution is only to remove the copper around the individual connector pins as shown in Fig. 16-11B, which keeps the signal current loop small.

### 16.3.6  Ground Fill

Ground fill, or ground pour, is a technique where copper is introduced into areas of the PCB signal layers that contain no traces. The intent is to reduce emissions and susceptibility by reducing field fringing from the signal traces and by providing some degree of shielding on the board. To be effective, the fill must be connected to the existing ground structure on the board at many places. If not properly grounded, the copper fill can actually increase emissions and susceptibility, as well as crosstalk between traces. Small areas of fill and long skinny areas are exceptionally troublesome in this respect. Small fill areas should especially be avoided, because they do no good and can actually make things worse if not properly grounded.

If any copper fill is left ungrounded, then noise can be coupled to the isolated fill area and then capacitively couple to adjacent traces increasing crosstalk. Copper fill, not properly grounded, can also create an ESD problem. *Therefore, never have regions of ungrounded copper fill on a PCB.*

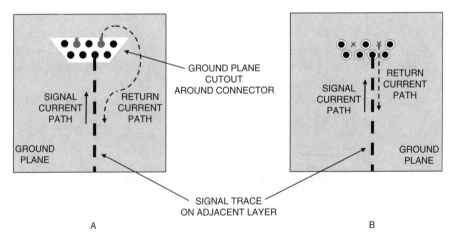

**FIGURE 16-11.** Connector area of the PCB with (*A*) a large area of the ground plane cut out for all the connector pins and (*B*) with only a small clearance hole in the ground plane around each individual connector pin.

Although often used with analog circuits on double-sided boards, copper fill is not recommended for high-speed digital circuits, because it can cause impedance discontinuities, which can lead to possible functional problems. On multilayer boards, the ground fill, if used, must be connected to the PCB ground plane at multiple points. If ground fill is used on multilayer boards, it should only be applied to the surface layers.

## 16.4 PCB LAYER STACKUP

PCB layer stackup (the ordering of the layers and the layer spacing) is an important factor in determining the EMC performance of a product. A good stackup will produce minimal radiation from the loops on the PCB (differential-mode emission), as well as the cables attached to the board (common-mode emission). However, a poor stackup will cause excessive radiation from both of these mechanisms.

The following four factors are important with respect to board stackup:

- The number of layers
- The number and types of planes (power and/or ground)
- The ordering or sequence of the layers
- The spacing between the layers

Usually not much consideration is given to these factors by the designer, except as to the number of layers. In many cases, however, the other three factors are of equal importance. Spacing between the layers is sometimes not even known by the PCB designer; it is left to the discretion of the PCB fabricator.

In deciding on the number of layers, the following should be considered:

- The number of signals to be routed and PCB cost.
- Clock frequency.
- Will the product have to meet Class A or Class B emission requirements?
- Will the PCB be in a shielded or unshielded enclosure?
- The EMC engineering expertise of the design team.

Often, only the number of signals to be routed and costs are considered. In reality, all the items are of critical importance and should be considered equally. If an optimum design is to be achieved, in the minimum amount of time and at the lowest cost the last item can be especially important and should not be ignored. For example, a design team with considerable EMC expertise may be able to do an acceptable design of a two-layer PCB, whereas a less-experienced team would be better off designing a four-layer board.

### 16.4.1    One- and Two-Layer Boards

One- and two-layer boards provide an EMC challenge to the PCB designer. These boards are selected primarily for cost considerations, not for EMC performance. The main EMC concern here is to keep loop areas as small as possible. One- or two-layer boards should only be considered when clock frequencies are less than 10 MHz. In this case, the third harmonic will be less than 30 MHz. The only advantage of a single-sided board is cost.

On one- or two-layer boards all critical signals (see Section 16.1.3) should be routed first, in order to guarantee optimum routing. Critical signals should be routed as short as possible, with an adjacent ground return trace. Clocks and buses should have a ground return trace on both sides of the signal traces or bus (see Section 12.2.2). Small damping resistors ($\approx 33\ \Omega$) should be placed in all clock outputs to reduce ringing. A small ground plane should be placed under the crystal or oscillator, and the crystal or oscillator case should be connected to it. On one- or two-layer boards, crystals are usually preferred to oscillators because they have much less harmonic energy.

On two-layer digital boards, the ground and power should be routed so as to form a grid (see Section 10.5.3). It is not unusual to see a 10- to 12-dB decrease in emissions when a ground grid is added to a two-layer digital board that previously did not have one.

Decouple the $V_{cc}$ on all clocked ICs by adding a small ferrite bead in series with the $V_{cc}$ line, which is located on the power supply side of the decoupling capacitor(s). Unused board areas should be filled with ground, but be sure that the fill is connected to the board's ground structure at multiple points and not left floating.

Consider dithering the clock to spread the clock energy out in the frequency spectrum and thereby reduce the peak amplitude of the emissions. See Section 12.2.3 on dithered clocks. Also use a *minimum* of two decoupling capacitors per IC (four capacitors on square packages); locate them on opposite sides of the IC to produce canceling loops for the transient power supply currents (see Section 11.5).

Many of the above techniques are applicable and can also be applied to multilayer PCBs.

One final method of reducing emissions on single- or double-sided boards is with the use of an image plane (German, Ott, and Paul, 1990; Fessler, Whites, and Paul, 1996). An image plane is a relatively large conductive metal plane located close to the board. This can be something as simple as aluminum or copper foil. Properly connected, it can reduce emissions not only from the PCB but also from the connected cables. For more details, see the listed references.

The two most important things that must be done to minimize emissions and susceptibility on two-layer boards are as follows:

1. To keep the loop area of critical signals (clocks, etc.) small
2. To grid the ground and power structures

### 16.4.2  Multilayer Boards

Multilayer boards (four or more layers) using ground and/or power planes provide significant reduction in radiated emissions over two-layer designs. A often–used rule of thumb is that a four-layer board will produce 20 dB or more, less radiation than a two-layer board—all other factors being equal. Boards containing planes are much better than those without planes for the following reasons:

- The planes allow signals to be routed in a microstrip (or stripline) configuration. These configurations are controlled impedance transmission lines that produce much less radiation than the random traces used on one- and two-layer boards.
- When the return current is on the adjacent plane, the loop area is reduced.
- The ground plane significantly decreases the ground impedance and hence the ground noise.

Although two-layer boards have been used successfully in unshielded enclosures at frequencies of 20 to 25 MHz, these cases are the exception rather than the rule, and it requires that the design team has a lot of EMC expertise. Above about 10 MHz, multilayer boards should normally be seriously considered.

*16.4.2.1 Multilayer Board Objectives.* When using multilayer boards, *six design objectives* should be kept in mind, as follows:

1. A signal layer should *always* be adjacent to a plane.
2. Signal layers should be tightly coupled (close) to their adjacent planes.
3. Power and ground planes should be closely coupled together.*
4. High-speed signals should be routed on buried layers located between planes. The planes can then act as shields and contain the radiation from the high-speed traces.†
5. Multiple-ground planes are very advantageous, because they will lower the ground (reference plane) impedance of the board and reduce the common-mode radiation.
6. When critical signals are routed on more than one layer, they should be confined to two layers adjacent to the same plane. As discussed, this objective has usually been ignored.

---

*In some designs it may be desirable not to use a power plane at all. Also special high-capacitance PCB laminates are available for use on power-ground plane sandwiches in order to improve IC power supply decoupling, as was discussed in Section 11.4.6.
†On a board with less than eight layers, this objective and objective #3 cannot both be satisfied simultaneously.

Most PCB designs cannot meet all six objectives, so a compromise is required. For example, one is often faced with the choice between close signal to return plane coupling (objective #2) and close power to ground plane coupling (objective #3).

Another choice is often between routing signals adjacent to the same plane (objective #6), or shielding signal layers by burying them between planes (objective #4). If the number of board layers permits, then one or the other of these objectives should be satisfied. From both an EMC and a signal integrity point of view, it is usually more important to have the return current flow on a single plane than to bury the signal layers between planes.

*Objectives #1 and #2 should always be achieved and not compromised.*

Many excellent board stackups only satisfy four or five of the six design objectives, which is perfectly acceptable. Seldom will a practical PCB satisfy all six objectives. An eight-layer PCB is the fewest number of layers that can be used to achieve five of the six above objectives. On four- and six-layer boards, some of the above objectives will always have to be compromised. Under those conditions, the designer will have to determine which objectives are the most important to the design at hand.

The above paragraph should not be construed to mean that a good EMC design cannot be achieved on a four- or six-layer board, because it can. It only indicates that only four of the six objectives can be met simultaneously, and some compromise will be necessary.

Another desirable objective, from a mechanical point of view, is to have the cross section of the board symmetrical (or balanced) to prevent warping. For example, on an eight-layer board, if layer two is a plane, then layer seven should also be a plane. Another issue to be considered is odd or even number of layers. Although odd layer count PCBs can be manufactured, it is usually simpler and therefore less expensive to manufacture boards with an even number of layers. All the configurations presented here will use symmetrical, or balanced, construction with an even number of layers. If non symmetrical, or odd number layer count construction is allowed, additional stackup configurations are possible.

### 16.4.2.2 Four-Layer Boards.

Four-layer boards are used to improve EMC performance and signal integrity over that of two-layer boards. They provide no additional routing layers, although they do remove the power and ground traces from the signal layers.

A common four-layer board configuration, consisting of two signal layers and two planes, is shown in Fig. 16-12* (power and ground planes may be reversed). It consists of four uniformly spaced layers with internal power and ground planes. The two external trace layers usually have orthogonal routed

---

*In the PCB stackup drawings that follow, the power and ground planes are shown with a heavy line for emphasis. This is not intended to infer that they are made from thicker copper than the signal layers.

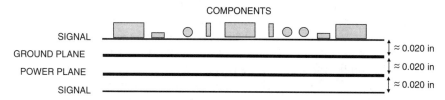

**FIGURE 16-12.** A common four-layer board configuration. This configuration satisfies only one of the six objectives.

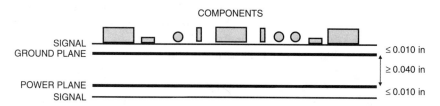

**FIGURE 16-13.** Improved four-layer board layer spacing. This configuration satisfies two of the six objectives.

traces. On a 0.062-in-thick board, the spacing between layers is approximately 0.020 in.

Although this configuration is significantly better than a two-layer board, it has a few less than ideal characteristics, and the stackup only satisfies objective #1. Because the layers are equally spaced, there is a large separation between the signal layers and the current return planes. There is also a large separation between the power and ground planes. On a four-layer board, both of these deficiencies cannot be corrected simultaneously.

With normal PCB construction techniques, there is not sufficient interplane capacitance between the adjacent power and ground planes to provide effective decoupling below about 500 MHz. The decoupling, therefore, will have to be taken care of by other means (such as proper use of decoupling capacitors as discussed in Chapter 11), therefore, the signal layers and the planes should be placed close together. The advantages of tight coupling between the signal (trace) layers and the current return planes will more than outweigh the disadvantage caused by the additional loss in interplane capacitance between the power and ground planes.

Therefore, one of the simplest ways to improve the EMC performance of a four-layer board is to space the signal layers as close to the planes as possible ($\leq$ 0.010 in), and to use a large core ($\geq$ 0.040 in) between the power and ground planes as shown in Fig. 16-13. This has three advantages and few disadvantages.

The first advantage is that signal loop areas are smaller and therefore produce less differential-mode radiation. For the case of 0.005-in spacing (trace-layer to plane-layer), the signal loop area will decease by a factor of four with respect to the equal spacing configuration. Because differential-mode (loop) radiation is directly proportional to loop area, it will decrease by 12 dB compared with a stackup with equally spaced layers, at no additional cost.

The second advantage is that tight coupling between the signal trace and the ground plane reduces the ground plane impedance (inductance), hence reducing the common-mode radiation from the cables connected to the board. The empirical data in Fig. 10-19 shows that the plane inductance will decrease from about 0.13 nH/in to 0.085 nH/in. as the spacing changes from 0.020 in. to 0.005 in. This is a 35% reduction in inductance. Because differential-mode logic currents flowing through the ground inductance cause the ground noise voltage, the ground noise will also decrease by the same 35%. This voltage is the excitation voltage for the common-mode currents on the cables; therefore, the currents will also be reduced by this factor. The radiation from a cable is directly proportional to the common-mode current in that cable; hence, the cable radiation will decrease by the same 35% or slightly less than 4 dB.

The third advantage is that close trace-to-plane coupling will decrease the crosstalk between adjacent traces. For a fixed trace-to-trace spacing, the crosstalk is proportional to the square of the trace height* (Eq. 10-15). Therefore, the crosstalk will decrease by a factor of 16 (or 24 dB) when the trace height is reduced from 0.020 to 0.005 in. *This is one of the simplest, least costly, and most overlooked methods of reducing radiation and crosstalk on a four-layer PCB.* The configuration in Fig. 16-13 satisfies objectives #1 and #2.

If the power plane shown in Figs. 16-12 or 16-13 is split to accommodate different dc voltages, it is important to restrict routing on the bottom signal layer so that traces do not cross the split in the plane. If some traces must cross the split, stitching capacitors should be located close to where the traces cross the split to provide a lower impedance return current path.

The overwhelming majority of four-layer boards have stackups as described above, with the two signal layers on the outside and the two planes in the center. The stackup shown in Fig. 16-13 is satisfactory for most applications of four-layer boards. However, other possibilities have also been used successfully.

Taking a slightly nonconventional approach, the signal layers and the plane layers could be reversed, producing the stackup shown in Fig. 16-14A. The major advantage of this stackup is that the planes on the outer layers provide shielding to the signal traces on the inner layers. The disadvantages are that the ground plane may be cut up considerably with component mounting pads on a high-density PCB. This can be alleviated somewhat, by reversing the planes and placing the power plane on the component side, and the ground plane on the solder side of the board. Second, some designers do not like to have an exposed power plane, and third, the buried signal layers make board rework difficult if not impossible. This stackup satisfies objectives #1, #2, and #4.

Two of the above three problems can be alleviated with the stackup shown in Fig. 16-14B, where the two outer planes are ground planes, and power is routed

* The spacing between the trace layer and the adjacent plane.

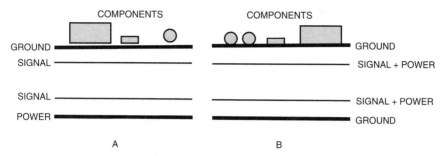

**FIGURE 16-14.** Four-layer board with signal traces on the inside layers and the planes on the outside layers. The configuration of (*A*) satisfies three of the six objectives, whereas the configuration of (*B*) satisfies four of the six objectives.

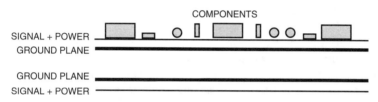

**FIGURE 16-15.** A four-layer board with two internal ground planes and no power plane. This configuration satisfies three of the six objectives.

as a trace on the signal layers. The power should be routed as a grid, using wide traces, on the signal layers. Two added advantages of this configuration are that (1) the two ground planes produce a much lower ground impedance and hence less common-mode cable radiation and (2) the two ground planes can be stitched together around the periphery of the board to enclose all the signal traces in a Faraday cage. This configuration satisfied objectives #1, #2, #4, and #5 while using only four layers.

A fourth possibility, which is not commonly used but performs well, is shown in Fig. 16-15. This is similar to Fig. 16-13, but with the power plane replaced with a ground plane, and the power routed as a trace on the signal layer. This stackup overcomes the rework problem associated with the Fig. 16-14 configurations, and it still provides for the low ground impedance as a result of the two ground planes. The planes, however, do not provide any shielding. The Fig. 16-15 configuration satisfies objectives #1, #2, and #5 but not objectives #3, #4, or #6. A well-known personal computer peripheral manufacturer has used this stackup successfully for many years.

So, as can be observed, more options are available for a four-layer board than you might have originally thought. It is possible to satisfy four of the six objectives with only four-layers. The configurations of Figs. 16-13, 16-14B, and 16-15 all can perform well EMC wise.

*16.4.2.3 Six-Layer Boards.* Most six-layer boards consist of four signal routing layers and two planes. From an EMC perspective, a six-layer board is preferred over a four-layer board because it is easy to shield high-frequency signals by placing them on buried layers between planes, or to provide for orthogonally routed signal layers that are referenced to the same plane.

One stackup that *should not be used* on a six-layer board is the one shown in Fig. 16-16. The planes provide no shielding for the signal layers, and two of the signal layers (1 and 6) are not adjacent to planes. The only time this arrangement works even moderately well is if all the high-frequency signals are routed on layers 2 and 5 and only low-frequency signals, or better yet no signals at all (just mounting pads and test points), are located on layers 1 and 6. In this configuration, any unused area on layers 1 and 6 should be provided with "ground fill" and tied into the primary ground plane, with vias, at as many locations as possible. This configuration satisfies only objective #3.

With six layers available, the principle of providing two buried layers for high-speed signals (as was done in the four-layer board of Fig. 16-14) is easily implemented as shown in Fig. 16-17. In addition to the high-speed signal routing layers (layers 3 and 4), this configuration also provides two surface layers for routing low-speed signals.

This is a common six-layer stackup, and it can be effective in controlling emissions. This configuration satisfies objectives #1, #2, and #4, but not #3, #5, and #6. Its main drawback, and not a serious one, is the separation of the power and ground planes. Because of this separation, there is no significant power to ground inter-plane capacitance. Therefore, the decoupling must be designed carefully to overcome this limitation.

Not nearly as common, but also a good performing stackup for a six-layer board, is shown in Fig. 16-18. This has the same ordering of planes as Fig. 16-17,

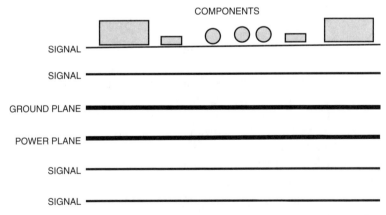

**FIGURE 16-16.** A six-layer PCB stackup that is *not recommended.* This configuration uses six layers and only satisfies one of the six objectives.

COMPONENTS

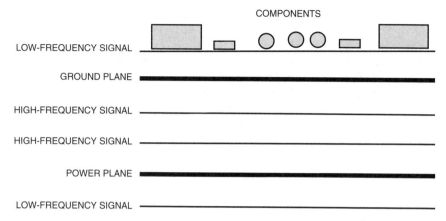

LOW-FREQUENCY SIGNAL

GROUND PLANE

HIGH-FREQUENCY SIGNAL

HIGH-FREQUENCY SIGNAL

POWER PLANE

LOW-FREQUENCY SIGNAL

**FIGURE 16-17.** A common and effective six-layer PCB stackup that provides shielding for the high-frequency signal layers. This configuration satisfies three of the six objectives.

COMPONENTS

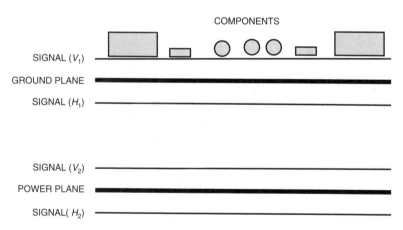

SIGNAL ($V_1$)

GROUND PLANE

SIGNAL ($H_1$)

SIGNAL ($V_2$)

POWER PLANE

SIGNAL( $H_2$)

**FIGURE 16-18.** Six-layer PCB stackup in which orthogonal routed signal layers reference the same plane. This configuration satisfies three of the six objectives.

but it assigns the layers differently and in many cases can provide better EMC performance than the stackup of Fig. 16-17.

In Fig. 16-18, $H_1$ indicates the horizontal routing layer for signal 1, and $V_1$ indicates the vertical routing layer for signal 1. $H_2$ and $V_2$ represent the same for signal 2. This configuration has the advantage that orthogonal routed signals always reference the same plane. The disadvantage is that the signals on layers one and six are not shielded. Therefore, the signal layers should be placed close to their adjacent planes, and the desired board thickness is made up by the use of a thicker center core. Typical layer spacing for this board might be 0.005 in/0.005 in/0.040 in/0.005 in/0.005 in.* This configuration satisfies objectives #1, #2, and #6, but not #3, #4, or #5.

If in the stackup shown in Fig. 16-18 multiple dc voltages are required, and the power plane is split into separate isolated voltage sections, then all critical signals must be routed only on layers 1 and 3 adjacent to the solid ground plane. Signals that do not cross the split in the power plane can be routed on layers 4 and 6. However, if one of the dc voltages were routed as a trace on a signal layer, this problem could have been avoided.

It is easier to achieve good EMC performance with a six-layer board than with a four-layer board. Six-layer boards also have the advantage of four signal routing layers instead of being limited to just two, as well as allowing for the possibility of using two ground planes. The configurations of Figs. 16-17 and 16-18 both perform well, the difference being that Fig. 16-17 provides for shielding of two of the high-frequency signal layers, whereas Fig. 16-18 allows the pairs of orthogonal routed layers to reference the same plane. Figure 16-17 would often be preferred if the product was in an unshielded enclosure (because the high-frequency signal traces are shielded by the outer planes), whereas the configuration of Fig. 16-18 might be preferred if the product were in a shielded enclosure.

### 16.4.2.4 Eight-Layer Boards.

An eight-layer board can be used to add two more routing layers or to improve EMC performance by adding two more planes. Although there are examples of both cases, most eight-layer board stackups are used to improve EMC performance rather than add additional routing layers. The percentage increase in cost of an eight-layer board over a six-layer board is less than the percentage increase in going from four to six layers, which makes it easier to justify the cost increase for improved EMC performance. Therefore, most eight-layer boards (and all the ones that we will concentrate on here) consist of four signal-routing layers and four planes.

An eight-layer board with six routing layers is definitely *not* recommended, no matter how you decide to stack up the layers. If you need six routing layers, you should be using a ten-layer board. Therefore, an eight-layer board can be thought of as a six-layer board with optimized EMC performance. Although many stackups are possible, I will only discuss a few that have proven themselves by providing excellent EMC performance.

The basic stackup of an eight-layer board with good EMC performance is shown in Fig. 16-19. This configuration is popular and satisfies five out of six of the original objectives; it does not satisfy objective #6. All signal layers are adjacent to planes, and all the layers are closely coupled together. The high-speed signals are buried between planes; therefore, the planes provide shielding

---

* Actual layer spacing may vary from these numbers depending on several factors, such as desired total board thickness and copper thickness used for the layers.

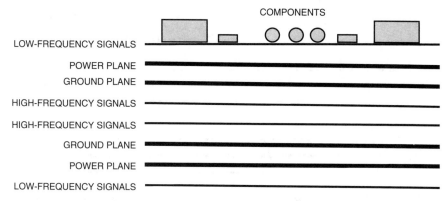

**FIGURE 16-19.** A common eight-layer PCB stackup with excellent EMC performance. This configuration satisfies five of the six objectives.

to reduce the emissions from these signals. In addition, the board uses multiple ground planes, which decreases the ground impedance.

For best EMC performance and signal integrity, when critical high-frequency signals change layers (e.g., from layer 4 to 5 in the case of Fig. 16-19) a ground-to-ground via should be added between the two ground planes near the signal via. This provides a path adjacent to the signal via for the return current.

The stackup in Fig. 16-19 can be further improved by using some form of embedded PCB capacitance technology, as discussed in Section 11.4.6, for layers 2 and 3 as well as 6 and 7. This approach will provide a significant improvement in the high-frequency decoupling and may allow the use of significantly fewer discrete decoupling capacitors.

If a design requires two dc voltages (e.g., 5 V and 3.3 V), the stackup of Fig. 16-19 should be considered. Each of the two power planes can be assigned a different voltage. This results in a design with two solid voltage planes and avoids the necessity of a split power plane and its associated problems.

Another excellent eight-layer configuration is shown in Fig. 16-20. This configuration is similar to the six-layer configuration of Fig. 16-18 but includes two outer layer ground planes. With this arrangement, all routing layers are buried between planes and are therefore shielded. In addition, orthogonal routed, high-frequency signals reference the same plane.

Although not as common as the stackup of Fig. 16-19, this excellent configuration also satisfies five of the six objectives presented previously; it does not satisfy objective #3. Typical layer spacing for this configuration might be 0.010 in/0.005 in/0.005 in/0.020 in/0.005 in/0.005 in/0.010 in.

The 0.010 in spacing between layers 1 and 2 is used so that most of the return current from the signals on layers 2 and 4 will return on the closer ground plane, in this case layer 3. Similarly in the case of layers 5 and 7, most of the signal current will return on layer 6. When a signal layer is located between two planes,

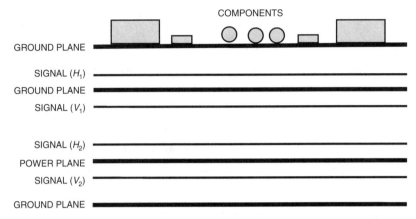

**FIGURE 16-20.** An excellent eight-layer PCB stackup where orthogonal routed signals reference the same plan, and high-frequency signal routing layers are shielded by the outer layer ground planes. This configuration satisfies five of the six objectives.

one twice as far away as the other, 67% of the current will return on the close plane and only 33% of the current on the farther plane (see Table 10-3).

An even better layer spacing for the stackup of Fig. 16-20 would be 0.015 in/ 0.005 in/0.005 in/0.010 in/0.005 in/0.005 in/0.015 in. In this case, the far plane is three times farther away from the signal layers than the near plane and the return current will divide 75% on the close plane and 25% on the far plane (Table 10-3).

Another possibility for an eight-layer board is to modify Fig. 16-20 by moving the planes to the center as shown in Fig. 16-21. This has the advantage of having a tightly coupled power-ground plane pair at the expense of not being able to shield the traces.

This is basically an eight-layer version of Fig. 16-18 with the addition of a tightly coupled power-ground plane pair at the center. Typical layer spacing for this configuration might be 0.006 in/0.006 in/0.015 in/0.006 in/0.015 in/ 0.006 in/0.006 in. The 0.006-in layer spacing allows tight coupling between the signal layers and their respective return planes as well as tight coupling between the power and ground plane, which will improve the decoupling above 500 MHz. This configuration satisfies objectives #1, #2, #3, #5, and #6, but not #4. This is an excellent performing configuration with good signal integrity and is often preferred over the stackup of Fig. 16-20 because of the tightly coupled power/ground planes. The stackup in Fig. 16-21 can be improved even more by using some form of embedded PCB capacitance technology for layers 4–5 to improve the high-frequency decoupling. This is one of my favorite configurations for high-frequency signals.

For high-frequency signals (with harmonics above 500 MHz) on a board in a shielded enclosure, the stackup of Fig. 16-21 would usually be preferred. For lower frequency and/or a product in an unshielded enclosure, the stackup

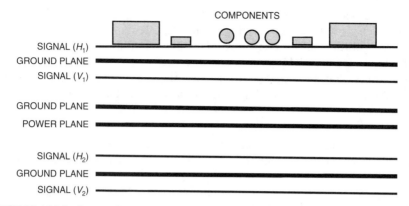

**FIGURE 16-21.** An excellent eight-layer PCB stackup with good signal integrity and EMC performance. This configuration provides a tightly coupled power-ground plane pair at the center of the board. This configuration satisfies five of the six objectives.

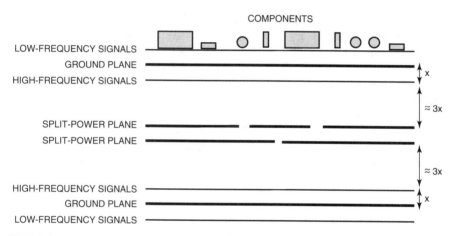

**FIGURE 16-22.** An acceptable, if not ideal, eight-layer board stackup with four signal layers and two split power planes. This configuration satisfies four of the six objectives.

of Fig. 16-20 might be preferred because it provides shielding for the signal layers.

Note that all three of the above eight-layer boards satisfy five of the six objectives.

Figure 16-22 shows an acceptable, if not ideal, eight-layer board that can be used if split-power planes are required. It has two split-power planes and four routing layers. Typical layer spacing for this stackup might be 0.006 in/0.006 in/0.015 in/0.006 in/0.015 in/0.006 in/0.006 in. Because the split-power planes are three times as far away from the inner signal layers (layers 3 and 6) than the ground planes (layers 2 and 7), 75% of the signal return current will be on the ground planes and only 25% on the split-power planes (Table 10-3). This will provide a 6-dB reduction in the detrimental effect of the split-power plane. This configuration satisfies objectives #1, #2, #4, and #5, but not #3 or #6.

There is very little EMC advantage to using a board with more than eight layers. More that eight layers are usually only used when additional signal routing layers are required. If six routing layers are needed, then a 10-layer board should be used.

***16.4.2.5 Ten-Layer Boards.*** Ten-layer boards usually have six signal layers and four planes. Having more than six signal layers on a 10-layer board is *not recommended.*

High layer count boards (10 plus) require thin dielectrics (typically 0.006 in or less on a 0.062 in thick board) and therefore they automatically have tight coupling between all adjacent layers and satisfy objectives #2 and #3. When properly stacked and routed, they can meet five or even all six of the objectives, and will have excellent EMC performance and signal integrity.

A common and nearly ideal stackup for a 10-layer board is shown in Fig. 16-23. The reason that this stackup has such good performance is the tight coupling of the signal and return planes, the shielding of the high-speed signal layers, the existence of multiple ground planes, as well as a tightly coupled power/ground plane pair in the center of the board. The high-frequency decoupling performance can be improved even more by the use of some form of embedded PCB capacitance technology for layers 5 and 6. High-speed signals normally would be routed on the signal layers buried between the planes (layers 3–4 and 7–8 in this case).

The common way to pair orthogonally routed signals in this configuration would be to pair layers 1 and 10 (carrying only low-frequency signals), as well as pairing layers 3 and 4, and layers 7 and 8 (both carrying high-speed signals).

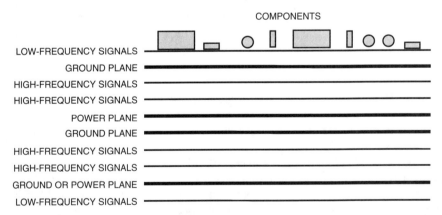

**FIGURE 16-23.** A commonly used and nearly ideal 10-layer PCB stackup. This configuration satisfies five of the six objectives.

By pairing signals in this manner, the planes on layers 2 and 9 provide shielding to the high-frequency signal traces on the inner layers. In addition, the signals on layers 3 and 4 are isolated (shielded) from the signals on layers 7 and 8 by the center power/ground plane pair. For example, high-speed clocks might be routed on one of these pairs, and high-speed address and data buses routed on the other pair. In this way, the bus lines are protected against being contaminated with clock noise by the intervening planes.

Where critical signals transition from one layer to another, decoupling capacitors or plane-to-plane vias, whichever is appropriate, should be added to reduce the return current discontinuity that would otherwise occur (see Section 16.3.3). This configuration satisfies five of the six original objectives for multilayer boards, it does not satisfy objective #6.

Another possibility for routing orthogonal signals on the 10-layer board shown in Fig. 16-23 is to pair layers 1 and 3, layers 4 and 7, and layers 8 and 10. In the case of layer pairs 1 and 3 as well as 8 and 10, this has the advantage of routing orthogonal signals with reference to the same plane. The disadvantage, of course, is that if layers 1 and/or 10 have high-frequency signals on them, then no inherent shielding is provided by the PCB planes. Therefore, these signal layers should be placed close to their adjacent plane (which occurs naturally in the case of a 10-layer board).

Each of the two 10-layer routing configurations discussed above have many advantages with few disadvantages, the primary difference being how the orthogonal routed signals are paired. Either stackup will provide good EMC and signal integrity performance if laid out carefully.

The stackup in Fig. 16-23 can be improved by using some form of embedded PCB capacitance technology for layers 5 and 6, which improves the high-frequency power/ground plane decoupling.

Figure 16-24 is another possible stackup for a 10-layer board. This configuration gives up the closely spaced power/ground plane pair. In return, it provides three signal-routing-layer pairs shielded by the ground planes on the outer layers of the board, and isolated from each other by the internal power and ground planes. All signal layers are shielded and isolated from each other in this configuration. The stackup of Fig. 16-24 is very desirable if you have few low-speed signals to put on the outer signal layers (as in Fig. 16-23) and most of your signals are high speed.

One consideration with this stackup relates to how badly the outside ground planes will be cut up by the component mounting pads and vias on a high-density PCB. This issue has to be addressed and the outside layers carefully laid out. This configuration satisfies objectives #1, #2, #4, and #5, but not #3 or #6.

Figure 16-25 shows yet another possible stackup for a 10-layer board. This stackup allows the routing of orthogonal signals adjacent to the same plane, but in the process has to give up the closely spaced power/ground planes. This configuration is similar to the eight-layer board shown in Fig. 16-20, with the addition of the two outer low-frequency routing layers.

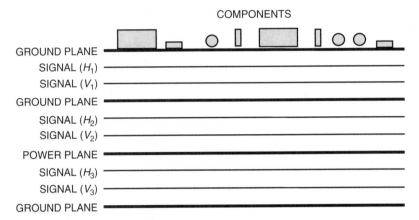

**FIGURE 16-24.** A 10-layer PCB stackup that provides three signal layer pairs, which are shielded and isolated from each other. This configuration satisfies four of the six objectives.

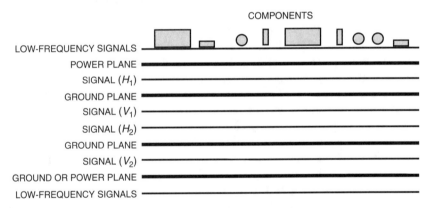

**FIGURE 16-25.** Ten-layer PCB stackup in which orthogonal routed signals reference the same plane. This configuration satisfies five of the six objectives.

The configuration in Fig. 16-25 satisfies objectives #1, #2, #4, #5, and #6, but not #3.

The stackup in Fig. 16-25 can be improved even more by replacing layers 2 and 9 each with a pair of embedded PCB capacitance layers (thereby satisfying objective #3). This, however, effectively converts it to a 12-layer board.

Figure 16-26 shows a 10-layer board that meets all six of the originally stated objectives. The drawback, however, is that it only has four signal routing layers. This configuration provides excellent performance, both from an EMC and a signal integrity perspective. The stackup in Fig. 16-26 can be improved

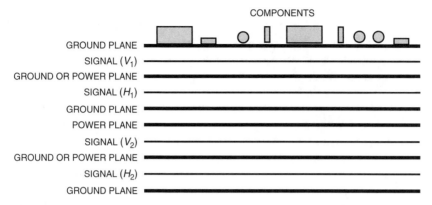

**FIGURE 16-26.** A 10-layer PCB stackup that satisfies all six of the objectives. However, it only has four signal routing layers.

with the use of some form of embedded PCB capacitance technology for layers 5 and 6.

***16.4.2.6 Twelve and More Layer Boards.*** High layer-count boards contain many planes; hence, the problems caused by split power planes can usually be avoided because there are enough planes that a different power plane can be assigned to each voltage.

An excellent stackup for a 12-layer PCB that satisfies all six of the original objectives is shown in Fig. 16-27. This is basically the 10-layer board of Fig. 16-25 with the addition of two more planes to satisfy objective #3. The performance of the stackup in Fig. 16-27 can be improved by the use of some form of embedded PCB capacitance technology for layers 2 and 3 as well as layers 10 and 11.

If a design uses a multiplicity of dc voltages that will require two split power planes, the stackup of Fig. 16-28 should be considered. In this configuration, the split-power planes are isolated from the signal layers by the solid-ground planes. Therefore, none of the signal layers are adjacent to the split-power planes, and the concern about signals crossing over a split plane is eliminated. The stackup of Fig. 16-28 has six routing layers and satisfies five of the original six objectives. As shown, it does not satisfy objective #6.

***16.4.2.7 Basic Multilayer PCB Structures.*** As demonstrated by the examples in this chapter, many times the PCB designer is faced with the choice of either shielding critical signal layers by burying them between planes (objective #4) or of routing critical signals on two layers that are adjacent to the same plane (objective #6).

Although contrary to popular practice, I believe that significant evidence exists to show that for high-frequency circuitry, good EMC performance, and

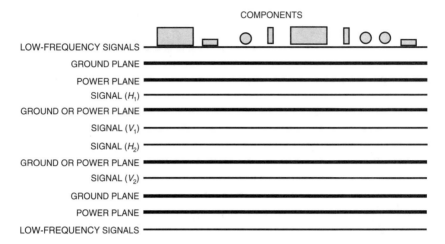

**FIGURE 16-27.** A 12-layer PCB that has 6 signal-routing layers and satisfies all six of the original objectives.

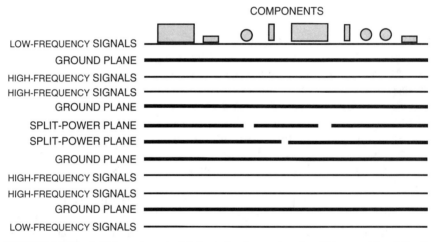

**FIGURE 16-28.** A 12-layer board with 6 routing layers that should be considered if split power planes are required. The two split-power planes are isolated from the signal layers by two solid-ground planes. This configuration satisfies five of the six objectives.

signal integrity, routing critical signals on layers that are adjacent to the same plane should take precedence over shielding critical signal layers by burying them between planes.* Both the EMC performance and the signal integrity of high-speed PCBs will be improved by this approach (Archambeault, 2002,

---

* This is especially true as frequencies become higher and higher, and signal integrity becomes more of an issue. For the case of constant impedance transmission lines, referencing different planes produces an impedance discontinuity at the location of the transition.

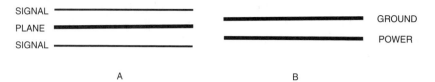

**FIGURE 16-29.** The two basic building blocks for multilayer boards. (*A*) Two signal layers adjacent to a plane and (*B*) a power-ground plane pair.

p. 191). Routing a signal on layers adjacent to the same plane significantly reduces the current return path inductance, because most PCB designers cannot, or do not, provide plane-to-plane vias adjacent to the signal trace vias, as was discussed in Section 16.3.3.

This then suggests a general procedure for determining an optimum stackup for high-layer-count, high-speed digital logic boards. The basic stackup should consist of multiples of the two basic structures that consist two signal layers adjacent to a plane (signal-plane-signal) as shown in Fig. 16-29A as well as adjacent power and ground plane pairs as shown in Fig. 16-29B. These two structures can then be combined in multiple ways to form PCBs with six or more layers.

For example, the six-layer stackup shown in Fig. 16-18 consisted of two sets of the building blocks from Fig. 16-29A, whereas the eight-layer PCB stackup shown in Fig. 16-21 used two of the basic building blocks of Fig. 16-29A, combined with the building block in Fig. 16-29B.

Figure 16-30 shows a 12-layer board with eight routing layers based on using four of the basic building blocks shown in Fig. 16-29A. This stackup does not have adjacent power and grounds, so it only satisfies five of the six original objectives.

Adding the basic building block from Fig. 16-29B to the center of the board shown in Fig. 16-30 produces a 14-layer board that meets all the design objectives.

### 16.4.3  General PCB Design Procedure

The previous sections have discussed various ways to stackup high-speed, digital logic PCBs with 4 to 14 layers. A good PCB stackup reduces radiation, improves signal quality, and helps aid in the decoupling of the power bus. No one stackup is best; several viable options are available in each case and some compromising of objectives is usually necessary.

Table 16-2 summarizes the number of routing layers and the number of planes for various common layer-count PCBs. Other combinations are possible, but these are the most common. From the table, one can visualize a pattern developing with respect to why more layers should be used.

Note that in the case of boards with eight or more layers, five or in some cases all six of the multilayer board objectives can be satisfied.

In addition to the number of layers, the type of layer (plane or signal), and the ordering of the layers, the following factors are also important in determining the EMC performance of the board:

- The layer spacing
- The assigning of signal layer pairs for orthogonal routing of signals
- The assignment of signals (clock, bus, high speed, low frequency, etc.) to which signal-routing-layer pairs

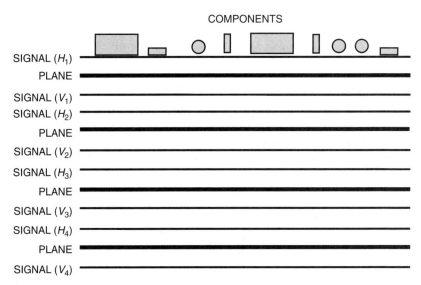

**FIGURE 16-30.** A 12-layer board with 8 routing layers constructed from the basic building block shown in Fig. 16-29A. This configuration satisfies five of the six objectives.

**TABLE 16-2.  Summary of PCB stackup options.**

| Number of layers | Routing layers | Planes | Reason for using | Number of objectives satisfied |
|---|---|---|---|---|
| 2 | 2 | 0 | Low cost | 0 |
| 4 | 2 | 2 | Improved EMC and SI performance | 1 to 4 |
| 6 | 4 | 2 | Improved EMC and SI, as well as two additional routing layers | 3 |
| 8 | 4 | 4 | Improved EMC and SI performance | 5 |
| 10 | 6 | 4 | Two additional routing layers | 4 or 5 |
| 12 | 6 | 4 | Improved EMC and SI performance | 5 or 6 |
| 12 | 8 | 4 | Two additional routing layers | 5 |
| 14 | 8 | 6 | Improved EMC and SI performance | 6 |

This discussion on board stackup has assumed a standard 0.062-in-thick board, with symmetrical cross section, an even number of layers, and conventional via technology. If blind, buried, micro-vias, nonsymmetrical boards, or odd number layer count boards are considered, other factors come into play and additional board stackups not only become possible but also desirable in many cases.

The following represents the general steps required in creating a PCB stackup:

- Determine the number of signal routing layers required
- Determine how to handle multiple dc voltages
- Determine the number of power planes required for the various system voltages
- Determine whether multiple voltages will be on the same power plane layer, thereby requiring a split-plane, and routing restrictions on the adjacent layers
- Assign each signal layer pair to a solid reference plane as shown in Fig. 16-29A
- Pair power and ground planes as shown in Fig. 16-29B
- Determine the ordering of the layers
- Determine the spacing between layers
- Define any necessary routing restrictions

Following the guidelines presented in this chapter will produce better PCBs and avoid many of the most common EMC problems associated with boards. All the stackups discussed, with the exception of Fig. 16-16, will provide good-to-excellent EMC performance.

The avoiding of current return path discontinuities is probably the most important, and often overlooked, principle involved in good PCB design. *Think, where are the return currents flowing?*

See Chapter 17 for additional information on the layout and routing of mixed-signal PCBs.


**SUMMARY**

- The six main objectives of multilayer board stackup are as follows:
  - Every signal layer should be next to a plane.
  - Tight coupling should exist between signal layers and their adjacent reference planes.
  - Tight coupling should exist between power and ground planes.
  - High-frequency signals should be shielded by burying them between planes.
  - Multiple ground planes are desirable.
  - Orthogonally routed signals should reference the same plane.

- To meet five or more of the above objectives, an eight or more layer board is required.
- Use multilayer boards with power and ground planes whenever possible.
- Give a lot of consideration to component placement and orientation.
- Ten percent of the circuitry, the critical signals, causes 90% of the problems.
- High spectral content, repetitive waveshape circuits are critical signals. These include:
  - Clocks
  - Buses and
  - Repetitive control signals
- The signal speed (spectral content) of a signal is proportional to
  - The fundamental frequency
  - The reciprocal of the rise/fall time
  - The magnitude of the current
- If possible, route high-frequency orthogonal signals adjacent to the same plane.
- Compromise is needed with most board stackup configurations.
- Many multilayer board configurations exist that provide good-to-excellent EMC performance.
- Planes are your friends; slots are your enemies.
- Route critical digital signals away from the periphery of the board and the I/O area.
- Multiple dc voltages can be dealt with using one or more of the following approaches:
  - Split the power plane and live with the routing restrictions that approach creates.
  - Use a separate solid power plane for each voltage.
  - Use a power island on a signal layer for some of the voltages.
  - Route some of the voltages as a trace on a signal layer.
  - As a last resort, use stitching capacitors when a trace must cross a split-power plane.
- Critical signals should be routed on no more than two layers, and these layers should be adjacent to the same plane.
- From an EMC and signal integrity perspective, routing critical signals on layers that are adjacent to the same plane should take precedence over shielding the signal layers by burying them between planes, in particular for boards contained in shielded enclosures.
- Connect circuit ground to the chassis, through a very low inductance connection, in the I/O area of the board.

## PROBLEMS

16.1 The signal speed (spectral content) of a signal is proportional to what parameters?

16.2 Where should the circuit ground be connected to the chassis ground?

16.3 Where, and how, should the metallic backshell on I/O connectors be connected?

16.4 What are the two most important objectives to follow in laying out double-sided PCBs?

16.5 What are the three most common causes of return current path discontinuities on a multilayer PCB?

16.6 a. Name three reasons why slots in power/ground planes should be avoided?
    b. If slots exist in power or ground planes, what routing restriction must be adhered to?

16.7 What two of the six multilayer PCB objectives should always be followed?

16.8 For the nonsymmetrical, cross-section, eight-layer PCB shown in Fig. P16-8, which of the basic multilayer board design objectives does the stackup satisfy?

16.9 For the stackup of Fig. 16-19, what would be the disadvantage of reversing the power and ground planes on layers 6 and 7?

16.10 Which of the six multilayer PCB objectives does the eight-layer stackup shown in Fig. 16-20 not satisfy?

16.11 How many of the basic multilayer PCB design objectives does the 12-layer PCB stackup shown in Fig. P16-11 satisfy?

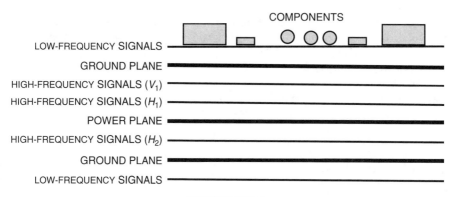

**FIGURE P16-8.**

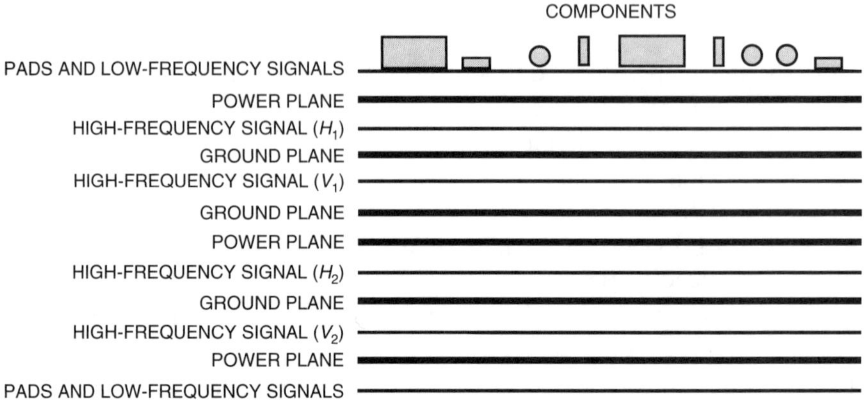

COMPONENTS

PADS AND LOW-FREQUENCY SIGNALS

POWER PLANE
HIGH-FREQUENCY SIGNAL ($H_1$)
GROUND PLANE
HIGH-FREQUENCY SIGNAL ($V_1$)
GROUND PLANE
POWER PLANE
HIGH-FREQUENCY SIGNAL ($H_2$)
GROUND PLANE
HIGH-FREQUENCY SIGNAL ($V_2$)
POWER PLANE
PADS AND LOW-FREQUENCY SIGNALS

**FIGURE P16-11.**

16.12  One of the dc voltages on a PCB uses a "power island" located on the top layer of the PCB.

  a.  Does the use of this power island require the designer to specify any specific routing restrictions with respect to the board layout?

  b.  What if the power island is located on the bottom layer of the board?

16.13  Draw a PCB stackup that is not in the book. Which of the basic design objectives does your stackup satisfy?

## REFERENCES

Archambeault, B. *PCB Design for Real-World EMI Control.* Boston, MA: Kluwer Academic Publishers, 2002.

Fessler, J. T., Whites, K. W., and Paul, C. R. "The Effectiveness of an Image Plane in Reducing Radiated Emissions." *IEEE Transactions on Electromagnetic Compatibility,* February 1996.

German, R. F., Ott, H. W., and Paul, C. R. "Effect of an Image Plane on Printed Circuit Board Radiation. *1990 IEEE International Symposium on Electromagnetic Compatibility,* August 21–23, 1990.

Paul, Clayton R. *Introduction to Electromagnetic Compatibility,* 2nd ed. New York Wiley, 2006.

Smith, D. C. *Routing Signals Between PWB Layers—Part 2, An Emission Example,* 2006. Available at www.dsmith.org. Accessed January 2009.

## FURTHER READING

Archambeault, B. "Effects of Routing High-Speed Traces Close to the PCB Edge." *Printed Circuit Design & Fab,* January 2008.

Bogatin, E. "An EMC Sweet 16." *Printed Circuit Design & Manufacture*, April 2006.

Ott, H. W. *PCB Stackup, Parts 1 to 6*, Henry Ott Consultants, 2002–2004. Available at http://www.hottconsultants.com/tips.html. Accessed January 2009.

Ritchey, L. W. *Right the First Time, A Practical Handbook on High Speed PCB and System Design,* vol. 1, Speeding Edge. May 2003. Available www.speedingedge.com. Accessed April 2009.

# 17  Mixed-Signal PCB Layout

The design and layout of a mixed-signal printed circuit board (PCB) can be a challenging task, the solution to which is not well addressed in most engineering literature. Mixed-signal PCB problems usually involve either one of two situations. One involves digital logic circuits that interfere with sensitive low-level analog circuits [often audio or radio frequency (rf)], and the second involves high-power motor and relay driver (noisy analog) circuits that interfere with both digital and analog circuits.

In the discussion that follows, keep in mind two basic principles of electromagnetic compatibility (EMC). One is that currents should be returned to their source as locally and compactly as possible, that is, through the smallest possible loop area. The second is that a system should have only one reference plane. If the current is not returned locally and compactly, it creates a loop antenna. If a system has two reference planes, it creates a dipole antenna.* Both are undesirable results.

The key to determining the optimum mixed-signal board layout is understanding how and where the ground return currents actually flow. Most PCB designers only think about where the signal current flows (obviously on the signal trace) and ignore the path taken by the return current.

The lowest impedance signal return path for high-frequency currents will be in a plane directly under the signal trace, as was discussed in Section 3.2. Slots in ground planes interrupt this optimum ground return current path, which produces a large ground impedance (inductance) and an increased ground plane voltage drop.

## 17.1  SPLIT GROUND PLANES

Before continuing, let us define the basic problem that we are trying to solve. It is not that the analog circuits might interfere with the digital logic. Rather, it is the possibility that the *high-speed digital logic might interfere with the low-level analog circuits*. This concern is legitimate. We want to make sure that the digital

---

* See Appendix D for more information on dipole antennas.

*Electromagnetic Compatibility Engineering,* by Henry W. Ott
Copyright © 2009 John Wiley & Sons, Inc.

ground currents do not flow in the analog ground plane: hence, the often-heard recommendation to split the ground plane into analog and digital sections.

However, if the ground plane is split and traces run across the split as shown in Fig. 17-1A, what will be the current return path? Assuming that the two planes are connected together somewhere, usually at a single point, the return current will have to flow in a large loop. High-frequency currents that flow in large loops produce radiation and create high ground inductance. Low-level analog currents that flow in large loops are susceptible to electromagnetic field pickup. Both of the above are undesirable results.

If the ground plane must be split and traces run across the split, first connect the planes together at one location, forming a bridge, as shown in Fig. 17-1B. Then, routing all the traces so that they cross at this bridge will provide a current return path directly underneath each of the traces, which produces a small loop area.

Other acceptable ways of passing a signal over a split plane are with opto-isolators, magnetoresistive isolators, or transformers. In the first case, the only thing crossing the split in the plane is light; in the second two cases, a magnetic field crosses the split. Another possibility is with a true differential signal, where the signal flows down one trace and returns on the other trace. This approach, however, is not as good as the other three approaches.*

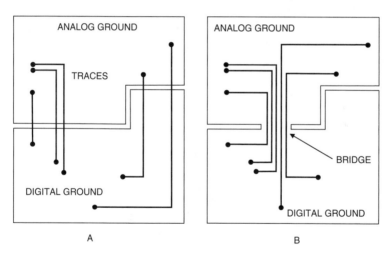

**FIGURE 17-1.** (*A*) Signal traces crossing over a split between the analog and digital ground planes. (*B*) Signal traces crossing over a bridge between the analog and digital ground planes.

---

* In differential signaling, the ground path can still matter, especially in the case of high-frequency signals. We usually assume that in differential signaling, the signal current returns on the second trace. Although a useful simplification, this is not completely true. Even in differential signaling, current flows on the plane adjacent to each conductor (due to the fields that are present), as if they were two independently routed, single-ended traces. However, under the proper conditions, these two ground plane currents cancel.

An exceptionally bad configuration occurs when the ground plane is split and the two grounds are kept separate all the way back to the system's single "star ground" point, which is located at the power supply as shown in Fig. 17-2. In this case, the return current for a trace crossing over the split will have to flow all the way back to the power-supply ground as shown in the figure—a really big loop! In addition, this produces a dipole antenna, which consists of the analog ground and the digital ground planes (which are at different rf potentials) connected together with the long power-supply ground wires.

## 17.2   MICROSTRIP GROUND PLANE CURRENT DISTRIBUTION

To address the above concern, it is helpful to understand a little more about the characteristics of high-frequency currents. Assuming a multilayer PCB, high-frequency currents will return on the plane closest to the signal trace, because this is the lowest impedance (lowest inductance) path. For the case of a microstrip line (a trace above a plane), the return current flows on the adjacent plane regardless of what that plane is, power or ground; see Section 10.6.1.1. The current will spread out in the plane as was shown in Fig. 10-9, but otherwise it will follow the trace.

Table 17-1 shows the percentage of the ground plane current contained within a distance $\pm\ x/h$ from the center of a microstrip trace, where $x$ is the horizontal distance from the trace centerline, and $h$ is the height of the trace

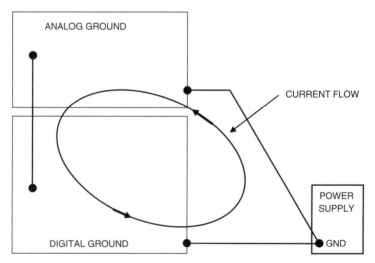

**FIGURE 17-2.** A poor layout with a trace crossing over a split between analog and digital ground planes, which are connected together at only one point—the power supply ground terminal.

**TABLE 17-1.   Percentage of Ground-Plane Current Contained within a Distance of ±x/h from the Center of a Microstrip Trace.**

| Distance from Center of Trace | % of Current | Ground-Plane Current[a] Reduction at a Distance > x/h |
|---|---|---|
| $x/h = 1$ | 50% | 12 dB |
| $x/h = 2$ | 70% | 16 dB |
| $x/h = 3$ | 80% | 20 dB |
| $x/h = 5$ | 87% | 24 dB |
| $x/h = 10$ | 94% | 30 dB |
| $x/h = 20$ | 97% | 36 dB |
| $x/h = 50$ | 99% | 46 dB |
| $x/h = 100$ | 99.4% | 50 dB |
| $x/h = 500$ | 99.9% | 66 dB |

[a] Increased by 6 dB to account for current on only one side of the trace.

above the ground plane (see Fig. 10-8). The numbers were calculated by integrating Eq. 10-13 between $\pm x$ for a 0.005-in-wide trace 0.010 in above the plane, but similar results will be obtained for other dimensions. Also listed in the table is the reduction (in dB) of ground plane current at a distance $> x/h$ from the center of the trace.

For example, if a microstrip trace is located 0.010 in above a plane, 97% of the return current will be contained in the portion of the plane that is within $\pm 0.200$ in of the trace centerline.

From the above, we can conclude that if the digital signal traces are routed properly, the digital-ground currents have no desire to flow through the analog portion of the ground plane and corrupt the analog signal. Figure 17-3 shows a digital logic trace on a split ground plane mixed-signal board and a representation of its return current path. Why is it then necessary to split the ground plane to prevent the digital return current from doing something that it does not want to do in the first place? The answer is that it is not necessary!

Therefore, I prefer the approach of using only one ground plane and *partitioning* the PCB into digital and analog sections. Analog signals must be routed *only* in the analog section of the board (on all layers). Digital signals must be routed *only* in the digital section of the board (on all layers). If this is done properly, the digital return currents will not flow in the analog section of the ground plane but will remain under the digital signal traces as shown in Fig. 17-4. Analog-to-digital converters can then be positioned to straddle the analog-digital partition as shown in Fig. 17-8.

Notice from comparing Figs. 17-3 and 17-4 that the digital logic ground current follows the same path whether or not the ground plane is split. What causes an analog noise problem is when a digital signal trace is routed in the analog section of the board, or vice versa. This is shown in Fig. 17-5. The digital ground currents now do flow in the analog section of the ground plane. Note, however, that this problem is not the result of not having split the ground

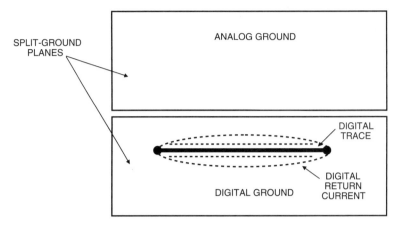

**FIGURE 17-3.** A digital logic trace and its associated ground return current on a split-ground plane PCB. Note, the digital return current remains close to and under the trace.

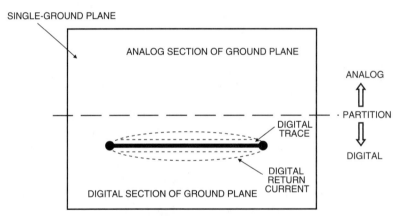

**FIGURE 17-4.** A digital logic trace and its associated ground return current on a partitioned, single ground plane mixed-signal PCB. Note, the digital return current still remains under the trace.

plane; rather, the problem is the result of the improper routing of the digital logic trace. The fix should be to route the digital logic trace, properly not to split the ground plane.

A PCB with a single ground plane, partitioned into analog and digital sections, and discipline in routing can usually solve most otherwise difficult mixed-signal layout problems, without creating any additional issues caused by splitting the ground plane. Component placement and partitioning are, therefore, critical to a good mixed-signal layout. If the layout is done properly, the digital ground currents will remain in the digital section of the board and will

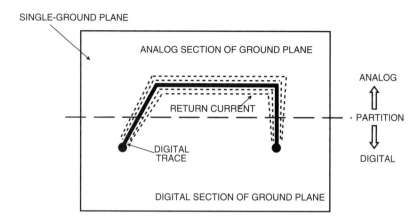

**FIGURE 17-5.** An improperly routed digital logic trace. The digital return current now flows through the analog section of the ground plane.

not interfere with the analog signals. The routing, however, must be checked carefully to ensure that the above routing restrictions are adhered to 100%! It only takes one improperly routed trace to destroy an otherwise perfectly good layout. Autorouting of a mixed-signal PCB, more often than not, results in a layout disaster; hence, manual techniques often have to be used.

## 17.3 ANALOG AND DIGITAL GROUND PINS

Another issue is where and how to connect the analog and digital ground pins of a mixed-signal integrated circuit (IC). Interestingly, most analog to digital (A/D) converter manufacturers, while suggesting the use of split ground planes, state in their data sheet, or application note, that the *AGND* and *DGND* pins must be connected externally to the same low impedance ground plane with minimum lead length.* Any extra impedance in the *DGND*-to-*AGND* connection will couple noise into the analog circuit through the stray capacitive coupling internal to the IC as shown in Fig. 17-6.

The $V_{noise}$ shown in Fig. 17-6 results from a transient current flowing through the internal inductance of the *DGND* lead frame, as well as any transient current flowing through the inductance of the external ground connection between *AGND* and *DGND*. The PCB designer, however, only has control over the inductance of the external ground connection. Their recommendation is then to connect both the *AGND* and the *DGND* pins of the A/D converter to the analog ground plane (also see Section 17.5).

---

* The reason for this is because most A/D converters do not have their analog and digital grounds connected together internally. Therefore, they rely on the external connection between the *AGND* and *DGND* pins to provide this connection.

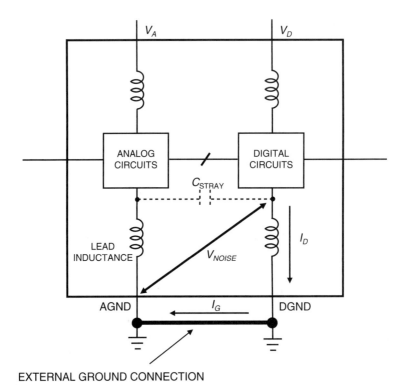

**FIGURE 17-6.** Simplified, internal model of an A/D (or D/A) converter, showing the noise voltage ($V_{noise}$) produced by the digital ground current.

All the above requirements can easily be satisfied if the system contains only one A/D converter. The ground plane can be split with the analog and digital sections connected together at one point under the converter, as shown in Fig. 17-7. The bridge between the two ground planes should be made the size of the IC, and no traces should be routed across the split in the plane. This is typical of the IC manufacturers' demo board.

What happens, however, if instead of a single A/D converter, your system has multiple converters? If the analog and digital ground planes are connected together under each converter, the planes are connected together at multiple points, and are no longer split. If the planes are not connected together under each converter, how do you satisfy the requirement that the *AGND* and *DGND* pins must be connected together through a low impedance?

A much better way to satisfy the requirement of connecting *AGND* and *DGND* pins together, and not creating additional problems in the process, is to use only one ground plane. The ground plane should be partitioned into analog and digital sections as shown in Fig. 17-8. This layout satisfies the requirement of connecting the analog and digital ground pins together through a low

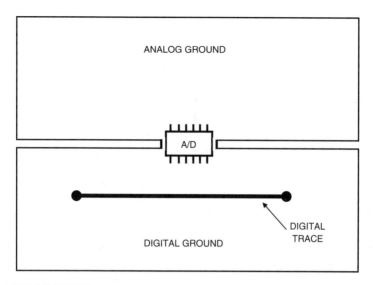

NOTE: NO TRACES ARE ALLOWED TO CROSS THE SPLIT IN THE GROUND PLANE.

**FIGURE 17-7.** An acceptable layout of a mixed-signal PCB with a single A/D converter and a split ground plane.

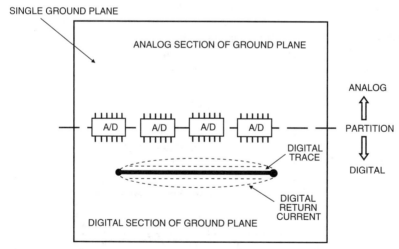

**FIGURE 17-8.** A properly partitioned mixed-signal PCB with multiple A/D converters and a single ground plane.

impedance plane, as well as meeting the EMC concerns of not creating any unintentional loop or dipole antennas.

As was eloquently stated by Terrell and Keenan (1997, p. 3–18) in *Digital Design for Interference Specifications*: "Thou Shalt Have But One Ground Before Thee."

## 17.4 WHEN SHOULD SPLIT GROUND PLANES BE USED?

Should split-ground planes ever be used? I can think of at least three instances where they would be appropriate. The instances are as follows:

- Some medical equipment with low leakage current requirements (10 μA)
- Some industrial process control equipment where the outputs are connected to noisy, high-power electromechanical equipment
- Possibly when a PCB is improperly laid out to begin with

In the first two cases listed above, signals that cross the split in the ground plane are usually optical or transformer coupled, which satisfy the requirement of no traces crossing the split in the ground plane. Notice that in these two cases, the ground plane was not split to protect the analog circuits from the digital ground currents, but for other externally imposed reasons.

The last case, however, is of more interest for the current discussion. It can be demonstrated clearly that if a mixed-signal board is poorly laid out, its performance can be improved by using a split ground plane.

Consider the situation that was shown in Fig. 17-5, where a high-speed digital trace was routed over the analog section of the board—a clear violation of the partition rules. Because the digital return current flows under the signal trace, it will flow in a portion of the analog ground plane. Splitting the ground plane in this case will improve the functional performance of the PCB by constraining the digital return current to the digital ground plane as shown in Fig. 17-9. This however, will increase the radiated emission from the board as the result of the larger loop area that exists between the signal trace and the

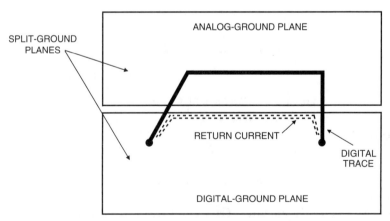

**FIGURE 17-9.** A mixed-signal PCB with an improperly routed digital logic trace and a split ground plane. The digital return current is now confined to the digital ground plane.

return current path. It will also increase the impedance of the ground plane, thereby increasing the radiation from cables connected to the board. In this case, the real problem is the improper routing of the high-speed digital trace, not the fact that the ground plane was not split. Two wrongs do not make a right! A better solution would have been to route the digital signal trace properly in the first place, and not to split the ground plane.

Remember the key to a successful PCB layout is *partitioning* and the use of *routing discipline*, not the splitting of the ground plane. It is almost always better to have only a single reference plane (ground) for your system.

The following two things that should be avoided if you do split planes:

- Overlapping the planes
- Routing traces across the split

Overlapping the planes will increase the interlayer capacitance, which will decrease the high-frequency isolation, and isolation is the reason for splitting the planes in the first place. Traces crossing over the split will have increased crosstalk to adjacent traces, and it can increase the radiated emission by 20 dB or more.

If analog and digital ground planes are split, low-capacitance, oppositely poled, Schottky diodes are often connected between the two planes to limit the direct current (dc) voltage differential to a few hundred millivolts. This can be important to prevent damage to mixed-signal ICs that have connections to both planes. Schottky diodes are used because of their low forward voltage drop (typically 300 mV) and their low capacitance. The low diode capacitance will minimize the high-frequency coupling between the planes.

The "Chain Saw "Test: On a split-plane board, conceptually you should be able to cut through the board at the split and not cut any traces or planes. If that is the case, the split plane board was most likely properly laid out and routed.

## 17.5   MIXED SIGNAL ICs

The layout information provided by A/D converter manufacturer's data sheets is usually only applicable to a simple system that contains one A/D converter (e.g., their demo board). Their recommended approach is not normally applicable for multiple A/D or digital-to-analog (D/A) converter systems or multiple board systems.

Analog-to-digital converters and digital-to-analog converters are mixed-signal ICs that have both analog and digital ports. Much confusion exists with respect to the proper grounding and decoupling of such devices. Digital and analog engineers often tend to view these devices differently.

A/D and D/A converters, as well as most other mixed-signal ICs, *should be considered to be analog components*. They are analog ICs with a digital section, not digital ICs with an analog section. The labels *AGND* and *DGND* on the IC pins refer to where these pins are connected internally, and it is not intended to imply where or how they should be connected externally. These pins should almost always be connected together and referenced, as well as decoupled, to the analog ground plane. Of course, if you use only a single ground plane, instead of split ground planes, the point becomes moot.

The exception to the above rule is some large DSPs that contain a large amount of digital processing. These devices draw large transient currents from the digital supply and should have their *AGND* pin tied to analog ground, and their *DGND* pin tied to digital ground, unless the data sheet specifies otherwise. These devices are specifically designed to have a high of degree of noise immunity between the internal analog and digital circuitry.

Figure 17-10 is a simplified generic block diagram of a mixed-signal system that consists of some analog circuitry, a mixed-signal IC, and some digital circuitry. If a difference in ground potential existed on the board, where would be the least harmful place to have it?

- Between grounds *A* and *B*
- Between grounds *B* and *C*
- Between grounds *C* and *D*

Ground noise between grounds *A* and *B* could have a detrimental effect on the low-level, sensitive analog circuits. Similarly, ground noise between grounds *B* and *C* could affect the analog to digital (or digital to analog) conversion taking place in the mixed-signal IC. The place that the noise would have the

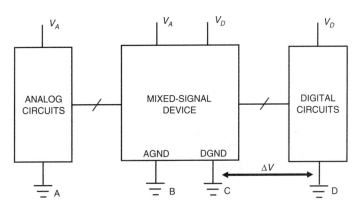

**FIGURE 17-10.** Grounding of a generic mixed signal system. Grounds A, B, and C should be connected to the analog ground plane, and ground D should be connected to the digital ground plane.

least detrimental effect is between the two digital interfaces, or between grounds *C* and *D*, as indicated by the $\Delta V$ in Fig. 17-10. This results from the fact that digital circuits have a larger inherent immunity than analog circuits. Therefore, grounds *A*, *B*, and *C* should all be connected to the analog ground plane, and ground *D* should be connected to digital ground plane.

### 17.5.1  Multi-Board Systems

Another way to isolate the digital ground from the analog ground is to place the digital circuits on one PCB and the analog circuits on another PCB. The question then is on which board should the A/D or D/A converters be mounted? The answer is simple if we remember that mixed-signal ICs are analog devices. The A/D or D/A converters should be mounted on the analog board as shown in Fig. 17-11. This approach (1) allows the *AGND* and *DGND* pins to be connected together, through the low impedance of the analog ground plane; (2) provides the shortest path for the sensitive analog circuitry; (3) isolates the analog circuits from the digital ground currents, and (4) applies any existing ground differential between the two boards, to the converter's digital inputs (or outputs,) where it is less of a problem than if applied to the low-level analog inputs (or outputs).

## 17.6  HIGH-RESOLUTION A/D AND D/A CONVERTERS

The use of a single solid ground plane, properly partitioned and routed (as discussed above), is usually adequate for most low-to-moderate resolution A/D converters (8, 10, 12, 14, or even 16 bit). For higher resolution systems (18 bits and up), even more ground noise voltage isolation *may* be required for

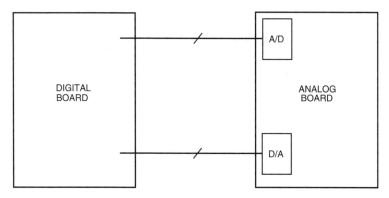

**FIGURE 17-11.** In a multiboard mixed-signal system, the A/D and/or D/A converter should be located on the analog board.

adequate performance. These converters often have minimum resolution voltages in the single digit microvolt range or less.

One conservative design approach is to keep the analog ground plane noise voltages smaller than the minimum analog signal level of concern. In the case of an A/D (or D/A) converter, the smallest resolvable signal voltage level, the least significant bit (LSB), is a function of the number of bits and the full-scale reference voltage of the converter. The lower the reference voltage and the larger the number of bits, the smaller the minimum resolvable signal voltage will be. Table 17-2 lists resolution voltages versus the number of bits for an A/D converter using a 1-V reference. These resolution levels can be scaled for other reference voltages by multiplying the resolution by the appropriate factor. For example, if the converter uses a 2-V reference, then multiply the resolution numbers in the table by two.

The dynamic range numbers listed in Table 17-2 are the ratio of the maximum signal (the reference voltage) to the minimum signal (the LSB) expressed in decibels. They will remain the same regardless of the reference voltage. It is interesting to note that most actual signals have dynamic ranges of less than 100 dB. Live music, for example, may have a dynamic range of as much as 120 dB. However, when recorded on a compact disk (CD) the dynamic range is limited to about 90 dB.

An estimation of the required ground noise voltage isolation can be obtained by assuming a digital ground noise voltage of 50 mV, which is representative of a PCB with a good layout. Also assume that it is desirable to limit the analog ground noise voltage to 5 μV. The ratio of these two numbers is a factor of 10,000 to 1 or 0.01%. This is equivalent to 80 dB of ground noise isolation.

We know from the previous discussion and Table 17-1 that for a microstrip trace, most of the return current will flow under or close to the trace. If the digital trace is kept 0.25 in or more away from the analog partition, then 99% of the digital return current will remain in the digital section of the PCB, assuming the trace is 0.005 in from the ground plane (an $x/h$ ratio of 50). However, a small amount of digital-ground current ($<1\%$) may still flow in the

**TABLE 17-2.    Converter Resolution Voltages for a 1-V Reference[a].**

| Number of Bits | Resolution | Dynamic Range |
| --- | --- | --- |
| 8 | 4 mV | 48 dB |
| 10 | 1 mV | 60 dB |
| 12 | 240 μV | 72 dB |
| 14 | 60 μV | 84 dB |
| 16 | 15 μV | 96 dB |
| 18 | 4 μV | 108 dB |
| 20 | 1 μV | 120 dB |
| 24 | 0.06 μV (60 nV) | 144 dB |

[a]The resolution can be scaled for other reference voltages (e.g., for a 2-V reference, multiply the resolution numbers by two).

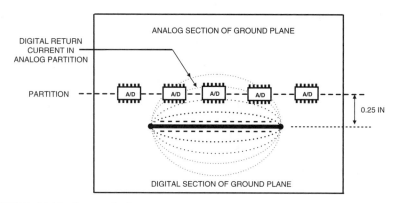

**FIGURE 17-12.** On a single ground plane, properly partitioned board, a small percentage (<1%) of the digital ground current may flow in the analog portion of the ground plane.

analog ground plane as shown in Fig. 17-12. As indicated in the previous paragraph, a current of only 0.1% or even 0.01% of the total digital ground current may cause a problem if it flows through the analog ground plane. A current of 0.01% or less would require a current reduction of 80 dB or more, whereas a current of 1% is equivalent to a reduction of only 40 dB.

From Table 17-1, it can be observed that separating the analog ground from the digital trace by more than an $x/h$ ratio of 50 provides little additional reduction in ground current. Therefore, the very small percentage of digital logic ground current present beyond an $x/h$ ratio of 50 *may* still present a noise problem for some high-resolution converters.

### 17.6.1  Stripline

Two possible solutions to this problem are available, short of splitting the ground plane. One is to run the digital logic traces in a stripline configuration, because for a stripline the return current does not spread out as far as it does for a microstrip line. This was discussed in Section 10.6.1.2 and is shown in Fig. 10-12.

Table 17-3 lists the percentage of the ground-plane current contained within a distance $\pm x/h$ from the center of a stripline trace, where $x$ is the horizontal distance from the trace centerline and $h$ is the height of the trace above the ground plane (see Fig. 10-11). The numbers were calculated by integrating Eq. 10-16 between $\pm x$ for a 0.005-in-wide trace 0.010 in above the plane, but similar results will be obtained for other dimensions. Also listed in the table is the reduction (in decibels) of ground-plane current at a distance more than $x/h$ from the center of the trace.

Notice that in the stripline case, 99% of the current will be contained within a distance of $\pm 3$ times the trace height, whereas for microstrip, the distance would be $\pm 50$ times the trace height—more than an order of magnitude

**TABLE 17-3.**   **Percentage of Ground-Plane Current Contained within a Distance of $\pm x/h$ from the Center of a Stripline Trace.**

| Distance From Center of Trace | % of Current | Ground-Plane Current[a] Reduction at a Distance $> x/h$ |
|---|---|---|
| $x/h = 1$ | 74% | 24 dB |
| $x/h = 2$ | 94% | 36 dB |
| $x/h = 3$ | 99% | 52 dB |
| $x/h = 5$ | 99.95% | 78 dB |
| $x/h = 10$ | 99.9999756% | 144 dB |

[a] Increased by 6 dB to account for current on only one side of the trace and increased by another 6 dB to account for the fact that there are two planes.

reduction in the current spread. For stripline, the amount of current beyond $\pm 10$ times the trace height is only 0.0000244%.

Therefore, if a digital stripline is 0.05 in or more away from the analog partition, virtually 100% of the digital return current will remain in the digital section of the PCB—assuming the trace is 0.005 in from the ground plane. For a trace height of 0.010 in the digital trace would only have to be kept 0.10 in away from the analog partition.

### 17.6.2   Asymmetric Stripline

For reasons of cost, stripline is seldom used on digital logic boards because it requires two planes for each signal layer, but asymmetric stripline is common. In the asymmetric stripline case, two orthogonally routed (this minimizes the coupling between the layers) signal layers are placed between two planes. For either signal layer, one plane is a distance $h$ away from the trace, whereas the other plane is a distance $2h$ away (see Fig. 10-14). In the case of asymmetric stripline, as well as stripline and microstrip, the reference planes or plane can be either power or ground as was discussed in Section 10.7. Figure 17-13 is a log-log plot of the normalized ground plane current density, on one side of the trace centerline, for a microstripline, a stripline, and an asymmetrical stripline as a function of $x/h$, for a 0.005-in-wide trace 0.020 in above a reference plane. The asymmetric stripline plot is for the case where $h_2 = 2 h_1$. As can be observed, the ground plane current distribution for asymmetrical stripline is much closer to that of stripline than that of microstrip. Therefore, we can conclude that *asymmetrical stripline behaves similar to stripline.*

Figure 17-13 clearly shows the advantage of stripline (or asymmetric stripline) over microstrip line, with respect to confining the spread of the return current. High-resolution converters should have no problem when used on a single ground plane, mixed-signal board, as long as the digital signals are routed as stripline (or asymmetric stripline) and are kept at least 5 to 10 times the trace height away from the analog–digital partition boundary. It is

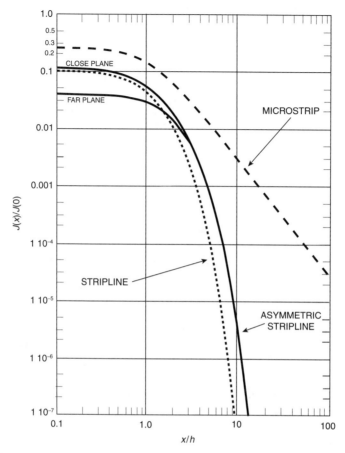

**FIGURE 17-13.** Log-log plot of normalized ground plane current density for microstrip, stripline, and asymmetric stripline, versus $x/h$. The asymmetric stripline plot is for the case where $h_2 = 2\,h_1$.

interesting to note that for asymmetrical stripline, the current at a distance from the centerline greater than $x/h = 3$ is the same in both planes.

### 17.6.3  Isolated Analog and Digital Ground Planes

A second approach to the high-resolution converter problem is to divide the board into separate isolated analog and digital ground plane regions, all still solidly connected to the digital ground plane under each of the A/D converter as shown in Fig 17-14. This approach provides additional ground noise isolation for high-resolution A/D converters while maintaining a single contiguous ground plane for the system. For the layout shown in Fig. 17-14, the

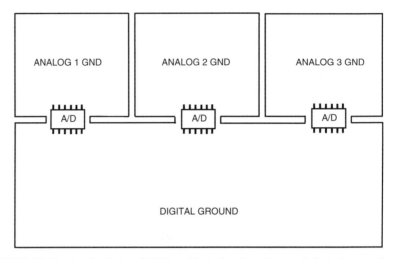

**FIGURE 17-14.** A mixed-signal PCB, with isolated analog and digital ground plane regions. This approach will provide additional noise isolation for high-resolution A/D converters while maintaining a single contiguous ground plane.

digital ground-plane current cannot flow in the analog portion of the ground, because no current loop is possible between the digital and analog planes.

Notice that even in this case, the analog and digital ground planes are not split; they are all connected together to form a single contiguous ground plane. Also remember that when using this approach no traces, on any layer, can cross over the isolating slots in the plane.

## 17.7   A/D AND D/A CONVERTER SUPPORT CIRCUITRY

Often in a mixed-signal PCB design, considerable effort is expended on the layout of the PCB to control the ground noise, whereas other aspects of the design that equally affect the performance are ignored.

### 17.7.1   Sampling Clocks

In a high-precision sampled data system, a low-jitter, noise-free, sampling clock is essential. Any jitter in the sampling clock varies the point in time at which the signal waveform is sampled, which produces an amplitude error in the sampled signal. This error is equal to the clock jitter times the rate of change of the signal being sampled. This is shown in Fig. 17-15. If a waveform is sampled at irregular intervals, as the result of clock jitter, then the original waveform cannot be correctly reconstructed after the fact.

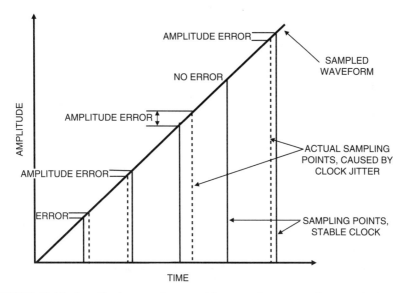

**FIGURE 17-15.** Amplitude error is created by sampling a waveform at the incorrect time, as a result of sampling clock jitter.

When the analog signal is reconstructed in a D/A converter with a stable clock, the waveform will be in error (see Fig. 17-16), as the result of the signal having been sampled at the wrong point in time; therefore, the samples represent incorrect amplitudes. Clock jitter effectively raises the noise floor of the system similar to the effect of the ground noise discussed previously.

The effect of sampling clock jitter on an A/D converter's signal-to-noise ratio (SNR) is given by the following equation (Kester, 1997; Nunn, 2005):*

$$ \text{SNR} = 20 \log\left(\frac{1}{2\pi f t_j}\right). \tag{17-1} $$

The SNR in Eq. 17-1 is expressed in decibels and is that for a perfect A/D converter (no additional noise sources other than the clock jitter), with a sine wave input of frequency $f$, which has a sampling clock with a root mean square (rms) jitter of $t_j$.

For example, a perfect A/D converter with a 100-MHz input signal and a sampling clock with 1 ps of rms jitter would have an SNR of no better than 64 dB. Comparing this result with the converter's dynamic range data listed in Table 17-2, one can conclude that a converter that operates under the above conditions cannot achieve a resolution greater than 10 bits. This result is

---

* Note that there is a typo in Nunn's paper, and the equation is incorrect as printed there.

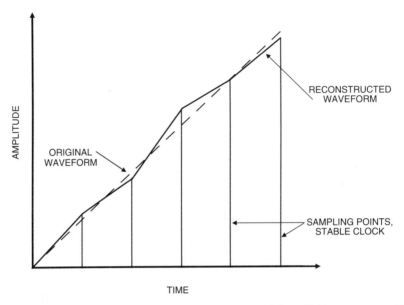

**FIGURE 17-16.** Waveshape of Fig. 17-15 reconstructed by a D/A converter with a stable clock. The solid line is the reconstructed waveform, and the dashed line is the original waveform.

startling! A 10-ps clock jitter would limit the converter resolution to 8 bits. To resolve the LSB of a 20-bit A/D converter sampling a 100-MHz signal, the clock would have to have a jitter less than 1.5 femtoseconds.* Therefore, we conclude that clock jitter can be a serious problem when sampling high-frequency signals.

Even when sampling a relatively low-frequency signal, clock jitter can be a major source of noise. Sampling a 1-MHz signal in a 20-bit A/D converter with a sampling clock having 5 ps of rms jitter limits the SNR to 90 dB. This means that the 20-bit converter cannot achieve a resolution greater than 14 bits. Even when sampling a 500-kHz signal, a converter under these conditions could not achieve a resolution greater than 16 bits.

As can be concluded from the above examples, in many cases, the error caused by the sampling clock jitter is far greater than that produced by any ground noise voltage that might exist.

### 17.7.2   Mixed-Signal Support Circuitry

Proper layout and grounding of the mixed-signal IC support circuitry is crucial to the noise performance of the system. To minimize loading and output current, each digital output of a mixed-signal device should feed only one load.

---

*That is $1.5 \times 10^{-15}$ s.

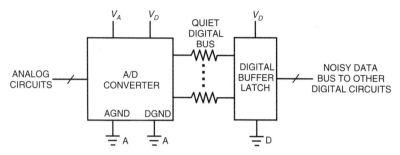

**FIGURE 17-17.** A/D converter with digital output buffers and/or resistors, used to minimize loading on the output drivers and minimize digital noise fed back to the converter.

In high-resolution converters, it is also a good idea to connect the digital output to an intermediate buffer register, which is located adjacent to the converter, to isolate it from the noisy digital data bus as shown in Fig. 17-17. The buffer serves to minimize loading (minimizing the required output current) on the converter's digital outputs and prevents the system data bus from coupling noise back into the converter's analog input through the converter's stray internal capacitance. In addition, series output buffer resistors (100 to 500 $\Omega$) could also be used in addition to, or in place of, the buffer to minimize the loading on the digital drivers, which reduces the transient currents in the converter's output.

Any ground noise between the converter and an external reference voltage or sampling clock can affect the converter's performance. Therefore, the reference voltage and sampling clock should both be referenced to the analog ground plane, not the digital ground plane. The A/D converter sampling clock should be located in the analog section of the PCB, and it should be isolated from the noisy digital circuits, and be grounded and decoupled to the analog ground.

If the sampling clock must be located on the digital ground plane, possibly because it is derived from the digital system clock, it should be transmitted to the A/D or D/A converter differentially, or through a transformer to reject any noise voltage between the two grounds.

Figure 17-18 shows a properly grounded mixed-signal converter and its support circuitry, where *A* represents analog ground and *D* represents digital ground.

## 17.8   VERTICAL ISOLATION

The discussion so far involved partitioning or separating the analog and digital circuitry in the plane of the board, which is often referred to as horizontal isolation. It is also possible to isolate or separate the analog and digital circuitry in the vertical, or Z-axis, of the board. For example, in the case of a double-sided

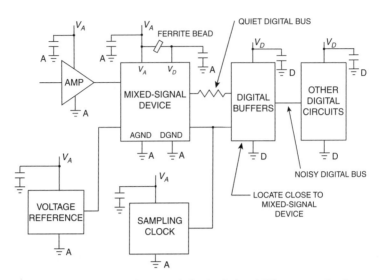

**FIGURE 17-18.** Properly grounded mixed-signal IC support circuitry.

surface-mount board, the digital circuits could be routed on the top layers of the board and the analog circuits routed on the bottom layers of the board. The primary reason for using this approach is to reduce the size of the product. For example, this approach is used commonly in cell phones. The analog components on one side of the PCB should be isolated from the digital components on the other side of the board with a plane (ground or power).

Some limitations of the vertical isolation approach include the following: The digital power supply can induce noise into the analog signal traces through the vias where the signals change layers. The magnitude of this coupling will depend on the effectiveness of the digital power supply decoupling and on the location of the vias involved. Through-vias will also provide coupling between the analog and the digital layers. The portion of the via that extends into the opposite partition of the board will act as a small stub antenna and can pick up, or radiate, high-frequency energy. This action is often referred to as Z-axis coupling (King, 2004). The power and ground plane antipads (clearance holes) will also provide some leakage (coupling) at high frequencies. These items can all be minimized or eliminated by the use of blind and buried vias.* A typical eight-layer, vertically isolated, mixed-signal PCB stackup is shown in Fig. 17-19.

Cell phones are good examples of the effective use of vertical isolation on a mixed-signal PCB. They are sold in the consumer market in large quantities where price and small size are important. The analog circuitry is on one side of the board, and the digital circuitry is on the other side of a typically eight-layer

---

*A blind via is one that does not go all the way through the board. A buried via is one that interconnects inner layers and does not extend to either outer surface of the board.

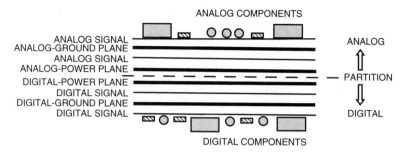

**FIGURE 17-19.** A typical stackup for a vertically isolated mixed-signal PCB.

board. Micro via technology, with blind and buried vias, is used to minimize coupling and reduce the size of the product.

## 17.9 MIXED-SIGNAL POWER DISTRIBUTION

### 17.9.1 Power Distribution

Mixed-signal boards may use separate power for the analog and digital circuits. This often results in a split-power plane. A split-power plane is acceptable, provided the routing is such that no traces on any layer adjacent to the power plane cross over the split.* Any traces that cross over the power-plane split must be routed on a layer adjacent to a solid ground plane. In many circumstances, a split-power plane can be avoided by routing some of the power, usually the analog power, as a trace on a signal layer rather than as a plane.

Analog power can be obtained by several different methods, including the following:

- A separate power supply
- A voltage regulator, off of the digital power
- A filter, off of the digital power

If a voltage regulator is used, a linear regulator is preferred to a switching regulator, especially in the case of a high-resolution converter, because it will provide an output without any additional switching noise. If a filter is used, it can consist of a single filter off the digital power to supply all the analog circuitry, or individual filters at each IC between the analog and digital power pins, or in some cases both.

---

* Because the power plane will be the return current path for the signal current on the adjacent trace layer, the return current path will be interrupted if the plane is split.

Because the digital circuitry draws the largest magnitude currents and has the largest current transients, the power connector should be located in the digital partition of the board. Power can then be fed directly to the digital circuits and then filtered or regulated to power the analog circuits, as shown in Fig. 17-20.

The analog power regulator or filter should be placed to straddle the partition between the analog and digital sections of the board, which is similar to the A/D converters.

### 17.9.2 Decoupling

For most mixed-signal ICs, the digital supply should be decoupled to the analog ground. The digital decoupling capacitor, however, must be connected *directly* to the *DGND* pin of the IC, to minimize digital currents in the analog ground plane. This is shown in Fig. 17-21.

The digital power pin of the mixed-signal IC may be powered from the digital or analog supply. In either case, it should be isolated with a small impedance such as a ferrite bead or resistor (*Z* in Fig. 17-21). If the analog supply is used, this isolation helps keep the digital noise out of the analog supply. If the digital supply is used, this helps to keep the digital supply noise out of the converter's digital circuitry and decoupling capacitor.

Large digital signal processor (DSP) ICs with microprocessors and codecs, which contain a large amount of digital circuitry, are the exception to the above grounding scheme. They usually should have *AGND* and *DGND* separately grounded to their respective ground planes. These chips are usually designed with good noise isolation between the analog and digital circuitry. Check the IC data sheet for recommendations.

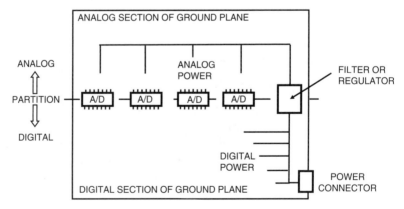

NOTE: KEEP THE LARGE POWER CURRENTS IN THE DIGITAL PARTITION OF THE BOARD.

**FIGURE 17-20.** Powering of a properly partitioned PCB from a single power supply.

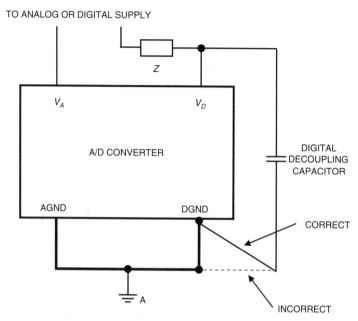

**FIGURE 17-21.** Proper decoupling of the digital supply on a mixed-signal IC. The decoupling capacitor must be connected directly to the *DGND* pin of the mixed-signal IC.

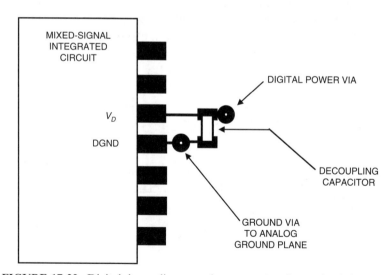

**FIGURE 17-22.** Digital decoupling capacitor mounting for a mixed-signal IC.

Figure 17-22 shows one acceptable layout for the digital decoupling capacitor on a mixed-signal IC. As shown the *DGND* pin has a via directly connected to the analog ground plane, which minimized the impedance between the *AGND* and *DGND* pins. The ground end of the decoupling capacitor is connected with a trace, as short as possible, directly to this via and hence the *DGND* pin. This layout keeps the transient decoupling capacitor ground currents out of the analog ground plane.

## 17.10   THE IPC PROBLEM

Industrial process control (IPC) equipment presents the mixed-signal designer with a slightly different problem than the one we have been discussing. Here, it is often the case that the noisy analog circuits that consist of motor, relay, and solenoid drivers interfere with the digital or low-level analog circuits. Because the motor, relay, and solenoid currents are low-frequency signals, their return currents will not take the path of least inductance and flow under the signal trace, but rather the currents will take the path of least resistance. All principles discussed so far are still applicable, but the application of them will be slightly different.

Following are some methods of dealing with the IPC problem:

- Use a return trace, not a plane, for the noisy analog signal (works well).
- Segment the ground plane and use a single bridge for traces that must cross over.
- Route noisy analog traces such that the return path of least resistance (direct path) does not pass through the digital or analog portion of the board.
- Split ground planes and use opto-isolators, transformers, or magnetoresistive isolators for signals that must pass over the split in the ground plane.

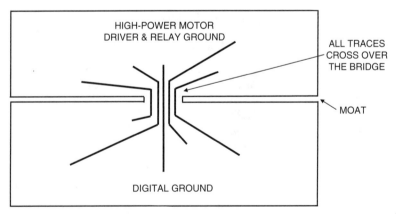

**FIGURE 17-23.** Example of a mixed-signal IPC board using a single bridge to connect the noisy analog ground to the digital ground, No traces can be routed over the split in the ground plane; all traces must be routed across the bridge.

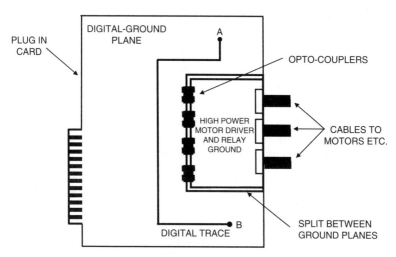

**FIGURE 17-24.** An example of a mixed-signal IPC board with a split ground plane and optocouplers to communicate across the split.

Figure 17-23 shows an example of the use of a segmented ground plane with a single bridge for signals that cross from one side to the other. This approach works well as long as no traces are routed across the gap, or moat, in the ground plane.

Figure 17-24 shows an example of a mixed-signal board using a split-ground plane approach with optocouplers used to communicate across the split. Note that the digital trace shown going from points $A$ and $B$ must be routed over the digital ground plane and not run directly from $A$ to $B$—in which case, it would cross the split in the ground plane.

## SUMMARY

- Partition mixed-signal PCBs with separate analog and digital sections.
- Straddle the partition with the A/D or D/A converters.
- Do not split the ground plane; use one solid ground plane under both analog and digital sections of the board.
- If for some reason ground or power planes are split, do not run any traces across the split on an adjacent layer.
- Traces that must go over a power plane split must be on a layer adjacent to a solid ground plane.
- Route digital signals only in the digital section of the board; this rule applies to all layers.
- Route analog signals only in the analog section of the board; this rule applies to all layers.

- Think about where and how the ground return currents are actually flowing.
- The key to a successful mixed-signal PCB layout is a single ground plane, proper partitioning, and routing discipline.
- Use stripline or asymmetric stripline for the digital logic signals on high resolution converter (greater than 18-bits) boards.
- For high-resolution A/D converters, you may want to divide the board into separate isolated analog and digital ground regions, each solidly connected to the digital ground under each A/D or D/A converter.
- A/D and D/A converters, as well as most other mixed-signal ICs, should be considered analog devices with a digital section, not digital devices with an analog section.
- For multiboard mixed-signal systems, the A/D and/or D/A converter should be mounted on the analog board.
- The designation *AGND* and *DGND* on the pins of a mixed-signal IC refers to where the pins are connected internally, and it does not imply where or how they should be connected externally.
- On most mixed-signal ICs, both the *AGND* and *DGND* pins should be connected to the analog-ground plane.
- On mixed-signal ICs, the digital decoupling capacitor should be connected directly to the digital ground pin.
- Most mixed-signal systems can be laid out successfully with a single contiguous ground plane.
- In high-resolution sampled data systems, a low jitter sampling clock is essential.
- Proper mixed-signal support circuitry layout and grounding is also essential for optimum system performance.

## PROBLEMS

17.1  What are the keys to a successful mixed-signal PCB layout?

17.2  Ninety-nine percent of the return current will be within what distance, on either side, of a digital logic trace that is located 0.010 in from a reference plane (or planes)?
   a. When the trace is a microstrip line.
   b. When the trace is a stripline.

17.3  What two things should be avoided if one splits a ground plane on a mixed-signal PCB?

17.4  How many reference planes are needed for a mixed-signal PCB?

17.5  What are two ways of dealing with high-resolution A/D converters?

17.6  What do the labels *AGND* and *DGND* on the pins of an A/D or D/A converter refer to?

17.7  Should most mixed-signal ICs be considered digital devices or analog devices?

17.8  What is the maximum SNR and resolution (in bits) obtainable by an ideal A/D converter with a sampling clock with an rms jitter of 300 femtoseconds, or less when sampling a 60-MHz input signal?

17.9  What is the maximum allowable rms sampling clock jitter for an A/D converter to achieve a resolution of 24 bits when sampling a 1-MHz signal?

17.10  An audio signal is band-limited to 50 kHz and then sampled with a 16-bit A/D converter. What is the maximum allowable clock jitter to resolve the LSB?

17.11  Where should the *AGND* and *DGND* pins of most mixed-signal devices be connected?

17.12  What is the exception to the answer to Problem 17-11?

17.13  On a mixed-signal IC, where should the ground side of the digital decoupling capacitor be connected?

## REFERENCES

Kester, W. "A Grounding Philosophy For Mixed-Signal Systems." *Electronic Design, Analog Applications Issue*, June 23, 1997.

King, W. M. "Digital Common-Mode Noise: Coupling Mechanisms and Transfers in the Z-Axis." *EDN*, September 2, 2004.

Nunn, P. "Reference-Clock Generation for Sampled Data Systems." *High Frequency Electronics,* September 2005.

Terrell, D. L. and Keenan, R. K. *Digital Design for Interference Specifications*, 2nd ed. Pinellas Park, FL, The Keenan Corporation, 1997.

## FURTHER READING

Holloway, C. L. and Kuester, E. F. Closed-Form Expressions for the Current Densities on the Ground Plane of Asymmetric Stripline Structures. *IEEE Transactions on Electromagnetic Compatibility*, February, 2007.

Johnson, H. "ADC Grounding." *EDN*, December. 7, 2000.

Johnson, H. "Multiple ADC Grounding." *EDN*, February. 2001.

Johnson, H. "Clean Power." *EDN*, August. 2000.

Kester, W. *Mixed-Signal and DSP Design Techniques*, Chapter 10, (Hardware Design Techniques). Norwood, MA, Analog Devices, 2003.

Kester W. *Ask the Application Engineer—12, Grounding (Again)*, Analog Dialogue 26-2, Norwood, MA, Analog Devices, 1992.

Ott, H. W. "Partitioning and Layout of a Mixed-Signal PCB," *Printed Circuit Design*, June 2001.

# 18 Precompliance EMC Measurements

Much has been written about "legitimate" electromagnetic compatibility (EMC) measurements, that is, compliance measurements performed according to various EMC regulations using expensive and complex test facilities, such as an open area test site (OATS) or a large semianechoic chamber costing. In some cases, millions of dollars. However, little has been written about simple EMC tests that can be performed in the product development laboratory, with inexpensive equipment that will provide a good indication of the EMC performance of a product.

This chapter is about EMC measurements that can be made in the product development laboratory. These simple measurements require limited, relatively inexpensive equipment. Everything needed, including a spectrum analyzer, is available for less than a total of $30,000 [U.S. dollars (USD) as of 2008]. If you already have a spectrum analyzer, then a $15,000 budget would be adequate. The big advantage is that these tests can be done early in the design in the product designers' own laboratory. Although not as accurate as legitimate EMC measurements performed in a controlled environment test facility, the fact that they are simple, quick, and can be performed easily at your workbench, more than compensates for the accuracy degradation, especially when performed early in the design. I like to refer to these tests as "workbench EMC measurements."

The advantages of early EMC testing during the design phase of a product include the following:

- Increasing the probability of passing the final compliance test
- Minimizing the number of retests required for compliance at an EMC test laboratory
- Eliminating surprises late in the design (caused by EMC failures)
- Ensuring that EMC considerations are part of the design—not add ons

Data from EMC test laboratories indicate that 85% of the products submitted for final compliance testing fail the first time. By using the simple

*Electromagnetic Compatibility Engineering,* by Henry W. Ott
Copyright © 2009 John Wiley & Sons, Inc.

workbench EMC measurements described in this chapter, that statistic can be reversed such that 85% or more of the products pass regulatory compliance tests the first time.

## 18.1  TEST ENVIRONMENT

Radiated emission test facilities are carefully designed and constructed to control reflections. The objective is to have only one reflective surface, and that is the ground plane. An OATS does this by locating the facility in an open area with no metallic objects nearby. The one reflective surface consists of a metallic ground plane located at the site. A large semianechoic chamber,* with a 3- or 10-m measuring distance, accomplishes the same objective by having a metallic ground plane (the chamber floor) and using radio frequency (rf) absorber material (carbon loaded pyramidal cones and/or ferrite tiles) on the walls and ceiling to absorb rf energy and prevent reflections.

The workbench EMC measurement environment (your design laboratory) is just the opposite from that described above. It typically has many uncontrolled reflective surfaces, such as metal file cabinets, metal desks and lab benches, as well as possibly metal walls. Therefore, you do not want to perform a radiated emission test in this uncontrolled environment; rather, you need to measure some parameter that is proportional to the radiated emission, not the radiated emission directly.

What you definitely do not want to do is build a small-shielded room (nonabsorber loaded), place your product and a receiving antenna inside the room, and attempt to measure the radiated emissions. This approach maximizes the errors associated with such a test. The large reflections from the walls and ceiling will produce nulls and peaks in the radiated emission pattern, which can produce errors as large as $\pm 40$ dB (Cruz and Larsen, 1986).

Useful workbench EMC measurements must be made such that they are not affected (or at least minimally affected) by the uncontrolled environment in which the tests are being performed.

## 18.2  ANTENNAS VERSUS PROBES

We will not, with one exception (see Section 18.9), use antennas as part of our precompliance tests. Antennas are large (usually a significant fraction of a

---

* Radiated emission measurements in a fully anechoic chamber (no reflective surface on the floor), possibly at a 5-m measuring distance, are being seriously considered by the international EMC measurement standards committees. However, one possible problem with this methodology is that the limits would have to be changed, which would require new studies into the interference potential of products, because the test methods and limits are interrelated.

**FIGURE 18-1.** A common-mode current clamp (courtesy of Fischer Custom Communications, Inc.).

wavelength) in size, are sensitive to nearby reflections, and interact with surrounding metal objects.

Rather, we will use small probes that are much smaller than a wavelength, can be used close to surrounding metal objects, and are very insensitive to reflected rf energy. The probes that we use will be a few inches or smaller in size, compared with antennas, which have dimensions of many feet. For example, at 30 MHz, a tuned dipole antenna is 16.4 ft (5 m) long.

## 18.3 COMMON-MODE CURRENTS ON CABLES

By far, the most useful precompliance measurement that you can make is to measure the common-mode currents on all the cables attached to your product.

The radiation from a cable is directly proportional to the common-mode current on that cable as indicated by Eq. 12-7. The common-mode current is the unbalanced current (current not returned) on the cable. If this current is not returned on the cable, where does it go? Into radiation, that is where! In the case of intentional signals (differential-mode signals), the current flows down one wire of the cable and returns on an adjacent wire; hence, the net current is zero and the common-mode radiation is eliminated.

Because cables are always a major source of product radiation, measuring the common-mode current is one of the most useful tests that you can learn to do. The common-mode current can easily be measured with a high-frequency, clamp-on current probe (such as the Fischer Custom Communications Model F-33-1 shown in Fig. 18-1) and a spectrum analyzer. The current probe is about 2 3/4 in in diameter with a 1-in hole in the center for the cable. The test setup is

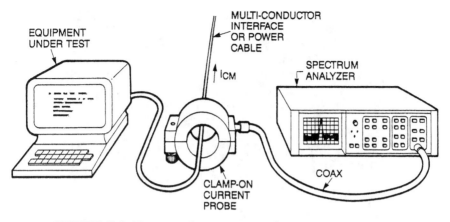

**FIGURE 18-2.** Test setup for common-mode current measurement.

shown in Fig. 18-2. The F-33-1 current probe has a flat frequency response from 2 to 250 MHz. The transfer impedance* of the current probe is 5 Ω ( + 14 dBΩ); therefore, a 1-μA current will produce a 5-μV output from the current probe.

Most common-mode cable radiation occurs below 250 MHz, so the bandwidth of the F33-1 probe is usually sufficient. However, if you want or need additional bandwidth, the F-61 common mode current clamp is useful up to 1 GHz, and it is more sensitive than the F-33-1 probe. Its frequency response is between ±1 dB from 40 MHz to 1 GHz. The transfer impedance of the F-61 probe is 18 Ω ( + 25 dBΩ). Therefore, a 1-μA current will produce an 18-μV output from the probe.

Make it a habit to measure the common-mode currents on *all* your cables! Do it early in the development process, on prototype models, while it is still easy to make a change to the design, and prior to performing final EMC compliance testing. If you fail the common-mode current test, then you will also fail the radiated emission test.

For a Class B product, the current must be less than 5 μA (15 μA for a Class A product). Use the above limits for cables that are 1 m long or longer. For cables shorter than 1 m, the allowable current is inversely proportional to the cable length. For example, for a $\frac{1}{2}$ -m- long cable, the maximum current would be 10 μA for a Class B product (30 μA for a Class A product).

---

* The transfer impedance $Z$ is the ratio of the probe output voltage $V$ to the current I in the cable being measured. Thus, $Z = V / I$ (in Ω) or $Z = 20 \log (V / I)$ (in dBΩ). The larger the transfer impedance, the more sensitive the probe.

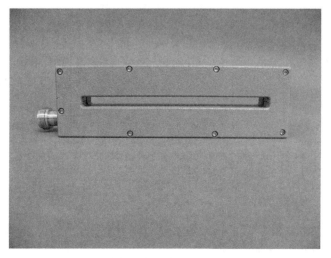

**FIGURE 18-3.** A common-mode current clamp designed for use with ribbon cables (courtesy of Fischer Custom Communications, Inc.).

Five microamps is equivalent to 14 dBμA. Using the F-33-1 probe, the output voltage for a current of 5 μA can be obtained by adding the current (in dB) to the probe transfer impedance (in dB) which gives the following:

$$V = 14 \text{ dB}\mu\text{A} + 14 \text{ dB}\Omega = 28 \text{ dB}\mu\text{V}. \tag{18-1}$$

Therefore, when measuring the current on a cable 1 m long or longer, the probe output voltage, read on a spectrum analyzer, should be less than 28 dBμV for the cable emissions to pass the Class B limit. For a Class A product, the voltage reading should be less than 38 dBμV.

This technique works equally well on shielded or unshielded cables. By the way, this technique is an excellent way to determine the effectiveness of your cable shield termination. If you use common-mode filters on your cables or ferrite cores to suppress common-mode radiation, the current probe measurement will indicate their effectiveness. Just measure the current before and after inserting the filter (or ferrite), or as you vary the way that the cable shield is terminated.

Special clamp-on current probes are also available for ribbon cables (see Fig. 18-3); however, in most cases, they are not needed. Just fold the ribbon cable so that it fits in the round hole of the standard current clamp shown in Fig. 18-1. This folding of the cable will have no effect on the common-mode current.

All cables should be measured regardless of their intended purpose. Measure the signal cables, the power cord [alternating current (ac) or direct current (dc)], fiber-optic cables, video monitor cables, input/output (I/O) cables, telecom cables, and any other cables that are attached to the product. If it is connected to the product, then it can be a source of common-mode radiation!

### 18.3.1   Test Procedure

What you want to do is reduce the common-mode current on each and every cable to below 5 µA (15 µA for a Class A product). The cables, however, may interact with each other. If you reduce the common-mode current on one cable, then it may increase on another cable.

Measure one cable at a time with the common-mode current clamp. Use common-mode filters, ferrite chokes, cable shields, and so on to reduce the current below the required limit; then go on to the next cable and do the same. When you get through all the cables, start over again, because the currents may have increased on some of the previously fixed cables. Keep this iterative process up until the currents on all the cables are under the limit. You may have to go through the process two or three times on each cable. When you are finished, the current on each cable should be under 5µA (15µA for a Class A device), and the cables should no longer present a radiated emission problem.

### 18.3.2   Cautions

Cables radiate energy coupled to them from the product as well as pickup energy from external sources such as the local FM or TV broadcast station. All measurements must, therefore, be validated to ensure that you are measuring what you think that you are measuring.* A simple validation test for the common-mode current measurement is to turn the product off and determine whether the signal read zero. If it remains, it is due to external pickup. FM radio stations commonly are picked up this way, so any signal in the 88 to 108 MHz (in the United States) band should be suspect, as possibly being from an external source. If you always perform the tests at the same location, you learn the frequencies of all the local FM stations.

Standing waves can exist on the cables under test. Under these conditions, moving the current probe along the cable will detect the location of the maximum current. At the frequencies above 30 MHz, you only have to move the current probe about a meter (an arms length) to detect the maximum. If 'the cable is 30 m long, you do not have to slide the current clamp along the full length of the cable. In most cases, the maximum will occur with the probe close to the product under test.

Of all the various types of EMC measurements possible, the common-mode current measurement is the most useful and will provide the biggest payback in terms of dollars saved and time spent trying to achieve compliance with EMC regulations—*learn to do it, and do it often!*

---

* This is often referred to as a "null experiment," that is a test that should produce a zero result, and it is used to validate the instrumentation test setup.

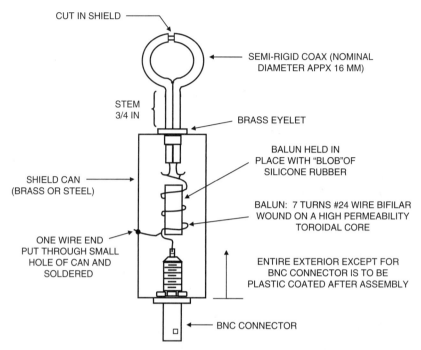

CUT IN SHIELD

SEMI-RIGID COAX (NOMINAL
DIAMETER APPX 16 MM)

STEM
3/4 IN

BRASS EYELET

BALUN HELD IN
PLACE WITH "BLOB"OF
SILICONE RUBBER

SHIELD CAN
(BRASS OR STEEL)

BALUN: 7 TURNS #24 WIRE BIFILAR
WOUND ON A HIGH PERMEABILITY
TOROIDAL CORE

ONE WIRE END
PUT THROUGH SMALL
HOLE OF CAN AND
SOLDERED

ENTIRE EXTERIOR EXCEPT FOR
BNC CONNECTOR IS TO BE
PLASTIC COATED AFTER ASSEMBLY

BNC CONNECTOR

**FIGURE 18-4.** Construction details for a shielded-loop magnetic field probe.

## 18.4   NEAR FIELD MEASUREMENTS

As indicated in Chapter 12, emissions from digital electronics can occur as either common- or differential-mode radiation. The above cable current measurement provides information about the common-mode radiation mechanism. What is needed next is some information on the differential-mode radiation mechanism. Differential-mode radiation is the result of currents flowing around loops on the printed circuit board (PCB). These current loops act as small-loop antennas that radiate magnetic fields. What you can do, therefore, is to measure the magnetic field close to the printed circuit board. This can be done with a small, shielded-loop probe similar to that shown in Fig. 2-29C.

Scott Roleson described the design and construction of such a loop probe in a 1984 EDN article (Roleson, 1984). The construction of this magnetic field probe is shown in Fig. 18-4. A commercial version of the probe described in the article is available from Fischer Custom Communications as the Model F-301 Shielded Loop Sensor (see Fig. 18-5). The loop is about 3/4 in in diameter and is useful over the frequency range of 10 to 500 MHz.

Although the probe can be calibrated in terms of the magnetic field, the results of the near-field measurement cannot be extrapolated to the far field to

**FIGURE 18-5.** Small split-shield magnetic field loop probe.

determine the magnitude of the radiation. For additional information on near and far fields, see Section 6.1. Therefore, measurements made with this probe are qualitative not quantitative. However, these measurements are still useful in indicating excessive sources of differential-mode radiation as well as in performing A and B comparisons, when a change or modification is made to the product.

As an alternative to the above magnetic field probe, a simple homemade probe can be constructed from a 50-$\Omega$ coaxial cable. One end of the cable should be formed into a $^3/_4$- to 1-in diameter loop with the cable center conductor soldered to the shield as shown in Fig. 18-6. The shield should be left unterminated at this end. Although not as good at rejecting electric fields as the "Roleson probe," its performance is usually adequate for most precompliance applications. The advantage is that this probe is inexpensive and easy to construct. Commercial probes using this construction are also available. For example, the EMCO (ETS Lindgren) 7405 near-field probe set or the A.R.A Technologies HFP-7410 probe set; both sets include both magnetic and electric field probes.*

### 18.4.1   Test Procedure

Scan the probe over the printed circuit board looking for "hot spots" (locations with strong magnetic fields). These fields will usually occur at frequencies that are harmonics of the clock.

---

* Electric field probes are usually not needed, or used, for precompliance testing.

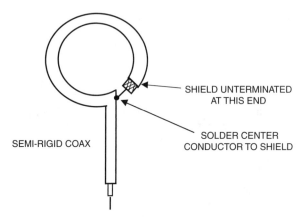

**FIGURE 18-6.** Simple magnetic field probe made from a coaxial cable.

When a hot spot is found, check the printed circuit board layout in that vicinity for violations of the good EMC practices discussed throughout this book. After making changes to the board, retest to confirm that the field has decreased in amplitude.

Many times you will find that it is the integrated circuit (IC) that is causing most of the emission. In this case, you should consider using a board-level shield over the component.

The probe can be held vertical or horizontal when making these tests. If the probe is vertical (plane of the loop perpendicular to the board), it will have to be rotated 0° to 90° to detect the maximum field strength. If the probe is used horizontal (plane of loop parallel to the board), it does not have to be rotated. I usually prefer to use the probe vertical, because it is easier to get between and around the various height components on a board.

### 18.4.2    Cautions

When comparing two magnetic field readings (such as before and after a fix), the probe must be held at the exact same distance from the board. This is extremely important because, in the near field, the magnitude of the magnetic field strength falls off as the cube of the distance from the source. The easiest distance to remember and repeat is zero. Therefore, I hold the probe vertical and touch it to the board or component (IC, etc.). This distance also gives the largest amplitude reading.

Because small-shielded magnetic field probes are insensitive to external fields, you can usually assume that what you observe is coming from the product under test. However, this assumption can be validated by moving the probe away from the board and verifying that the magnitude of the reading drops significantly, because the field strength decays as the cube of the distance from

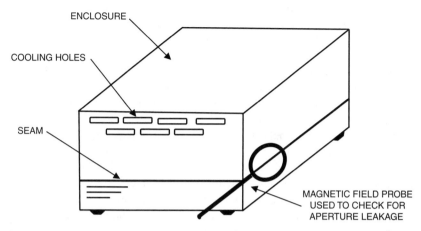

**FIGURE 18-7.** Using a magnetic field probe to check for leakage at the apertures of an enclosure.

the source. Of course another option is to turn the power OFF to the circuit under test and confirm that the reading disappears.

### 18.4.3  Seams and Apertures in Enclosures

A small-shielded magnetic field probe, as described above, can also be used to search for leakage at the seams and apertures of shielded enclosures.

Place the probe next to the shield with the plane of the loop parallel to the shield as shown in Fig. 18-7. Move the probe along the seam or aperture and search for the presence of a strong magnetic field.

In the case of an aperture, the maximum magnetic field strength will occur at the ends of the aperture with a null in the center. This is because the induced-shield currents (the currents produce the magnetic field) are maximum at the ends of the aperture, as was shown in Fig. 6-25.

The above two EMC precompliance measurements (common-mode current and magnetic field near the PCB) are the most important and are the ones that everyone should be performing on early prototype models and again before final compliance testing at a legitimate EMC test facility.

## 18.5  NOISE VOLTAGE MEASUREMENTS

Noise voltage measurements can be helpful in pinpointing the source of radiated emissions. Many of these measurements can be made with a standard X10 unbalanced scope probe and a high-frequency oscilloscope.

Some useful measurements to perform are clock waveforms, $V_{cc}$-to-ground noise voltage on ICs, ground differential noise voltage on PCBs, and common-mode voltage from I/O-ground to the chassis.

Clock waveform measurements are useful to detect ringing as well as, undershoot and/or overshoot on the waveform (see Fig. 10-2A). If any of these conditions exist, they usually can be eliminated by inserting a small resistor or ferrite bead in series with the signal output to damp out the transient (see Fig. 10-2B). A value of 33 Ω is usually a good starting value. Then, vary the value until you find the smallest value that will eliminate the problem.

$V_{cc}$-to-ground noise measurements are useful to determine the effectiveness of the power supply decoupling. When viewed on an oscilloscope, a noise spike will appear whenever the signal switches. These measurements can also be performed by using a spectrum analyzer as the readout device, instead of an oscilloscope. This provides additional information on the decoupling effectiveness versus frequency, and it can be very enlightening.

Measuring the common-mode noise voltage between the PCB ground in the I/O area (where the cables terminate on the PCB) and the chassis will indicate the common-mode noise voltage that excites the cables and causes them to radiate. The common-mode cable current measurement (discussed previously) is a similar, and a better, test for cable radiation; however, this test provides an additional option.

### 18.5.1   Balanced Differential Probe

Measuring the differential noise voltage between the ground pins of different ICs will provide an indication of the quality of the PCBs ground structure (see Table 10-2). Because you are measuring between two points on the ground, this measurement should be performed using a balanced differential probe rather than a standard unbalanced probe (see Section E.4 in Appendix E).

Most instrument manufacture's make high-impedance active [field-effect transistor (FET) input] differential probes for their equipment. The problem with these probes is that they are often very expensive, fragile (easily damaged as a result of their FET input, if an overvoltage is applied), and have a limited bandwidth. These problems can all be overcome if one realizes that high-input impedance is not required for ground-to-ground noise voltage measurements. Input impedances of a few hundred to a thousand ohms are more than adequate in this application. Figure 18-8 shows the construction details for a homemade X10, balanced, passive differential-probe that works well in this application and has a 1000-Ω input impedance (Smith, 1993). This probe is actually an application of the principle of the use of coaxial cables in a balanced circuit as was shown in Fig. 4-7. Note that the cable shield should not be connected to the circuit at the probe tip end.

When using an oscilloscope with this balanced probe, the inputs are fed to two channels of the scope, one of them inverted and the two inputs added. If

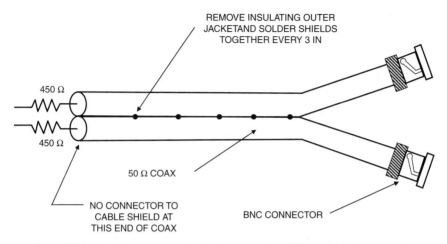

**FIGURE 18-8.** Construction details for a passive differential voltage probe.

this passive differential probe is used with a spectrum analyzer, however, only one input is available. Therefore, the subtraction of the two signals must be done before applying the signal to the spectrum analyzer. This can easily be accomplished with a high-frequency 180° combiner (such as a Mini-Circuits Model ZFSCJ-2-1). This combiner has a frequency response from 1 to 500 MHz, and it has about 4 dB of insertion loss that has to be accounted for when measuring the voltage. Figure 18-9 shows such a homemade passive X10

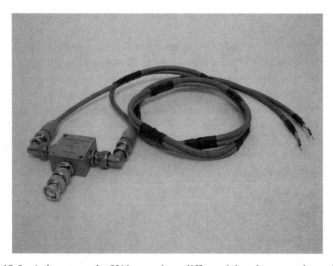

**FIGURE 18-9.** A home-made X10, passive, differential voltage probe, with a 180° combiner.

differential probe, with the 180° combiner. This probe works well up to 500 MHz and can be modified to work up to a gigahertz if necessary. The gigahertz version, however, becomes a X20 probe.

If only one of the two cables of the balanced probe is made, it can be used as an unbalanced dc to 500-MHz probe with a 500-Ω input impedance.

### 18.5.2   DC to 1-GHZ Probe

What limits the frequency response of the above probe to 500 MHz, in addition to the combiner, is the fact that the input impedance of the oscilloscope is not truly 50 Ω; rather it is 50 Ω in parallel with the scope's input capacitance (typically 3 to 10 pF). Above 500 MHz, this capacitance causes a reflection to occur on the probe's cable, which reflects a signal back to the probe tip. Because the probe tip resistance is 450 Ω, not 50 Ω, the signal reflects again. This reflected wave, bouncing back and forth on the cable, will produce standing waves and will cause the measurement to be in error. The solution is to place a 50-Ω termination at the probe tip (between the center conductor and the shield) on the cable side of the 450-Ω resistor. To reduce any inductance in series with this shunt resistor, the termination is usually constructed from four 200-Ω resistors in parallel. Doug Smith describes the construction detail for a similar unbalanced probe on his website (Smith, 2004). For the balanced differential probe, use a 450-Ω tip resistor instead of the 976-Ω resistor suggested in the construction detail, leave the ground lead off, and modify both tips of the balanced probe as described. For this balanced probe configuration, a wider bandwidth combiner, than discussed above, will also have to be used if the probe is connected to a spectrum analyzer.

### 18.5.3   Cautions

When making high-frequency noise measurements, make sure that the input impedance of the oscilloscope is set to 50 Ω, not 1 MΩ.

When making high-frequency noise voltage measurements on PCBs, lead dress is important to guarantee that the leads do not pick up stray magnetic fields from the PCB. The leads should be dressed perpendicular to the PCB (see Section E.4 in Appendix E).

### 18.6   CONDUCTED EMISSION TESTING

Precompliance conducted emission tests can be performed easily using a line impedance stabilization network (LISN); see Section 13.1.1. The conducted emission precompliance tests can be performed similarly to the way they are specified in the regulations.

Figure 18-10 shows the conducted emission test setup specified in the Federal Communications Commission (FCC)/International Special Committee on

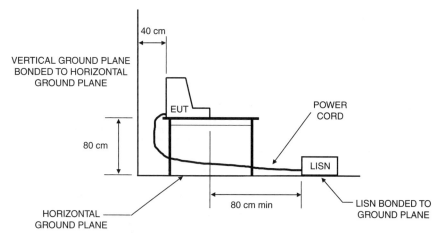

**FIGURE 18-10.** Conducted emission test setup as per the FCC/CISPR regulations.

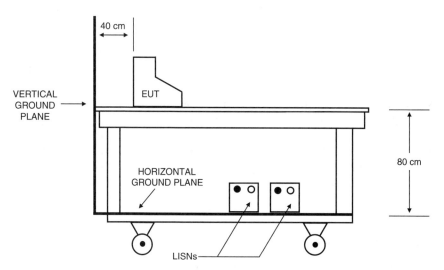

**FIGURE 18-11.** Precompliance conducted emission test setup on a laboratory cart. The cart must be nonmetallic.

Radio Interference (CISPR) regulations. For precompliance, the test is often performed without the ground planes. A better approach, however, is to set up the conducted emission test on a nonmetallic laboratory bench or cart as shown in Fig. 18-11. This includes the ground planes, although they are not as large as specified in the regulations.

Two LISNs are required, one connected to each side of the ac power line. Hence, many manufacturers package two LISNs in one enclosure and provide a method to switch the measuring port from one side of the line to the other.

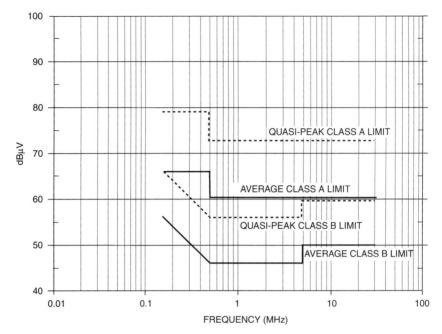

**FIGURE 18-12.** FCC/CISPR quasipeak and average conducted emission limits.

When the measuring port is switched a 50-Ω termination is automatically placed on the unused port.

The FCC and the European Union require that both quasipeak (narrowband) and average (broadband) conducted emission measurements be made. The respective limits are shown in Fig. 18-12. The average limits are 13 dB lower than the peak limits for a Class A product and 10 dB lower for a Class B product.

### 18.6.1 Test Procedure

Set up the equipment under test (EUT) at the appropriate distance from the two ground planes (if used). Plug the EUT into the LISN and the LISN into the power line. Connect the spectrum analyzer to the voltage measuring port on the LISN. Set the spectrum analyzer for a 10-kHz resolution bandwidth using a peak or quasipeak detector.

Measure the common-mode voltage from each side of the line to ground (hot to ground and neutral to ground) and compare the results with the regulatory limits. Repeat the measurement with the spectrum analyzer detector function set to average.* Conducted emission measurements are usually performed from

---

* If the average limit is met when using the peak or quasipeak detector, the EUT shall be considered to have met the limits, and the average detector measurement is not required.

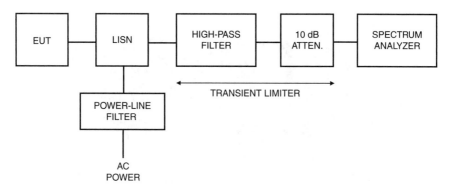

**FIGURE 18-13.** Block diagram of the conducted emission instrumentation test setup, including a high-pass filter and 10-dB attenuator.

150 KHz to 30 MHz with the unused LISN's measuring port always terminated with a 50-Ω resistor.

### 18.6.2 Cautions

Noise can be coupled into the LISN from the external power line. Therefore, a validation test should be performed to confirm that you are measuring what you think that you are. A simple validation test for conducted emission measurements is to turn the product OFF to determine whether the signal goes away. If it remains, it is most likely because of external power line noise. Adding a power-line filter between the LISN and the external ac power line can help to reduce any externally coupled noise.

Although the spectrum analyzer can be connected directly to the LISN's measuring port, this approach could possibly damage the spectrum analyzer or cause incorrect readings. Other equipment connected to the ac power line can cause large transient voltages that could damage the sensitive input of the spectrum analyzer. In addition, a large 50/60-cycle component will be present in the measured signal; this can overload the input circuitry of the spectrum analyzer, which also causes the reading to be in error.

Therefore, although not required by the regulations, it is prudent to add a high-pass filter (which will suppress the 60-Hz component of the signal by 60 dB or more) and a 10-dB external attenuator to absorb any transients present on the power line. A block diagram of this test setup is shown in Fig. 18-13. The high-pass filter and the 10-dB attenuator, as well as an additional diode limiter (to clip any large transient voltages), are available as a single component as shown in Fig. 18-14 (e.g., an Agilent 11947 Transient Limiter). Do not forget to scale the limit line on the spectrum analyzer by 10 dB to account for the external attenuator. The spectrum analyzer should also be set for an internal 10-dB attenuation to provide additional protection.

**FIGURE 18-14.** A transient limiter used for conducted emission testing.

The two most common problems associated with conducted emission testing are power-line frequency overload of the spectrum analyzer and not terminating the unused measuring port with 50 Ω.

### 18.6.3   Separating C-M From D-M Noise

The conducted emission test procedures just described measures the total noise, which consists of both common- and differential-mode components. When diagnosing conducted emission problems, it is helpful to distinguish between the common- and the differential-mode emissions. This is desirable, because different power supply components affect differential-mode noise currents than those that affect common-mode noise currents; see Section 13.2. Similarly, different components of the power-line filter suppress differential-mode noise, whereas other components suppress common-mode noise; see Section 13.3. Knowing which mode is predominant in a product's conducted emission spectrum provides a clue as to which components in the power supply and/or power-line filter need to be changed or modified. It basically divides the problem in half.

Figure 18-15 shows a power supply connected to a LISN, with both common-mode and differential-mode noise currents emanating from the power supply. The two 50-Ω resistors represent the LISN. The noise voltage on the phase side of the LISN will be

$$V_P = 50\,(I_{CM} + I_{DM}). \qquad (18\text{-}2)$$

The noise voltage on the neutral side of the LISN will be

$$V_N = 50\,(I_{CM} - I_{DM}). \qquad (18\text{-}3)$$

Adding the phase and neutral voltages gives

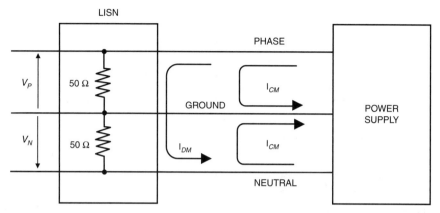

**FIGURE 18-15.** A power supply connected to a LISN, showing both common- and differential-mode noise currents.

$$V_P + V_N = 50\,(2I_{CM}) = 2V_{CM}. \tag{18-4}$$

Subtracting the phase and neutral voltages gives

$$V_P - V_N = 50\,(2I_{DM}) = 2V_{DM}. \tag{18-5}$$

Therefore, we can determine the common-mode and differential-mode noise voltages by adding or subtracting the two LISN voltages.

The addition and/or subtraction of the phase and neutral voltages, however, cannot be done after the measurement has been made, because the measurement is just a magnitude and does not provide any phase information. Therefore, what is required is a network that adds (or subtracts) the two voltages before they are measured—and the phase information is lost.

Figure 18-16 shows a conducted emission test setup that includes the addition of such a network. If the network adds the phase and neutral voltage, it is called a differential-mode rejection network, and if it subtracts the two voltages, it is called a common-mode rejection network. Such a network must satisfy three requirements, as follows:

- Add (or subtract) the phase and neutral voltages
- Attenuate the resultant voltage by 6 dB (reduce it to half its value)
- Provide a 50-Ω termination for each of the LISN outputs

***18.6.3.1 Differential Mode Rejection Network.***   One simple network for doing this, which consists of only five resistors, is shown in Fig. 18-17 (Nave, 1991, p. 94). This network does the addition of the two input voltages so that it is a

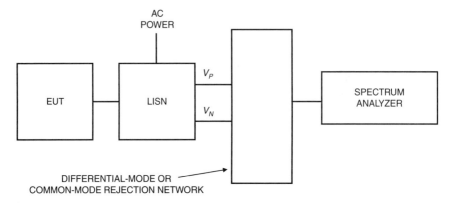

**FIGURE 18-16.** Conducted emission test setup, with the addition of a differential-mode or common-mode rejection network.

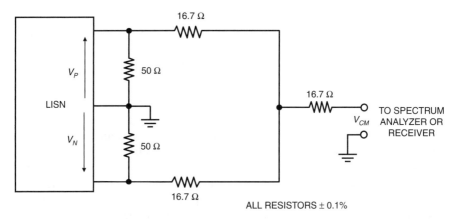

**FIGURE 18-17.** A differential-mode rejection network.

differential-mode rejection network. To achieve a differential-mode rejection of 50 dB or more, 0.1% resistors must be used and they must be laid out carefully to minimize parasitics. In use, the conducted emission test can be done first without the network, to obtain the total noise voltage (common mode plus differential mode). Then, the test can then be repeated with the differential-mode rejection network in place to obtain the common-mode noise voltage. The difference between the two is the differential-mode noise voltage.

*18.6.3.2 Switchable-Mode Rejection Network.* Another network that can be used to separate the common-mode from the differential mode emission is shown in Fig. 18-18 (Paul and Hardin, 1988). In this network, both the phase

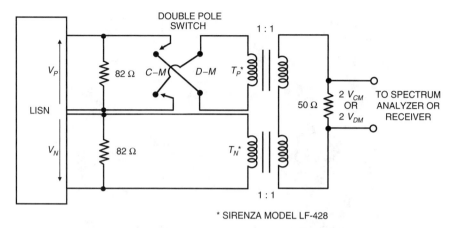

**FIGURE 18-18.**  Combination differential- and common-mode rejection network.

and neutral signals from the LISN are fed to the primaries of a two wide bandwidth (10 kHz to 50 MHz) transformers. The subtraction of the two signals is accomplished by connecting the two transformer secondaries in series. Addition is accomplished by reversing the polarity of one of the primary signals with a double-pole transfer switch. This network has the advantage of easily switching between the common- and differential-mode noise measurements by just flipping a switch.

This network, however, does not provide the required 6-dB attenuation of the output; hence, the output is twice the desired common-mode or differential-mode noise voltage. This limitation can be overcome easily by increasing the allowable limit on the spectrum analyzer by 6 dB. It is also easy enough to add a 6-dB, 50-$\Omega$ attenuator between the output of the network and the spectrum analyzer (my preference). If the setup of Fig. 18-13, which already has a 10-dB attenuator, is used simply apply only a 4-dB correction to the spectrum analyzer's limit line. A commercial version of this network that is available from Fischer Custom Communications is shown in Fig. 18-19.

## 18.7  SPECTRUM ANALYZERS

The most expensive piece of test equipment required for precompliance testing is the spectrum analyzer. However, most of the major instrument manufacturers (Agilent, Tektronix, Anritsu, IFR, etc.) make small portable spectrum analyzers priced in the $10,000 to $15,000 price range (USD in 2008) that are more than adequate for this application (see Fig. 18-20). In addition, other manufacturers have available spectrum analyzers, costing $10,000 or less with limited capability, some of which are also applicable for precompliance testing.

**FIGURE 18-19.** A commercial version of the combination differential- and common-mode rejection network used for conducted emission testing (courtesy of Fischer Custom Communications, Inc.).

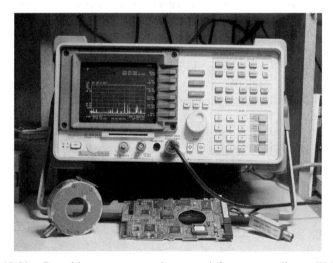

**FIGURE 18-20.** Portable spectrum analyzer used for precompliance EMC testing. Also shown are a clamp on common-mode current probe (left front) and a shielded loop magnetic field probe (right front).

The frequency range of the spectrum analyzer should be at least 1 GHz, with a max-hold mode. Max-hold allows the spectrum analyzer to run for a period of time (e.g., 5 or 10 s) and record the peak signal amplitudes occurring during that time period. The input impedance should be 50 $\Omega$. Some spectrum analyzers intended for use in the TV industry have 75-$\Omega$ input impedances.

The spectrum analyzer should have both a peak and average detector function. Most precompliance measurements will be performed with a peak detector, although conducted emission measurements must be performed with both a peak and an average detector. A quasipeak detector, although desirable, is not a necessity for precompliance measurements.

## 18.7.1 Detector Functions

*18.7.1.1 Peak Detector.* The default detector for most spectrum analyzers is a peak detector. Spectrum analyzers use a simple envelope detector (a series diode followed by a shunt capacitor) at the output of the intermediate frequency (IF) amplifier. The time constant of this detector is such that the voltage follows the peak amplitude of the detected IF signal.

Most precompliance measurements are made using the peak detector because it is much faster than measurements made with a quasipeak or average detector. Peak measurements will always be equal to or higher than those made with the other detectors. Therefore, if all the signals are below the regulatory limit when measured with the peak detector, the product will also be below the limit when measured with the other detectors, and no more testing is necessary.

*18.7.1.2 QuasiPeak Detector.* Most radiated and conducted emission limits are based on the use of a quasipeak detector. A quasipeak detector weighs signals according to their pulse repetition rates, which is a way of assessing the interference potential to AM radio. A quasipeak detector uses a filter at the output of the envelope detector. The filter has a charge rate that is much faster than its discharge rate. Therefore, the higher the repetition rate of the signal, the more the filter capacitor charges and the higher the output of the quasipeak detector. The quasipeak detector will always read less than or equal to the readings obtained with the peak detector. For continuous signals (nonpulsing), quasipeak and peak detectors will always read the same. The characteristics of the quasipeak detector are described in CISPR 16-1-1 (2006).

A quasipeak detector is a nice feature to have but is not essential for precompliance testing. For many products, the peak detector and quasipeak detector will give the same reading. For those products where this is not true, the peak detector will read higher than the quasipeak detector, which will thereby provide an additional degree of margin.

*18.7.1.3 Average Detector.* An average detector, which is only required for conducted emission tests, displays the average value of the detected envelope. In an average detector, the output of the envelope detector passes through a filter whose bandwidth is much less than the resolution, or IF, bandwidth of the spectrum analyzer. This filters the high-frequency components at the output of the envelope detector producing an average value of the signal.

**TABLE 18-1.   Precompliance Spectrum Analyzer Minimum Specifications**

| Parameter | Specification |
|---|---|
| Frequency range | 100 kHz to1 GHz |
| Resolution (IF) bandwidth | 9 or 10 kHz and 100 or 120 kHz |
| Detector functions | Peak and average[a] |
| Sensitivity | < 20 dB μV at 100-KHz Bandwidth |
| Display | Free run and max-hold modes |
| Input impedance | 50 Ω |

[a]Quasipeak, optional.

For more information on various spectrum analyzer detector functions, see Schaefer (2007).

Table 18-1 lists the minimum specifications for a precompliance spectrum analyzer.

### 18.7.2   General Test Procedure

Connect the signal source or probe to the spectrum analyzer's input. Set the appropriate frequency range and resolution bandwidth (100 or 120 kHz for the common-mode current clamp and the magnetic field probe measurements, and 9 or 10 kHz for conducted emission testing). Set the detector function to peak. For conducted emissions, measurements must be made with both peak and average detector functions. Set the vertical axis to dBμV, with 10 dB per major division scale. Set any internal attenuator to 0 dB for common-mode current and magnetic field measurements and to 10 dB for the conducted emission measurements. Set the reference level (top of the screen) appropriately in order to display a trace. A typical level for precompliance measurements (radiated or conducted) is 70 to 80 dBμV.

Some older or limited-capability spectrum analyzers can only indicate dBm* instead of dBμV. The relationship between dBm and dBμV is 107 dB:

$$\mathrm{dBm} = \mathrm{dB\mu V} - 107\,\mathrm{dB}. \qquad (18\text{-}6)$$

The spectrum analyzer should be able to point the screen display to a plotter to provide a hard copy of the measurements. Some spectrum analyzers (e.g., Agilent) can also point the screen trace to a portable ink-jet printer. Most can also save the display to internal memory, so that it can be recalled later and compared with the then active trace.

---

* Decibels with respect to a 1-mw reference (see Appendix A).

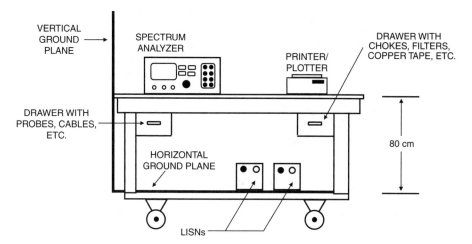

**FIGURE 18-21.** EMC crash cart containing all the equipment required for precompliance testing.

## 18.8  EMC CRASH CART

Rather than having all the equipment needed for workbench EMC measurements scattered all over the laboratory, where it may or may not be easy to find when needed, it can be combined all together in one place, in what I call an "EMC Crash Cart." This can easily be accomplished by starting with the conducted emission test cart of Fig. 18-11 and making some modifications to it, by adding a couple of drawers, as shown in Fig. 18-21.

The cart can be made of wood, plastic, or fiberglass, but not metal. The cart can be rolled to wherever it is needed, and it has everything required to perform the precompliance tests and apply fixes to the equipment if necessary. The spectrum analyzer and plotter, or printer, can be placed on the top of the cart. One of the two drawers can be used for all the necessary test equipment, such as follows:

- Common-mode current clamp
- Magnetic field loop probe
- Balanced voltage probe
- Transient limiter
- Common-mode and differential-mode rejection networks, etc.
- Coaxial cables, attenuators, etc.
- Small hand tools.

The second drawer can be used to hold EMC mitigation components that might be necessary to reduce the emissions; see Section 18.8.1.

For most precompliance measurements, the spectrum analyzer and plotter will remain on the cart. However, when performing a conducted emission test, remove the spectrum analyzer and plotter from the crash cart, and place the equipment under test (assuming that it is a table top size product) on the cart, 40 cm from the vertical ground plane.

### 18.8.1  Mitigation Parts List

The following items and parts are useful in fixing and/or isolating EMC problems in equipment and should be included in the crash cart drawer:

- Ferrite cores to reduce common-mode emissions from cables. Most of the ferrite manufacture's have kits available with an assortment of various ferrites for cable noise suppression.
- Aluminum foil and copper tape to improve shielding of enclosures and cables.
- Copper braid to improvise cable shields and to use as grounding straps.
- Small metal cable clamps for improving cable shield terminations. Add copper tape to plastic cable clamps, to make them conductive, if metal clamps are not available.
- AC power-line filters, as well as filter-pin connectors in common configurations such as for DB-9, DB-25, RJ-11, and RJ-45 connectors.
- AC safety, agency-listed, power-line capacitors to use for X- and Y-capacitors in power-line filters, 1000 pf to 2 μF.
- Connector backshell grounding clips and EMI gaskets for common connectors, used to ground the connector backshell directly to the chassis.
- An assortment of conductive EMI gaskets and spring fingers to provide electrical conductivity across enclosures seams.
- Resistors, 10 to 1000 Ω, to dampen clock oscillations, etc.
- Small ceramic capacitors, 470 pF to 0.1 μF, and small ferrite beads with 50-to 100-Ω impedance.
- Sandpaper to remove paint and nonconductive coatings from enclosures.

By using a crash cart, everything needed to perform the necessary precompliance measurements and apply fixes to the product will be readily available in one location. This will save time and prevent you from having to search all over the laboratory for what you need. This approach will also encourage you to use the equipment, because you do not have the excuse of not knowing where all the equipment is!

# 18.9   ONE-METER RADIATED EMISSION MEASUREMENTS

## 18.9.1   Test Environment

Radiated emission measurements could be performed as part of precompliance testing, if a way can be found to overcome the uncontrolled reflection problem discussed in Section 18.1.

To accomplish this, the following two things must be done:

1. Reduce the amplitude of the reflected field
2. Maximize the amplitude of the desired signal coming directly from the product under test

The reflected energy can be minimized by placing the EUT as far away from any reflecting surfaces (metal) as possible. Therefore, you will need to find some open space in which to perform the test. An indoor test site could possibly be an unused conference room, a break room, the cafeteria, or some open warehouse space. An outdoor test site could be an empty parking lot or open field. If possible, try to find a location where any large metal objects are at least 3 m away from both the EUT and the antenna. Indoor sites are always preferred because of the possibility of adverse weather conditions associated with outdoor sites.

To maximize the desired signal, the test antenna should be placed 1 m away from the product, instead of the 3- or 10-m test distance used at a commercial EMC test facility. If all reflective objects are kept at least 3 m away from both the antenna and the EUT, a 1-m measuring distance will usually keep the reflected signal 15 dB, or more, below the desired signal's level.*

A nonconductive turntable is also required to rotate the product to determine the maximum emissions. The basic test setup is shown in Fig. 18-22.

Radiated emission tests should not be used in place of the previously described precompliance tests but possibly can be used in addition to them.

## 18.9.2   Limits for 1-m Testing

In the far field, radiated emission limits can be scaled inversely proportional to the measuring distance. Therefore, if the measuring distance were decreased from 3 to 1 m, the limit would have to increase by a factor of three, or 10 dB. However, when testing at 1 m, several other complex factors come into play, which include product and antenna size, and the simple inverse distance extrapolation may not be accurate.

---

*This assumes far-field conditions with a $1/r$ falloff in signal strength with distance ($r$ being the radius, or distance involved), and it is a conservative approach because the fields may actually fall off at a faster rate; see Section 6.1.

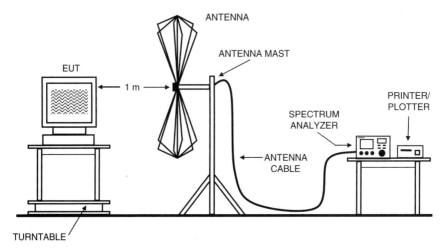

**FIGURE 18-22.** A simple 1-m radiated emission test setup.

Experience with 1-m testing shows that a 6-dB correction factor is closer to reality (Curtis, 1994). This will be somewhat conservative and will provide sufficient accuracy for precompliance testing. Figure 18-23 shows the FCC and CISPR radiated emission limits extrapolated (using the 6-dB correction factor) for 1-m testing.

### 18.9.3   Antennas for 1-m Testing

Small broadband antennas are the best choice for 1-m EMC testing. Most users test with a biconical antenna (from 30 to 200 MHz) and a small log periodic antenna (from 200 to 1000 MHz). Although far from ideal, this technique has worked well in many cases.

***18.9.3.1 Cautions.***   A word of caution is in order with respect to the use of a log periodic antenna. The active element of a log periodic will change with frequency. If the center of the log periodic antenna's boom is placed 1 m from the equipment under test, then the high-frequency element will be closer than 1 m and the low-frequency element will be farther than 1 m. Therefore, the smallest (shortest boom) log periodic possible should be used for these 1-m tests. This, however, means a less-sensitive antenna. This same phenomena occurs when testing at a legitimate EMC test facility; however, because the test distance is either 3 or 10 m, the percentage change of the position of the active element on the log periodic, with respect to the measuring distance, is much smaller than in the case of 1-m testing.

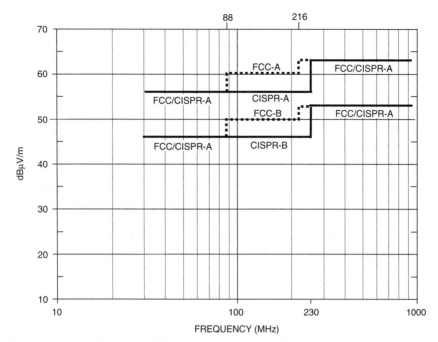

**FIGURE 18-23.** FCC and CISPR radiated emission limits extrapolated for 1-meter testing.

Another problem with 1-m testing is the effect of large antenna elements. The problem is that the beam width of the radiation from the product may not illuminate the complete antenna. The calibration antenna factors used to convert field strength to voltage output, however, were derived assuming uniform illumination of the complete antenna.

***18.9.3.2 Ideal Antenna for 1-m Testing.***  The ideal antenna for 1-m precompliance testing would have:

1. A wide bandwidth, 30 to 1000 MHz
2. Be physically small, less than 12 in. long
3. Be very sensitive, have an antenna factor* less than 6 dB
4. Have a reasonably flat frequency response

---

*The antenna factor (*AF*) units of 1/m is the ratio of the incident field strength *E* to the voltage *V* at the antenna terminals. Thus, $AF = E/V$, or $V = E/AF$. Therefore, the smaller the antenna factor, the more sensitive the antenna.

The requirement for a physically small antenna and a very sensitive antenna are contradictory. Small antennas, at frequencies below 300 MHz, are usually inefficient and therefore not very sensitive. An antenna that satisfies all the above criteria does not exist. However, it is possible to come close with an active antenna.

***18.9.3.3 Active Antennas.***   The advantages of an active antenna are small size, broad bandwidth, high sensitivity (small antenna factor), and flatter frequency response (the amplifier gain can be shaped to compensate for the falloff of the normal antenna response). Of course, active antennas also have some disadvantages, such as concerns over amplifier stability, nonlinear distortion, and dynamic range (overload with a strong signal). The first two problems can be overcome by proper active device selection and good amplifier design. The third problem can be overcome by monitoring the output level (on the spectrum analyzer) over the bandwidth of the amplifier to be sure that it is always below the saturation level at all frequencies.

None of the major antenna manufacturers produce an antenna that satisfies the above criteria. However, some antennas have existed on the market from smaller companies, which come close to satisfying the above criteria. To the best of my knowledge, none are still available.

An active antenna that I have been using since 1997, with good success, for 1-m precompliance testing was made by a company called ETA Engineers in Tucson, Arizona. However, the antenna is no longer available. The antenna is an active antenna, called a Model 100 Bowtop Antenna (see Fig. 18-24). The antenna is basically half of a bow-tie antenna, therefore a monopole, with a capacitive top hat. The antenna's specifications are as follows: The maximum dimension of the antenna is 7.5 in; the useful frequency range is from 30 MHz to 700 MHz; and the antenna factor over that frequency range varies from +6 to –3.5 dB. This antenna satisfied all of the above ideal requirements (small size, high sensitivity, and flat frequency response) for a 1-m precompliance antenna, except its frequency response does not go to 1 GHz. This antenna works well, and I use it whenever I perform precompliance, 1-m radiated emission tests.

Note: When performing 1-m EMC testing, it is not necessary to vary the antenna height, as is required when performing 3- or 10-m testing. The antenna height search is done to guarantee that the ground plane reflected wave and the direct wave from the equipment under test are in phase, and therefore add together. In the case of 1-m testing, the reflection point usually occurs beyond the antenna location, and it is not therefore detected by the antenna. Tests, however, must be performed with both horizontal and vertical polarization of the antenna. When performing the test, rotate the turntable and manipulate the cables for maximum emissions.

It is also interesting to note that a 1-m indoor radiated emission test facility is used by the FCC (with biconical and log periodic antennas) to test randomly selected products for EMC compliance. If the 1-m data indicate that the

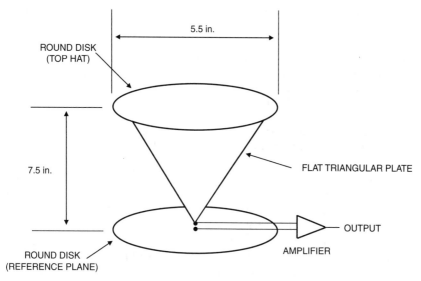

**FIGURE 18-24.** An active bowtop antenna.

product passes, the device is approved and no more testing is done. However, if the 1-m data indicate that the product fails, the test is repeated on the Commission's 3-m OATS test site to confirm the failure.

A more thorough discussion of EMC testing and troubleshooting techniques is presented in the book, *Testing for EMC Compliance* (Montrose and Nakauchi, 2004).

## 18.10   PRECOMPLIANCE IMMUNITY TESTING

In this chapter, the discussion so far has covered simple EMC tests that relate to a product's emissions. Although the FCC does not have requirements that pertain to immunity, both the European Union and the military do. Also, medical equipment, automotive equipment, and avionics have to meet immunity requirements. This section therefore discusses some simple precompliance tests that relate to the radiated, conducted, and transient immunity of a product.

### 18.10.1   Radiated Immunity

One cannot indiscriminately radiate rf energy in an open area, across the frequency spectrum from 30 MHz to 1 GHz, without posing a significant risk of causing interference to licensed radio services and incurring the wrath of the FCC or other regulatory authorities. Evidently, this was not obvious to everyone, because in 1996 the FCC felt it necessary to issue Public Notice 63811, which said in part (FCC, 1996):

> Laboratories are alerted that measurements of the radio frequency (RF) immunity of electronic equipment may not be performed on open area test sites without a station license. . . . Parties wishing to perform RF immunity tests on an open area site are reminded that they must be conducted in a manner that confines the RF energy to the immediate area, such as a shielded room or anechoic chamber, to prevent interference to communications. Failure to do so is a violation of Section 301 of the Communications Act of 1934, as amended.

However, performing precompliance radiated immunity testing in a small shielded room is not a viable option, for the same reasons that it was not viable for performing radiated emission tests, that is, the problem of reflections causing large uncontrollable peaks and nulls in the radiated field pattern.

Therefore, some other method must be found to do precompliance-radiated immunity testing. Two possibilities exist. One would be to use a low-power broadband source of radiation and to place it close to the EUT. The second would be to use a narrowband radio transmitter that is presently authorized by the FCC for use by the public, license free, such as those for the Family Radio Service or Citizen Band (CB) radios. This approach, however, only allows testing at certain specific spot frequencies.

A convenient source of broadband-radiated fields is any small motor that uses brushes to make connections to the commutator, such as an electric drill or a Dremel®* motor tool. The arcing that occurs at the brushes is a source of low-level broadband emissions. Figure 18-25 shows the radiated emission spectrum of

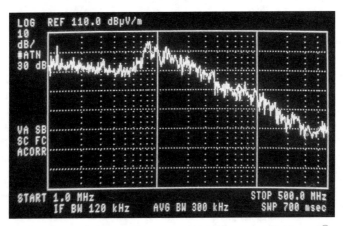

**FIGURE 18-25.** Emission spectrum, in the 1-to-500 MHz range, of a Dremel® motor tool.

---

* Registered trademark of Robert Bosch Corp, mount Prospect, IL.

a Dremel® motor tool. The emission was measured at a distance of 30 cm using a bowtop antenna (Fig. 18-24). As can be observed, the emission is relatively constant from 1 to 10 MHz and then falls off at a rate of 20 dB per decade up to 500 MHz. In the 1- to 10-MHz range, the level is about 90 dB μV/m (0.03 μV)

A simple test that will provide an indication of the susceptibility of a PCB to radiated fields is to hold the Dremel® motor tool about 1 in above the PCB and move it across the board while observing any malfunction of the product. The fields produced by this test are fairly low, a few V/m at this distance, but will detect overly susceptible circuits. This is a good test to perform early in the design phase of a product. If the product cannot pass this test, it is overly susceptible to rf fields and should be fixed even if it does not have to meet any regulatory immunity requirements. This is also a good way to detect poor PCB layout, especially in the case of analog circuits on single or double sided board. Most digital circuits can easily pass this test.

Small, handheld, portable transmitters can also be used for rf immunity testing at specific spot frequencies. CB radios, Family Radio Service transmitters, cordless phones, and cell phones are all possibilities. Table 18-2 lists the approximate frequency, power, and field strength at a 1-m distance from many unlicensed radio services transmitters available in the United States. Similar radio services are available in other countries.

Electric field strength as a function of power and distance can be approximated by Eq. 14-2. A convenient number to remember is that the electric field strength 1 m from a 1-w transmitter will be about 5.5 V/m. This value can be scaled for other conditions as the square root of the power and as the reciprocal of the distance.

Note: Cell phones automatically adjust their transmitted power as a function of the received signal strength; therefore, it is difficult to know what the radiated power is in any individual case. Because various models of low-power radio transmitters can have specifications that differ from Table 18-2, it is

**TABLE 18-2.  Specifications for Various Low-Power Transmitters.**

| Service | Approximate Frequency | Maximum Power | Approximate V/m at 1 m |
|---|---|---|---|
| Citizen band | 27 MHz | 5 W | 12.3 V/m |
| Family radio | 465 MHz | 500 mW | 3 V/m |
| Cell phones | 830 MHz/1.88 GHz | 600 mW | 6 V/m |
| Cordless phones | Various[a] | 200 mW | 2.5 V/m |
| GMRS[b] | 465 MHz | 1 to 5 W | 5.5 to 12.3 V/m |

[a]Most present-day (in 2006) cordless phones operate in one of the following frequency bands: 400 MHz, 900 MHz, 2.4 GHz, or 5.8 GHz.
[b]General Mobile Radio Service (GMRS) requires an FCC license; however, many companies already have a license for the use of the GMRS. These transmitters are the walkie-talkies that you often see security people using when patrolling the premises of buildings and malls.

prudent to check the product's specifications with respect to power and frequency before using it for precompliance testing.

To perform a precompliance test, place the transmitter an appropriate distance (for the field strength desired) from the EUT and/or its cables and initiate a brief transmission while observing the response of the product.

### 18.10.2   Conducted Immunity

Noise can be created on the dc or ac power line by connecting a "Chattering Relay" across the line. This consists of a relay connected across the power line in series with its own normally closed contact as shown in Fig. 18-26. When power is turned ON, the relay will pull-in; when it does, the normally closed contact will open and the relay will drop out. This process will continue until power is removed from the relay, thus, the name chattering relay. When the relay drops out, a large inductive kick voltage will be produced across the power line in accordance with the equation:

$$V = - L \, dI / dt. \tag{18-7}$$

This voltage is typically 10 to 100 times the power-line voltage, and it can be in the range of hundreds or even thousands of volts. The voltage can be controlled (limited) by adding two back-to-back Zener diodes across the relay, as shown in Fig. 18-26. By mounting the relay in an enclosure with different values of breakdown diodes connected across the relay, through a rotary switch, the magnitude of noise voltage on the power line can be varied by the switch setting. This test is good for determining susceptibility of the EUT to noise on the power line. The test is also realistic, because many power lines have relays or solenoids connected to them.

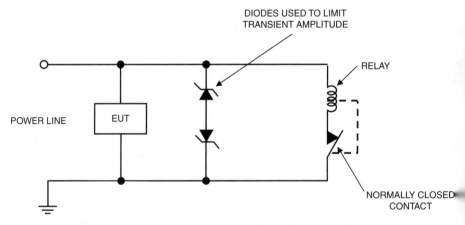

**FIGURE 18-26.** Circuit for a chattering relay, with diode voltage limiter.

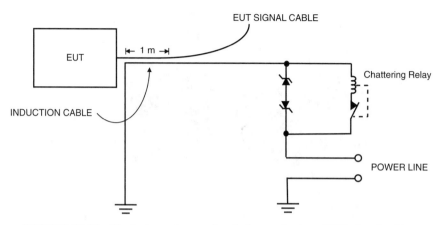

**FIGURE 18-27.** Chattering relay used to induce noise into EUT signal cables.

The chattering relay can also be used to induce a noise voltage into signal cables as shown in Fig. 18-27. The chattering relay is powered by a separate power source through a separate cable. About a 1-m of this cable is run parallel, and taped, to the signal cable under test. The noise current from the relay circuit will magnetically induce a noise voltage into the cable under test. Switching in diodes having different breakdown voltage values, into the relay circuit, will vary the magnitude of the induced test voltage.

Conducted rf immunity can also be simulated by holding a small handheld, portable transmitter (as discussed in Section 18.10.1) close to the cable being tested.

### 18.10.3   Transient Immunity

*18.10.3.1 Electrostatic Discharge.*   Precompliance electrostatic discharge (ESD) testing can be performed using a small portable, battery-powered, ESD simulator as shown in Fig. 18-28. One such device is Thermo Scientific's MiniZap®*.

An ESD simulator is the second most expensive piece of precompliance test equipment—after a spectrum analyzer. Prices will typically range from $5000 to $10,000 (USD as of 2008). A "poor man's" ESD generator can, however, be made from an inexpensive piezoelectric gas barbecue lighter.[†] The output voltage from such a device is not repeatable or controlled but will be in the 10-

---

* Registered trademark of Thermo Fisher Scientific, Inc., Waltham, MA.
[†] Another commercial device, the Zerostat antistatic gun, is a piezoelectric device that can also be used for crude ESD testing. The device, which was originally developed for removing static from vinyl phonograph records, will generate a few kilovolts. When the trigger is pulled, a positive voltage is generated; when the trigger is released, a negative voltage is generated. The cost is about $100 (USD in 2008). If your product cannot pass an ESD test with this device, it is overly sensitive and should be fixed.

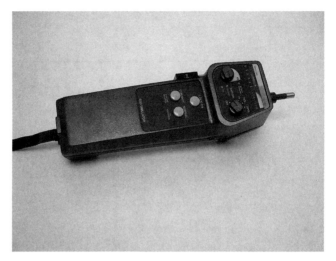

**FIGURE 18-28.** A small battery powered ESD simulator.

to 15-kV range and is can provide a crude ESD test. I have used one many times, in a pinch, when I did not have my regular ESD simulator available.

An ESD generator that is used for precompliance testing should be capable of performing both contact and air discharge, with positive and negative voltages, and it should have a variable voltage range of 1 to 10 kV. The generator should simulate the human body model with a 150-pF and 330-Ω R–C network. The ability to apply both single and repetitive pulses, at the rate of 1 or 10 pulses per second, is also desirable.

A typical ESD test setup for small tabletop equipment is shown in Fig. 18-29. Discharges should be made directly to any accessible surface of the EUT as well as to the edges of the vertical and horizontal coupling planes. Tests should be performed with both positive and negative polarity of the test voltage.

***18.10.3.2 Electrical Fast Transient.*** An ESD simulator can also be used to induce transients into the cables of the EUT. Take a wire with one end grounded and tape it to the cable under test for a length of 1 m and apply the ESD pulses to the open end of the conductor as shown in Fig. 18-30. Setting the generator to produce repetitive pulses of 10 or 20 pulses per second will provide a simulation of the electrical fast transient (EFT) test.

Assuming a 0.5 coupling coefficient between the wire and the cable under test (a reasonable assumption), the ESD simulator should be set for twice the desired EFT voltage. For example, to simulate a 2-kV EFT transient, set the ESD simulator for a repetitive 4-kV discharge.

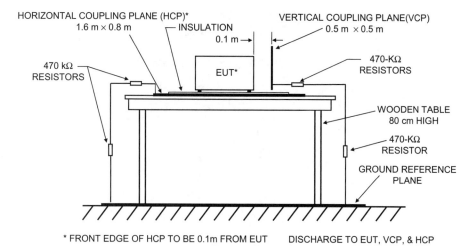

**FIGURE 18-29.** A typical ESD test setup for table-top equipment.

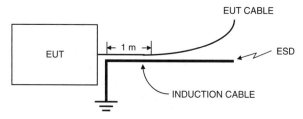

**FIGURE 18-30.** ESD induced cable transients.

## 18.11 PRECOMPLIANCE POWER QUALITY TESTS

Two unique emission tests required by the European Union are the harmonics and flicker tests as described in Section 1.7.2. The harmonic requirement limits the harmonic content of the current drawn by the product from the ac power line, up to the 40th harmonic. The major problem occurs with nonlinear loads such as switching power supplies, fluorescent lamps, and variable speed motor drives, all of which draw current rich in harmonics (see Section 13.9). The flicker requirement limits the transient ac power line current (inrush current) drawn by the product. Because of the finite impedance of the power line, large transient currents produce a fluctuation of the line voltage that result in the flickering of lights.

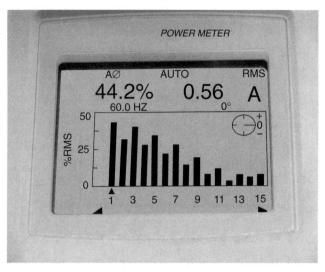

**FIGURE 18-31.** Harmonic current from of a small switching power supply, displayed on a harmonic meter.

### 18.11.1 Harmonics

The harmonic current drawn by a product can be measured easily using a small handheld power meter/harmonic analyzer. Figure 18-31 shows the harmonic distortion of a small switching power supply displayed on such a meter. The harmonic limit for Class A equipment, which includes most products with the exception of personal computers, TV sets, portable tools, and lighting equipment, are listed in Table 18-3 for odd and even harmonics, up to the 20th harmonic.

For odd harmonics above the 19th the requirement is that the current does not exceed $2.25/n$, where $n$ is the harmonic number. For even harmonics greater than the 20th, the current limit is $1.84/n$. These limits apply up to the 40th harmonic.

**TABLE 18-3.** European Union's Harmonic Current Limits for Class A Equipment, up to the 20th harmonic.

| Harmonic Number | Current (A) | Harmonic Number | Current (A) |
|---|---|---|---|
| 1 | ----- | 2 | 1.08 |
| 3 | 2.30 | 4 | 0.43 |
| 5 | 1.14 | 6 | 0.30 |
| 7 | 0.77 | 8 | 0.23 |
| 9 | 0.40 | 10 | 0.18 |
| 11 | 0.33 | 12 | 0.15 |
| 13 | 0.21 | 14 | 0.13 |
| 15 | 0.15 | 16 | 0.12 |
| 17 | 0.13 | 18 | 0.10 |
| 19 | 0.12 | 20 | 0.09 |

Personal computers and TV sets must meet the Class D limits that are more severe and are based on the power rating of the device. Class D products must also not exceed the values shown in Table 18-3.

If you do not have a harmonic analyzer, then a crude approach to estimating the harmonic distortion is to look at the input current waveform on an oscilloscope. You can easily detect distortion of 5% or more by just visually observing the waveform. If the current waveform looks approximately sinusoidal, the product will most likely pass the harmonic emission test. If the waveform looks something like Fig. 13-4, it almost surely will fail. If the waveform is distorted but not as bad as Fig. 13-4, it is hard to tell whether the product will comply. The slower the pulse rise time, and the more the pulse is spread out over the cycle, the more likely the product will be compliant.

### 18.11.2 Flicker

The flicker standard requires that a large number of measurements be made, and then a statistical analysis performed on the data (involving the magnitude, time duration, waveshape, and repetition rate of the disturbance) to determine the flicker factor. This cannot easily be performed as a precompliance test.

One portion of the flicker requirements, however, can be measured, and that is the maximum inrush current. This can be measured with a clamp-on current meter having an inrush current measuring capability, such as the Fluke 330 Series True RMS Clamp Meters.

The flicker standard has three levels of compliance, each of which is applicable to a different product class. Level A is applicable to any product that does not meet the description of Level B or Level C. Level A is the most stringent requirement. Level B is applicable to products that are manually switched or automatically switched at more than two times per day. A Level C product is one that is attended when in use (e.g., a hair dryer or vacuum cleaner) or is automatically switched no more than two times per day. Level C is the least restrictive requirement.

The maximum ac power line voltage dip allowed for a Level A product is 4%, for Level B it is 6%, and for Level C it is 7%. The power-line source impedance of the flicker test set is $0.4 + j\,0.25\ \Omega$. The magnitude of this impedance is $0.472\ \Omega$. Using this impedance, the maximum allowable inrush current can be determined for the three levels of the test. The results for a 240-V ac power line are tabulated in Table 18-4.

**TABLE 18-4. European Union's Maximum Peak Inrush Current Requirements.**

| Test Level | Maximum Relative Voltage Change | Peak Inrush Current (A) |
|---|---|---|
| A | 4% | 20.3 |
| B | 6% | 30.5 |
| C | 7% | 35.6 |

For a 120-V power line, the current levels would be one half those listed in Table 18-4.

## 18.12   MARGIN

Although not directly related to precompliance testing, the subject of test margin is an important subject that should be discussed. Because this is the only chapter related to EMC testing, I decided to include the information here. The discussion in this section relates primarily to legitimate EMC compliance testing, not to precompliance testing.

Legally, all that is required is that the product complies with the applicable EMC requirements. Why then do you need to have margin? The two primary reasons are (1) that products vary, and (2) that a degree of measurement uncertainty always exists.

### 18.12.1   Radiated Emission Margin

Radiated emission measurement uncertainty has five sources. The first source is the quality of the test site (OATS or anechoic chamber), which includes the quality of the ground plane and how well the reflections are controlled, and it can easily produce an uncertainty of $\pm$ 2 dB.

The second source is the accuracy and calibration of the equipment, which includes the cables. The measuring instrument, a spectrum analyzer or radio receiver, probably has an accuracy of no better than $\pm 1.5$ dB; similarly, the antenna calibration usually has an accuracy no better than $\pm 2$ dB. What about the loss of the 50 ft of cable between the antenna and the spectrum analyzer? Was the cable loss measured, or taken from a table in a handbook? If measured, was that before or after the cable was sitting on the ground for a year or more, and being stepped on by a few people? Let us allow $\pm 0.5$ dB for this. If a preamplifier is used, then it probably contributes another $\pm 1.5$ dB of uncertainty.

Root mean square (rms) summing the above tolerances gives an uncertainty of approximately $\pm 3.6$ dB, with no allowance provided for product variation or the other issues discussed below.

The third source is the test procedure. Was the turntable rotated for a maximum reading, and then a vertical height search with the antenna done to determine the peak? Or was the antenna vertical height search performed to determine the maximum reading, and then the product was rotated on the turntable to determine the peak? The results can be different!

The fourth source is the test setup. What length cables were connected to the product, and how were they routed? What was the exact placement of the interconnected equipment being tested.

The fifth source is the operator technique. For example, how much time was spent, at each frequency, manipulating cables and rotating the product looking for the maximum emission?

One study of the reproducibility of radiated emission measurements among seven, 10-m open field test sites showed that most of the readings, but not all, over a frequency range of 30 to 400 MHz were within $\pm 4$ dB of the average (Kolb, 1988).

ANSI C63-4 requires that radiated emission test sites be validated by making normalized site attenuation* (NSA) measurements, at frequencies from 30 to 1000 MHz. To be considered acceptable, the measured NSA must be within $\pm 4$ dB of the theoretical NSA for an ideal site (ANSI C63-4, 2003). Therefore, two sites could have readings as much as 8 dB apart from each other and still be considered acceptable.

As a result of the above uncertainties, many companies require their products to have a 6-dB margin against the regulatory limit. For final compliance testing, a 4- to 6-dB margin is probably reasonable, with 6 dB being preferred. For precompliance measurements, however, the margin should be even greater, typically 6 to 8 dB.

### 18.12.2   Electrostatic Discharge Margin

ESD is inherently a statistical process; see Section 15.11. The product has a probability of failing. If the probability is small, and you have not performed a thousand or more discharges, then you may not detect the failure. The European Union's ESD standard only requires 10 discharges at each test point, which is not a statistically valid sample.

Digital equipment also has "time windows"; see Section 15.11. The equipment performs different functions at different times, such as reading from the hard drive, recording to the hard drive, reading random-access memory (RAM), outputting data to a printer, and so on. If only one of these functions is susceptible, did the performance of the function coincide with the discharge?

The environment also affects the ESD test results. Such things as humidity, temperature, and altitude can produce different test results. The routing of the ground strap on the ESD gun will affect the results, because different positions will produce different fields resulting from the large discharge currents.

The results are also dependent on the testing technique. For air discharge tests, the speed at which the ESD simulator's tip approaches the product can affect the result. Similarly, the exact angle that the simulator is positioned with respect to the product can produce different results. Last but not least, is the variation in ESD susceptibility of different test samples.

To overcome some of these variabilities and to improve confidence in the results, a degree of margin should be built into the precompliance test. For example, the European Union's ESD standard requires that products

---

*Site attenuation is a measure of the path loss between two antennas separated by a specified distance and height above a flat reflecting surface (ground plane).

pass a 4-kV contact discharge and an 8-kV air discharge. A degree of margin can be obtained by testing the product to 5-kV contact, and 10-kV air discharges, when performing precompliance tests.

## SUMMARY

- Precompliance EMC measurements should be performed early in the design stage and prior to final regulatory compliance testing.
- The most useful and important precompliance test is the measurement of the common-mode currents on *all* the cables.
- The second most important precompliance test is the scanning of the PCB with a magnetic field probe.
- Magnetic field probes should also be used to check for leakage at the seams and apertures of shielded enclosures.
- Noise voltage measurements can be helpful in pinpointing the source of radiated emission problems.
- Some useful noise-voltage measurements include clock waveforms, $V_{cc}$-to-ground voltage on ICs, ground noise voltage between ICs, and common-mode voltage between the I/O ground and the chassis.
- All the equipment necessary for precompliance EMC testing should be assembled together in one place, such as an EMC Crash Cart.
- If radiated emission testing is to be done, as part of the precompliance testing, it should be done in an open area (away from metal objects) and at a test distance of 1 m.
- Precompliance radiated emission testing should not be done in place of the common-mode cable current measurement, but possibly in addition to it.
- Precompliance conducted-emission tests can easily be done, in your own laboratory, using an LISN and a spectrum analyzer.
- Differential- and common-mode conducted emission can easily be distinguished from each other by using a common-mode or differential-mode rejection network, when performing the conducted emission precompliance test.
- Radiated immunity tests can be performed, at spot frequencies, by using a small handheld transmitter such as a Personal Radio Service or Citizen'sBand radio, or by using an electric drill or Dremel® tool as the noise source.
- Chattering relays are useful for inducing transient noise onto an ac or dc power line.
- ESD testing can be performed with a small battery-powered ESD generator or an inexpensive piezoelectric gas barbecue lighter.

- The ideal 1-m antenna for precompliance radiated emission testing should have a bandwidth from 30 to 1000 MHz, should be no bigger than 12 in, and have a small antenna factor of no more than 6 dB.
- A small *active* broadband antenna is preferred for 1-m testing; however, they are not readily available commercially.
- For 1-m radiated emission testing, the antenna height does not have to be varied; however, both horizontal and vertical polarization must be tested.
- When doing 1-m emission testing, rotate the product on a turntable and manipulate the cables to find the maximum emission.
- To extrapolate a 3-m radiated emission limit for 1-m testing, increase the allowable level by 6 dB.
- *Do not* perform radiated emission testing inside a small-shielded room (nonanechoic room).
- Radiated emission uncertainty is in excess of $\pm 4$dB.
- Products should have
  - 4- to 6-dB margin against the regulatory limit when doing final compliance testing
  - 6- to 8-dB margin for precompliance testing

## PROBLEMS

18.1 Why is a small-shielded enclosure not suitable for radiated emission measurements?

18.2 What will be the output voltage from an F-33-1 common-mode current clamp when it is placed around a cable with 100 μA of common-mode current?

18.3 When using the F-33-1 current clamp to measure the common-mode current on a 3-m-long cable, what is the maximum allowable reading in dBm for a Class B product?

18.4 At what voltage level (in dBμV) should the limit line be placed on a spectrum analyzer when using an F-61 common-mode current clamp on a 1/3-m long cable in order to pass FCC Class A radiated emission?

18.5 a. What is a null experiment?
b. What is the purpose of the null experiment?

18.6 What would be a good null experiment to perform when making a $V_{cc}$-to-ground noise measurement on an IC, using a balanced differential probe?

18.7 When performing common-mode current tests on cables of a product, how can you be sure that the measured current is indeed coming from the product under test, and not being picked up from some external source?

18.8  Are the following EMC precompliance measurements quantitative or qualitative?

    a. Common-mode current clamp measurements

    b. Magnetic field loop probe measurements

    c. Noise voltage measurements

    d. Conducted emission measurements

18.9  What are two uses for a small magnetic-field loop probe?

18.10  What precompliance EMC measurements provide information on the following:

    a. The differential-mode radiated emission from the product

    b. The common-mode radiated emission from the product

    c. Waveform distortion of the ac power current drawn by the product

18.11  What are four types of useful precompliance noise voltage measurements?

18.12  Name four disadvantages of commercial, active, differential probes?

18.13  a. When performing a conducted emission test, what three functions does a transient limiter provide?

    b. Why is a transient limiter used for conducted emission measurements?

18.14  What are the three common types of detectors found on spectrum analyzers?

18.15  Assume that 1-m radiated emission precompliance testing is being performed in a location where the closest metallic object is 5 m away from both the product and the antenna. How many decibels below the desired signal will the reflected signal be?

18.16  What is the European Union's harmonic current requirement applicable to the 33rd harmonic for a Class A product?

18.17  What are two ways to produce an rf field for precompliance radiated immunity testing?

18.18  What are the five sources of radiated emission measurement uncertainty?

## REFERENCES

ANSI C63-4. *American National Standard for Methods of Measurement of Radio-Noise Emissions from Low-Voltage Electrical and Electronic Equipment in the Range of 9 kHz to 40 GHz*, 2003.

CISPR 16-1-1. *Specification for Radio Disturbance and Immunity Measuring Apparatus and Methods Part 1-1: Radio Disturbance and Immunity Measuring Apparatus–*

*Measuring Apparatus,*International Special Committee on Radio Interference (CISPR), 2006.

Cruz, J. E. and Larsen, E. B. *Assessment of Error Bounds for Some Typical MIL-STD461/462 Types of Measurements.* NBS Technical Note 1300, October 1986.

Curtis, J. "Toil and Trouble, Boil and Bubble: Brew Up EMI Solutions at Your Own Inexpensive One-Meter EMI Test Site." *Compliance Engineering*, July/August 1994.

FCC. Public Notice 63811, *Conditions For The Use of Outdoor Test Ranges For RF Immunity Testing.* July 3, 1996.

Kolb, L., *Reproducibility of Radiated EMI Measurements.* Pan Alto, CA, Hewlett Packard, April 15, 1988.

Montrose, M. I. and Nakauchi, E. M. "Testing for EMC Compliance," New York, IEEE Press, Wiley Intersciences, 2004.

Nave, M. J. *Power Line Filter Design for Switched-Mode Power Supplies.* New York, Van Nostrand Reinhold, 1991.

Paul, C. R. and Hardin, K. B. "Diagnosis and Reduction of Conducted Noise Emissions." *1988 IEEE International Symposium on Electromagnetic Compatibility*, Seattle, WA, August 2–4, 1988.

Roleson, S. "Evaluate EMI Reduction Schemes with Shielded-Loop Antennas." *EDN*, May 17, 1984.

Schaefer, W. "Narrowband or Broadband Discrimination with a Spectrum Analyzer or EMI Receiver." *Conformity*, December 2007.

Smith, D. C. *High Frequency Noise and Measurements in Electronic Circuits.* New York, Van Nostrand Reinhold, 1993.

Smith, D. C. "DC to 1GHz Probe Construction Plans." 2004. Available http://www.emcesd.com/1ghzprob.htm. Accessed April 2009.

## FURTHER READING

AN-150. *Spectrum Analyzer Basics*, Agilent Technologies, 2004.

AN-1328. *Making Precompliance Conducted and Radiated Emissions Measurements with EMC Analyzers*, Agilent Technologies, 2000.

Gerke, D. and Kimmell, W. *EDN The Designer's Guide to Electromagnetic Compatibility*, Chapter 13 (EMI Testing: If You Wait to the End, It's Too Late). St. Paul, MN, Kimmel Gerke Associates, 2001.

*LISN-UP™ Application Note.* Fischer Custom Communications, 2005.

Roleson, S. "Using Field Probes as EMI Diagnostic Tools." *Conformity*, March 15, 2007.

Roleson, S. "Field Probes as EMI Diagnostic Tools." *Conformity*, September 2006.

Smith, D. C. "Build It! Magnetic Field Probe." *Conformity* , June 2003.

# APPENDIX A
# The Decibel

One of the most commonly used, but often misunderstood, terms in the field of electrical engineering is the decibel, which is abbreviated dB. The decibel is a logarithmic unit used to express the *ratio* of two powers. It is defined as follows:

$$dB = 10 \log \frac{P_2}{P_1}. \qquad (A\text{-}1)$$

The decibel is not an absolute quantity; it is always a *ratio* of two quantities. The unit can be used to express power gain ($P_2 > P_1$); or power loss ($P_2 < P_1$) — in the latter case, the result will be a negative number.

The decibel actually comes from a logarithmic unit of measurement called the *bel*, which is named after Alexander Graham Bell. One bel is defined as a power ratio of 10 (or 10 times the power). It was originally used to measure acoustic sound power ratios in telephony. The bel is a large unit, so the decibel, which is 1/10 of a bel, is more commonly used.

Being a logarithmic unit, it compresses the result and allows us to easily measure or plot quantities that cover a large dynamic range. This feature can be useful in the case of electromagnetic compatibility (EMC) measurements. For example, a voltage ratio of 1,000,000 to 1 could be expressed as 120 dB.

## A.1   PROPERTIES OF LOGARITHMS

Since the decibel involves logarithms, it is appropriate to review some of the properties of logarithms. The *common logarithm* (log) of a number is the power to which 10 must be raised to equal that number.* Therefore, if

$$y = \log x, \qquad (A\text{-}2)$$

---

* The *natural logarithm* (ln) of a number is the power to which $e$ (approximately 2.718) must be raised to equal the number. The common logarithm and the natural logarithm are related by log x = (log e) (ln x), or log x = 0.4343 ln x.

---

*Electromagnetic Compatibility Engineering*, by Henry W. Ott
Copyright © 2009 John Wiley & Sons, Inc.

then

$$x = 10^y. \tag{A-3}$$

Some useful facts and identities relating to logarithms are as follows:

$$\log 1 = 0,$$
$$\log \text{ of numbers} > 1 \text{ are positive,}$$
$$\log \text{ of numbers} < 1 \text{ are negative,}$$

$$\log(ab) = \log a + \log b, \tag{A-4}$$

$$\log(a/b) = \log a - \log b, \tag{A-5}$$

$$\log a^n = n \log a. \tag{A-6}$$

As can be observed from the equations above, multiplication of numbers becomes the addition of their logs, and division of numbers becomes the subtraction of their logs. This can be useful for EMC measurements. For example, let us assume that we want to know the voltage that would be measured by a spectrum analyzer connected by a long cable to an antenna exposed to an electromagnetic field. We would have to multiply the incident field strength by the antenna factor, and then multiply it by the loss of the cable connecting the antenna to the measuring instrument, and multiply it by the gain/loss of any amplifier or attenuator used. If all these numbers are expressed in decibels, however, all we have to do is add/subtract their gain or losses—a much simpler task.

## A.2   USING THE DECIBEL FOR OTHER THAN POWER MEASUREMENTS

Although the decibel is defined with respect to power, it has become common practice to also use it to express voltage and/or current ratios; in which case, it is defined as follows:

$$\text{dB voltage} = 20 \log(V_2/V_1), \tag{A-7}$$

or

$$\text{dB current gain} = 20 \log \frac{I_2}{I_1}. \tag{A-8}$$

These equations are only correct when both voltages, or both currents, are measured across equal value impedances. Common usage, however, is to use the definitions in Eqs. A-7 and A-8 regardless of the impedance levels.

The relationship between voltage gain and power gain can be determined by referring to Fig. A-1.The power into the amplifier is

$$P_1 = \frac{V_1^2}{R_1}. \tag{A-9}$$

The power out of the amplifier is

$$P_2 = \frac{V_2^2}{R_2}. \tag{A-10}$$

The power gain $G$ of the amplifier, expressed in dB is

$$G = 10 \ \log \frac{P_2}{P_1} = 10 \ \log \left[ \left( \frac{V_2}{V_1} \right)^2 \frac{R_1}{R_2} \right]. \tag{A-11}$$

Using the identities of Eqs. A-4 and A-6, Eq. A-11 can be rewritten as

$$G = 20 \ \log \frac{V_2}{V_1} + 10 \ \log \frac{R_1}{R_2}. \tag{A-12}$$

Comparing Eq. A-12 with A-7 shows that the first term of the power gain is the voltage gain, as defined in Eq. A-7. If $R_1 = R_2$, the second term in Eq. A-12 equals zero and both the voltage gain and the power gain, expressed in dB, are numerically equal. The values of resistances $R_1$ and $R_2$ must be known, however, to determine the power gain from a given voltage gain.

In a similar manner, the power gain of the circuit in Fig. A-1 can be expressed as

$$G = 20 \ \log \frac{I_2}{I_1} + 10 \ \log \frac{R_2}{R_1}. \tag{A-13}$$

Notice that in this case, the resistance ratio in the second term is the reciprocal of that in Eq. A-12.

**Example A-1.** A circuit has a voltage gain of 0.5, an input impedance of 100 $\Omega$, and a load impedance of 10 $\Omega$. From Eq. A-7, the voltage gain in decibels is $-6$ dB. From Eq. A-12

$$\text{dB power gain} = -6 + 10 \ \log \frac{100}{10} = 4 \ \text{dB}. \tag{A-14}$$

Therefore in this case, the power gain in decibels is positive, whereas the voltage gain in decibels is negative.

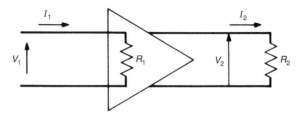

**FIGURE A-1.** Circuit for comparing power gain and voltage gain.

## A.3  POWER LOSS OR NEGATIVE POWER GAIN

Let us compute the power gain from point 1 to point 2 for the case where the power at point 2 is less than the power at point 1. The power gain in decibels is

$$G = 10 \log \frac{P_2}{P_1}. \tag{A-15}$$

To express the power ratio $P_2/P_1$ as a number greater than 1, we can rewrite Eq. A-15 as

$$G = 10 \log \left(\frac{P_1}{P_2}\right)^{-1} \tag{A-16}$$

From the identity of Eq. A-6 this becomes

$$G = -10 \log \frac{P_1}{P_2}. \tag{A-17}$$

Therefore, power loss is indicated by a negative decibels power gain.

## A.4  ABSOLUTE POWER LEVEL

The decibel may also be used to represent an absolute power level by replacing the denominator of Eq. A-1 with a reference power $P_0$, such as 1mW. This gives

$$\text{Number of dB(absolute)} = 10 \log \frac{P}{P_0}. \tag{A-18}$$

Equation A-18 represents the absolute power level above or below the reference power $P_0$. In this case, the user must know the reference power, which is usually expressed by adding letters to the abbreviation dB. For example, dBm is used to

**TABLE A-1. Reference Levels for Various dB Units**

| Unit | Type unit | Reference | Use | Remarks |
|------|-----------|-----------|-----|---------|
| dBa | Power | $10^{-11.5}$W | Noise | Measured with F1A weighting |
| dBm | Power | 1 mW | | |
| dBrn | Power | $10^{-12}$W | Noise | Reference noise power |
| dBrnc | Power | $10^{-12}$W | Noise | Measured with "C-message" weighting |
| dBspl | Sound Pressures | 20 μPa$^a$ | Acoustics | |
| dBu | Voltage | 0.775 V$^b$ | Audio | |
| dBV | Voltage | 1 V | | |
| dBmV | Voltage | 1 mV | | |
| dBμV | Voltage | 1 μV | | |
| dBμV/m | Field Strength | 1 μV/m | Electromagnetic fields | |
| dBw | Power | 1 W | | |

$^a$SPL is an abbreviation for Sound Pressure Level. Zero SPL is considered to be the threshold of hearing at 1 kHz and is equivalent to 20 μ Pa. This was the original use of the decibel.
$^b$A dBu is equal to a dBm if the impedance is 600 Ω. In other words, 0.775V across 600 Ω equals 1 mW.

signify a reference power of 1 mW. Table A-1 lists some of the more commonly used decibel units and their reference level and abbreviations.

In EMC measurements, we usually talk in terms of dB μV (or dB μV/m), which means the reference is a microvolt (or a microvolt/m). If the measured signal were 40 dB μV, then it would represent a 100-μV signal. An 80-dB μV signal would be a 10,000-μV signal.

Remember, decibels *are always ratios of numbers*, never an absolute quantity. So to say that the voltage gain of an amplifier is 22 dB makes sense. It is 20 times the log of the ratio of the output voltage divided by the input voltage. However, to say that the signal level out of the amplifier is 40 dB makes no sense, because decibels are not an absolute measurement. We could, however, say that the signal level out of an amplifier is 40 dB mV. The signal then is 40 dB greater than a millivolt, or 100 mV.

The decibel can therefore be properly used in only the following two situations:

- First, when it is obvious that we are talking about the ratio of two numbers, such as *the gain of the amplifier* or *the attenuation of the filter*.
- Second, if we specify the *reference* to which a single number(or measurement) is being compared, in which case the decibel term requires a subscript.

## A.5  SUMMING POWERS EXPRESSED IN DECIBELS

It is often necessary to determine the sum of two powers when the individual powers are expressed in decibels with respect to some reference power level (eg., dBm). The individual powers could always be converted to absolute power, added, and converted back to decibels, but this is time consuming. The following procedure can be used when combining such terms.

$Y_1$ and $Y_2$ are two power levels expressed in decibels above or below a reference power level $P_0$. $P_1$ and $P_2$ represent the absolute power levels corresponding to $Y_1$ and $Y_2$, respectively. Let us also assume that $P_2 \geq P_1$. From Eqs. A-18 and A-3 we can write

$$\frac{P_1}{P_0} = (10)^{Y_1/10} \tag{A-19}$$

and

$$\frac{P_2}{P_0} = (10)^{Y_2/10}; \tag{A-20}$$

therefore

$$\frac{P_1}{P_2} = (10)^{(Y_1 - Y_2)/10}. \tag{A-21}$$

Let us define the difference ($D$) between the two powers, expressed in decibels as

$$D = Y_2 - Y_1. \tag{A-22}$$

Therefore

$$P_1 = P_2 (10)^{-D/10}. \tag{A-23}$$

Adding $P_2$ to both sides gives

$$P_1 + P_2 = P_2 \left(1 + 10^{-D/10}\right). \tag{A-24}$$

Expressing the sum of the powers $P_1$ and $P_2$ in terms of decibels referenced to $P_0$ gives

$$Y_T = 10 \log \left(\frac{P_1 + P_2}{P_0}\right). \tag{A-25}$$

This can be rewritten as

$$Y_T = 10 \log(P_1 + P_2) - 10 \log P_0. \tag{A-26}$$

Substituting from Eq. A-24 for $P_1 + P_2$ gives

$$Y_T = 10\log\left[P_2\left(1 + 10^{-D/10}\right)\right] - 10\log P_0. \qquad \text{(A-27)}$$

$$Y_T = 10\log\left(\frac{P_2}{P_0}\right) + 10\log\left(1 + 10^{-D/10}\right). \qquad \text{(A-28)}$$

The first term in Eq. A-28 represents $Y_2$, the larger of the two individual powers expressed in terms of decibels. The second term represents how much $Y_2$ must be increased when the two powers are added.

The sum of two powers expressed in dB is therefore equal to the larger power increased by

$$10\log\left(1 + 10^{-D/10}\right) \qquad \text{(A-29)}$$

where $D$ equals the difference in decibels between the two original powers. The maximum value of this expression is 3 dB and occurs when $D = 0$. Therefore, if two equal powers are summed, the magnitude increases by 3 dB. Values of this expression as a function of $D$ are tabulated in Table A-2.

**TABLE A-2   Sum of Two Powers Expressed in Decibels.**

| Amount $D$ by Which Two Powers Differ (dB) | Amount by Which Larger Quantity Is Increased to Obtain Sum (dB) |
|---|---|
| 0 | 3.00 |
| 0.5 | 2.77 |
| 1 | 2.54 |
| 1.5 | 2.32 |
| 2 | 2.12 |
| 3 | 1.76 |
| 4 | 1.46 |
| 5 | 1.19 |
| 6 | 0.97 |
| 7 | 0.79 |
| 8 | 0.64 |
| 9 | 0.51 |
| 10 | 0.41 |
| 11 | 0.33 |
| 12 | 0.27 |
| 15 | 0.14 |
| 20 | 0.04 |

# APPENDIX B
# The Ten Best Ways to *Maximize* the Emission from Your Product

As an electromagnetic compatibility (EMC) consultant, I am not sure that this appendix is necessary. My experience indicates that many product designers already know about these techniques and practice them regularly. But as an aid to those of you new to the business, or to those who would like a review, I present it to help you quickly become proficient, and be on par with your more experienced colleagues. Following is a list of the 10 best ways to *maximize* the emission from your product.

1. The *clock* is usually the best place to start *maximizing* emission. Pick the highest frequency clock and the fastest rise time possible. Clock frequencies above 100 MHz with sub-nanosecond rise times are especially desirable. You should always route a clock around the board at a frequency higher than necessary and then divide the frequency down at the load. When viewed on an oscilloscope, clocks with sub-nanosecond rise time's look like ideal square waves, just what your digital guru likes to see, not those wimpy trapezoidal waves produced by slow rise time clocks.

2. *Clock routing* is important. Make sure that the layout is such that the clock trace is as long as possible. Also, route it as far away from any ground planes, power planes, and other ground or return traces as possible. When doing component placement on a printed circuit board (PCB) orient and locate all integrated circuits (ICs) to maximize the length of the clock runs. Routing clocks along the edge of the PCB and near or through the input/output (I/O) area is also very desirable.

3. Make sure that you have plenty of *slots* in your ground and power planes. The extra copper removed will decrease the weight of your printed circuit boards as well as force the logic return currents to flow through a larger loop, which thereby increases the inductance of the ground and power

*Electromagnetic Compatibility Engineering,* by Henry W. Ott
Copyright © 2009 John Wiley & Sons, Inc.

planes. The increased ground impedance will also increase the ground voltage that excites the I/O cables, thereby increasing their radiation.

4. If you have *split-ground planes* (e.g., analog ground and digital ground), make sure that you route a lot of high-frequency clock traces (or other high-frequency signals such as buses) across the gap between the planes. This approach will force the ground return currents to flow in big loops. If you do a really good job of this, you can force the return currents to flow all the way back to the power-supply ground terminal, where you have provided the only single-point connection between the two ground planes. These same principles also apply to split-power planes.

5. To provide ineffective high-frequency *decoupling* for your digital logic, place a single 0.1-uF capacitor (a 0.01-uF capacitor will do if you run out of 0.1-uF capacitors) somewhere in the vicinity of the IC. This decoupling method is the same as that used on digital logic ICs for the last 40 plus years, so it must still be the correct approach. Of course, if low cost is an objective, then you could leave the decoupling capacitors off completely, or use one capacitor for every three to five digital logic ICs. What ever you do, do not use more than one capacitor per IC. After all, multiple capacitors are expensive and take up additional board space and make routing more difficult.

6. *Unshielded I/O cables* should not have any common-mode filters or ferrite chokes on them. Ferrite cores produce large bulges in the cable and look strange. Besides, most users do not know what they are or why they are there. If you do use filters, then make sure that the printed circuit board layout maximizes the parasitics—the capacitance across a series element and the inductance in series with a shunt element. Also, locate the filter far away from where the cable enters or leaves the enclosure. This way, the I/O signal can pass through the filter and then be routed through the high-frequency logic area of the board, before exiting the enclosure.

7. If *shielded cables* are used for I/O signals, then terminate the cable shields with long pigtails. A 3-or 4-in. pigtail is probably sufficient, but 6 to 8 in. will be even more effective. One especially effective method is to connect the pigtail to a pin of the connector. Then, you can add an additional pigtail to the pin inside the product and terminate the shield to the circuit ground. Better yet, do not terminate the cable shield at all. This last approach will also reduce the manufacturing costs of the product.

8. The *logic ground* probably should be connected to the chassis somewhere. One method is to connect the edge of the printed circuit board farthest from the I/O cables to the chassis. This method will guarantee that the maximum ground voltage on the printed circuit board will be available to excite the I/O cables, which will cause them to radiate effectively. If only a single point connection is to be made to the chassis, then make sure it is made with a long thin wire or trace. This circuit ground to chassis connection, combined with the above-mentioned techniques (poor cable

shield termination, and ineffective filtering of unshielded cables), is very effective in *maximizing* the emission from the I/O cables.

9. If the product is in a *metal enclosure*, then many things can be done to *maximize* the emission. First of all, make sure that all seams are covered with a heavy coat of paint or other nonconductive finishes. This helps to resist corrosion and improves appearance. If you must use a conductive finish in the seam, then make sure that the mechanical design is such that little or no pressure is applied to push the surfaces together and make electrical contact. Also, maximize the number of seams and make each one as long as possible. Large cooling holes are another effective way to *maximize* the emission.

10. If a *power-line filter* is used, then make sure it is mounted a long way from where the power cord enters the enclosure. The power cord can then be run via a long circuitous route to the power line filter inside the enclosure. Also, the ground connection to the filter should be via a long wire. Whatever you do, do not connect the metal case of the filter directly to the chassis. Another of my favorite tricks is to tie or lace the filter's input and output power cables together. This technique makes for neat looking wiring.

Many other possible things could be done as part of the design process to *maximize* the emission, but the above 10 examples are a good place to start. Using the approach described in this appendix, you should be able to increase the emission from your product by at least 20 to 40 dB.

If, however, after building and testing the prototype, you should decide that you would like to *actually sell the product*, then you may want to consult an EMC engineer. He or she should be able to help you incorporate the necessary fixes into the product, after the fact, so that it will pass the regulatory compliance tests. *Good luck!!!*

# APPENDIX C
# Multiple Reflections of Magnetic Fields in Thin Shields

Consider the case of a magnetic field with a wave impedance $Z_1$ incident on a thin shield of characteristic impedance $Z_2$, as shown in Section 6.5.6, Fig. 16-14. Because the shield is thin and the velocity of propagation is large, the phase shift through the shield can be neglected. Under these conditions, the total transmitted wave can be written as

$$H_{t(\text{total})} = H_{t2} + H_{t4} + H_{t6} + \cdots. \qquad (\text{C-1})$$

From Eqs. 6-10 and 6-15 we can write

$$H_{t2} = \frac{2Z_1 H_0}{Z_1 + Z_2}(e^{-t/\delta})K, \qquad (\text{C-2})$$

where $K$ equals the transmission coefficient at the second interface from medium 2 to medium 1 (Eq. 6-17).

We can now write for $H_{t4}$,

$$H_{t4} = \frac{2Z_1 H_0}{Z_1 + Z_2}(e^{-t/\delta})(1 - K)(e^{-t/\delta})(1 - K)(e^{-t/\delta})K, \qquad (\text{C-3})$$

which reduces to

$$H_{t4} = \frac{2Z_1 H_0}{Z_1 + Z_2}(e^{-3t/\delta})(K - 2K^2 + K^3). \qquad (\text{C-4})$$

Consider the case of a metallic shield where $Z_2 \ll Z_1$. Then $K \ll 1$, $K^2 \ll K$, $K^3 \ll K$, and so on. The total transmitted wave can then be written as

$$H_{t(\text{total})} = 2H_0 K(e^{-t/\delta} + e^{-3t/\delta} + e^{-5t/\delta} + \cdots). \qquad (\text{C-5})$$

*Electromagnetic Compatibility Engineering,* by Henry W. Ott
Copyright © 2009 John Wiley & Sons, Inc.

The infinite series in brackets in Eq. C-5 has the limit*

$$e^{-t/\delta} + e^{-3t/\delta} + e^{-5t/\delta} + \cdots = \frac{\text{cosech}(t/\delta)}{2} = \frac{1}{2\sinh(t/\delta)}. \tag{C-6}$$

Substituting Eq. 6-17 for $K$ and Eq. C-6 for the infinite series in Eq. C-5, gives

$$H_{t(\text{total})} = \frac{4H_0 Z_2}{Z_1}\left[\frac{1}{2\sinh(t/\delta)}\right], \tag{C-7}$$

or

$$\frac{H_0}{H_{t(\text{total})}} = \left(\frac{Z_1}{4Z_2}\right)2\sinh\left(\frac{t}{\delta}\right). \tag{C-8}$$

Shielding effectiveness is 20 times the log of Eq. C-8, or

$$S = 20\log\frac{Z_1}{4Z_2} + 20\log\left[2\sinh\left(\frac{t}{\delta}\right)\right]. \tag{C-9}$$

Replacing $Z_1$ with the impedance of the wave at the shield $Z_w$, and replacing $Z_2$ with the shield impedance $Z_s$, gives

$$S = 20\log\frac{Z_w}{4Z_s} + 20\log\left[2\sinh\left(\frac{t}{\delta}\right)\right]. \tag{C-10}$$

The first term of Eq. C-10 is the reflection loss $R$, as defined by Eq. 6-22. To calculate the correction factor $B$, we must put Eq. C-10 into the form of Eq. 6-8. The second term of Eq. C-10 must therefore be equal to $A + B$. Thus, we can write

$$B = 20\log\left[2\sinh\left(\frac{t}{\delta}\right)\right] - A. \tag{C-11}$$

Substituting for $A$, from Eq. 6-12a, gives

$$B = 20\log\left[2\sinh\left(\frac{t}{\delta}\right)\right] - 20\log e^{t/\delta}. \tag{C-12}$$

---

*Standard Mathematical Tables,* 21st edition, p. 343 (Chemical Rubber Co., 1973).

Combining terms

$$B = 20 \log \left[ \frac{2 \sinh(t/\delta)}{e^{t/\delta}} \right], \tag{C-13}$$

expressing the sinh $(t/\delta)$ as an exponential, gives the correction factor $B$ as

$$B = 20 \log \left[ 1 - e^{-2t/\delta} \right]. \tag{C-14}$$

Figure 6-15 is a plot of Eq. C-14 as a function of $t/\delta$. Note that the correction factor $B$ is always a negative number, which indicates that less shielding is obtained from a thin shield because of the multiple reflections.

Table C-1 lists values of $B$ for very small values of $t/\delta$, which are not shown in Fig. 6-15.

**TABLE C-1.  Reflection Loss Correction Factor ($B$) for Thin Shields.**

| $t/\delta$ | $B(\mathrm{dB})$ |
|---|---|
| 0.001 | −54 |
| 0.002 | −48 |
| 0.004 | −42 |
| 0.006 | −38 |
| 0.008 | −36 |
| 0.01 | −34 |
| 0.05 | −20 |

# APPENDIX D
# Dipoles for Dummies (as well as for all the rest of us without a Ph.D. in electromagnetics)

Why is this appendix on antenna theory in a book on electromagnetic compatibility engineering? Because an understanding of some basic antenna theory is helpful for all electrical engineers, especially those involved in electromagnetic compatibility (EMC). After all, if a product radiates, or is susceptible, to electromagnetic energy, it is an antenna even if you call it something else such as a microprocessor, an integrated circuit (IC), a printed circuit board (PCB), a power cord, or an RS-232 cable.

An important characteristic to remember about antennas is reciprocity. Reciprocity means that if a structure (antenna) radiates well, then it will also pick up energy well, and vice versa. What prevents an antenna from radiating will also prevent an antenna from picking up energy. Therefore, the same techniques can be used to solve both emission and susceptibility problems.

## D.1  BASIC DIPOLES FOR DUMMIES

A dipole is a basic antenna structure that consists of two straight collinear wires (arms or poles) as depicted in Fig. D-1. The first thing to notice about a dipole is that it has two parts, hence, the term "di" in its name.

How can we explain the fact that it is possible to drive current into a dipole when the ends are open, and we, therefore, do not have a closed loop? The simplest way to resolve this seeming dilemma, without getting involved with electromagnetic field theory, is to consider the parasitic capacitance between the two arms (poles) of the antenna as the return current path, as shown in Fig. D-2. At high frequency, this capacitance will represent a low impedance. Current through this uncontrolled parasitic capacitance represents radiation.

Therefore, *a dipole requires two parts to radiate* and *the amount of radiation will be proportional to the dipole current*. Note also from Fig. D-2

*Electromagnetic Compatibility Engineering,* by Henry W. Ott
Copyright © 2009 John Wiley & Sons, Inc.

**FIGURE D-1.** Dipole antenna.

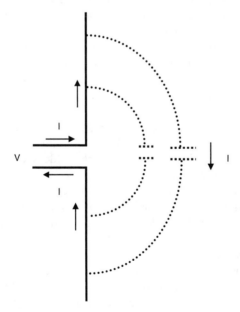

**FIGURE D-2.** Current in a dipole antenna flows through the capacitance between the arms (poles).

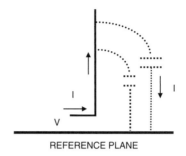

**FIGURE D-3.** Monopole antenna, showing the arm (pole) and the reference plane.

that *a dipole does not require a "ground" to work*, just capacitance between the two arms.

A good analogy is to consider what happens when we clap our hands. Clapping launches an acoustic wave, whereas a dipole launches an electromagnetic wave. It takes two hands to clap, just like it takes two arms for a dipole to radiate.

What about a monopole; does it only need one part to radiate? The answer is no; a monopole also needs two parts. A monopole is just a dipole cut in half, as we shall see. The second part is normally a reference plane located below the one arm (pole) as shown in Fig. D-3. If you do not provide the reference plane (which is the second part of the antenna), then the monopole will find something to work against, usually the largest metal object around. The current path for a monopole is through the parasitic capacitance betweeen the one arm of the monopole and the reference plane as shown in Fig. D-3. Note that the reference plane does not have to be a plane, and it does not have to be grounded. Any metal object with capacitance to the pole will do just fine, regardless of its shape. Other examples of monopole antennas configurations are shown in Fig. D-4.

Note that even in the case of the monopole, "no ground" is needed to operate.

Referring to our clapping hands analogy again, consider the case where you are asked to clap with one hand in your pocket. You take your free hand and find something to clap against such as your knee, a desk, a table, or wall; that is exactly how a monopole works.

Therefore, *the way to make an antenna (dipole or monopole) is to have a radio frequency (rf) potential between two pieces of metal*. The capacitance between the two pieces of metal will provide the current return path.

*The way to prevent radiation is to connect the two halves of the antenna together so that they are at the same potential*. By the way, it does not matter what potential the halves are at, as long as a potential difference does not exist between them.

Returning to the clapping analogy, if you put your two hands together and wrap duct tape around them (equivalent to holding the two arms of the dipole at the same potential), then you no longer can separate your hands and clap.

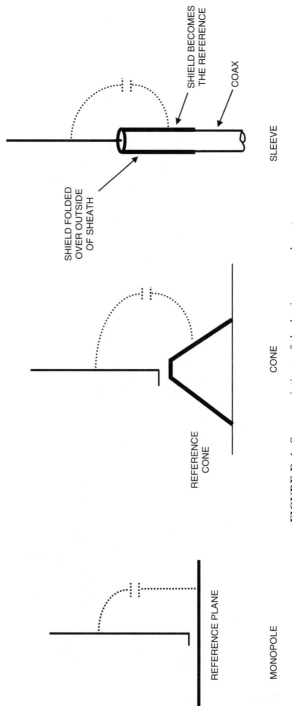

**FIGURE D-4.** Some variations of the basic monopole antenna.

SHIELD FOLDED
OVER OUTSIDE
OF SHEATH

SHIELD BECOMES
THE REFERENCE

COAX

SLEEVE

REFERENCE
CONE

CONE

REFERENCE PLANE

MONOPOLE

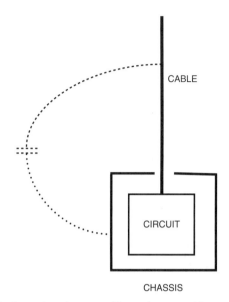

**FIGURE D-5.** A product in a metallic enclosure, with an attached cable.

So what does all of this have to do with EMC, you ask? It turns out quite a bit! Let us consider a simple product in a metallic enclosure with a single cable attached as shown in Fig. D-5. To make it more interesting, let us assume that we put the product on a rocket and launch it into space so that it is now orbiting the earth. I think under these circumstances we can bypass a discussion of how we should "ground" the product!

However, if a potential difference exists between the chassis and the cable, we have produced a monopole antenna (the cable is the arm of the monopole and the chassis is the reference plane), and the cable will radiate. This potential difference is referred to as a common-mode voltage.

Because we do not want a difference of potential to exist between the cable and chassis, how we connect the internal circuit to the chassis becomes important. *The internal circuit's reference (often called "circuit ground") should be connected to the chassis as close to where the cable terminates as possible* to minimize the voltage between the cable and chassis. This connection must provide a low impedance at rf frequencies. Any impedance between the circuit reference and the chassis will produce a voltage drop and will cause the assembly to radiate. In practice, this ground-to-chassis connection is often made with poorly placed metal standoffs and can be of considerable impedance. Seldom is the connection optimized for EMC purposes. This connection and how it is made is *critical* to the EMC performance of the product (see Section 3.2.5).

A second possibility to reduce the cable radiation is to place capacitors between all the cable conductors (even the ones we call ground) and the chassis, to short out the rf potential between the cable and chassis.

Third, we could use a common-mode choke (ferrite core) on the cable to raise the cable's common-mode impedance and hence to reduce the cable current produced by the common-mode voltage between the chassis and cable.

Last, but not least, we could shield the cable and terminate the shield properly (360° connection, see Section 2.15) to the chassis. In this case, the cable effectively does not leave the enclosure. You can think of the cable shield as just an extension of the chassis, and how well it does or does not behave in this manner is a strong function of the shield-to-chassis connection.

Note in the above example it matters not what potential the chassis is at with respect to the earth or any other arbitrary reference, only the difference of potential between the chassis and the cable matters.

Now, let us take our product out of orbit and bring it back to earth. Does it matter how we ground the chassis to some external reference, such as the earth or the power line ground? No, not from an EMC perspective! The same criteria, as when we were in orbit, still applies, our only requirement is to have no common-mode voltage between the cable and the chassis.

## D.2  INTERMEDIATE DIPOLES FOR DUMMIES

OK, so now you know something about how dipole and monopole antennas work. Let us look into the matter a little more and determine the current distribution along the length of a monopole antenna. The same result will also be applicable to a dipole if we apply the result to each of the dipole arms individually.

Let us assume that we drive a current $I$ into the base of the monopole antenna as shown in Fig. D-6. The current at the top of the antenna must be zero. Therefore, the current varies from $I$ at the base to zero at the top. If the antenna is short compared with a quarter of a wavelength (e.g., less than 1/10 wavelength), the current distribution will be linear from base to tip. If the

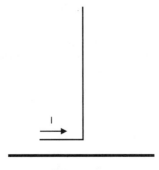

REFERENCE PLANE

**FIGURE D-6.** Current feeding a monopole antenna.

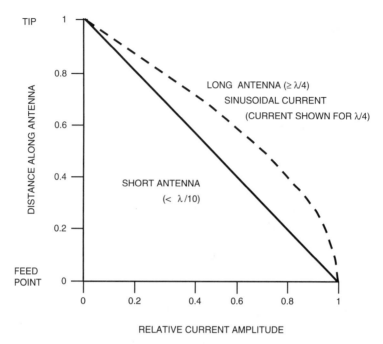

**FIGURE D-7.** Current distribution on a monopole antenna.

antenna is long, then the current distribution will be sinusoidal. This is show in Fig. D-7.

Clearly, the antenna does not radiate uniformly along its entire length. The bottom millimeter will radiate the maximum, and the top millimeter will hardly radiate at all. The average current will be 0.5 $I$ for a short antenna, and 0.637 $I$ for a quarter wavelength long antenna. Compared with an ideal antenna (one having uniform current along its entire length), a short dipole will produce half as much radiation, and a quarter wavelength long dipole will produce 64% as much radiation.

This leads directly to the concept of effective length or effective height of an antenna. If the effective length (in meters) is multiplied by the incident electric field strength (in V/m), it will give the voltage picked up by the antenna. For an ideal (uniform current distribution) dipole or monopole, the effective length will be equal to the actual length of the antenna. For a short dipole (monopole), however, the effective length will be equal to one half its actual length.

How could we make the antenna more efficient? By increasing the average current! That means forcing more current up to the top of the antenna. Because the current is flowing through the parasitic capacitance between the antenna arm and the reference plane, we have to increase the capacitance from the top of the antenna to the reference plane.

Figure D-8 shows what is often called a "top hat" or capacitive loaded antenna. By adding a large piece of metal at the top, we can increase the

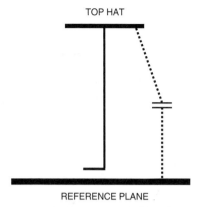

**FIGURE D-8.** Capacitive loaded monopole antenna.

capacitance from the top of the antenna to the reference plane and thereby increase the current flowing up to the top of the antenna. The "top hat" can be a metal disk, radial wires, or a metal sphere. It does not matter what shape it is as long as it increases the capacitance from the top of the monopole to the reference plane.

As was mentioned before, the same approach can be used with a dipole, only in the dipole case we must apply the "top hat" to the ends of both arms. The resulting antenna is then often referred to as a "dumbbell" antenna; see Fig. D-9.

Therefore, we can conclude that *adding metal (capacitance) to the end(s) of a dipole or monopole antenna will increase its radiation efficiency.* Because the tophat adds capacitance, it will also decrease the resonant length (see Section D.3.2) of the antenna, because the antenna will now have some current at the top. This can be deduced by studying Fig. D-16.

Again you can ask the question, what does this all have to do with EMC? What it has to do with EMC is that you want to make sure you do not configure your product in such a way that you produce a "top hat" antenna.

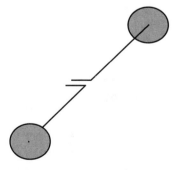

**FIGURE D-9.** Capacitive loaded dipole antenna.

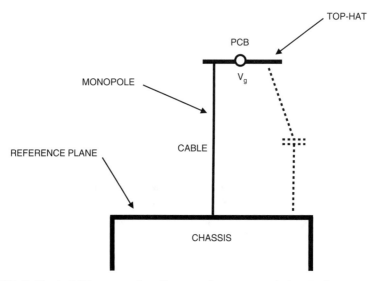

**FIGURE D-10.** A PCB mounted a distance above a metal chassis forms a top-hat antenna.

Consider the product depicted in Fig. D-10, which consists of a PCB connected to the end of a long cable and mounted a significant distance above a metal chassis. We have just created a "top hat" antenna and the structure will radiate efficiently. The cable is the monopole, the chassis is the reference plane, and the PCB is the "top hat." Therefore, *when a PCB is mounted in a product with a metal chassis, it should be mounted as close to the chassis as possible and have its reference (ground) connected to the chassis.*

A similar situation exists when a daughterboard is mounted above a PCB as shown in Fig. D-11. This is not as bad as the case shown in Fig. D-10, because the dimensions are much smaller, however, it can also, on occasion, be a problem. The solution is also simpler in this case; just connect the daughter-board ground to the motherboard ground through multiple metal standoffs or by some other means.

Figure D-12 shows an interesting example. It is similar to the example of Fig. D-5, but this time the product is in a plastic enclosure instead of a metal box. In

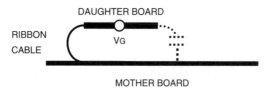

**FIGURE D-11.** A daughter board mounted above a PCB forms a top-hat antenna.

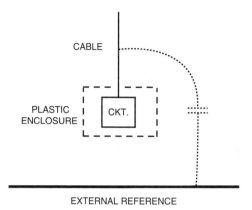

**FIGURE D-12.** A product in a plastic enclosure, with an attached cable.

this situation, the product does not provide a reference plane for the monopole to work against, so that the monopole will have to find something external to the product to act as its reference plane. This reference plane could be the actual ground (earth) or a metal table, file cabinet, or other metal object in the vicinity. In each location where the product is placed, the reference plane could be something different. Under these circumstances, how do you eliminate the common-mode radiation?

In this case, you may be better off by intentionally providing the other half of the antenna (the reference plane) as part of the product, instead of letting it find something different in each installation. One way to do this would be to add a metal plate, as shown in Fig. D-13, to the bottom of the plastic enclosure and then short out the monopole to this plate. This plate does not have to be thick or heavy (metal foil will do), but it should be large to have maximum

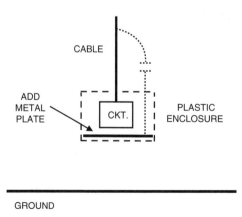

**FIGURE D-13.** A product in a plastic enclosure with the addition of a metal reference plate.

capacitence to the pole. At this point, the question is usually asked, how large? The answer is simple, as large as the enclosure will allow.

## D.3   ADVANCED DIPOLES FOR DUMMIES

If you followed the discussion so far, you now know quite a bit about dipoles and are ready to move to the advanced class. Here we will determine the impedance of a dipole antenna. The impedance is important because it affects the ability to couple energy into, or extract energy from, the antenna structure.

### D.3.1   Impedance of a Dipole

Referring to Fig. D-2, we can determine that one of the elements of the antenna impedance will be a capacitor. We also know that the wire arms of the antenna must have inductance, and this inductance will be in series with the capacitance.

If the antenna radiates, then energy will be lost, and this lost energy must be accounted for in our model. The only component that will dissipate power is a resistor, so let us add that to our model in series with the inductor and capacitor. *Therefore, the equivalent circuit for a dipole antenna becomes a series R, L, C network* as shown in Fig. D-14. We call the resistance $R_R$ the "radiation resistance" because it represents the energy that is lost as radiation.

From Fig. D-14, we observe that a dipole is actually a series resonant circuit. At frequencies below resonance, the impedance will be capacitive, at frequencies above resonance it will be inductive, and at resonance it will be resistive.

The impedance of a monopole is one half that of a dipole. This can be deduced by studying Fig. D-15. Figure D-15A shows a dipole antenna and its impedances. If we cut the dipole in Fig. D-15A in half and add a reference plane, along the cut then we form a monopole antenna as shown in Fig. D-15B. The monopole inductance and resistance will be one half that of the dipole, and the capacitance will be twice that of the dipole.

### D.3.2   Dipole Resonance

Refering to Fig D-14 we can conclude that below the resonant frequency the input impedance to the antenna will be large ($> 1000$ $\Omega$) because of the

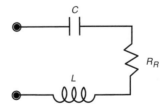

**FIGURE D-14.** The impedance of a dipole is a series R-L-C circuit.

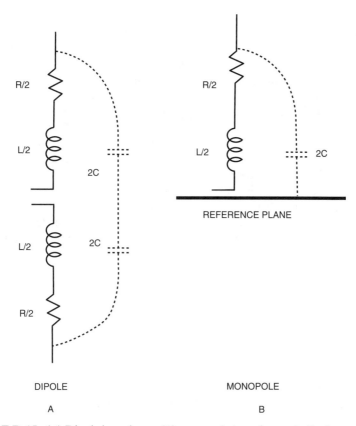

**FIGURE D-15.** (*A*) Dipole impedance, (*B*) monopole impedance. A dipole cut in half is equivalent to a monopole.

impedance of the capacitor. Above the resonant frequency, the impedance will also be large ($> 1000\ \Omega$) because of the impedance of the inductor. At resonance, however, the impedance will be low (around $70\ \Omega$ for a dipole and $35\ \Omega$ for a monopole) because, at resonance, the inductive reactance will cancel the capacitive reactance, which leaves just the radiation resistance.

It will be difficult for the common-mode voltage (or any other voltage for that matter) to drive much current into the antenna when the input impedance is large. However, it will be easy to drive current into an antenna at resonance, when the impedance is low. Therefore, dipole (monopole) resonance is important with respect to EMC. *At the resonant frequency, it is much easier to couple energy into or extract energy from the antenna, and it will therefore be a more efficient radiator or receptor of electromagnetic energy.*

As it turns out, the resonant frequency of a dipole (or monopole) is related to its length. Resonance will occur when the length of one of the antenna arms (elements) is one quarter wavelength. Therefore, a dipole will be resonant when its overall length is equal to one half a wavelength, and a monopole will be resonant when its length is one quarter wavelength.

Why this is so can best be understood by recalling the discussion relating to Fig. D-7, with respect to the current in a monopole. At any point in time, the current distribution along the length of a conductor will be sinusoidal as shown in Fig. D-16A. The required boundary condition for an antenna element is that the current at the tip be zero. Figure D-16B shows various length antenna elements placed such that the current at the tip will be zero. As can be observed, the current at the base will be maximum when the element is a quarter of a wavelength long. The highest current point also represents the lowest impedance point; hence, this represents the resonant length.

If the antenna element is shorter than a quarter wavelength, then the current at the base will be lower; hence, a higher impedance and the element will be below resonance. If the antenna element is longer than a quarter wavelength, then the current at the base will also be lower; hence, a higher impedance and the element will be above resonance.

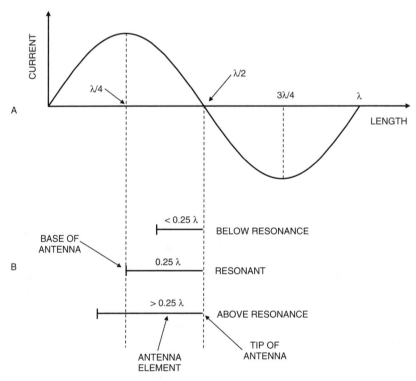

**FIGURE D-16.** (A) Current distribution along the length of a conductor. (B) Current distribution on various length antenna elements.

Additional resonances will also occur when the element length is equal to odd multiples of a quarter wavelength. This can be deduced from Fig. D-16 by extending the sine wave of current to the left (Fig D-16A) and then moving the base of the antenna (Fig D-16B) to the next current peak. At these frequencies, the cable radiation will also be increased because of the resonance.

### D.3.3   Receiving Dipole

Figure D-17A shows a dipole antenna exposed to an electric field $E$. Figure D-17B shows the equivalent circuitry for the receiving dipole of Fig. D-17A, where $Z_A$ represents the dipole impedance (Fig. D-14) and $R_L$ is the load impedance. If the dipole has an effective length of $L_e$, then the voltage induced into the antenna when exposed to the electric field $E$ will be

$$V_i = L_e E, \tag{D-1}$$

regardless of the length of the antenna or whether the antenna is resonant.

The voltage across the load $R_L$ at the terminals of the antenna will be equal to

$$V_L = \left(\frac{R_L}{R_L + Z_A}\right) V_i = \left(\frac{R_L}{R_L + Z_A}\right)(L_e E). \tag{D-2}$$

From Fig. D-14, we know that the impedance $Z_A$ is equal to

$$Z_A = R_R + X_L - X_C, \tag{D-3}$$

which is frequency dependent. The antenna impedance $Z_A$ will be large at frequencies above and below resonance, and it will be small at resonance, where the second and third terms cancel, which will leave just the radiation resistance $R_R$. Therefore, $V_L$ will be a maximum at resonance (where $Z_A$ is small) and will decrease at frequencies above and below resonance (where $Z_A$ is large).

Notice from the above three equations, and Fig. D-17B, that the voltage induced into the antenna $V_i$ is frequency independent, whereas the voltage $V_L$ at the antenna terminals is frequency dependent. Therefore, the problem is not that a nonresonant antenna will not pick up a voltage, rather the problem is that the picked up voltage cannot be coupled out of the antenna because of the large magnitude of the antenna impedance $Z_A$, when the antenna is not resonant.

### D.3.4   Theory of Images

Let us compare the performance of the dipole and monopole as radiators. For example, let us assume that we measure the field at a point in space $d$ meters

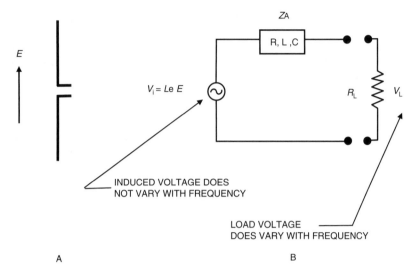

A                                                                    B

**FIGURE D-17.** (A) Receiving dipole, (B) equivalent circuit of a receiving antenna.

from, and at a 45° angle from the axis of a dipole shown in Fig. D-18A. How will this compare with the field measured at a similar point the same distance away from a monopole of one half the length and carrying the same current as the dipole as shown in Fig. D-18B.

To answer this question, we can use the theory of images. The easiest way to understand the theory of images is to consider something that we are all

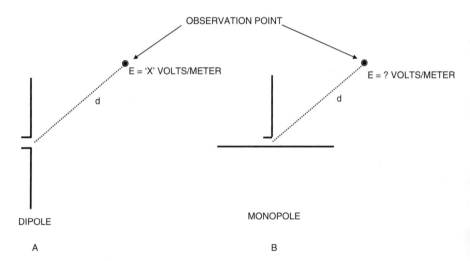

A                                                                    B

**FIGURE D-18.** (A) Transmitting dipole, (B) transmitting monopole.

familiar with—a mirror. After all, light is electromagnetic energy similar to what we have been discussing, except of a much higher frequency.

If you look into a mirror (a reflective surface), what do you see? You see yourself! If you move three steps back from the mirror, what does your image do? It also moves three steps back. Therefore, a mirror produces an image of the object in front of it, and that image is located as far behind the mirror as the object was in front of the mirror.

The same thing happens with a monopole above a reference plane (a reflective surface). The reference plane produces an image of the monopole as far below the plane as the antenna is above the plane. Saying it slightly differently, the field produced at any point in space, in the upper hemisphere, by a conductor perpendicular to a reflective plane is equivalent to the field that would be produced by the original conductor plus a second identical conductor located an equal distance below the plane, as the original conductor is above the plane, but without the plane present. Figure D-19 shows this equivalence. Therefore, the monopole is equivalent to the dipole in the upper hemisphere.

The answer to our original question is then, that as long as we limit ourselves to considering the fields in the upper hemisphere, *the monopole and the dipole produce the exact same field at the observation point.*

### D.3.5  Dipole Arrays

Dipoles do not have to be used individually; rather, they are often combined in various ways to modify their radiating or receiving characteristics. Two common dipole arrays are a Yagi and a log periodic.

A Yagi antenna consists of a single driven dipole, a number of slighter shorter parasitic dipoles called directors located in front of the active dipole, and a single slightly longer dipole called a reflector located behind the active

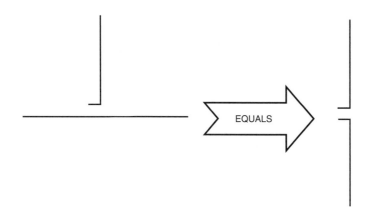

**FIGURE D-19.** Dipole–monopole equivalence.

dipole, as shown in Fig. D-20. The objective of a Yagi is to increase the gain of the antenna. Because an antenna is a passive structure, the only way to obtain gain (i.e., more energy in one direction) is to take the energy from somewhere else, that is, reduce the energy in another direction, which decreases the beamwidth and increases the directivity of the antenna. An optimized Yagi antenna will provide about 10 dB of gain over a dipole. Yagi's are commonly used as very high frequency (VHF) TV antennas.

A log periodic antenna is an array of driven elements that decreases in length and spacing as shown in Fig. D-21. The log periodic is fed from the front with the feed-line crisscrossing between antenna elements. The purpose of a log periodic antenna is to produce an antenna that will operate efficiently over a

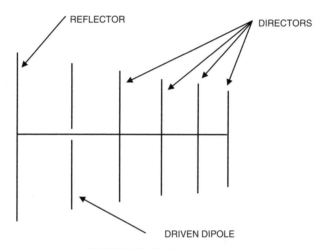

REFLECTOR                    DIRECTORS

DRIVEN DIPOLE

**FIGURE D-20.**  Yagi antenna.

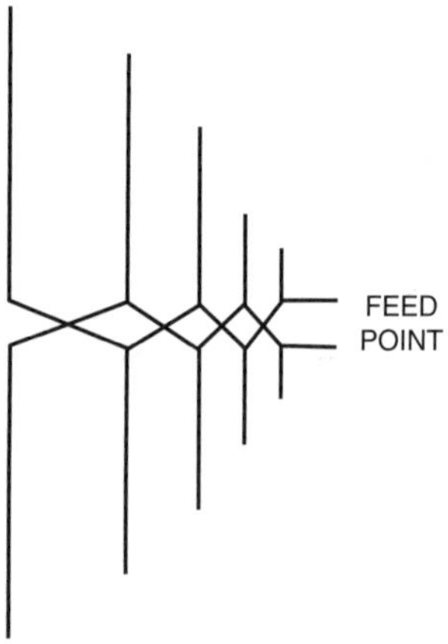

FEED
POINT

**FIGURE D-21.**  Log periodic antenna.

large frequency range. At different frequencies, a different dipole becomes the active element. Because the impedance of a dipole is large at frequencies off of resonance, the nonresonant dipoles draw little or no current from the feed line.

A log periodic antenna has fairly uniform impedance and radiation pattern from the resonant frequency of the longest element to the resonant frequency of the shortest element. A single log periodic antenna is often used for EMC testing over the frequency range of 300 to 1000 MHz.

### D.3.6   Very High-Frequency Dipoles

At very high frequencies ($>1$ GHz), the size of a resonant dipole is small; at 3 GHz it is equal to 2 in. Therefore, it will pick up or radiate little energy because from Eq. D-1, the induced voltage is equal to the incident field strength $E$ times the effective length $L_e$, which is less then the actual length of the antenna. If the dipole is made larger, then its impedance will be high, and we will not be able to couple energy to it or extract energy from it. What can be done to solve this dilemma? How about using a small resonant dipole with a big reflector behind it to concentrate a lot of energy at the focal point, where we will locate the small dipole as shown in Fig. D-22. This increases the field strength $E$ at the dipole, which increases the voltage that it picks up. We just made a commonly used satellite-receiving antenna.

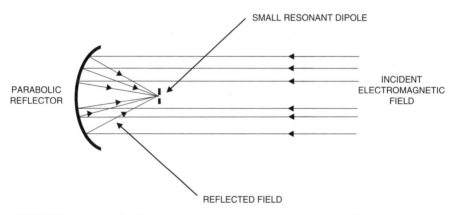

**FIGURE D-22.** A small dipole with a parabolic reflector makes an effective, very high-frequency antenna.

## SUMMARY

- A dipole (or monopole) antenna requires two parts.
- The magnitude of the radiation will be proportional to the dipole (or monopole) current.

- A dipole (or monopole) antenna does not require a ground to work.
- A monopole is just a dipole in disguise.
- The way to make a dipole (or monopole) antenna is to have an rf potential between two pieces of metal.
- The way to prevent radiation is not to have a potential difference between the two halves of the antenna.
- The internal circuit reference (ground) of a product should be connected to the chassis as close to where the cable enters/leaves the product as possible.
- The effective length (effective height) of an antenna is defined as the ratio of the voltage induced into the antenna (not the terminal voltage), to the magnitude of the incident electric field.
- Adding metal (capacitance) to the end(s) of a dipole or monopole antenna will increase its radiation efficiency.
- A PCB should be mounted as close to a metal chassis as possible, and have its ground connected directly to that chassis.
- A plastic enclosure for an electronic product should contain a metal reference plane.
- The equivalent circuit of a dipole (or monopole) is a series R–L–C circuit.
- The impedance of a monopole is one half that of a dipole.
- At the resonant frequency, it is much easier to couple energy into an antenna and, therefore, it will be a more efficient radiator (or receptor) at this frequency.
- Antenna resonance occurs when the element(s) of the antenna are one quarter wavelength long.
- Multiple resonances will occur at frequencies that are odd-numbered harmonics of a quarter wavelength frequency.
- A monopole and dipole both radiate the same field.

## FURTHER READING

German, R. F. and Ott, H. W. *Antenna Theory Simplified.* One-Day Seminar, Henry Ott Consultants, 2003.

Iizuka, K. "Antennas for Non-Specialists." *IEEE Antennas and Propagation,* February 2004.

# APPENDIX E
# Partial Inductance

Inductance is an important concept to understand when analyzing electromagnetic compatibility (EMC) issues in electronic systems. However, inductance is not well understood. As a result, there is considerable misunderstanding and confusion about the meaning of inductance as well as its calculation and/or measurement. One possible interpretation of inductance is depicted in Fig. E-1. If Fig. E-1 is not the correct interpretation, just what is?

## E.1  INDUCTANCE

When current flows in a conductor, a magnetic flux $\phi$ is produced around the conductor as shown in Fig. E-2. If the current increases, the magnetic flux increases proportionally. Inductance is the constant of proportionality between current and magnetic flux. This can be written as

$$\phi = LI, \tag{E-1a}$$

where $\phi$ is the magnetic flux produced by the current $I$ and $L$ is the inductance of the conductor. Solving Eq. E-1a for the inductance $L$ gives

$$L = \frac{\phi}{I}. \tag{E-1b}$$

Inductance comes in many flavors, which include self-inductance, mutual inductance, loop inductance, and partial inductance. It is important to understand the differences between these. This difference revolves around what magnetic flux $\phi$ is used in Eq. E-1b when calculating the inductance.*

---

* In this appendix, we are only interested in calculating the external inductance of conductors. The external inductance involves only the fields external to the conductor (as shown in Fig. E-2), not the fields internal to the conductor itself. At high frequencies, the internal inductance is negligible, and the only inductance of significance is the external inductance; see Section 5.5.1.

---

*Electromagnetic Compatibility Engineering,* by Henry W. Ott
Copyright © 2009 John Wiley & Sons, Inc.

**FIGURE E-1.** Milli and Henry's in-duck-dance (Courtesy of Otto Buhler).

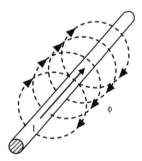

**FIGURE E-2.** Magnetic fields surrounding a current-carrying conductor.

Equation E-1a points out why EMC engineers are always emphasizing the importance of reducing the inductance of signal and ground conductors. If these conductors have inductance, they will have a magnetic flux $\phi$ surrounding them, and that flux is proportional to the inductance. Where there is an uncontained magnetic flux, there will be radiation. Therefore, the more inductance a conductor has, the more noise it will generate, as is clearly shown in Fig E-3.

**FIGURE E-3.** More in-duck-dance produces more noise (Courtesy of Kathryn Whitt).

## E.2   LOOP INDUCTANCE

To produce a magnetic flux $\phi$, there must be current flow, and to have current flow, there must be a current loop. This often leads to the erroneous conclusion that inductance can only be defined for the case of a complete loop. Weber (1965) even states; "It is important to observe that inductance of a piece of wire not forming a loop has no meaning." We will see shortly, however, that this is not true.

The magnetic flux density at a distance $r$ from a current-carrying conductor can be determined using the Biot-Savart Law and is equal to (Eq. 2-14 in Section 2.4).

$$B = \frac{\mu I}{2\pi r},$$ (E-2)

for $r$ greater than the radius of the conductor. $B$ is the magnetic flux density ($\phi$/ unit area), $\mu$ is the magnetic permeability, $I$ is the current in the conductor, and $r$ is the distance or radius from the conductor to the point at which the magnetic flux density is being determined.

The *self-inductance* of a loop is equal to

$$L_{loop} = \frac{\phi_T}{I}$$ (E-3)

where $\phi_T$ is the total magnetic flux penetrating the surface area of the loop, and $I$ is the current in the loop.

The *mutual inductance* between two loops 1 and 2 is equal to

$$M_{12} = \frac{\phi_{12}}{I_1},$$    (E-4)

where $\phi_{12}$ is the magnetic flux produced by loop 1, which penetrates the surface area of loop 2, and $I_1$ is the current, in loop 1, which produced the magnetic flux.

It is interesting to note, by comparing Eqs. E-3 and E-4, that the maximum value of the mutual-inductance is equal to the self-inductance. This is true because the self-inductance is equal to the total magnetic flux $\phi_T$ divided by the current that produced it, whereas the mutual inductance is equal to some of the magnetic flux $\phi_{12}$ divided by the current that produced it. The maximum value of some of something is all of it. Therefore, one can write

$$M_{12} \leq L_{loop}.$$    (E-5)

### E.2.1   Inductance of a Rectangular Loop

The inductance can be easily calculated for only a few simple geometries. This emphasizes an important point, that although the theory of inductance is simple, the calculation of inductance is often complex.

Consider the rectangular loop shown in Fig. E-4, with sides having lengths $a$ and $b$, and carrying a current $I$. The magnetic flux penetrating the surface area ($S = a\,b$) of the loop as the result of the current in the left-hand current-carrying segment can be obtained by summing, by means of an integral, the flux in the small area $dS$ from $r$ equals $r_1$ to $a$, which gives

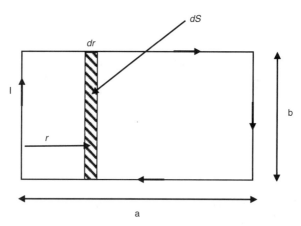

**FIGURE E-4.** A rectangular loop.

$$\phi = BS = \int_{r_1}^{a} \frac{\mu I dS}{2\pi r},$$

(E-6)

where $r_1$ is the radius of the current-carrying conductor. The surface area $dS$ of the small segment of the loop located a distance $r$ from the left-hand conductor is

$$dS = bdr.$$

Substituting this for $dS$ in Eq. E-6 gives

$$\phi = \frac{\mu Ib}{2\pi} \int_{r_1}^{a} \frac{1}{r} dr = \frac{\mu Ib}{2\pi} \ln \frac{a}{r_1}.$$

(E-7)

By symmetry, the magnetic flux that penetrates the area of the loop as the result of the current in the right-hand segment of the loop will also be equal to Eq. E-7.

Similarly we can write the magnetic flux penetrating the area of the loop as the result of the current in the top segment of the loop as

$$\phi = \frac{\mu Ia}{2\pi} \int_{r_1}^{b} \frac{1}{r} dr = \frac{\mu Ia}{2\pi} \ln \frac{b}{r_1}.$$

(E-8)

Again by symmetry, the magnetic flux that penetrated the surface area of the loop as the result of the current in the bottom segment of the loop will also be equal to Eq. E-8.

The total magnetic flux that penetrates the loop will therefore be twice that of Eq. E-7 plus twice that of Eq. E-8 or

$$\phi_T = \frac{\mu Ib}{\pi} \ln \frac{a}{r_1} + \frac{\mu Ia}{\pi} \ln \frac{b}{r_1},$$

(E-9)

and the inductance of the rectangular loop is equal to

$$L_{loop} = \frac{\phi_T}{I} = \frac{\mu}{\pi} \left[ b \ln \frac{a}{r_1} + a \ln \frac{b}{r_1} \right]$$

(E-10)

Equation E-10 neglects the fringing of the magnetic fields at the corners of the loop. A more accurate equation for the inductance of a rectangular loop, which includes the effect of fringing, is given by Eq. E-20.

The loop inductance represented by Eq. E-10 can be placed anywhere around the loop; *its location in the loop cannot be uniquely determined.* Therefore, from a

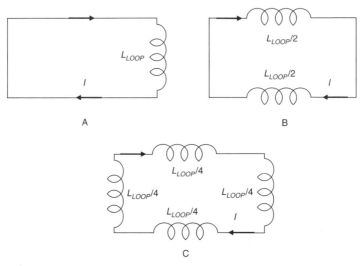

**FIGURE E-5.** The loop inductance can be placed anywhere around the loop.

loop inductance perspective, the circuits of Figs. E-5A, E-5B, and E-5C all behave the same. Which, if any, of these models is correct? Possibly none! That question cannot be answered with just knowledge of the loop inductance. So how can one determine the inductance of just one segment of a loop?

For example, let us say that you want to determine the inductance of just the ground conductor of your circuit to calculate the ground noise voltage. Or possibly you want to determine the inductance of just the power trace on a printed circuit board (PCB) to be able to determine the magnitude of the voltage dip that will occur when an integrated circuit (IC) switches and draws a large transient current. Knowledge of the loop inductance is not helpful in either of these cases.

For the case of a square loop, it may be reasonable to assume that one quarter of the inductance is in each of the four segments of the loop, which is similar to that shown in Fig. E-5C. But what about the case when the loop is not square or the conductors are not the same length, or diameter? For example, one conductor is a 26-Ga wire (or a narrow trace on a PCB) and the other conductor is a large ground plane. In these cases, the inductance of the individual conductors (ground plane and trace) cannot be determined from knowledge of the loop inductance. However, one can determine a unique inductance for each segment of a loop using the theory of partial inductance.

## E.3   PARTIAL INDUCTANCE

The theory of partial inductance is a powerful concept that is important to understand, because it allows one to define a unique inductance associated with only part of a loop. This approach allows one to explain the phenomena of ground bounce and power rail collapse. Ground bounce, or ground voltage, is created when a transient current flows through the partial inductance of a ground bus or plane. Power rail collapse, or power voltage dip, occurs when a transient current flows through the partial inductance of the power bus or plane. Without the theory of partial inductance, these concepts cannot be explained, because the inductance of a segment of a loop cannot otherwise be uniquely determined.

Ruehli (1972), expanding on the work of Grover (1946), has shown *that a unique inductance can be attributed to a segment of an incomplete loop*. As was the case with loop inductance, both partial self-inductances and partial mutual inductances exist.

### E.3.1  Partial Self-Inductance

Vital to the understanding of partial inductance is the ability to define the surface area over which the magnetic flux density must be summed to determine the value of the magnetic flux to be used in Eq. E-1b when calculating the partial inductance.

For the case of an isolated segment of a current-carrying conductor, Ruehli has shown that the partial self-inductance flux area is the surface area bounded on one side by the conductor segment, on one side by infinity, and on the other two sides by straight lines perpendicular to the conductor segment as shown in Fig. E-6.

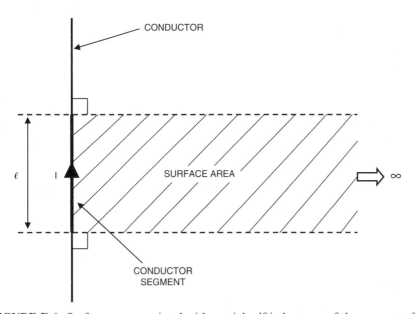

**FIGURE E-6.** Surface area associated with partial self-inductance of the segment of a conductor.

*Therefore, the partial self-inductance of a conductor segment is the magnetic flux penetrating the surface area between the conductor segment and infinity, divided by the current in the conductor segment.*

The magnetic flux that penetrates the surface area shown in Fig. E-6 is equal to the following surface integral:

$$\phi = \int_S \bar{B} \cdot d\bar{S}. \tag{E-11}$$

The partial self-inductance of a conductor segment of length $\ell$ and radius $r_1$ can therefore be written as

$$L_p = \frac{\mu I}{2\pi} \int_{r_1}^{\infty} \frac{1}{r} dr. \tag{E-12}$$

Equation E-12 cannot be evaluated directly because of the infinite limit of integration. However, because the magnetic flux density $\bar{B}$ is equal to the curl of the vector magnetic potential $\bar{A}$, we can write, $\bar{B} = \bar{\nabla} x \bar{A}$, and by using Stokes' theorem, the integral of Eq. E-11 over the surface area $S$ can be transformed into a line integral of the vector magnetic potential $\bar{A}$ over the circumference $C$ of the surface area. Therefore,

$$\phi = \int_S \bar{B} \cdot d\bar{S} = \int_C \bar{A} \cdot d\bar{l}. \tag{E-13a}$$

At first glance, this equation does not seem to solve the problem of the infinite integral, because the circumference of the surface area is also infinite. The circumference of the surface area has four sides: one along the conductor, two sides perpendicular to the conductor, and one side parallel to the conductor at infinity, see Fig E-6.

It can easily be demonstrated, however, that the line integral of the vector magnetic potential only has to be taken along the side of the loop adjacent to the conductor. The vector magnetic potential $\bar{A}$ is oriented in the direction of the current $I$ on the conductor, as shown in Fig. E-7. Because the vector magnetic potential $\bar{A}$ is equal to zero at infinity, the integral along that side of the surface area will be zero. The two sides of the area perpendicular to the conductor are at right angles to $\bar{A}$; hence, the line integral of $\bar{A} \cdot d\bar{l}$ along these paths is also equal to zero. Therefore, the integration over the circumference of the surface area reduces to just the integration, from point $a$ to point $b$, over the side of the area adjacent to the conductor. Therefore, Eq. E-13a reduces to

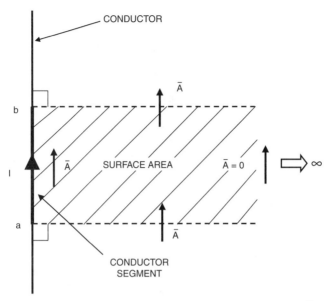

**FIGURE E-7.** Direction of the vector magnetic potential $\bar{A}$.

$$\phi = \int_a^b \bar{A} \cdot d\bar{l},$$
(E-13b)

and this integral is fininte.

To determine the vector magnetic potential $\bar{A}$ and transform it into a form that can be evaluated involves considerable mathematical manipulation beyond the scope of this book (see Ruehli, 1972), which demonstrates, as previously stated, that the theory of inductance may be simple but the actual calculation of the inductance is often complex.

For a round conductor segment of length $\ell$ and radius $r_1$, Grover (1946) gives the *partial self-inductance* as

$$L_p = \frac{\mu l}{2\pi} \left[ \ln \frac{2l}{r_1} - 1 \right]$$
(E-14)

where $\mu$ is the permeability of free space and is equal to $4\,\pi \times 10^{-7}$.

### E.3.2   Partial Mutual Inductance

The *partial mutual inductance* between segments of two arbitrary conductors can be determined in a manner similar to that used above to obtain the partial self-inductance of a conductor. For this case, Ruehli has shown that the

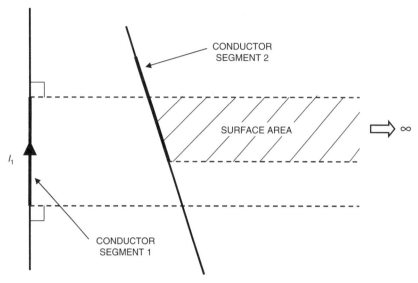

**FIGURE E-8.** Surface area associated with the partial mutual inductance of two conductors segments.

partial mutual inductance flux area is the surface area bounded on one side by the conductor segment 2, on another side by infinity, and on the two remaining sides by two straight lines perpendicular to conductor segment 1, as shown in Fig. E-8.

Figure E-8 shows the partial mutual inductance flux area associated with two coplanar, nonparallel, and offset conductor segments. Note, it is not necessary for the segments to be coplanar, but the analysis is simplified if they are.

*Therefore the partial mutual inductance between the two conductor segments is the ratio of the magnetic flux penetrating the surface area between the second conductor segment and infinity, divided by the current $I_1$ in the first conductor segment.*

Consider the case shown in Fig. E-9 of two coplanar parallel conductor segments, separated by a distance $D$. Calculating the flux produced by the current $I_1$ that penetrates the partial mutual inductance surface area (the surface area between conductor 2 and infinity), and dividing by the current $I_1$, gives the partial mutual-inductance between the two conductor segments as

$$L_m = \frac{\mu l}{2\pi} \int_D^\infty \frac{1}{r} dr, \qquad \text{(E-15)}$$

where $\ell$ is the length of the current-carrying conductor segment and $D$ is the spacing between the conductors.

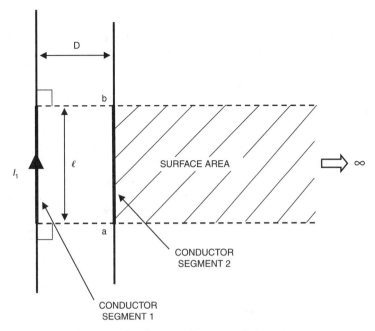

**FIGURE E-9.** Example of two coplanar parallel conductor segments.

Again, the infinite integral cannot be directly evaluated but can be converted by Stokes' theorem into a line integral of the vector magnetic potential $\bar{A}$ over the circumference of the surface area. As was the case for the partial self-inductance, the integration of $\bar{A}$ only has to be performed between points $a$ and $b$ over the side of the surface area adjacent to conductor segment 2, because the integration over the other three sides equals zero. This evaluation once again involves considerable mathematical manipulation beyond the scope of this book (see Ruehli, 1972).

For two identical and parallel round conductor segments of length $\ell$ and separated by a distance $D$, Grover (1946) gives the *partial mutual inductance* in terms of the following infinite series

$$L_{p12} = \frac{\mu l}{2\pi}\left[\ln\frac{2l}{D} - 1 + \frac{D}{l} + \frac{1}{4}\frac{D^2}{l^2} + \cdots\right].\qquad\text{(E-16)}$$

If $D \ll \ell$, Eq. E-16 reduces to

$$L_{p12} = \frac{\mu l}{2\pi}\left[\ln\frac{2l}{D} - 1\right].\qquad\text{(E-17)}$$

### E.3.3   Net Partial-Inductance

The *net partial inductance* $L_{np}$ of any conductor segment is equal to the partial self-inductance of that segment, plus or minus the partial mutual inductance from all nearby current-carrying conductors. The sign of the mutual inductance will depend on the direction of current flow. If the current flow in the two conductor segments is in the same direction, then the sign of the partial mutual-inductance term will be positive. If the currents in the two conductor segments are in opposite directions, then the sign will be negative. The partial mutual inductance between orthogonal conductor segments will be zero.

    If a loop is composed of a number of segments, and the net partial inductances (both self and mutual) of each segment are summed, then the result will be the loop inductance. Therefore, *the loop inductance can be determined from the partial inductances, but the partial inductances cannot be determined from the loop inductance.* Hence, the theory of partial inductance is the more basic, or fundamental, concept. Loop inductance is just a special case of the more general theory of partial inductance.

### E.3.4   Partial Inductance Applications

*E.3.4.1 Rectangular Loop.*   Consider the case of the rectangular loop shown in Fig. E-10, where $r_1$ is the radius of the conductor, $a$ is the length of one side, and $b$ is the length of the other side. Considering each side to be a conductor segment, the net partial inductance of the loop will be equal to

$$L_{loop} = \left(L_{p11} - L_{p31}\right) + \left(L_{p22} - L_{p42}\right) + \left(L_{p33} - L_{p13}\right) + \left(L_{p44} - L_{p24}\right), \quad \text{(E-18)}$$

where $L_{pxx}$ is the partial self-inductances of each conductor segment and $L_{pyx}$ is the partial mutual inductances of each conductor segment.

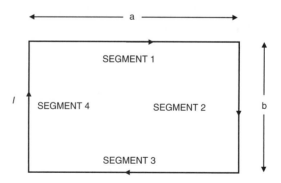

**FIGURE E-10.** A rectangular loop, with four conductor segments.

Substituting Eq. E-14 for each of the partial self-inductances and Eq. E-17 for each of the partial mutual-inductances in Eq. E-18 gives for the inductance of the rectangular loop

$$L_{loop} = \frac{\mu}{\pi}\left[b\ln\frac{a}{r_1} + a\ln\frac{b}{r_1}\right]. \tag{E-19}$$

Equation E-19 is identical to the loop inductance equation derived as Eq. E-10. Both of these equations ignore the magnetic field fringing that occurs at the corners of the loop.

Grover (1946) gives the following more accurate equation for the inductance of a rectangular loop:

$$L_{loop} = \frac{\mu}{\pi}\left[a\ln\frac{2a}{r_1} + b\ln\frac{2b}{r_1} + 2\sqrt{a^2 + b^2} - a\sinh^{-1}\frac{a}{b}\right.$$
$$\left. -b\sinh^{-1}\frac{b}{a} - 2(a+b) + \frac{\mu}{4}(a+b)\right]. \tag{E-20}$$

**Example E-1:** For the rectangular loop shown in Fig. E-10, let $a = 1$ m, $b = 0.5$ m, and $r_1 = 0.0001$ m. The net partial inductance calculation (Eq. E-19) gives the loop inductance as 5.25 µH. Grover's equation (Eq. E-20) gives the loop inductance as 4.97 µH. The difference is apparently the result of fringing at the corners of the loop.

It is interesting to note that, substituting the infinite integral Eq. E-12 for each of the partial self-inductances and the infinite integral Eq. E-15 for each of the partial mutual inductances in Eq. E-18, as well as using the definite integral

$$\int_{x_1}^{x_2}\frac{dx}{x} = \ln\frac{x_2}{x_1}, \tag{E-21}$$

and doing a significant amount of mathematical manipulation, Eq. E-18 reduces to Eq. E-19. This is true because all the infinite terms in the partial self-inductance and the partial mutual-inductance equations cancel each other, and therefore these terms do not have to be evaluated. This derivation is similar to the procedure that will be used in deriving Eq. E-33 from Eq. E-32 in Section E.3.5.

*E.3.4.2 Two Unequal Diameter Parallel Conductors.* Now, let us examine the case of two closely spaced, but unequal diameter, conductors as shown in Fig. E-11. Assume that the length $\ell$ of each conductor is much greater than the spacing $D$ of the conductors. Conductor 1 has a radius $r_1$ and conductor 2 has a radius $r_2$.

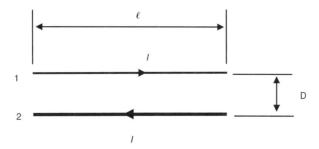

**FIGURE E-11.** Two equally spaced parallel conductors of unequal diameter.

The net partial inductance of conductor 2 is equal to

$$L_{np2} = L_{22} - L_{12} \tag{E-22}$$

Substituting Eq. E-14 for $L_{22}$ and Eq. E-17 for $L_{12}$ gives for the net partial inductance of conductor 2

$$L_{np2} = \frac{\mu l}{2\pi} \left[ \ln \frac{2l}{r_2} - \ln \frac{2l}{D} \right]. \tag{E-23}$$

Equation E-23 points out an important fact; if two conductors that carry equal and opposite currents are moved closer together, then the net partial inductance of the conductors will decrease, because the partial mutual inductance, which is the second term in Eq. E-23, increases. This method is a practical way to reduce inductance—*place conductors that carry equal and opposite currents close together.*

The above clearly demonstrates that if conductor 2 is the ground conductor and conductor 1 is the signal conductor, *then the ground inductance is a function not only of the characteristics of the ground conductor but also of the spacing between the ground and signal conductor*—the closer that the signal conductor is to the ground, the lower the ground inductance. This will be demonstrated by the ground plane measurements described in Section E-4, the results of which are shown in Fig. E-19.

### E.3.5   Transmission Line Example

Let us use the theory of partial inductance to calculate the per-unit-length inductance $L$ of an infinitely long transmission line that consists of two identical round conductors of radius $r_1$ and spaced a distance of $D$ apart as shown in Fig. E-12. By using an infinite transmission line, we can ignore any effects that occur at the ends of the line.

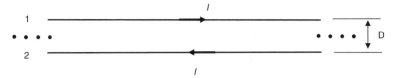

**FIGURE E-12.** An infinite two-wire transmission line.

The net partial inductance of the transmission line will be

$$L = \left(L_{p11} - L_{p21}\right) + \left(L_{p22} - L_{p12}\right). \tag{E-24a}$$

Because of symmetry, $L_{p11} = L_{p22}$ and $L_{p21} = L_{p12}$, therefore,

$$L = 2\left(L_{p11} - L_{p21}\right). \tag{E-24b}$$

Substituting Eq. E-14 for $L_{p11}$ and Eq. E-17 for $L_{p12}$ gives

$$L = \frac{\mu l}{\pi}\left[\ln\left(\frac{2l}{r_1}\right) - 1 - \ln\left(\frac{2l}{D}\right) + 1\right], \tag{E-25}$$

dividing by $l$ and expanding the terms, gives the *per-unit-length loop inductance* of a two-wire transmission line as

$$L = \left(\frac{\mu}{\pi}\right)\ln\frac{D}{r_1}, \tag{E-26}$$

where $\mu = 4\pi \times 10^{-7}$ H/m.

To check our answer, let us compare Eq. E-26 to the result obtained by using the standard transmission line equations. In Chapter 5, we showed that the per-unit-length inductance of a transmission line was equal to Eq. 5-30

$$L = \frac{\sqrt{\varepsilon_r}}{c}Z_0 \tag{E-27}$$

where $c$ is the speed of light and $Z_0$ is the characteristic impedance of the line.

From Eq. 5-18b, we know that the characteristic impedance of a transmission line that consists of two parallel round conductors is equal to

$$Z_0 = \frac{120}{\sqrt{\varepsilon_r}}\ln\left[\frac{D}{r_1}\right]. \tag{E-28}$$

Substituting Eq. E-28 for $Z_0$ in Eq. E-27 gives the inductance of the transmission line as

$$L = \frac{120}{c} \ln\left(\frac{D}{r_1}\right). \tag{E-29}$$

The speed of light is equal to

$$c = \frac{1}{\sqrt{\mu\varepsilon}} = \frac{120\pi}{\mu}. \tag{E-30}$$

Substituting Eq. E-30 into Eq. E-29 gives for the inductance of the transmission line

$$L = \left(\frac{\mu}{\pi}\right) \ln\frac{D}{r_1}, \tag{E-31}$$

which is identical to what we derived (Eq. E-26) by using the theory of partial inductance.

The inductance of the transmission line of Fig. E-12 could also have been calculated by substituting the infinite integral Eqs. E-12 and E-15 for the partial self-inductance and partial mutual inductances, respectively, in Eq. E-24b, which gives

$$L = \frac{\mu l}{\pi}\left[\int_{r_1}^{\infty} \frac{1}{r}\,dr - \int_{D}^{\infty} \frac{1}{r}\,dr\right]. \tag{E-32}$$

Evaluating the integrals using the identity of Eq. E-21 and dividing by $\ell$ to obtain the per-unit-length inductance gives

$$L = \frac{\mu}{\pi}\left[\ln\frac{\infty}{r_1} - \ln\frac{\infty}{D}\right] = \frac{\mu}{\pi}[\ln\infty - \ln r_1 - \ln\infty + \ln D] = \left(\frac{\mu}{\pi}\right)\ln\frac{D}{r_1}, \tag{E-33}$$

which again is the same as Eq. E-26.

## E.4   GROUND PLANE INDUCTANCE MEASUREMENT TEST SETUP

The discussion above on partial inductance leads naturally to an understanding of, and method for measuring, the inductance of a segment of a conductor (e.g., a PCB ground plane or trace). The voltage developed across any conductor is a function of the current in that conductor, as well as the current in all nearby conductors. The latter results from the mutual inductance between the conductors.

The magnitude of the inductive voltage drop across a conductor segment will be proportional to the rate-of-change of current through that segment. This was expressed in Eq. 10-1 and is repeated here.

$$V = L\frac{di}{dt},$$  (E-34a)

or because $L = \phi/I$

$$V = \frac{d\phi}{dt}.$$  (E-34b)

Equation E-34b is Faraday's Law.

Notice that whereas the voltage across a resistance is proportional to the current through that resistance, the voltage across an inductance is proportional to the rate of change of current through the inductance.

When using Eq. E-34a to calculate the voltage developed across a conductor, the inductance $L$ to use in the equation depends on what voltage is desired. If one wishes to calculate the voltage across a complete loop, then the loop inductance $L_{loop}$ would be used. Alternatively, to calculate the voltage drop across only a segment of a loop, the net partial inductance $L_{np}$ of that segment would be used. A similar discussion would apply as to what magnetic flux $\phi$ to use in Eq. E-34b.

Rewriting Eq. E-34a for the voltage $V_S$ developed across only a segment of a loop gives

$$V_S = L_{np}\frac{di}{dt},$$  (E-35)

where $L_{np}$ is the net partial inductance of that segment and $di/dt$ is the rate of change of current through that segment.

Skilling (1951, pp. 102–103) shows that this voltage can be measured with a meter connected to the ends of the conductor segment, provided the leads of the meter are run perpendicular to the segment and extend out to a great distance from the segment. This configuration is necessary to insure that the magnetic fields that surround the current segment being measured do not interact with the meter leads.

Figure E-13 shows the correct and the incorrect way to connect a meter to a current-carrying conductor segment to measure the inductive voltage drop across that segment. Figure E-13A shows the test leads run parallel to the current-carrying conductor. In this case, the meter leads pick up an error voltage caused by magnetic field coupling from the current-carrying conductor. Figure E-13B shows the test leads run perpendicular to the current-carrying conductor. In this case, the leads do not pick up an error voltage. Therefore, the meter leads should be routed perpendicular to the current-carrying conductor

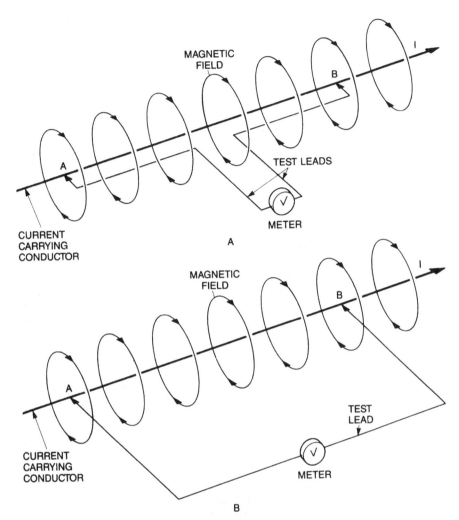

**FIGURE E-13.** Test set up to measure the voltage drop between points *A* and *B* on a segment of a current-carrying conductor. (A) Test leads parallel to current-carrying conductor pick up an error voltage due to the current-carrying conductor's magnetic field. (B) Test leads perpendicular to current-carrying conductor do not pick up an error voltage.

for a great distance, ideally to infinity. In practice, if the test leads are routed perpendicular to the conductor for a reasonable distance, then the coupling will be negligible as the result of the drop off of the magnetic flux density with distance (see Eq. E-2).

Another way to look at this is that the test leads must not cross the surface area associated with the partial self-inductance, see Fig. E-6. If the test leads are routed along the circumference of the surface area, then the only contribution to the integral of the vector magnetic potential will be along the conductor being measures, as indicated in Fig. E-7 and Eq. E-13b.

Solving Eq. E-35 for the net partial inductance gives

$$L_{np} = \frac{V_s}{di/dt} \tag{E-36}$$

It can be observed from Eq. E-36 that if the voltage drop $V_s$ across a segment of a trace or plane can be measured, then the net partial inductance of the that segment can be determined by dividing the measured voltage by the rate of change of current through the segment. The rate of change of the current can be determined by measuring the voltage across a terminating resistor placed at the end of the signal trace.

When measuring ground plane noise voltages, consideration must be given to (1) the bandwidth of the instrumentation, (2) the high-frequency common-mode rejection ratio (CMRR) of the instrumentation, and (3) the dress, or routing, of the leads from the measuring instrument to the circuit under test.

If ground plane voltage measurements are to be made, then they must be made with a wide-bandwidth oscilloscope and with a wide-bandwidth differential probe with a good high-frequency CMRR, at least 100:1 at the frequencies involved. Instrumentation with a large bandwidth is needed to measure the high-frequency components of the noise.

A differential probe with high CMRR is required when measuring the voltage drop across a segment of a current-carrying conductor, since the oscilloscope ground will be at a different potential than either of the probe tips. This potential difference can cause a high-frequency common-mode current to circulate in the instrumentation system, which can interfere with the measurement, hence the necessity for the high CMRR differential probe.

The ground inductance measurements discussed in Section 10.6.2.1 were made with a 500-MHz bandwidth digital oscilloscope and the homemade, 10:1 ratio coaxial, differential probe shown in Fig. 18-8. The test was performed on a $3 \times 8$ in double-sided PCB with a single trace on the top of the board, and a full ground plane on the bottom, as was shown in Fig. 10-17. The trace was 6 in long, 0.050 in wide, and terminated with a 100-$\Omega$ resistor. Boards of various thicknesses were used to vary the height of the trace above the ground plane. A cross-section of the test printed circuit board configuration is shown in Fig. E-14.

Test points for measuring the ground voltage drop were located underneath the trace every inch along the ground plane. *Because of the skin effect, the trace return current is present only on the top of the ground plane.* Therefore, the ground plane voltage must be measured from the trace side of the PCB, where it is much harder to measure, as shown in Fig. E-15. The balanced differential probe's test leads were spaced 1 in apart and were routed perpendicular to the PCB for a distance of 2 in as shown in Fig. E-16.

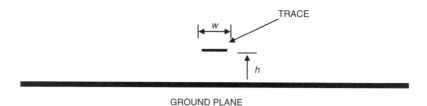

**FIGURE E-14.** Cross section of test printed circuit board used to measure ground plane inductance.

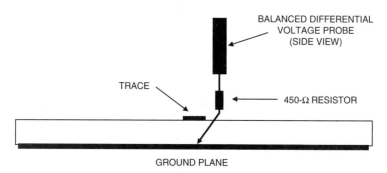

**FIGURE E-15.** Side view of a balanced differential probe, making a ground voltage measurement.

To determine the partial self-inductance of the ground plane, time domain measurements were made by driving the trace with a square wave having an amplitude of 3 V and a rise time (10% to 90%) of 3 ns. Frequency domain measurements could also have been performed using the same technique, except that the trace would be excited with a sine wave instead of a square wave.

Figure E-17 shows the overall instrumentation configuration for both time and frequency domain measurements. For the time domain measurements, a home-brew 74HC240 square-wave oscillator was used as the signal source.

The output of the 180° combiner (which had a 500-MHz bandwidth and a 4-dB insertion loss) provided a single-ended output that equals the difference between the two differential input signals. Because the measured ground plane voltage drop is usually very small and the balanced differential probe has a loss of 20 dB (10:1 ratio), a 25-dB radio frequency (rf) amplifier (having a 1.3 GHz bandwidth) was used between the combiner and the spectrum analyzer or oscilloscope. The combination of the probe loss and the combiner loss, plus the amplifier gain, provided an overall gain of +1 dB for the instrumentation setup.

Figure E-18 shows the waveshape of the typical ground noise voltage. A ground noise pulse occurs on every transition of the square wave signal. On

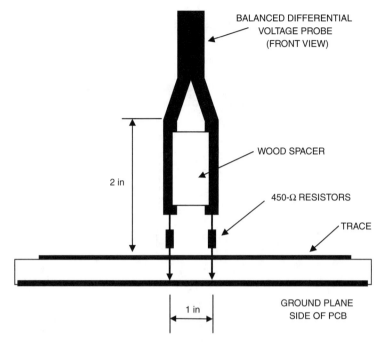

**FIGURE E-16.** Front view of a balanced differential probe, making a ground voltage measurement.

the low-to-high transition, a positive ground noise voltage pulse occurs, and on the high-to-low transition a negative ground noise pulse occurs.

The result of the net partial inductance ground plane measurements were shown in Fig. 10-19.

Although complex, Holloway and Kuester (1998) were able to calculate the net partial inductance of a ground plane. Their results, however, are not expressed in a closed-form equation; rather, they are in the form of a complex integral equation that can only be evaluated by numerical methods.

Figure 10-20, which is repeated here as Fig. E-19 (without the 7-mil height data point) compares the net partial inductance values measured in Chapter 10, using the measurement technique described here, with the theoretical values calculated by Holloway and Kuester. The geometry (shown in Fig. E-14) had a trace width of 50 mils, and trace heights of 16, 32, and 60 mils. As shown in Fig. E-19, the results correlate reasonably well for trace heights from 10 to 60 mils.

## E.5  INDUCTANCE NOTATION

Partial inductances are often indicated by using a subscript $p$ with the inductance term, for example, by using the symbol $L_p$. Loop inductance is

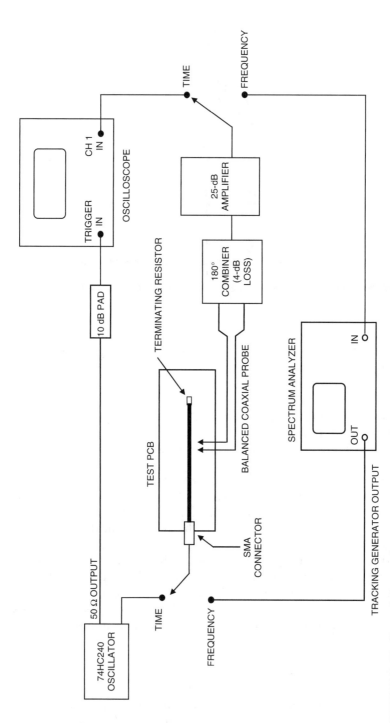

**FIGURE E-17.** Ground plane voltage measurement test instrumentation setup for both time domaine and frequency domaine measurement.

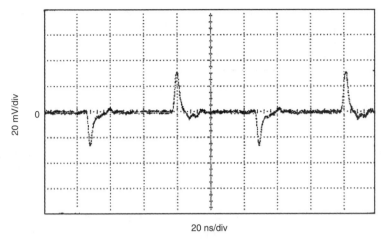

20 ns/div

**FIGURE E-18.** Waveshape of measured ground plane noise voltage.

often distinguished from partial inductance by using no subscript, an $\ell$ subscript, or the word *loop* as the subscript, for example, $L_{loop}$.

Usually the type of inductance being referred to is, or should be, obvious from the use of the inductance term. For example, if one refers to the inductance of a complete current path, they are talking about loop inductance. However, if one is discussing the inductance of only a segment of a loop, such

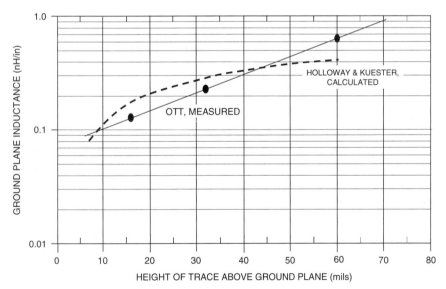

**FIGURE E-19.** Measured versus theoretical ground plane inductance, as a function of trace height. This is the same as Fig. 10-20 less the 7 mil data point. The problem with this data point was discussed in Section 10.6.2.3.

as the ground inductance, they are referring to the partial inductance. There-fore, subscripts are not used in the chapters of this book (except for this appendix), to distinguish between loop inductance and partial inductance. The use of the term should make the distinction clear.

## SUMMARY

This appendix has shown that a unique inductance can be attributed to a segment of a loop. Both partial self-inductances and partial mutual induc-tances exist and can be calculated for individual conductor segments. This tool is very powerful and extremely useful. It is what allows one to determine such things as ground noise voltage and power rail dip.

The validity of this approach was demonstrated by calculating the induc-tance of various conductor configurations, which include a square loop, two parallel but unequal diameter conductors, and an infinite transmission line, using the theory of partial inductance and comparing the results to well-known results calculated by other means. In all cases, the results were identical.

A practical case where the theory of partial inductance must be used is a PCB with a narrow signal trace adjacent to a large ground plane. In this case, the inductance of the individual conductors (signal trace and ground plane) that make up the loop cannot be determined without the theory of partial inductance, and as demonstrated in Chapter 10, the magnitude of the two partial inductances (trace versus plane) are about two orders of magnitude apart.

We have also demonstrated that the inductance of a loop can be calculated by determining the magnetic flux that penetrates the loop area, or by summing all the partial self-inductances and mutual inductances of the segments of the loop.

Another important point to keep in mind is that the loop inductance can be derived from knowledge of the partial inductances, but the partial inductances cannot be derived from knowledge of the loop inductance.

Finally, we have demonstrated that the partial inductances and the voltage drop associated with these partial inductances can be measured, and we have shown the appropriate method of measurement, and that the measured partial inductances values for ground planes show good correlation with the theore-tical values calculated by Holloway and Kuester.

## REFERENCES

Grover, F. W. *Inductance Calculations*. New York, NY, Dover, 1946. Reprinted in 1973 by the Instrument Society of America, Research Triangle Park, NC.

Holloway, C. L. and Kuester E. F. "Net Partial Inductance of a Microstrip Ground Plane." *IEEE Transactions on Electromagnetic Compatibility*, February 1998.

Ruehli, A. "Inductance Calculations in a Complex Integrated Circuit Environment," *IBM Journal of Research and Development*, September 1972.

Skilling, H. H. *Electric Transmission Lines*. New York, McGraw-Hill, 1951.

Weber, E. *Electromagnetic Theory*. Mineola, NY, Dover, 1965.

## FURTHER READING

Hoer, C. and Love, C. "Exact Inductance Equations for Rectangular Conductors with Applications to More Complicated Geometries." *Journal of Research of the National Bureau of Standards-C, Engr. Instrum.*, April–June, 1965.

Paul, C. R. "What Do We Mean by 'Inductance'? Part I: Loop Inductance." *IEEE EMC Society Newsletter*, Fall 2007.

Paul, C. R. "What Do We Mean by 'Inductance'? Part II: Partial Inductance." *IEEE EMC Society Newsletter*, Winter 2008.

# APPENDIX F
# Answers to Problems

*Note:* For some of these problems, a unique solution may not exist. Therefore, solutions other than the ones listed here may in some cases also be acceptable.

## CHAPTER 1

### Problem 1.1

Noise is any signal present in the circuit other than the desired signal. Interference is the undesirable effect of noise.

### Problem 1.2

a. Yes.
b. No, it falls under the power of less than 6 nW exemption.

### Problem 1.3

a. No, it falls under the test equipment exemption.
b. Yes, even equipment exempted from meeting the technical standards must still comply with the non-interference requirement.

### Problem 1.4

a. The manufacturer or importer.
b. The user.

### Problem 1.5

a. The FCC's.
b. The European Union's.

*Electromagnetic Compatibility Engineering,* by Henry W. Ott
Copyright © 2009 John Wiley & Sons, Inc.

c. The FCC's.
d. The European Union's.

**Problem 1.6**

a. 216 to 230 MHz.
b. 5.5 dB.

**Problem 1.7**

a. 150 kHz to 30 MHz.
b. 30 MHz to 40 GHz.

**Problem 1.8**

a. The equipment must be constructed to ensure that any electromagnetic disturbance it generates allows radio and telecommunication equipment and other apparatuses to function as intended, and the equipment must be constructed with an inherent level of immunity to externally generated electromagnetic disturbances.
b. The EMC Directive, 2004/108/EC (which superceded the original directive, 89/336/EEC).

**Problem 1.9**

By publication in that country's official journal.

**Problem 1.10**

The European Union has immunity requirements, and the FCC has no immunity requirements.

**Problem 1.11**

Harmonics and flicker.

**Problem 1.12**

The generic standards for residential/commercial/light industrial environments; specifically EN 61000-6-3 for emissions and EN 61000-6-1 for immunity.

**Problem 1.13**

No, the product must be compliant with the EMC directive, not necessarily the standards. In the European Union, the standards are not legal documents but the directives are.

### Problem 1.14

1. A declaration of conformity.
2. A technical construction file.

### Problem 1.15

FCC Part 15 B: Legal.
MIL-STD-461E: Contractual.
2004/108/EC, EMC Directive: Legal.
RTCA/DO-160E for avionics: Contractual.
GR-1089 for telephone network equipment: Contractual.
TIA-968 for telecom terminal equipment: Legal, because FCC Part 68 requires compliance with it.
SAE J551 for automobiles: Contractual.

### Problem 1.16

United States: The Federal Register.
Canada: The Canada Gazette.
European Union: The Official Journal of the European Union.

### Problem 1.17

No, the Food and Drug Administration (FDA) not the FCC regulates medical equipment.

### Problem 1.18

1. Noise source.
2. Coupling channel.
3. Receptor.

### Problem 1.19

Frequency, amplitude, and time.

### Problem 1.20

a. Magnesium.
b. Nickel (passive).

### Problem 1.21

The zinc.

## CHAPTER 2

### Problem 2-1

a. 2.5 V.
b. 314 mV.
c. 15.7 mV.

### Problem 2-2

a. 187 mV.
b. 12.6 mV.
c. 628 μV.

### Problem 2-3

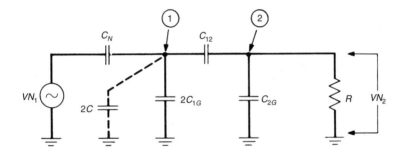

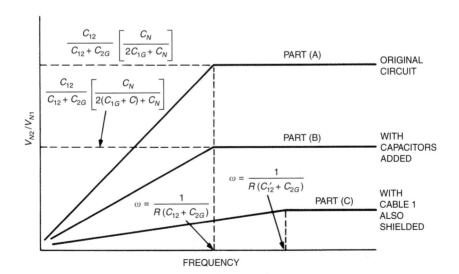

$$\frac{C_{12}}{C_{12} + C_{2G}} \left[ \frac{C_N}{2C_{1G} + C_N} \right]$$

PART (A)    ORIGINAL CIRCUIT

$$\frac{C_{12}}{C_{12} + C_{2G}} \left[ \frac{C_N}{2(C_{1G} + C) + C_N} \right]$$

PART (B)    WITH CAPACITORS ADDED

$$\omega = \frac{1}{R(C'_{12} + C_{2G})}$$

$$\omega = \frac{1}{R(C_{12} + C_{2G})}$$

PART (C)    WITH CABLE 1 ALSO SHIELDED

$V_{N2}/V_{N1}$

FREQUENCY

An equivalent circuit of the noise coupling is shown above. A reasonable assumption to simplify the problem is that $2C_{1G} \gg C_{12}$.

a. The asymptotic plot of $V_{N2}/V_{N1}$ is shown as the top curve in the above plot.
b. If capacitors $C$ are added, then it has the effect of increasing $C_{1G}$, which decreases the maximum coupling while keeping the breakpoint constant. This is shown as the middle curve in the above plot.
c. Shielding cable 1 reduces $C_{12}$ to a value $C'_{12}$. This lowers the maximum coupling further and also increases the break frequency. This is shown as the bottom curve in the above plot. A second effect of shielding is to increase capacitance $C_{1G}$, which decreases the coupling even more.

**Problem 2-4**

377 µV.

**Problem 2-5**

The magnetic field pickup is of opposite polarity on either side of where the input traces crossover, which thereby cancels the noise voltage.

**Problem 2-6**

$$M = \frac{\mu}{2\pi} \ln \left( \frac{b^2}{b^2 - a^2} \right)$$

**Problem 2-7**

a. 57.4 nH/m.
   12.9 nH/m.
   5.62 nH/m.
b. 36 mV/m.

**Problem 2-8**

The mutual inductance is less than, or equal to, the self-inductance of the circuit producing the magnetic flux.

**Problem 2-9**

a. 1/r.
b. $1/r^2$.

## Problem 2.10

By combining the equivalent circuits shown in Figs. 2-9A and 2-9B, we can determine the answer.

a. 25 mV across one of the resistors and 0 V across the other resistor.
b. The two noise currents add in one of the terminating resistors and cancel in the other terminating resistor.
c. The terminating resistor that had 25 mV across it will now have 0 V, and the terminating resistor that had 0 V across it will now have 25 mV.

## Problem 2.11

From Eq. 2-2, we know that the capacitive (electric field) pickup is a function of the terminating resistor $R$. If the two conductors of a twisted pair have different terminating resistances, then they will each pickup a different voltage, and a net noise voltage will appear between them. Only if the terminating resistances are equal will they pickup the same voltage—hence, the net noise voltage between them will be zero.

## Problem 2.12

a. 100% outside the shield, and 0% inside the shield.
b. 0% outside the shield, and 100% inside the shield.

## CHAPTER 3

### Problem 3.1

Safety grounds.

### Problem 3.2

False, power grounding will only affect the common-mode noise.

### Problem 3.3

By using a ground plane or grid (ZSRP).

### Problem 3.4

a. The current $I_g$.
b. The impedance $Z_g$.

### Problem 3.5

a. 10 to 15 $\Omega$.
b. 25 $\Omega$.

### Problem 3.6

$$V_B = (I_1 + I_2 + I_3)Z_1 + (I_2 + I_3)Z_2.$$

### Problem 3.7

To reduce the inductance between the enclosure and ground.

### Problem 3.8

105 nH.

### Problem 3.9

29 MHz. Do not forget to divide the inductance of each grounding strap by four, because there are four straps in parallel.

### Problem 3.10

The treatment of the interconnect (I/O) signals and cables.

### Problem 3.11

In a central system with extensions, the remote elements are not grounded locally and they receive their power from the central system, whereas in a distributed system the remote elements are all powered and grounded locally.

### Problem 3.12

1. Avoid the loop.
2. Tolerate the loop.
3. Break the loop.

### Problem 3.13

1. Isolation transformer.
2. Opto-coupler.
3. Common-mode choke.

**Problem 3.14**

It violates the requirements of the NEC, and it is unsafe.

**Problem 3.15**

a. 90.4 Hz.
b. 10.8 dB at 60 Hz.
   20 dB at 180 Hz.
   24 dB at 300 Hz.

## CHAPTER 4

**Problem 4.1**

From Figs. 4-4 and 4-5 we can write,

$$V_{dm} = \left[ \frac{R_L}{R_L + R_S} - \frac{R_L}{R_L + R_S + \Delta R_S} \right] V_{cm} = \left[ \frac{R_L \Delta R_S}{(R_L + R_S + \Delta R_s)(R_L + R_s)} \right] V_{cm}$$

From  Eq. 4-5

$$CMRR = 20 \log \left( \frac{V_{cm}}{V_{dm}} \right) = 20 \log \left[ \frac{(R_L + R_S + \Delta R_s)(R_L + R_S)}{R_L \Delta R_S} \right]$$

**Problem 4.2**

300 μV.

**Problem 4.3**

a. 60 dB.
b. 89.5 dB.
c. 106 dB.

**Problem 4.4**

6 dB.

**Problem 4.5**

As large as possible.

### Problem 4.6

a. 9.4 kΩ.
b. 57.5, or 35 dB.
c. 137.4 kΩ.
d. 1.15, or 1.2 dB.
e. 34 dB.

### Problem 4.7

a. 21, or 26 dB.
b. 60 dB.

### Problem 4.8

The instrumentation amplifier will have a CMRR 34 dB greater than the differential amplifier.

### Problem 4.9

When the source impedance is small and the load impedance is large, or vice versa.

### Problem 4.10

a. Greater than the sum of the source and load impedances.
b. Less than the parallel combination of the source and load impedances.

### Problem 4.11

1. The source impedance is not normally known.
2. The load impedance is not normally known.
3. The filter must not affect the differential-mode signal.

### Problem 4.12

To the enclosure, or chassis, ground.

### Problem 4.13

By proper layout.

### Problem 4.14

The characteristic impedance $Z_0$ of the power distribution system.

## Problem 4.15

a. 34.5 mV.
b. 1.26 Ω.
c. 630 mV.

## CHAPTER 5

## Problem 5.1

a. Dielectric material.
b. Frequency.

## Problem 5.2

a. Aluminum and tantalum electrolytics.
b. Paper and film.
c. Mica and ceramic.

## Problem 5.3

a. Ceramic.
b. Mica.
c. Multilayer ceramic.

## Problem 5.4

The inductance is inversely proportional to the log of the diameter.

## Problem 5.5

At 0.2 MHz, $R_{ac}/R_{dc} = 1.33$.
At 0.5 MHz, $R_{ac}/R_{dc} = 1.96$.
At 1.0 MHz, $R_{ac}/R_{dc} = 2.66$.
At 2.0 MHz, $R_{ac}/R_{dc} = 3.65$.
At 5.0 MHz, $R_{ac}/R_{dc} = 5.63$.
At 10 MHz, $R_{ac}/R_{dc} = 7.85$.
At 50 MHz, $R_{ac}/R_{dc} = 17.23$.

## Problem 5.6

a. 0.172 m Ω/m.
b. 16.57 m Ω/m.

### Problem 5.7

a. From Eq. 5-6, $R = \rho/A$, $\rho = 1.724 \times 10^{-8}$ Ω-m.
   For $d \gg \delta$, $A = \pi\, d\, \delta$.
   For copper, $\delta$ is given in Eq. 5-12.
   $R_{ac} = \frac{\rho}{A}$. Substituting the above for $\rho$ and $A$ gives Eq. 5-9b.
b. $d \sqrt{f} \geq 0.66$, for d in meters.

### Problem 5.8

The inductive reactance is directly proportional to frequency, and the ac resistance is proportional to the square root of the frequency.

### Problem 5.9

a. $\delta = \dfrac{wt}{2(w + t)}$.

b. $\delta = \dfrac{t}{2}$.

c. The break frequency occurs at a frequency where the conductor thickness equals twice the skin depth and therefore the high-frequency current will flow across the full cross section of the conductor. Hence, one would expect the ac and dc resistances to be equal at this frequency.
d. The slope is proportional to the $\sqrt{f}$, or 10 dB/decade.

### Problem 5.10

Answers to parts a, b, and c are in the following table:

| Conductor | Cross Section Area | $R_{dc}$ | $R_{ac}$ @ 10 MHz | L |
|---|---|---|---|---|
| Round | 0.05 in² | 13.6 μΩ/in | 1.04 mΩ/in | 14 nH/in |
| Rectangular | 0.05 in² | 13.6 μΩ/in | 0.690 mΩ/in | 12.6 nH/in |

d. Both conductors have the same cross-sectional area and dc resistance. However, the rectangular conductor has 34% less ac resistance and 10% less inductance than the round conductor.

### Problem 5.11

$X_L = 8.69$ Ω/in.
$R_{ac} = 0.139$ Ω/in.

**Problem 5.12**

Any two of the following:

1. Only has one conductor.
2. Cannot propagate dc.
3. Does not use the TEM mode of propagation.

**Problem 5.13**

6 ns.

**Problem 5.14**

96 nH/ft.

**Problem 5.15**

50 $\Omega$.

**Problem 5.16**

The dielectric constant.

**Problem 5.17**

$L = 6$ nH/in.
$C = 2.4$ pF/in.

**Problem 5.18**

a. 50 $\Omega$.
b. 1.24 radians or 71°.

**Problem 5.19**

For a dielectric constant of 4.5 (Table 4-3):
$\alpha_{ohmic} = 0.084$ dB/in. Neglecting the loss in the reference planes.
$\alpha_{dielectric} = 0.293$ dB/in.
The total attenuation $= 0.377$ dB/in.

**Problem 5.20**

1. Use a longer cylindrical core.
2. Use multiple turns on the core.

## CHAPTER 6

**Problem 6.1**

| | |
|---|---|
| Silver: | $|Z_s| = 3.6 \times 10^{-5} \, \Omega.$ |
| Brass: | $|Z_s| = 7.2 \times 10^{-5} \, \Omega.$ |
| Stainless: | $|Z_s| = 5.8 \times 10^{-3} \, \Omega.$ |

**Problem 6.2**

| Frequency (kHz) | Skin Depth (in) | Absorption Loss (dB) |
|---|---|---|
| 0.1 | 0.51 | 1.1 |
| 1.0 | 0.16 | 3.3 |
| 10 | 0.05 | 10.6 |
| 100 | 0.02 | 33.4 |

**Problem 6.3**

Using nonferrous material would require a shield thickness of 1.2 in or more; this is impractical.

Using steel, however, the shield would have to be 0.12 in thick; this is considerably more reasonable.

For a high-permeability material such as mumetal, the shield thickness would only have to be 0.05 in thick; this may be the best solution.

**Problem 6.4**

a. 138 dB.
b. 138 dB.

**Problem 6.5**

24 dB.

**Problem 6.6**

133 dB.

**Problem 6.7**

218 dB.

**Problem 6.8**

Greater than 313 MHz.

**Problem 6.9**

| Thickness (in) | Absorption Loss (dB) |
| --- | --- |
| 0.020 | 2.11 |
| 0.040 | 4.22 |
| 0.060 | 6.34 |

**Problem 6.10**

0.6 in (1.5 cm).

**Problem 6.11**

35 dB.

**Problem 6.12**

161 dB.

**Problem 6.13**

1 in (2.54 cm).

**Problem 6.14**

1. A conductive surface finish.
2. Adequate pressure.

**Problem 6.15**

92 dB.

**Problem 6.16**

348 MHz.

**Problem 6.17**

183 MHz.

**CHAPTER 7**

**Problem 7.1**

Arc discharge, because the required voltage is much lower than required for glow discharge.

**Problem 7.2**

A resistor of 43 to 400 Ω in series with a capacitor of 0.075 μF or larger across the load, or the contact, ($R = 270$ Ω, $C = 0.1$ μF would be a good choice).

**Problem 7.3**

Approximate waveshapes are shown in the following figures:

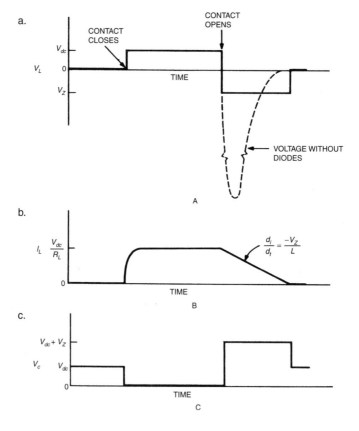

**Problem 7.4**

a. A value of $R$ between 60 and 240 $\Omega$, and a value of $C \geq 0.1$ μF ($R = 100$ $\Omega$, $C = 0.22$ μF would be a good choice).
b. $C \geq 0.35$ μF.

**Problem 7.5**

$C \geq 0.1$ μF.
$R \geq 600$ $\Omega$.
A diode with a voltage rating $> 24$ V and a current rating $> 100$ mA.

**CHAPTER 8**

**Problem 8.1**

283 nV.

**Problem 8.2**

a. 0.91 μV.
b. 1.01 μV.

**Problem 8.3**

8.33 nV/$\sqrt{\text{Hz}}$.

**Problem 8.4**

a. 179 μV.
b. 179 μV.

**Problem 8.5**

10 nV/$\sqrt{\text{Hz}}$

**Problem 8.6**

400 pA.

**Problem 8.7**

4 kHz.

**Problem 8.8**

$56.6 \text{ pA}/\sqrt{B}$.

**Problem 8.9**

a.  $4 \mu\text{V}$.
b.  $2 \mu\text{V}$.

**Problem 8.10**

$39 \mu\text{V}$.

## CHAPTER 9

**Problem 9.1**

Let:  $N_o$ = Noise power output of device.
$N_i$ = Noise power input to device.
$G$ = Power Gain of device.
$S_i$ = Signal power input to device.
$S_o$ = Signal power output of device.

From Eq. 9-1,

$$F = \frac{N_o}{GN_i}.$$

Multiplying the numerator and denominator by $S_i$ gives

$$F = \frac{N_o S_i}{GS_i N_i} = \frac{N_o S_i}{S_o N_i} = \frac{S_i / N_i}{S_o / N_o}.$$

**Problem 9.2**

a.  Bipolar: $V_{nd} = 38 \text{ nV}/\sqrt{\text{Hz}}$.
b.  FET: $V_{nd} = 70 \text{ nV}/\sqrt{\text{Hz}}$.

Therefore the bipolar transistor produces the least equivalent input device noise.

**Problem 9.3**

$S_o/N_o = 14.9$ dB.

**Problem 9.4**

a. $NF = 5.4$ dB.
b. $R_s = 300$ kΩ.
   $NF = 4.0$ dB.

**Problem 9.5**

a. Turns ratio $= 100$.
b. $NF = 0.5$ dB.
c. $NF = 27.9$ dB.
d. SNI $= 556$.

**Problem 9.6**

a. 5µV.
b. The inherent noise immunity of FM allows operation at a lower signal-to-noise ratio, and the 75-Ω transmission line has less thermal noise than a 300-Ω system. Also, the smaller bandwidth allows less noise into the system.

**Problem 9.7**

$NF = 3$ dB.

**Problem 9.8**

a. $F = 1.11$, $R_s = 26{,}500$ Ω.
b. $F = 1.25$, $R_s = 572$ Ω.

**Problem 9.9**

$NF = 6$ dB.

**Problem 9.10**

The total input noise voltage $V_{nt}$ is given by Eq. 9-30.
The input thermal noise voltage $V_t$ is given by Eq. 9-4.

The noise power output (total or thermal) is equal to the appropriate noise voltage squared times the power gain $G$ of the device divided by the source resistance $R_s$.

F = Total Noise Power Output/Noise Power Output Due to Thermal Noise. Therefore,

$$F = \frac{(V_{nt})^2}{(V_t)^2} = \frac{4kTBR_s + (V_n)^2 + (I_nR_s)^2}{4kTBR_s} = 1 + \frac{1}{4kTB}\left[\frac{V_n^2}{R_s} + I_n^2 R_s\right].$$

### Problem 9.11

Total output noise power equals $N_o = \dfrac{V_{nt}^2}{R_s}(G)$, where $V_{nt}$ is given by Eq. 9-30, and $G$ is the power gain of the device.

Total output signal power is given by $S_o = \dfrac{V_s^2}{R_s}(G)$

Therefore,

$$\frac{S_o}{N_o} = \frac{V_s^2 G R_s}{R_s\left[4kTBR_s + V_n^2 + (I_nR_s)^2\right]G} = \frac{(V_s)^2}{(V_n)^2 + (I_nR_s)^2 + 4kTBR_s}.$$

## CHAPTER 10

### Problem 10.1

A grid. A plane is not the correct answer. Although a plane performs well, it is not the "basic" structure.

### Problem 10.2

a. Resistance, at dc or low frequencies.
b. Inductance, at high frequencies.

### Problem 10.3

a. 70%.
b. 94%.

### Problem 10.4

a. Proportional to the square of the height.
b. Inversly proportional to the square of the spacing between the traces.

## Problem 10.5

a.  75%.
b.  25%.

## Problem 10.6

Because of the skin effect, the return current from the signal on the trace will only exist on the top of the plane. The voltage on the top of the ground plane will be different than the voltage on the bottom of the ground plane.

## Problem 10.7

a.  300 m$\Omega$/ in.
b.  0.0024 in.

## Problem 10.8

1. The decoupling capacitor.
2. The parasitic trace and load capacitance.

## Problem 10.9

1. There are two signal current loops. In one, the current flow is CW, and in the other the current flow is CCW; hence, the radiation from the two loops tend to cancel each other.
2. The two reference planes providing shielding to any fields that do radiate.

## Problem 10.10

a.  See figure below.
b.  Parasitic trace capacitance.
c.  Decoupling capacitor.

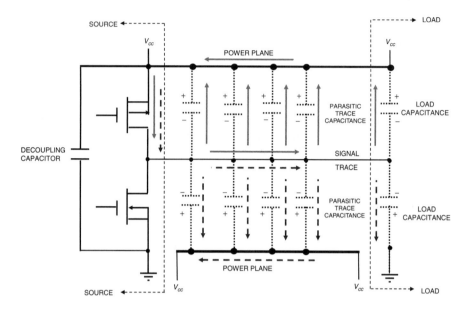

## CHAPTER 11

**Problem 11.1**

132 mA.

**Problem 11.2**

1. Radiation from the decoupling capacitor loop.
2. Decoupling currents produce a ground voltage drop, which excites cables connected to the system causing common-mode radiation.
3. Power bus noise couples to other ICs on the same power bus and then internally couples to I/O signal and power cables, which then radiate.

**Problem 11.3**

1. The capacitor.
2. The traces and vias.
3. The IC lead frame.

**Problem 11.4**

1. The number of capacitors used.
2. The inductance in series with each capacitor.

**Problem 11.5**

The total capacitance of all the capacitors used.

**Problem 11.6**

1. The capacitance must be large enough that its reactance is equal to or less than the low-frequency target impedance at the lowest frequency of interest.
2. The total capacitance must be large enough to supply the total transient current required by the IC (Eq. 11-12).

**Problem 11.7**

Two, because it will significantly reduce the radiated emission, both common mode and differential mode, from the decoupling capacitors.

**Problem 11.8**

1. Use equal value capacitors.
2. Spread the out around the IC.

**Problem 11.9**

1. The total capacitance multiplies up proportional to the number of capacitors used.
2. The total inductance divides down proportional to the number of capacitors used.
3. The absence of parallel or antiresonances.

**Problem 11.10**

1. The capacitance does not multiply up proportional to the number of capacitors, it is just the sum of the individual capacitances.
2. Parallel or antiresonances will be present.

**Problem 11.11**

An impedance spikes resulting from the parallel resonance (antiresonance) between the two capacitor networks will be present.

**Problem 11.12**

a. 34.
b. 0.05 µF.

**Problem 11.13**

a. 50 mΩ from 20 to 318 MHz, then increasing 20 dB/decade above that frequency.
b. 200 capacitors.
c. 800 pF for each capacitor.
d. Yes, providing it does not increase the inductance of the capacitor package.

**Problem 11.14**

0.01 μF.

**Problem 11.15**

4.5 nF.

**CHAPTER 12**

**Problem 12.1**

a. A small loop.
b. A dipole or monopole.

**Problem 12.2**

Differential-mode: Rising 40 dB/decade up to the $1/\pi\, t_r$ frequency, then flat beyond that.
Common-mode: Rising 20 dB/decade up to the $1/\pi\, t_r$ frequency, then falling at a rate of 20 dB/decade beyond that.

**Problem 12.3**

64 mA.

**Problem 12.4**

1. Canceling loops.
2. Dithered clocks.
3. Shielding.

**Problem 12.5**

The small dipole.

**Problem 12.6**

a. 13 dB μV/m.
b. 13 dB μV/m at 10 MHz, rising to 33 dB μV/m at 100 MHz, and flat from there to 350 MHz.
c. 8 dB margin at a frequency of 88 MHz.

**Problem 12.7**

a. 369 MHz.
b. 20 dB/decade.

**Problem 12.8**

They both will be the same.

**Problem 12.9**

Differential-mode, due to the $f^2$ term in the emission equation.

**ProblemS 12.10**

800 μV/m.

**Problem 12.11**

4 in (10.2 cm).

**Problem 12.12**

6.4 μA.

**CHAPTER 13**

**Problem 13.1**

a. From Fig. 13-8,

$$V_{CM} = 25I_{CM} = \frac{25V(f)}{R_{LISN} + \dfrac{1}{j2\pi f C_P}}.$$

For

$$R_{LISN} \ll \frac{1}{j2\pi fC_P}.$$

$$|V_{CM}| = 50\pi fC_P V(f).$$

b. $R_{LISN} \ll 1/(2\pi f C_P)$.
c. Letting $C_P = 500$ pF and $f = 500$ kHz (values all on the high side), the inequality of the answer to Problem 13-1b becomes $25 \ll 637$.

**Problem 13.2**

6.92 nF.

**Problem 13.3**

The current pulses in Fig. 13-38,

1. Are lower in amplitude.
2. Are wider (spread over a larger part of a cycle).
3. Have a lower harmonic content.

**Problem 13.4**

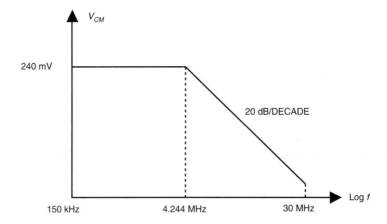

**Problem 13.5**

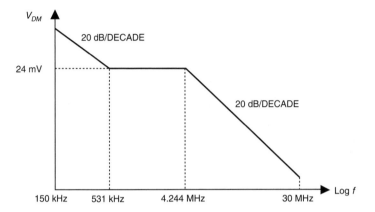

**Problem 13.6**

From Eq. 13-7, differential-mode.

**Problem 13.7**

1. It provides some differential-mode inductance, which improves the
   differential mode effectiveness of the filter.
2. Too much leakage inductance can cause the common mode-choke to
   saturate due to the power line current.

**Problem 13.8**

The parasitic capacitance across the common-mode choke, and the inductance in series with the Y-capacitors (capacitors to ground).

**Problem 13.9**

231 pF.

**Problem 13.10**

The ESL and ESR of the input ripple filter capacitor.

**Problem 13.11**

As can be observed from Fig. 13-9, the switching frequency.

## CHAPTER 14

**Problem 14.1**

0.88 V/m.

**Problem 14.2**

5.48 V/m.

**Problem 14.3**

6.2 V/m.

**Problem 14.4**

$$E = \sqrt{120 \cdot \pi \cdot P}.$$

**Problem 14.5**

194 V/m.

**Problem 14.6**

The unintentional detection (rectification) of high-frequency energy by a nonlinear element in a low-frequency circuit.

**Problem 14.7**

a. 2 μH.
b. 0.02 μF.

**Problem 14.8**

a. It will increase the attenuation.
b. It will decrease the attenuation.

**Problem 14.9**

0.9 in (2.3 cm). From Eq. 6-33.

**Problem 14.10**

72 V.

**Problem 14.11**

a.  20.1 Ω.
b.  1.1 Ω.

**Problem 14.12**

a.  12.12 A.
b.  20 A.
c.  1000 A.

**Problem 14.13**

a.  39.5 A.
b.  6.9 A.
c.  163.1 A.

**Problem 14.14**

5000 μF.

**Problem 14.15**

1.5 cycles.

**Problem 14.16**

a.  Yes, based on the CBEMA curve (Fig. 14-21).
b.  No, based on Fig. 14-21.
c.  Yes, based on Fig. 14-21.
d.  No.

**Problem 14.17**

Any combination of conditions falling into the upper right-hand portion of Fig. 14-21.

**CHAPTER 15**

**Problem 15.1**

No, only charged conductors can produce a discharge.

**Problem 15.2**

a. Yes.
b. No.

**Problem 15.3**

Polyester: negative.
Aluminum: positive.

**Problem 15.4**

a. No.
b. Yes.

**Problem 15.5**

a. Yes.
b. No, because charge is mobile on a conductor, the opposite charges will recombine.

**Problem 15.6**

a. 0.7 pF.
b. 84.6 pF.

**Problem 15.7**

10 pF.

**Problem 15.8**

22.5 pF.

**Problem 15.9**

If the spacing of the measurement electrodes are doubled, then the resistance between them will also double. However, to maintain the measurement over a square the length of the electrodes must also double, which provides twice as many parallel paths, hence the resistance will be cut in half. Combining the above two effects, the resistance remains the same regardless of the size of the square over which the measurement is made.

**Problem 15.10**

a. 4 V.
b. 0.2 V.

**Problem 5.11**

The chassis, metallic enclosure or ESD ground plate.

**Problem 5.12**

It reduces the magnitude of the ESD current, and hence the associated magnetic field, reducing the likelihood of transient upset or soft errors.

**Problem 15.13**

10 V.

**Problem 15.14**

40 A.

**Problem 15.15**

1. Program flow.
2. Validity of I/O data.
3. Validity of data in memory.

**Problem 15.16**

About 94%.

**CHAPTER 16**

**Problem 16.1**

1. Directly proportional to the fundamental frequency.
2. Directly proportional to the magnitude of the signal current.
3. Inversely proportional to the rise time.

**Problem 16.2**

In the I/O area of the board.

**Problem 16.3**

To the enclosure, with a 360° connection.

**Problem 16.4**

1. Keep loop areas of critical signals (clocks, etc.) very small.
2. Use a grid for the power and ground.

**Problem 16.5**

1. Slots or splits in planes.
2. Signal traces changing layers.
3. Ground plane cutouts (voids) around connectors.

**Problem 16.6**

a. 1. They cause return currents to flow through large loops.
   2. They increase the impedance of the planes.
   3. They increase the radiated emissions.
b. No high-frequency signal traces can cross over the slots on adjacent layers.

**Problem 16.7**

1. All signal layers adjacent to planes.
2. Signal layers tightly coupled (close) to their adjacent reference planes.

**Problem 16.8**

Objectives number 1, 2, 4, and 5.

**Problem 16.9**

You will not be able to use adjacent ground-to-ground vias where signal traces change from layers from 4 to 5.

**Problem 16.10**

#3 (closely spaced power and ground planes).

**Problem 16.11**

All.

**Problem 16.12**

a. No. As long as a signal layer is adjacent to a plane (objective #1), the second layer in the board will always be a plane. Therefore, none of the internal signal layers will be adjacent to the layer with the power island. The power island will, however, block some routing channels on the signal layer on which it is located.

b. Same answer as part a.

## CHAPTER 17

**Problem 17.1**

1. A single ground plane.
2. Proper partitioning.
3. Routing discipline.

**Problem 17.2**

a. $\pm 0.50$ in.
b. $\pm\ 0.030$ in.

**Problem 17.3**

1. Routing traces across the split.
2. Overlapping the planes.

**Problem 17.4**

It is almost always better to have only one single reference plane for your system.

**Problem 17.5**

1. Route the digital traces as stripline, or asymmetric stripline.
2. Use isolated analog and digital ground regions on the PCB, similar to that shown in Fig. 17-14.

**Problem 17.6**

Where the pins are connected internally. *AGND* to the internal analog ground, and *DGND* to the internal digital ground. They do not indicate to where the pins should be connected externally.

**Problem 17.7**

Analog devices.

**Problem 17.8**

$SNR$ = 78.9 dB.
Maximum resolution = 12 bits.

**Problem 17.9**

10 femtoseconds (0.010 ps).

**Problem 17.10**

50 ps.

**Problem 17.11**

Analog ground.

**Problem 17.12**

Large DSPs, and codecs, containing large amounts of digital circuitry.

**Problem 17.13**

Directly to the $DGND$ pin.

**CHAPTER 18**

**Problem 18.1**

Because of the reflective surfaces of the walls.

**Problem 18.2**

54 dBμV, or 500 μV.

**Problem 18.3**

−79 dBm.

## Problem 18.4

58 dBμV.

## Problem 18.5

a. An experiment the result of which should be zero.
b. To validate the test setup.

## Problem 18.6

Position the probe as it will be used for the measurement, and then short the two probe tips together—the voltage should read zero.

## Problem 18.7

Turn the product off, and see that the signal goes away.

## Problem 18.8

a. Quantitative.
b. Qualitative.
c. Quantitative.
d. Quantitative.

## Problem 18.9

1. Measuring the magnetic field in the vicinity of a PCB.
2. Measuring the leakage at the apertures in an enclosure.

## Problem 18.10

a. Magnetic fields close to the PCB.
b. Common-mode cable currents.
c. Harmonics.

## Problem 18.11

1. Clock waveforms.
2. $V_{cc}$-to-ground voltage.
3. Ground voltage differential.
4. Common-mode voltage, cable-to chassis.

**Problem 18.12**

1. They are expensive.
2. They are easily damaged.
3. They have limited bandwidths.
4. They must be supplied power.

**Problem 18.13**

a. 1. A low pass filter, to reject 60 Hz.
   2. A 10 dB attenuator, to attenuate transients present on the power line.
   3. A diode limiter, to clip very large transients on the power line.
b. To protect the input of the spectrum analyzer from overloads or damage.

**Problem 18.14**

1. Peak.
2. Average.
3. Quasi-peak.

**Problem 18.15**

20 dB.

**Problem 18.16**

$\leq$ 68 mA.

**Problem 18.17**

1. Use an electric drill or Dremel motor tool, to produce a broadband field.
2. Use a small handheld transmitter that is authorized for use by the public without a license, such as a Family Radio Service or Citizen Band radio.

**Problem 18.18**

1. The quality of the test site.
2. The accuracy of the measuring equipment.
3. The test procedure.
4. The test setup.
5. The operator.

# INDEX

---